A TEXTBOOK OF PH

A TEXTBOOK OF
PHYSICS

(Written strictly according to the sytlabi of Plus Two, Pre-University, Higher Secondary and Intermediate Classes)

PATITPABAN MISRA
Professor, Ravenshaw College
Cuttack

BIPIN BEHARI SWAIN
Reader, Ravenshaw College
Cuttack

PURNENDUNATH DAS
Reader, F.M. College
Balassore

VIKAS PUBLISHING HOUSE PVT LTD

VIKAS PUBLISHING HOUSE PVT LTD
576, Masjid Road, Jangpura, **New Delhi** - 110 014
Phones: 4314605, 4315313 • Fax: 91-11-4310879
E-mail: ***orders@vikas.gobookshopping.com***
Internet: ***www.vikaseforum.com***

First Floor, N.S. Bhawan, 4th Cross, 4th Main,
Gandhi Nagar, **Bangalore**-560 009 • Phone : 2204639

F-20, Nand Dham Industrial Estate, Marol,
Andheri (East), **Mumbai**-400 059 • Phone : 8502333, 8502324

A-26, 4th floor, Nelson Chambers, Nelson Manickam Road,
Aminjikarai, **Chennai**-600 029

Distributors:
UBS PUBLISHERS' DISTRIBUTORS LTD
5, Ansari Road, **New Delhi**-110 002
Ph. 3273601, 3266646 • Fax : 3276593, 3274261
E-mail: orders@gobookshopping.com • *Internet: www.gobookshopping.com*
- 10, First Main Road, Gandhi Nagar, **Bangalore**-560 009 • Ph. 2263901
- 6, Sivaganga Road, Nungambakkam, **Chennai**-600 034 • Ph. 8276355
- 8/1-B, Chowringhee Lane, **Kolkata**-700 016 • Ph. 2441821, 2442910
- 5-A, Rajendra Nagar, **Patna**- 800 016 • Ph. 672856, 656169
- 80, Noronha Road, Cantonment, **Kanpur**-208 004 • Ph. 369124, 362665

Distributors for Western India:
PREFACE BOOKS
223, Cama Industrial Estate, 2nd Floor,
Sun Mill Compound, Lower Parel (W), **Mumbai**-400 013

Reprint 2001

Printed at Hindustan Offset Printers, Delhi - 110 032

PREFACE

This book is designed to cover the syllabi of Plus Two, Pre-University, Higher Secondary and Intermediate examinations of all the states. In following the syllabus, we have sacrificed the logical development of subject matter, to some extent which we feel the teachers should be in a position to sort out. The units used in this book are S.I. Units, though other units also have been mentioned at places to acquaint the student with them. Simple use of calculus has been made wherever necessary. Adequate number of problems have been worked out to illustrate the underlying principle of the relevant text. At the end of each chapter, sufficient number of questions and problems, to elicit proper understanding of the text, have been included. We have deliberately avoided to include questions for conventional essay type answers.

Last but not the least, we have tried to present the subject matter in simple English and in a straightforward manner to reach the students fresh from secondary schools where teaching has been mostly in Oriya. It is for them to judge how far we have succeeded in our attempt. In spite of our efforts, errors might have crept into the book. We will be grateful to the teachers and the students if such errors are brought to our notice. Any suggestion for improvement of the book will be thankfully appreciated.

We are grateful to the authors and publishers of several excellent books and magazines from which we have drawn the material, to Sri S.P. Chattopadhyaya for typing the script, to Sri P. Sahoo, the artist who worked assiduously in drawing the diagrams.

AUTHORS

CONTENTS

4. OSCILLATORY MOTION AND WAVE MOTION 115—187

5. LIQUIDS 188—224

1

MEASUREMENT AND OTHER PRELIMINARIES

It is difficult to say when man learnt the art of measurement. The recurrence of the day and night and the change in season could not have escaped even the early man's attention. The change in the length of the shadow during the course of the day, the movement of the Sun during the day and of the moon and stars during the night in the sky certainly engaged the attention from early times. These facts perhaps gave the idea of passage of time. Sundials were used to reckon time in early days. Man also felt the need for measuring the lengths and distances. Naturally he used his own steps and the arm for this purpose.

With the growth of civilisation, when man became accustomed to use of numbers, the mathematical and astronomical studies developed. Thus gradually developed systems of measurement which should not only be accurate but should also be dependable. The 'Cubit' (length of the arm from elbow to the middle finger) and 'step' differs from person to person. With expanding frontiers of science and technology and development of communication between countries need for standardisation and acceptance of those standards for measurement over the world was felt. The systems in use for quantitative measurement of physical quantities and some other basic properties associated with physical quantities are discussed in this chapter.

1.1. Units of Measurement

Measuring any physical quantity means comparing it with a standard to determine its relationship to the standard. This standard is called a *Unit*. Thus unit may be defined as some specific magnitude of a measurable physical quantity in terms of which one may conveniently express other magnitudes of the same quantity. Thus all measurable physical quantities are expressed in terms of (*i*) some number and (*ii*) some units. For example, when it is said that the distance between two places is 200 kilometers it means the distance is 200 times that of a standard which is one kilometer. Thus one kilometer is the unit here and 200 is the number.

1.2. Fundamental and Derived Units

There are some units which are regarded as basic or *fundamental.* All other Units may be derived from them. Such units are called *derived* units. The meter is the unit of length and basing on it the unit of area is $(\text{Meter})^2$ and the unit of volume is $(\text{Meter})^3$.

Thus the unit of length is a fundamental unit and units of area and volume are derived units. Similarly the unit of velocity which is displacement (L) divided by time is (Meter/Sec). Thus from one or more fundamental units, other units can be derived by multiplication or division of the unit by itself or by others.

Conventionally, the units of mass (M), length (L) and time (T) are regarded as the three *fundamental units.*

However, with the growth of modern science and adoption of S.I. system of units four more basic units have been added *i.e.* the unit of current (I), the unit of temperature (T), the unit of luminous intensity (Cd), and the unit for amount of substance (mole). In addition there are 'dimensionless' units such as the radian (rad) for measurement of plane angle and steradian (sr) for measuring a solid angle.

In this chapter, we will confine ourselves to the fundamental units of mass, length and time only. Units concerning other quantities will be discussed in concerned chapters.

1.3. Systems of Units

A complete set of units, based on a system of fundamental units from which all derived units are obtained by multiplication and/or division of the fundamental units involved without introducing numerical factors is known as a *coherent* system of units.

As man learnt the art of measurement, different systems of units developed in different parts of the world. There are three major systems in use (1) F.P.S. system (British system), (2) C.G.S. system (French system) and (3) S.I. system (system International).

In F.P.S. system, the units of length, mass and time are Foot, Pound and Second respectively. The system derives its name from the initial alphabet of the basic units chosen. Similarly in the C.G.S. system the units are Centimeter, Gram and Second respectively. In the S.I. system Meter, Kilogram and Second are three basic units for length, mass and time respectively. Besides, Ampere degree Kelvin, candella, and mole are used as units for current, temperature, luminous intensity, and amount of substance respectively. The S.I. system of units will be used in this book giving reference to other systems wherever necessary.

1.4. Basic Standards for Units in S.I. System

The units chosen may be arbitary, but it should be well defined should not change with time or environment, should be accessible and above all, should be accepted internationally. We discuss below the standards chosen for units in S.I. system.

Meter. The 'International Prototype meter preserved at the International Bureau of Weights and Measures at Sevres, near Paris, was accepted as the standard of length. This standard is a bar of Platinum-iridium alloy. The distance between the two lines engraved on gold plugs near the ends of the bar, when the bar is at the temperature of melting ice is called one meter. This alloy is selected for its property of remaining unaffected with time and environment. Prototype of the standard meter bar were distributed to other bureaus of standards throughout the world. In India, it is available at the National Physical Laboratory, New Delhi.

This standard, however, is arbitary. Scientists continued the search for a natural standard whose value should not change with time; it should be accessible easily and accepted internationally. In 1960 the Eleventh General Conference of Weights and Measures redefined the meter in terms of wavelength of a certain orange line in the spectrum of the Krypton isotope of atomic mass 86. Each line in the spectrum of a gas has a specific wavelength. For any gas. the number of lines in the spectrum and their wavelengths will never change. The wavelengths of these lines can be accurately determined. Hence it was decided to use the wavelength of one of the spectral lines as reference for standard of length. Hence Krypton (86) a gas, which gives a very sharp well defined bright spectral line in the orange-red end of the spectrum was chosen. The meter is defined in terms of this standard as exactly 1,650,763.73 wavelengths of this line.

Alternately, λ for orange-red line of Krypton (86)

$$=6{,}067.802\times 10^{-10}\,\text{m}.$$

Kilogram. When the metric system was established in 1791, the gram was intended to be the mass of one cubic centimeter of pure water at 4°C. More precise measurement have shown that this is not exactly true. The prototype Kilogram is the accepted standard now.

The present standard for the unit of mass, the International Prototype Kilogram (kg) is a cylinder of Platinum-iridium alloy—Platinum 90 per cent and iridium 10 per cent—kept at the International Bureau of Weights and Measures in Sevres, France. This was adopted as the standard in 1889.

Second. Man used to depend on the position of the sun in the sky to have an idea of the time interval during the day. Similarly, the position of certain stars during night could be taken as a guide. But the apparent changes in the position of the sun and stars are caused due to the rotation of the earth about its axis.

The time interval between the noon of one day and the noon of the next day is defined as a day. But the length of the day varies as the earth moves in an elliptical orbit around the sun. The mean or average interval for a day over the year is 'mean solar day'. The day is divided into $24 \times 60 \times 60 = 86{,}400$ seconds. Thus a second was defined as 1/86,400 part of mean solar day. The second is the unit of time in all systems of units.

The astronomers, however, for precise calculation devised new standards of time called sideral time and ephemeries time, based on movement of distant stars and on planetary motions respectively. The second defined on the basis of these new standards were however, close to the mean solar second for ordinary calculations.

The search for a suitable time standard did not end here. It was known that a pendulum takes a fixed time for one complete oscillation. Could there be some natural phenomena like that? That undisturbed atoms emit or absorb radiation at a fixed frequency was known by then. This could be taken as a reference if the frequency is measured as accurately as possible keeping the atom undisturbed. Atomic clocks were built utilising this principle. The first clock built in 1948 used ammonia molecule which vibrates at a sharply defined frequency of $23{,}870 \times 10^6$ HZ. The latest model built in 1965 has utilised Cesium, 133 as the standard. Its accuracy is 5 parts in 10^{12}. This means, the likely variation in time in 6,000 years is around 1 second. The frequency assigned to Cesium standard was so carefully choosen that the second defined on this basis and the ephemeris second choosen on the basis of planetary motion were very close to each other. According to the new standard, 1 second is the time taken for 9,192,631,770 oscillations in Cesium 133 atom. Thus the Cesium oscillator with suitable equipment is the current laboratory standard for time.

It may be of interest to note that currently the mean solar second differs from the ephemeris second by about 130 parts in 10^{10}. Hence this is also the difference of mean solar second with the Cesium standard.

1.5. Multiples and Submultiples of the Units

The S.I. units discussed are some times found either too big or too small to express the magnitude of physical quantities involved. By putting suitable prefixes as given in Table 1.1. One can get the multiple or sub-multiple as desired. For example:

1 kilometer $= 1 \times 10^3$ meter $= 1000$ meters,
1 centimeter $= 1/100$ meter $= 1 \times 10^{-2}$ meter $= 0.01$ meter.
1 kilogram $= 1 \times 10^3$ gram $= 1000$ grams.
1 milligram $= 1 \times 10^{-3}$ gms $= 0.001$ gms.

TABLE 1.1. Metric Prefixes.

Prefix	*Symbol*	*Multiple or sub-multiple*
tera	T	$\times 10^{12}$
giga	G	$\times 10^{9}$
mega	M	$\times 10^{6}$
kilo	K	$\times 10^{3}$
deci	d	$\times 10^{-1}$
centi	c	$\times 10^{-2}$
mill	m	$\times 10^{-3}$
micro	μ	$\times 10^{-6}$
nano	n	$\times 10^{-9}$
pico	p	$\times 10^{-12}$
femto	f	$\times 10^{-15}$
atto	a	$\times 10^{-18}$

These prefixes can be used for multiples and sub-multiples of any unit, even though the unit is not in the S.I. system.

1.6. Inter Conversions of Units in Different Systems

The units in one system are related to similar units in other systems. There are simple conversion factors or equivalents as given in Table 1.2.

TABLE 1.2

	S.I. system	*C.G.S. system*	*F.P.S. system*
Length	1 meter $= 100$ cms	1 centimeter $= 10^{-2}$ meter $= 0.01$ meter	1 foot $= 30{\cdot}48$ cm $= 30{\cdot}48 \times 10^{-2}$m $= 0.3048$ meters
Mass	1 kilogram $= 1000$ gms	1 gm $= 10^{-3}$ kg $= 0.001$ kg	1 pound $= 453.6$ gms $= 453.6 \times 10^{-3}$kg $= .4536$ kg

1.7. Order of Magnitude of Distances and Methods for Measurement

Now that we have learnt about different systems of units, we will discuss in brief about the different range of distances we come across and units suitable for their measurement.

The radius of the earth is 6.4×10^6m; the distance between the earth and the moon is 3.85×10^8m. The mean distance of the sun from the earth is 1.50×10^{11}m. This distance called 1 astronomical unit appears comparatively large. But compared to other celestial bodies, the sun is quite close to the earth. The light from the sun takes about 8 minutes to reach the earth. But the light from the nearest star, outside the solar system, Alpha Centauri takes about 4.3 years to reach the earth. Such great distances are expressed in a special unit of length called *light year.* One 'light year' is equal to the distance travelled by light during one year. Taking the speed of light as 3×10^8 m/s, one light year is approximately 9.47×10^{15}m. Using this unit, Alpha Centauri is 4.3 light years away from us.

On a clear moonless night one can see a bright band of stars stretching north-south in the sky. This is the milkyway galaxy to which the sun belongs. There are about 100 billion stars in our galaxy. It is a flat almost circular disc about 100,000 light years across with a thickness of nearly 2,500 light years. Ours is not the only galaxy; there are nearly a billion of galaxies. The space between two neighbouring galaxies is empty and the distance between them may be several million light years. Our nearest galaxy—the great Nebulae in Andromeda—is situated at a distance of nearly thirty million light years.

Hubble observed a shift of the spectral line towards the red end of the spectrum of the light received from the distant galaxies. On the basis of Doppler's effect, he interpreted that all galaxies are receeding from each other. They are also moving away from the earth at speeds increasing with their distances from the earth. Assuming that the maximum speed of these galaxies cannot be more than that of light (according to the special Theory of Relativity) the theoretical limit of the size of the universe is estimated as 10 billions light years *i.e.*, approx 10^{26}m.

We have seen the highest range of distances. At the other extreme, we come across very small distances in the atomic and sub-atomic scale. The radius of hydrogen atom is 5×10^{-11}m. But its mass is concentrated at the central nucleus having a radius of 1.2×10^{-15}m. The rest of the space inside the atom is empty excepting the space occupied by one electron only. In between the two extreme limits, lie the usual distances we normally come across.

Methods of measurement. A single method of measurement cannot be used to measure such wide range of distances.

The length of a table can be measured with the help of a meter scale or the length of a room may be measured by a 10 meter tape. Similarly, the diameter of a narrow wire can be measured by a screw gauge. But this method of direct comparison cannot be applied in too large or too small distances. Indirect methods of measurement, therefore, becomes necessary.

The distance of nearer celestial objects such as moon, other planets, sun and nearer stars can be determined by the triangulation or parallax method, using the diameter of the earth's orbit as the base line. To illustrate the principle of this method, let us consider the simple case, shown in Fig. 1.1.

Let '*O*' be a celestial object whose distance from the earth is to be determined. For our reference, we choose a very distant star, say *S*, whose direction remains same at all positions of earth during its orbital motion. When the earth is at *A* in the orbit, let the observer measure the angle between the direction of the star *S* and the object *O* *i.e.* the angle *SAO* with the help of a sextant. After six months, when the earth has come to a point *B* which is diametrically opposite to the point *A* in the earth's orbit, the observer again measures the angle between *S* and *O* *i.e.* the angle *SBO*. The sum of the magni-

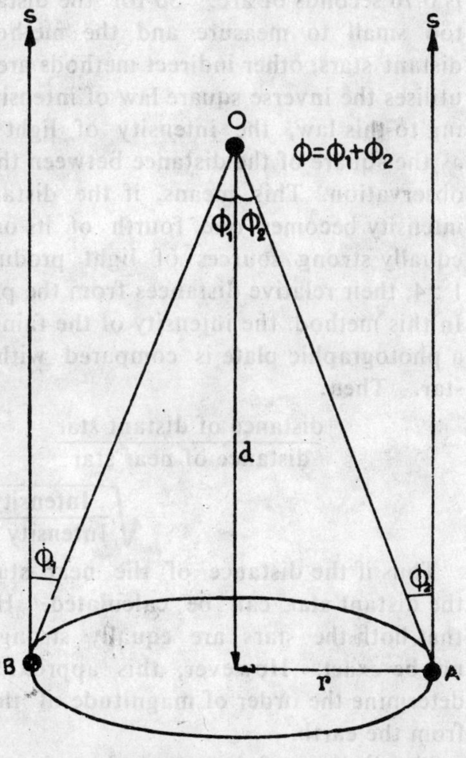

Fig. 1.1. Triangulation method for measuring large distances.

tudes of two angles gives the magnitude of the angle substended by the celestial object on the earth's orbital diameter AB.

It is known that in circular measures the angle ϕ radians

$$=\frac{\text{arc}}{\text{radius}}.$$

In Fig. 1.1 since the distance of O from the earth is quite large, BO, CO and AO are nearly equal to each other. So AB may be regarded as the arc of a circle of which O is the centre.

$$\text{Hence, } \phi \text{ (radians)} = \frac{\text{orbital diameter, } AB}{\text{distance of } O \text{ from earth}} = \frac{2r}{d}$$

or $\qquad d=2r\times\phi$ (radians).

Here r is the mean radius of the earth's orbit.

It is found, for $\phi=1$ second of arc, the distance $d=3.08\times10^{16}$m. This distance is called 'parsec' and is used as a unit for astronomical distances. The value of ϕ for the nearest star, Alpha Centauri is 0.76 seconds of arc. So for the distant star, this angle becomes too small to measure and the method fails. In case of very distant stars, other indirect methods are used. One such method utilises the inverse square law of intensity of illumination. According to this law, the intensity of light at a place varies inversely as the square of the distance between the light source and place of observation. This means, if the distance is increased two times, intensity becomes one fourth of its original value. Thus if two equally strong sources of light produce intensity at the ratio of 1 : 4, their relative distances from the place of observation is 1 : 2. In this method, the intensity of the faint image of a distant star on a photographic plate is compared with the intensity of a near-by star. Then,

$$\frac{\text{distance of distant star}}{\text{distance of near star}} = \sqrt{\frac{\text{Intensity due to near star}}{\text{Intensity due to distant star}}}$$

Thus if the distance of the near star is known, the distance of the distant star can be calculated. Here, of course, we assume that both the stars are equally strong sources of light; this may not be exact. However, this approximate method enables us to determine the order of magnitude of the distances of distant stars from the earth.

The distance of the moon have been measured by use of radar pulses. The radar pulses travel with the speed of light. The pulse directed to the moon is reflected by it and is received back 2.56

seconds after it is sent out. Thus light takes 1.28 seconds to reach the moon. Multiplying this time with the speed of light, the moon's distance can be determined. The same principle is used to measure the depth of sea by 'Sonar'. In Sonar, however, ultrasonic waves are used.

For measurement of small distances we use slide callipers, screw gauge and even a magnifying device such as an optical microscope. With the aid of suitably divided scale we can measure up to a micrometer by an optical microscope. Still smaller distances cannot be measured by optical microscope as we cannot see objects smaller than the wavelength of visible light which is around 5×10^{-7}m. But with the aid of electron microscope we can measure down to 10^{-8} m. The X-ray diffraction in a crystal enables us to go down the scale to about 10^{-10}m *i.e.* the diameter of hydrogen atom. The size of the atom can also be determined with the aid of Avogadro's hypothesis. The size of the gold nucleus was estimated to be of the order of 10^{-14}m by alpha particle scattering due to Sir Ernest Rutherford in 1911. The particle scattering technique has been used to estimate the effective size of elementary particles.

In Table 1.3 we have given the order of magnitudes of distances from very large to very small. Note that the distances at the two extreme ends vary by a factor of about 10^{41}.

1.8. Order of Magnitude of Time and Their Measurement

We have discussed earlier how the standard for measurement of time developed starting from the orbital motion of the earth and ending with the Cesium, 133 atom.

With a stop-watch we may measure time interval of the order of 10^{-1} seconds. Still lower intervals of time can be measured by indirect methods like comparing with period of electrical oscillation. Time interval of less than 10^{-12} second has been possible to measure with 'maser'. Many fundamental particles created during nuclear reactions have a life time as small as 10^{-24} second. Such life times are measured by indirect methods. It is observed that such a particle leaves a trace in suitable photographic emulsion before disintegrating. By dividing this length of the trace with the speed of the particle which is very nearly equal to that of light, we calculate the life time of the particle $\left(i.e.\ t=\frac{l}{c}\right)$.

On the other extreme of the time scale, we know that the life appeared on earth about few million years ago, the earth itself is 4-5

TABLE 1.3. Order of Maguitude of Distances.

Associated distances	*Order of magnitude in meters*	*Associated distances*	*Order of magnitude in meters*	*Associated distances*	*Order of magnitude in meters*
The limit of the Universe	10^{26}	Diameter of a cricket ground	10^{2}	Diameter of gold nucleus	10^{-14}
Distance to the nearest galaxy	10^{22}	Height of a man	10^{0}	Diameter of proton	10^{-15}
Diameter of milkyway galaxy	10^{20}	Width of your palm	10^{-1}	Radius of the sun	10^{9}
Distance of nearest star, the Alpha Centauri	10^{17}	Width of a finger	10^{-2}	Mean distance from the earth to moon	10^{8}
Mean distance of Pluto, the farthest planet, from sun	10^{13}	Thickness of a paper sheet	10^{-4}	Radius of the earth	10^{7}
Mean distance from the earth to sun	10^{11}	Diameter of red blood corpuscles	10^{-5}	Radius of the moon	10^{6}
		Size of a virus	10^{-8}	Distance from Delhi to Puri	10^{6}
		Diameter of hydrogen atom	10^{-10}	Height of Mount Everest	10^{5}

The order of magnitude We first express the quantity in terms of the nearest power of 10. The power or the exponent of 10, so obtained, is called the order of magnitude of the quantity as given below :

Quantity	*Expressed in nearest power of 10*	*Order of magnitude*
8	8×10^{1}	1
12	1.2×10^{1}	1
49	4.9×10^{1}	1
51	$\cdot 51 \times 10^{2}$	2
480	4.8×10^{2}	2
510	$.51 \times 10^{3}$	3

billion years old ; the age of the universe is beleived to be 5-6 billion years. Such large time intervals are estimated by measuring the content of some redioactive elements such as Uranium-238. Any natural radioactive substance reduces to half of its original mass after a fixed time interval known as half-life time. This data is used to estimate the age of the sample in which the radioactive element is traced. In estimating the age of fossils on the earth, a process involving measurement of content of C^{14} is used. In Table 1.4, we give the order of time magnitude of time interval for various events.

TABLE 1.4. Order of Magnitude of Time Intervals.

Associated event	*Order of magnitude of time in seconds*	*Associated event*	*Order of magnitude of time in seconds*
Age of universe	10^{18}	One minute	10^{2}
Age of earth	10^{17}	Time interval between heart beats/	10^{0}
Time elapsed since life appeared on earth	10^{16}	Time for one rotation of an electric fan	10^{-2}
Time since appearance of prehistoric animals	10^{15}	Time period of vibration of high frequency audible sound	10^{-4}
Time since appearance of earliest man	10^{13}	Life expectancy of an excited state of atom	10^{-8}
Age of Konark temple	10^{10}	Time for light to travel a distance of 1 cm.	10^{-11}
Human life expectancy	10^{9}	Time taken by the electron to revolve once around the nucleus of hydrogen atom	10^{-15}
Time for earth to revolve once around the sun	10^{7}	Time for light to travel a distance equal to the diameter of Proton	10^{-24}
Time for earth to complete one rotation about its axis	10^{5}		
Time taken by light to travel from sun to earth	10^{3}		

1.9. Dimensions

The alphabet or alphabets used to represent a particular unit may be called its *symbol*. There are various symbols in use for the same unit. However, generally the symbols M, L and T are used for mass, length, and time respectively.

The dimensions of a physical quantity refer to the fundamental unit or units contained in it. Any quantity which can be measured in mass units only, may be said to have the dimension of mass. This is expressed by the symbol (M). Similarly, any quantity which can be measured in length units only is said to have the dimension of length (L).

The derived units are based on the fundamental units and in many cases involve more than one fundamental units. In such a case the dimensions of such units are expressed in general as $K\,(M)^x\,(L)^y\,(T)^z$ where K is a pure number. x, y and z indicate how many times the particular unit is involved. Their values can be found from the difinition of the physical quantities involved, The power to which the fundamental units are raised to obtain the derived units are called the dimensions of the derived units. Following are some examples.

Area. The area of a square whose sides are $1m$ each is $1m \times 1m = 1\,m^2$

Thus the unit of area is the square of unit of length.

As the dimension of length is (L), the dimensions of area is $(L) \times (L) = (L)^2$. Thus area has two Dimensions in length.

Volume. The volume of a unit cube $= 1m \times 1m \times 1m = 1m^3$ *i.e.* the unit of volume is the cube of unit of length.

So the dimensions of volume is $(L)\,(L)\,(L) = (L)^3$. Thus volume has 3 dimensions in length.

Velocity. Velocity is the rate of displacement. Thus its unit is meter/sec. Its dimentions are

$$(L) \times \frac{1}{(T)} = \frac{(L)}{(T)} = (L)\,(T)^{-1}.$$

Acceleration. Acceleration is the rate of change of velocity. Its unit is meter/sec^2. Thus the dimensions are $(L)/(T)^2 = (L)\,(T)^{-2}$.

Momentum. Momentum by definition is mass $\times$ velocity.

Thus unit of momentum is the product of the unit of mass (kg) and the unit of velocity (m/sec) *i.e.* kg meter/sec. So it involves the dimensions of mass, length and time *viz.*

$$(L)(M)/(T) = (L)\,(M)\,(T)^{-1}.$$

Force. Force $=$ mass $\times$ acceleration.

Hence its unit is the product of unit of the mass (kg) and the unit of acceleration (m/sec^2). Thus the unit of force is kg. Meter/sec^2 and the dimensions are $(M)(L)/(T)^2$.

$$=(M)(L)(T)^{-2}.$$

Moment. Moment of a force or a couple involves the product of a force with a distance. Thus the unit of moment is the product of unit of force (kg. meter/sec^2) and the unit of length (meter).

Thus its unit is kg. meter^2/sec^2. and the dimensions are

$$(M)(L)^2/(T)^2=(M)(L)^2(T)^{-2}.$$

Work and energy. Work is derived by the product of a force and a distance. Hence its unit and dimensions are same as that of moment.

Power. It is defined as rate of doing work. Thus the unit will be the unit of work divided by the unit of time *i.e.*

$$\frac{\text{kg. meter}}{\text{sec}^2}.\ \text{meter/sec.}$$

The dimensions are $(M)(L)^2/(T)^3$

$$=(M)(L)^2(T)^{-3}$$

Pressure. It is calculated using the relation force/area. So its unit will be unit of force divided by unit of area *i.e.*

$$\frac{\text{kg. meter/sec}^2}{\text{meter}^2}$$

$$=\text{kg/meter. sec}^2.$$

So the dimensions are $(M)/(L)(T)^2=(M)(L)^{-1}(T)^{-2}$.

Numerical quantities have no dimensions. Quantities like specific gravity and angle in radians which are expressed in the form of a ratio of two quantities having the same units are dimensionless. The dimensions of a quantity will be same in all systems of units.

It is to be noted that two physical quantities having different dimensions cannot be added together. So the dimensions of each term of an equation must be same.

1.10. Use of Dimensional Analysis

1. The correctness of a relation involving number of physical quantities can be checked by use of dimensions. We take a relation,

$$V=V_i+at$$

Where V_i is the initial velocity of a body, V is the velocity acquired by the body after a time interval t and 'a' is the uniform acceleration with which the body is moving.

If the dimensions of the quantities on the left hand side of this equation are same as dimensions of the quantities on the right hand side then the equation is correct.

Putting the dimensions, $(L)(T)^{-1}=(L)(T)^{-1}+(L)(T)^{-2}.(T)$

$=(L)\ (T)^{-1}+(L)\ (T)^{-1}.$

So the dimensions of the terms on both sides are same and the equation is correct.

Another use to which dimensional analysis can be applied is to derive simple equations connecting physical quantities. One such example is discussed in Chapter 5 for deriving the Stoke's Law in Viscosity

Dimensional analysis is also used for conversion of units from one system to another.

Let us suppose that X_1 is a measure number at a physical quantity in a system of units, in which L_1, M_1, T_1 are the basic units of length, mass and time respectively. We want to find the corresponding measure number X_2 in a second system of units having L_2, M_2, T_2 as the basic units. If a, b and c are the dimensions in length, mass and time respectively of the physical quantity then the derived units of the quantity in the two systems are

$U_1 = L_1{}^a M_1{}^b T_1{}^c$ and $U_2 = L_2{}^a M_2{}^b T_2{}^c$

The measure number $X \propto 1/U$

so $X_1 U_1 = X_2 U_2$

Therefore, $X_1 L_1{}^a M_1{}^b T_1{}^c = X_2 L_2{}^a M_2{}^b T_2{}^c$

$$X_2 = X_1 \left(\frac{L_1}{L_2}\right)^a \left(\frac{M_1}{M_2}\right)^b \left(\frac{T_1}{T_2}\right)^c$$

Let us apply this relation to convert Newton to Dynes

[Force] = (L) (M) $(T)^{-2}$ so, a = 1, b = 1, c = –2

$$\text{Now } L_1/L_2 = \frac{\text{meter}}{\text{centimeter}} = 100$$

$$M_1/M_2 = \frac{\text{kilo gramme}}{\text{gramme}} = 1000 \qquad T_1/T_2 = \frac{\text{second}}{\text{second}} = 1.$$

So, $n_2 = 1 \times 100 \times 1000 \times 1 = 10^5$ dynes.

However, the dimensional analysis has many limitations. A physical relation concerning more than three variables cannot be derived by it. It does not give any information about the nature of numerical or dimensionless constants in it. Thus it cannot be applied to trigonometrical or exponential problems.

.11. Scalars and Vectors

Physical quantities involved in different fields of science and technology are classified as scalars and vectors.

A *Scalar* is some physical quantity which has magnitude but no direction. Thus to specify a scalar quantity completely only a number and a unit are necessary. Mass, distance, speed, temperature, energy, time and electrical charge are some such examples.

A *Vector* is a physical quantity which has both magnitude and direction. Hence to specify a vector quantity completely, its direction besides a number and the unit is to be mentioned. Displacement of a body is its change of position in a particular direction. Thus it is a vector quantity. Displacement, velocity, acceleration, force and momentum etc., are some examples of vector quantities.

Two scalars can be added by simple process of arithmetic as they involve magnitudes only. But the combined effect of two vectors is not found by simple arithmetic but by a process called vector addition. This process is essentially a geometrical operation as the direction of the vectors is to be taken into account. We will discuss this process in detail later.

Representation of vectors. (*a*) *In symbols.* A vector quantity is represented by a letter with an arrow at its top *i.e.* $\vec{\mathbf{a}}$. It is usually represented in the printed books by bold face type letter *i.e.* **a**. The magnitude of a vector is called its modulus and is indicated as mod **a** or | **a** |

(*b*) *Graphical.* To present a vector, a straight line is drawn with an arrow-head. The arrow-head indicates the direction of the vector and the length of the line is chosen to represent the magnitude of the vector. This is done by establishing a scale as is done for graph or map drawing. For example, we have to draw the vector diagram of a force 20 Newtons acting towards east and another force 30 Newtons directed to south. We choose a scale 10 Newton = 1 cm. Thus the first force is a line of length 2 cms, drawn west-east with the arrow-head at the eastern end, Fig 1.2 (*a*). Similarly, the other force is represented by 3 cm line, drawn north-south with the arrow-head at the southern end.

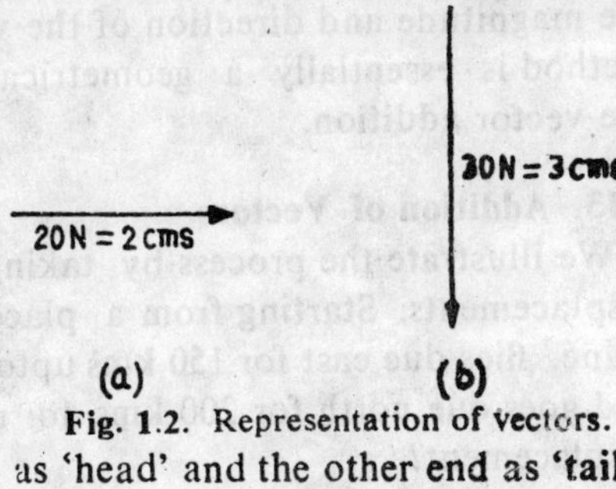

Fig. 1.2. Representation of vectors.

The arrow-head is referred to as 'head' and the other end as 'tail' in subsequent discussion.

Unit vector. A vector whose magnitude or modulus is unity is defined as a unit vector.

Null vector. A vector having modulus as Zero is called a null vector.

Equal Vectors. Two vectors are said to be equal when they have the same magnitude and direction.

Negative vectors. The vector having the same magnitude as the vector $\vec{\mathbf{a}}$, but in opposite direction is known as the negative vector of $\vec{\mathbf{a}}$ and denoted by $-\vec{\mathbf{a}}$.

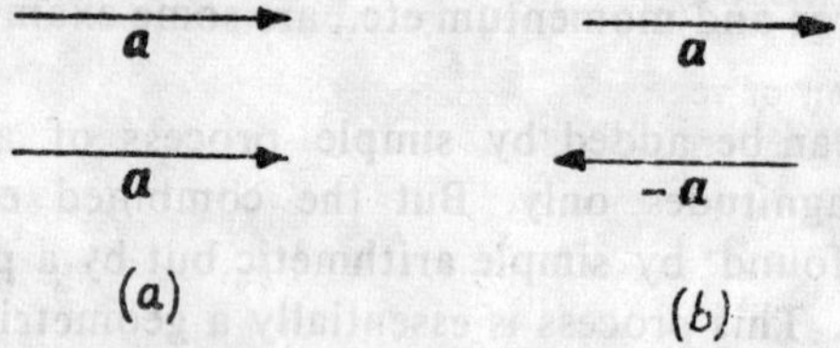

Fig. 1.3. Equal and negative vectors.

1.12. Resultant of Vectors

If a force acts on an object, the object moves in the direction of the force. But if two forces act simultaneously on the object, it is displaced in a direction which may be different from the direction of either of the forces acting on the object. This shows the possibility of finding a third force which could have caused the same displacement as the combined effect of two forces. This third force is called the *resultant force* of the two forces.

Any number of forces acting simultaneously on an object can be replaced by a single resultant force. Since force is a vector quantity, the resultant is not found by simple algebraic addition. The process called vector addition that takes into account both the magnitude and direction of the vectors is followed. Hence the method is essentially a geometrical process. We explain below the vector addition.

1.13. Addition of Vectors

We illustrate the process by taking the example involving two displacements. Starting from a place represented by '*O*', an aeroplane flies due east for 150 kms upto *A* and then changes direction and goes due north for 200 kms to reach *B*. What is the actual displacement?

A scale 1 cm=50 kms is chosen. A horizontal line *OA* 3 cms in lenth to represent 150 kms displacement, is drawn. A second line *AB*, as shown in the figure, 4 cms in length represent 200 kms displacement in the northward direction. By joining *O* to *B*.

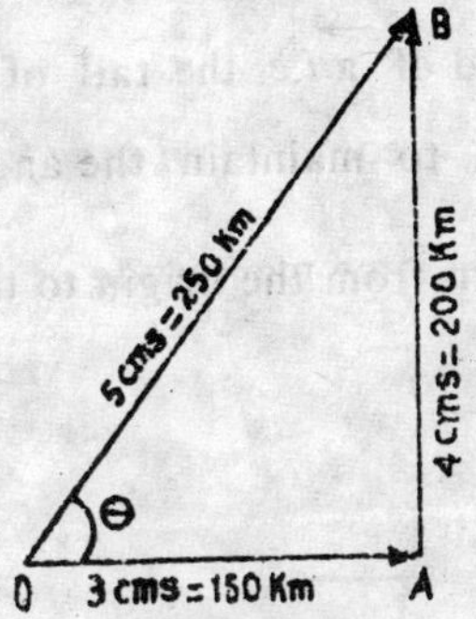

Fig. 1.4. Addition of vectors

the vector triangle OAB is completed. It is obvious from the figure, that the total displacement is given by OB. Here the length of the hypoteneous $OB=5$ cms which is equivalant to 250 kms. The direction of the resultant vector is known by measuring the angle AOB which is nearly 53° in this case. Thus the direction of the resultant displacement OB is 53° north of east.

The process illustrates the addition of two displacements to get resultant displacement and is applied for addition of any two vectors such as force, velocity etc. The vector addition is represented by the equation

$$\overrightarrow{\mathbf{OA}}+\overrightarrow{\mathbf{AB}}=\overrightarrow{\mathbf{OB}}.$$

It may be noted that $OA+AB \neq OB$.

To sum up, the following steps are followed in addition of two vectors $\overrightarrow{\mathbf{a}}$ and $\overrightarrow{\mathbf{b}}$.

(*i*) A suitable scale is chosen for the vector to be added.

(*ii*) The vector $\overrightarrow{\mathbf{a}}$ and $\overrightarrow{\mathbf{b}}$ are drawn according to this scale.

(*iii*) Starting from any arbitrary origin, the vector $\overrightarrow{\mathbf{a}}$ is drawn. The origin end of the vector is called 'tail' and the other end with the arrow is named 'head'.

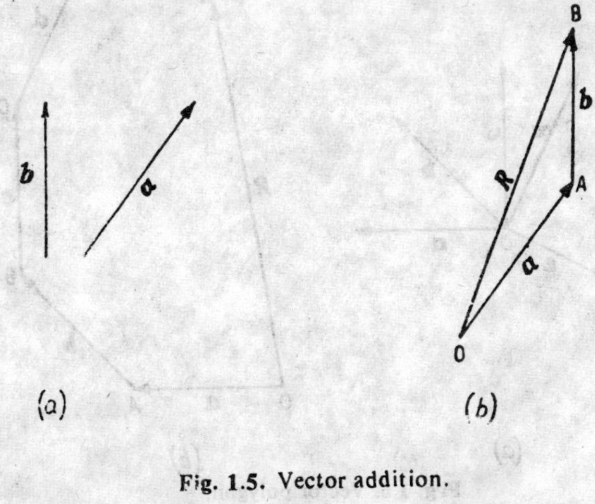

Fig. 1.5. Vector addition.

(*iv*) The vector $\vec{\mathbf{b}}$ is drawn from the head of $\vec{\mathbf{a}}$ *i.e.* the tail of $\vec{\mathbf{b}}$ starts from the head of $\vec{\mathbf{a}}$. Care is taken to maintain the angle between the two vectors.

(*v*) The resultant vector is a vector drawn from the origin to the head of the vector $\vec{\mathbf{b}}$

In this figure, Resultant $\vec{\mathbf{R}} = \vec{\mathbf{a}} + \vec{\mathbf{b}}$

1.14. Polygon of Vectors

For addition of more than two vectors, the same procedure is applied. But in place of triangle of vectors we will deal with polygon of vectors.

Vectors $\vec{\mathbf{a}}$, $\vec{\mathbf{b}}$, $\vec{\mathbf{c}}$, $\vec{\mathbf{d}}$ and $\vec{\mathbf{e}}$ as represented in the Fig. 1.6 (*a*) are to be added. After selecting the proper scale for these vectors, the vector $\vec{\mathbf{a}}$ is drawn from an origin *O*. The choice of $\vec{\mathbf{a}}$ is arbitary. We could start with any of the vectors. Tail of vector $\vec{\mathbf{b}}$ starts from the head of vector $\vec{\mathbf{a}}$. Similarly tail of vector $\vec{\mathbf{c}}$ starts from the head of vector $\vec{\mathbf{b}}$. In this 'head to tail' method all vectors are drawn *i.e.* $\vec{\mathbf{OA}}$, $\vec{\mathbf{AB}}$, $\vec{\mathbf{BC}}$, $\vec{\mathbf{CD}}$, $\vec{\mathbf{DE}}$. The vector joining the origin to the head of the last vector is the resultant vector. Here Restultant vector

$$\vec{\mathbf{R}} = \vec{\mathbf{OE}} \text{ [Fig. 1.6 (b)]}$$

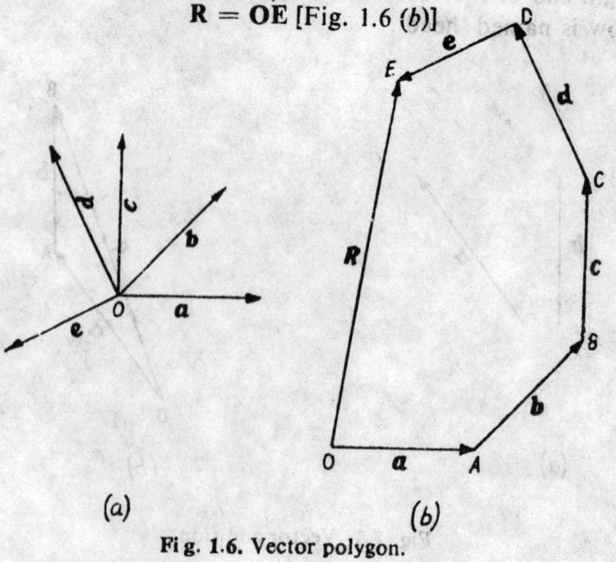

Fig. 1.6. Vector polygon.

Sometimes the last vector drawn comes back to the origin. In Fig. 1.7 such a case of a closed polygon is shown. The resultant vector in this case is zero ; the vectors have cancelled each other's effect.

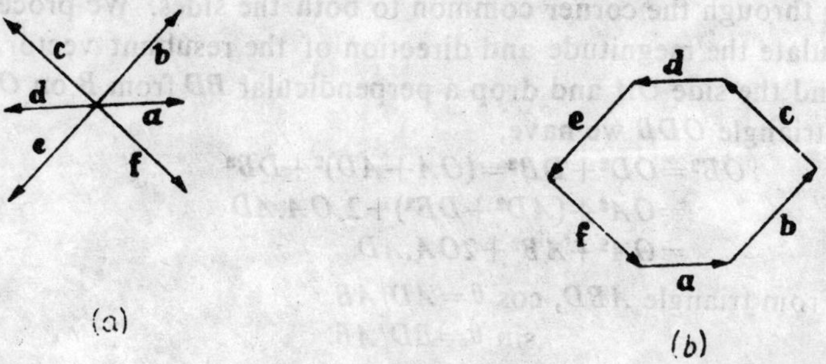

Fig. 1.7. Closed polygon.

1.15. Analytical Methods of Vector Addition

Addition of vectors can also be done by analytical methods as discussed below.

(*a*) *Vector triangle*. Refer to Fig. 1.4. Since OAB is a right-angled triangle, the magnitude of the resultant vector

$$OB=\sqrt{OA^2+AB^2}=\sqrt{3^2+4^2}=5 \text{ units}.$$

As 1 unit=50 kms, OB=250 kms.
From the triangle OAB,

$$\tan\theta=\frac{AB}{OA}=\frac{4}{3}=1.333$$

$$\theta=53°\ 6'\ 2''$$

Thus the resultant vector is determined.

(*b*) *Vector parallelogram*. Consider two vectors $\vec{\mathbf{a}}$ and $\vec{\mathbf{b}}$ represented by $\vec{\mathbf{OA}}$ and $\vec{\mathbf{OC}}$. The angle of inclination between them is θ. Complete the parallelogram $OABC$.

Since $AB=OC$, and they are parallel to each other, $\vec{\mathbf{AB}}$ and $\vec{\mathbf{OC}}$ are equal vectors.

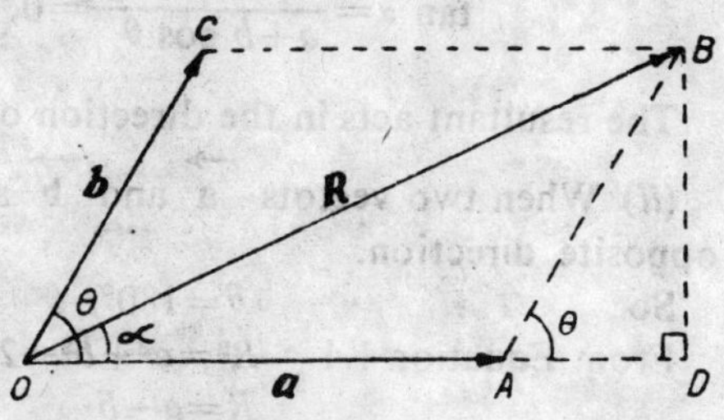

Fig. 1.8. Parallelogram of vectors.

Hence, the resultant $\vec{\mathbf{R}}$ $=\vec{\mathbf{OA}}+\vec{\mathbf{OC}}=\vec{\mathbf{OA}}+\vec{\mathbf{AB}}=\vec{\mathbf{OB}}$ *i.e.* the diagonal represents the resultant. This can be stated as follows:

If two vectors can be represented by the two adjacent sides of a parallelogram, in both magnitude and direction, the resultant of the vectors is given by the diagonal of the parallelogram, passing through the corner common to both the sides. We proceed to calculate the megnitude and direction of the resultant vector.

Extend the side OA and drop a perpendicular BD from B on OA. In the triangle ODB we have,

$$OB^2=OD^2+DB^2=(OA+AD)^2+DB^2$$
$$=OA^2+(AD^2+DB^2)+2\,OA.AD$$
$$=OA^2+AB^2+2OA.AD$$

From triangle ABD, $\cos\theta=AD/AB$

$\sin\theta=BD/AB$

Thus $AD=AB\cos\theta$ and $BD=AB\sin\theta$.

Substituting the value of AD.

$$OB^2=OA^2+AB^2+2\,OA.AB\cos\theta.$$

Replacing OB, OA and AB by R, a and b respectively,

$$R^2=a^2+b^2+2\,ab\cos\theta \quad ...(1.1)$$

Equation 1.1 gives the magnitude of the resultant vector. From the Fig. 1.8

$$\tan\alpha=\frac{BD}{OD}=\frac{BD}{OA+AD}=\frac{AB\sin\theta}{OA+AB\cos\theta}=\frac{b\sin\theta}{a+b\cos\theta} \quad ...(1.2)$$

Substituting the values of a, b and θ, we get the direction of the resultant with the help of a tangent table.

Special cases. (*i*) When two vectors $\vec{a}$ and $\vec{b}$ act in the same direction, so $\theta=0°$.

From Equation 1.1 $\quad R^2=a^2+b^2+2\,ab=(a+b)^2$.

Or $\quad R=a+b$.

Thus the magnitude of the resultant is equal to the sum of the magnitude of the vectors

$$\tan\alpha=\frac{b\sin\theta}{a+b\cos\theta}=0,\ \text{so},\ \alpha=0°.$$

The resultant acts in the direction of vectors $\vec{a}$ and $\vec{b}$.

(*ii*) When two vectots $\vec{a}$ and $\vec{b}$ act in the same line, but in opposite direction.

So $\quad \theta=180°$.

From Equation 1.1 $\quad R^2=a^2+b^2-2ab=(a-b)^2$

$$R=a-b$$

$$\tan\alpha=\frac{b\sin\theta}{a+b\cos\theta}=0$$

Thus $\quad \alpha=0°$

Hence the magnitude of the resultant is the difference of the magnitudes of vectors and will act along the line of vectors.

If $a > b$, the resultant will be in the direction of $\vec{\mathbf{a}}$.

If $b > a$, the resultant will be in the direction of $\vec{\mathbf{b}}$.

(*iii*) If vectors $\vec{\mathbf{a}}$ and $\vec{\mathbf{b}}$ act at right angles to each other.

So $$\theta = 90$$

$$R = \sqrt{a^2 + b^2}$$

and $$\tan \alpha = \frac{b \sin \theta}{a + b \cos \theta} = \frac{b}{a}$$

This relation is discussed in vector triangle also.

From the foregoing discussion it is evident that the resultant of two vectors ma be greater than, equal to, or less than either one of them depending on the angle between them.

The maximum value of the magnitude of resultant is the arithmetic sum of the magnitude of the two vectors and the minimum value is the arithmetic difference.

Example 1.1. Two vector of 5.0 units and 7.0 units make an angle of 60° with each other. What is the vector sum ?

Solution. The problem can be solved graphically by either vector triangle or parallelogram method.

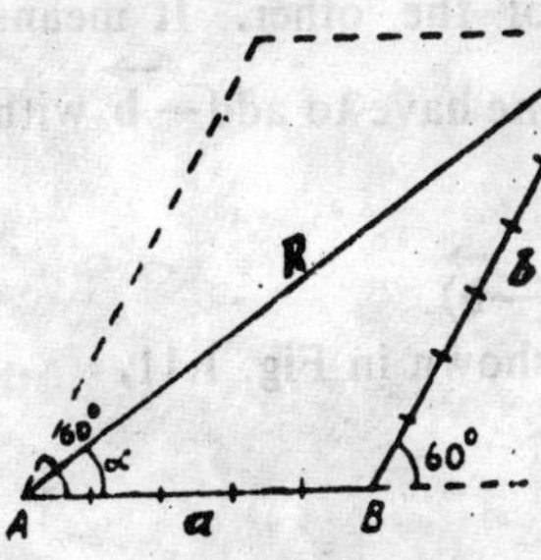

Fig. 1.9

Choosing a convenient scale draw a vector 5.0 units. At the end of $\vec{\mathbf{a}}$ draw the vector $\vec{\mathbf{b}}$ 7.0 units long so that there is a 60° angle between the direction of $\vec{\mathbf{a}}$ and $\vec{\mathbf{b}}$. The resultant vector $\vec{\mathbf{R}}$ is the line joining *A* to *C*. On measurement $\vec{\mathbf{R}}$ is found to be nearly 10.4 units and makes an angle 24°30′ with the vector $\vec{\mathbf{a}}$.

Applying analytical method the problem is solved with more accuracy.

Applying the Equation 1 1

$$R^2 = a^2 + b^2 + 2\,ab \cos \theta,$$

putting $a = 5.0$ units $b = 7.0$ units and $\theta = 60$

$$R^2 = (5.0)^2 + (7.0)^2 + 2\,(5.0)\,(7.0)\,(0.5)$$

$$R = 10.44 \text{ units.}$$

$$\tan \alpha = \frac{5 \times .8660}{7 + 5 \times .500} = 0.4558$$

$$\text{or } \alpha = 24°30'$$

Example 1.2. A boat moving at a speed of 4 km/h is required to cross a stream flowing at a speed of 3 km/h. If the boat were pointed directly to the opposite bank what would be its speed and direction while crossing the stream?

Solution. The boat will drift downward making an angle with the direct course. Drawing a vector diagram of the two velocities, we find the resultant is

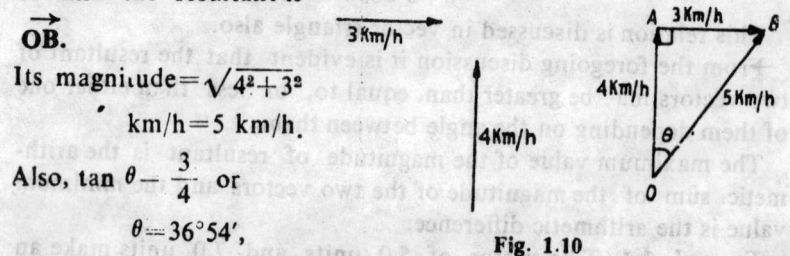

$\overrightarrow{\mathbf{OB}}$.

Its magnitude $= \sqrt{4^2 + 3^2}$

km/h = 5 km/h.

Also, $\tan \theta = \frac{3}{4}$ or

$\theta = 36°54'$,

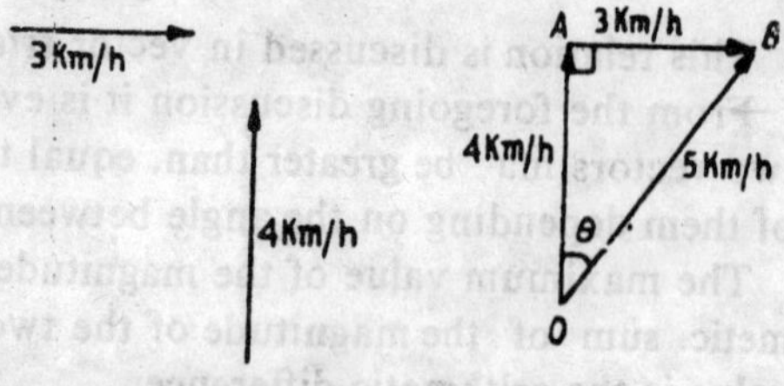

Fig. 1.10

So the boat moves making an angle of 53°6′ with the bank.

1.16. Subtraction of Vectors

The subtraction of a vector from another is caused by adding vectorially one vector to the negative of the other. It means to subtract a vector $\vec{\mathbf{b}}$ from a vector $\vec{\mathbf{a}}$ one have to add $-\vec{\mathbf{b}}$ with $\vec{\mathbf{a}}$.

Symbolically, $\vec{\mathbf{a}} - \vec{\mathbf{b}} = \vec{\mathbf{a}} + (-\vec{\mathbf{b}})$

The vector diagram of this statement is shown in Fig. 1.11,

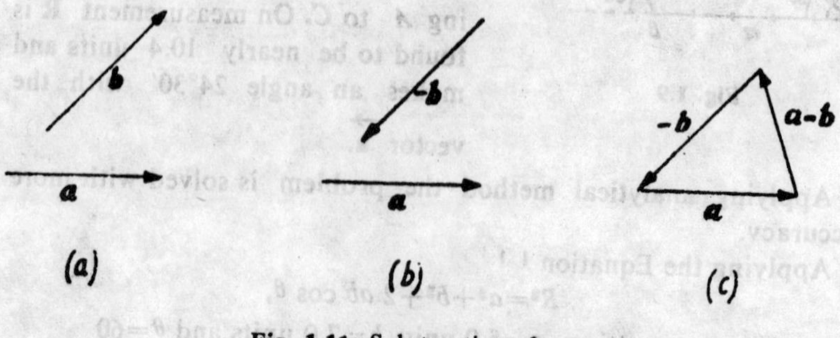

Fig. 1.11. Substraction of vectors.

It will of interest to note that in Fig. 1.8. (The Parallelogram of vectors) the vector represented by the diagonal $\vec{AC} = \vec{a} - \vec{b}$.

1.17. Properties of Vector Addition

The process of vector addition differs from scalar addition yet many rules of scalar addition are also applicable for vector addition.

(*i*) *Vector addition is commutative*. *OABC* is a vector parallelogram. Since the opposite sides of a parallelogram are equal and parallel, therefore,

$\vec{OA} = \vec{CB} = \vec{a}$ and

$\vec{AB} = \vec{OC} = \vec{b}$

Hence $\vec{OA} + \vec{AB} = \vec{OB}$

$= \vec{OC} + \vec{CB}$.

$\therefore \quad \vec{a} + \vec{b} = \vec{b} + \vec{a}$.

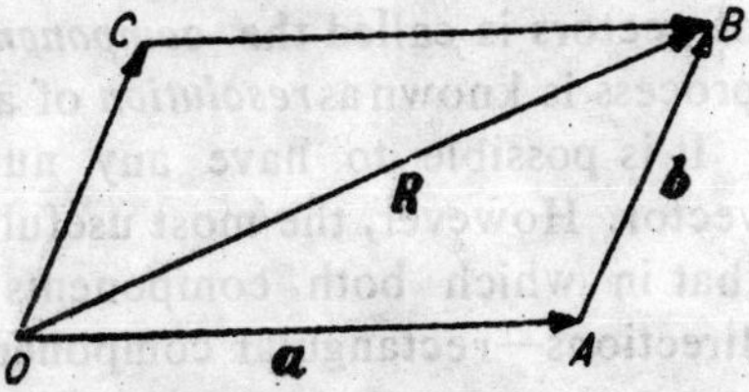

Fig. 1.12. Vector addition is commutative.

(*ii*) *Vector addition is associative.* By tail-head-tail...arrangement arrange the vectors as shown in Fig 1.13.

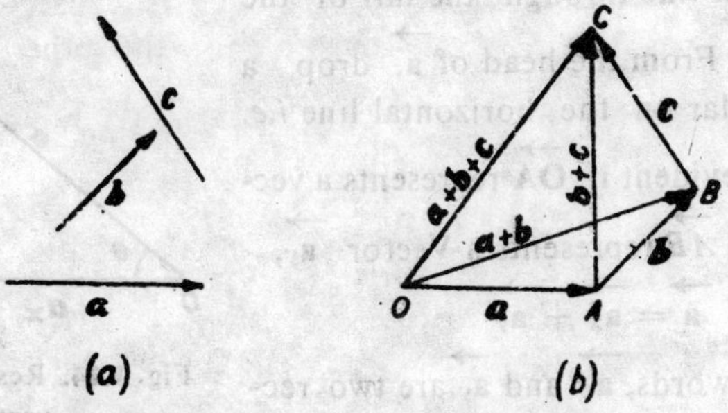

Fig. 1.13. Vector addition is associative.

Here $\qquad \vec{OA} = \vec{a}$, $\vec{AB} = \vec{b}$ and $\vec{BC} = \vec{c}$.

Join *O* to the head of vector $\vec{c}$.

$$\vec{OA} + \vec{AB} = \vec{OB} = \vec{a} + \vec{b}.$$

But $\qquad \vec{OB} + \vec{BC} = \vec{OC}$.

So, $\qquad (\vec{a} + \vec{b}) + \vec{c} = \vec{OC} \qquad \text{...}(i)$

Similarly $\vec{AB} + \vec{BC} = \vec{AC} = \vec{b} + \vec{c}$. But $\vec{OA} + \vec{AC} = \vec{OC}$.

Thus $\vec{\mathbf{a}} + (\vec{\mathbf{b}} + \vec{\mathbf{c}}) = \vec{\mathbf{OC}}$ ...(*ii*)

From (*i*) and (*ii*), $(\vec{\mathbf{a}} + \vec{\mathbf{b}}) + \vec{\mathbf{c}} = \vec{\mathbf{a}} + (\vec{\mathbf{b}} + \vec{\mathbf{c}})$.

Thus the vectors may be grouped in any order.

1.18. Resolution of a Vector

A number of vectors acting together can be replaced by a single resultant vector. The inverse process is also possible *i.e.* a given vector may be separated into a set of vectors whose effect when acting together is the same as that of the given vector. This set of vectors is called the *components* of the original vector and the process is known as *resolution* of a vector.

It is possible to have any number of components for a given vector. However, the most useful choice of components is usually that in which both components act in mutually perpendicular directions—rectangular components. Of the two, one is horizontal component or x-component and the other is vertical component or y-component.

A given vecter $\vec{\mathbf{a}}$ which makes an angle θ with the horizontal, Fig. 1.14, is to be resolved into two rectangular components. Draw a horizontal line through the tail of the vector $\vec{\mathbf{a}}$. From the head of $\vec{\mathbf{a}}$, drop a perpendicular on the horizontal line *i.e.* *BA*. It is evident if $\vec{\mathbf{OA}}$ represents a vector $\vec{\mathbf{a}}_x$ and $\vec{AB}$ represents a Vector $\vec{\mathbf{a}}_y$,

$$\vec{\mathbf{a}} = \vec{\mathbf{a}}_x + \vec{\mathbf{a}}_y$$

In other words, $\vec{\mathbf{a}}_x$ and $\vec{\mathbf{a}}_y$ are two rectangular components to which $\vec{\mathbf{a}}$ has been resolved.

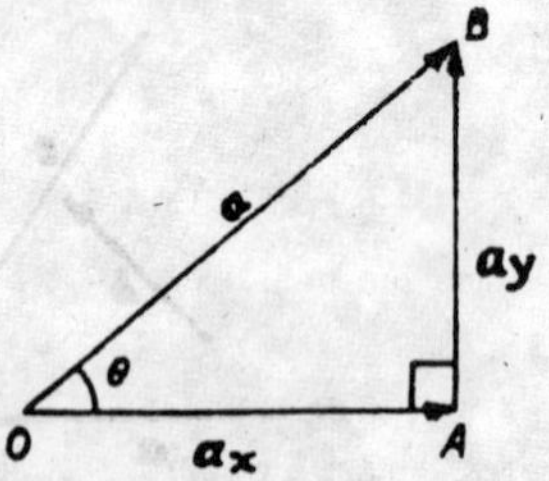

Fig. 1.14. Resolution of a vector.

In the triangle *OAB*,

$$\sin\theta = \frac{AB}{OB} \text{ and } \cos\theta = \frac{OA}{OB}$$

The magnitude of $\vec{\mathbf{a}}_x = OA = OB\cos\theta = a\cos\theta$.

$$\therefore \quad \vec{\mathbf{a}}_x = \vec{\mathbf{a}}\cos\theta \qquad ...(i)$$

Similarly the magnitude of $\vec{\mathbf{a}}_y = AB = OB\sin\theta = a\sin\theta$.

$$\therefore \quad \vec{\mathbf{a}}_y = \vec{\mathbf{a}}\sin\theta \qquad ...(ii)$$

Thus a single vector $\vec{a}$ is replaced by two vectors $\vec{a_x}$ and $\vec{a_y}$.

The advantages of resolution of a vector into its rectangular components will be evident when we deal with several vectors making various odd angles with each other. Each of these vectors can be replaced by two components in the horizontal and vertical directions. Thus there will be a set of *X*-components and another set of *Y*-components. All the *X*-component vectors can be simply added arithmetically thus replacing them by a single vector. Similarly all *Y*-component vectors added arithmetically can be replaced by a single vector. Thus large number of vectors are reduced to only two mutually perpendicular vectors. The resultant of these two vectors can be calculated easily and is the resultant of all vectors we started with.

Example 1.3. Four forces $\vec{a} = 10$ Newtons, $\vec{b} = 5$ Newtons, $\vec{c} = 6$ Newtons and $\vec{d} = 6$ Newtons are acting at a point in the manner shown in the Fig. 1.15. Determine thair resultant applying resolution method.

Solution. We consider vectors along *OX* and *OY* as +ve and vectors along *OX'* and *OY'* as negative. The two components to

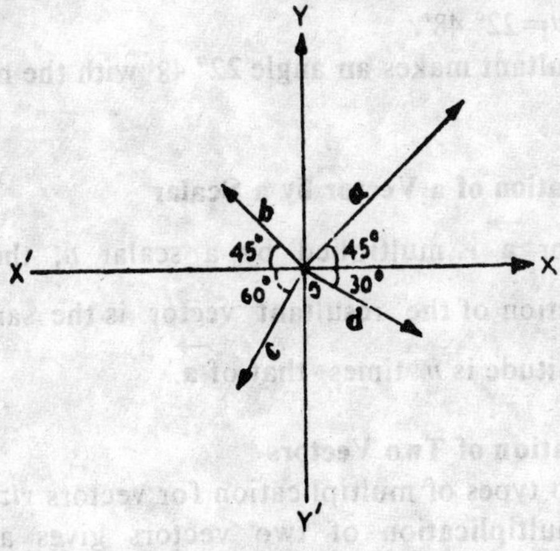

Fig. 1.15

which vector $\vec{a}$ is resolved are,

$$\vec{a_x} = \vec{a} \cos 45° \; N = 7.07 \; N$$

$$\vec{a_y} = \vec{a} \sin 45° \; N = 7.07 \; N$$

Similarly from

$$\vec{b}, \; \vec{b_x} = -\vec{b} \cos 45° \; N = -3.53 \; N$$

$$\vec{b}_y = \vec{b} \sin 45° \; N = 3.535 \; N$$

from $\vec{c}$,

$$\vec{c}_x = -\vec{c} \cos 60° \; N = -3.0 \; N$$

$$\vec{c}_y = -\vec{c} \sin 60° \; N = -5.196 \; N$$

from $\vec{d}$,

$$\vec{d}_x = \vec{d} \cos 30° \; N = 5.196 \; N$$

$$\vec{d}_y = -\vec{d} \sin 30° \; N = -3.00 \; N$$

Sum of all X-components

$$\vec{R}_x = \vec{a}_x + \vec{b}_x + \vec{c}_x + \vec{d}_x = 5.731 \; N$$

and sum of Y-components

$$= \vec{R}_y = \vec{a}_y + \vec{b}_y + \vec{c}_y + \vec{d}_y = 2.409 \; N$$

The resultant of the vector $\vec{R}_x$ and $\vec{R}_y$ is obtained by vector addition *i.e.* $\vec{R} = \vec{R}_x + \vec{R}_y$.

Magnitude $R = \sqrt{R^2_x + R^2_y}$

$$= \sqrt{(5.731 \; N)^2 + (2.409 \; N)^2}$$

$$= 6.217 \; N.$$

$$\tan \theta = \frac{BC}{AB} = 0.420$$

$$\theta = 22° \; 48'.$$

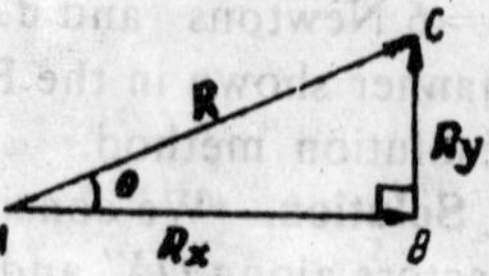

Fig. 1.16

Thus the resultant makes an angle 22° 48′ with the horizontal *i.e.* X-axis.

1.19. Multiplication of a Vector by a Scalar

When a vector $\vec{a}$ is multiplied by a scalar n, the resultant is $n\vec{a}$. The direction of the resultant vector is the same as that of $\vec{a}$ but the magnitude is n times that of $\vec{a}$.

1.20. Multiplication of Two Vectors

There are two types of multiplication for vectors *viz.*

(*i*) When multiplication of two vectors gives a scalar, the multiplication is called a *scalar or dot* product.

(*ii*) When multiplication of two vectors gives a third vector the multiplication is called a *vector or cross* product.

(*i*) *Scalar product*. The scalar product of two vectors $\vec{a}$ and $\vec{b}$ is written as $\vec{a} \cdot \vec{b}$ and is defined as $\vec{a} \cdot \vec{b} = |\vec{a}| \cdot |\vec{b}| \cos \theta$

$ab \cos \theta = ba \cos \theta = \vec{b} \cdot \vec{a}$. where θ is the angle between the direction of the two vectors.

Consider two vectors $\vec{\mathbf{a}}$ and $\vec{\mathbf{b}}$ as shown in Fig 1.17. Drop perpendiculars from A on OB and from B on OA.

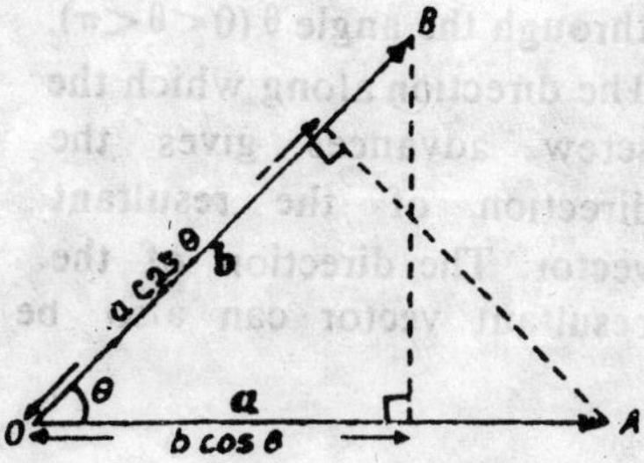

Fig. 1.17. Scalar product of two vectors.

Thus $\vec{\mathbf{a}}\ \ \vec{\mathbf{b}} = a\,(b\cos\theta)$

(magnitude of $\vec{\mathbf{a}}$) times (scalar projection of $\vec{\mathbf{b}}$ on $\vec{\mathbf{a}}$).

Also $\vec{\mathbf{b}}\cdot\vec{\mathbf{a}} = b(a\cos\theta)$ = (magnitude of $\vec{\mathbf{b}}$) time (scalar projection of $\vec{\mathbf{a}}$ on $\vec{\mathbf{b}}$).

Hence the scalar or dot product of two vectors is the product of the modulus of either vector and the projection of the other in its direction. Thus the dot product results in a scalar quantity. An example of scalar product is given by work which is a scalar quantity but derived by the scalar product of two vectors—Force and displacement.

Characteristics of dot product. (*i*) Since $\vec{\mathbf{a}}\cdot\vec{\mathbf{b}} = \vec{\mathbf{b}}\cdot\vec{\mathbf{a}}$; the scalar product is commutative.

(*ii*) If $\vec{\mathbf{a}}\cdot\vec{\mathbf{b}} = 0$, then either of the two vectors is a null vector or the vectors are mutually perpendicular *i.e.*

$$\theta = \frac{\pi}{2}.$$

(*iii*) If the vectors $\vec{\mathbf{a}}$ and $\vec{\mathbf{b}}$ are parallel, then $\vec{\mathbf{a}}\cdot\vec{\mathbf{b}} = \pm ab$ according as $\theta = 0$ or $\theta = \pi$.

(*iv*) $(m\vec{\mathbf{a}})\cdot(n\vec{\mathbf{b}}) = mn(\vec{\mathbf{a}}\cdot\vec{\mathbf{b}}) = mn\,\vec{\mathbf{a}}\cdot\vec{\mathbf{b}} = \vec{\mathbf{a}}\cdot mn\,\vec{\mathbf{b}}$ $= n\,\vec{\mathbf{a}}\cdot m\,\vec{\mathbf{b}}$ where m and n are scalar quantities. This is known as associative law.

(*ii*) **Vector product.** The vector product or cross product of two vectors $\vec{\mathbf{a}}$ and $\vec{\mathbf{b}}$ is a vector, written as $\vec{\mathbf{a}}\times\vec{\mathbf{b}}$ and is defined as $\vec{\mathbf{a}}\times\vec{\mathbf{b}} = |\vec{\mathbf{a}}|\times|\vec{\mathbf{b}}|\sin\theta\,\vec{\mathbf{n}}$ where θ is the angle between the two vectors $\vec{\mathbf{a}}$ and $\vec{\mathbf{b}}$ and $\vec{\mathbf{n}}$ is a unit vector normal to the plane containing $\vec{\mathbf{a}}$ and $\vec{\mathbf{b}}$. The direction of this resultant vector is obtained in the following way. Let a right-handed screw (Fig. 1.18)

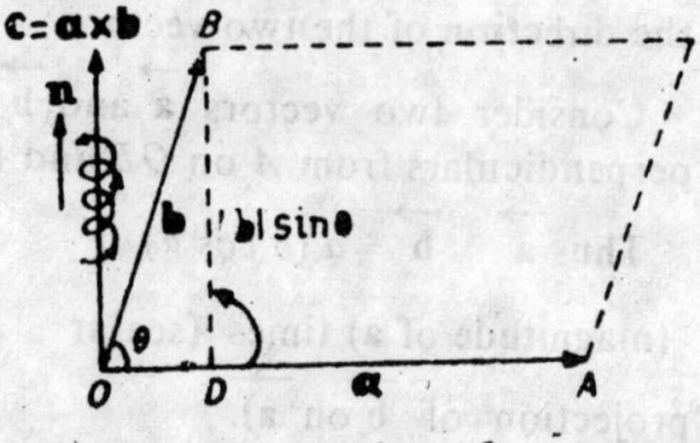

Fig. 1.18. Vector product of two vectors.

whose axis is perpendicular to the plane containing $\vec{a}$ and $\vec{b}$ be rotated from $\vec{a}$ to $\vec{b}$ through the angle θ (0<θ<π). The direction along which the screw advances gives the direction of the resultant vector. The direction of the resultant vector can also be found by a simple rule. Hold

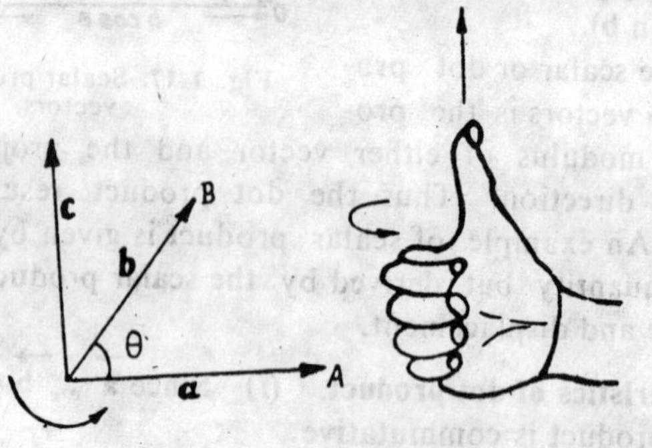

Fig. 1.19. Right hand rule.

your right hand in the manner shown in Fig. 1.19 with the thumbs erect and the fingers folded round. Let the direction of the thumb be perpendicular to the plane containing the vectors $\vec{a}$ and $\vec{b}$. If the fingers are folded in the direction in which $\vec{a}$ is to rotate to $\vec{b}$ through the angle θ (0<θ<π), then the thumb points in the direction of the resultant vector.

Thus the vector product of two vectors is a vector quantity.

Magnitude of $(\vec{a} \times \vec{b}) = |\vec{a} \times \vec{b}| = |\vec{a}|\ |\vec{b}| \sin\theta\ |\vec{n}|$

$= |\vec{a}|\ |\vec{b}| \sin\theta = ab \sin\theta = (OA)\ (OB) \sin\theta$ (Ref. Fig 1.19.)

=Area of the parallelogram in which a and b are the two adjacent sides.

Characteristics of vector product. (*i*) By reversing the order of the vectors, the sign of the product is reversed. To get the direction of the vector product $\vec{a} \times \vec{b}$, the sense of rotation of the right handed screw is from $\vec{a}$ to $\vec{b}$. But for getting the direction of the vector product $\vec{b} \times \vec{a}$, the sense of rotation of the right handed screw is from $\vec{b}$ to $\vec{a}$. Thus the direction of the vector product in this case is opposite to that in the other case.

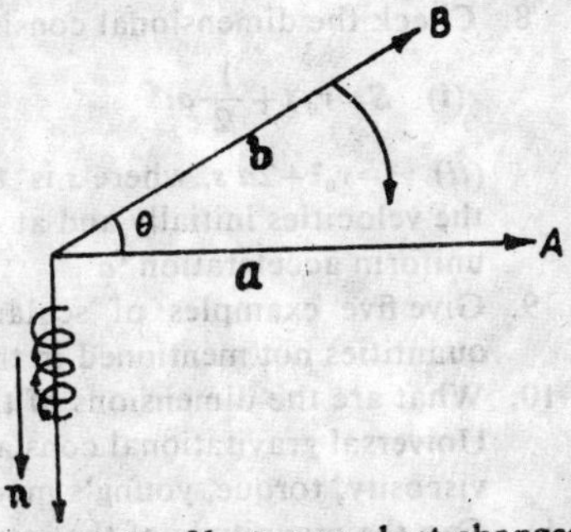

Fig. 1.20. Vector product changes the sign when the order of vectors in the product is reversed.
$a \times b = b \times a$

i.e. $\vec{a} \times \vec{b} = -\vec{b} \times \vec{a}$.

Hence vector product is not commutative. The student should note carefully the sense of rotation which is very important in this case.

(ii) The vector product is associative. If m is a scalar,

$$(m\vec{a} \times \vec{b}) = \vec{a} \times (m\,\vec{b}) = m\,(\vec{a} \times \vec{b})$$

(iii) If $\vec{a} \times \vec{b} = 0$, then either of two vectors is a null vector or vectors $\vec{a}$ and $\vec{b}$ are parallel.

When a charged particle carrying a charge $+q$ moves with a velocity $\vec{v}$ in a magnetic field of induction $\vec{B}$, the force $\vec{F}$ acting on it is given by $\vec{F} = q\,\vec{v} \times \vec{B}$ and the direction of $\vec{F}$ is perpendicular to the plane containing $\vec{B}$ and $\vec{v}$. This is an example of a vector product.

QUESTIONS

1. Give two examples to show that the S.I. is a coherent system of units.
2. What is the advantage in choosing the wavelength of a light radiation as a standard of length ? Is this standard temperature dependant?
3. Why is a Cesium oscillator a more accurate standard of time?
4. Arrange the following distances in decreasing order of magnitude: 1 foot, 1 dekameter, 1 micrometer, 1 mile, 3 yds, 2.5 m, 1 light year, 2 inches, 5 cms, 1 parsec, 1 Angstrom.
5. Does the measure of an angle depend upon unit of length?
6. What is the only fundamental unit still not defined by a natural standard?
7. Arrange the following masses in increasing order of magnitude: 1 mgm, 2 lbs, 1 kg, 11 quintols, 1 metric ton, 1 ton.

8. Check the dimensional consistencies in the following equations:

(*i*) $S = v_0 t + \frac{1}{2} at^2$

(*ii*) $v^2 = v_0^2 + 2a\,s$, where s is the distance moved in time t, v_0 and v are the velocities initially and at the time t when the body is moving with uniform acceleration 'a'.

9. Give five examples of scalar quantities and five examples of vector quantities not mentioned in this book.
10. What are the dimensions of the following: Universal gravitational constant, (G), angular velocity, surface tension, viscosity, torque, young's modulus of elasticity.
11. Can the magnitude of the resultant of two vectors be greater than the sum of magnitudes of individual vectors?
12. A vector is resolved along two mutually perpendicular directions in such a manner that the magnitude of both components are same. How are the components oriented with respect to the original vector?
13. A charged particle $+q$ is moving in a direction from bottom to the top of this page with a velocity $\vec{v}$ while a magnetic field of induction $\vec{B}$ is applied in a direction from left to right of the page. What will be the magnitude of the force acting on the charge? What would happen to the force if the direction of the field is reversed?
14. If $\vec{a} \cdot \vec{b} = \vec{b} \cdot \vec{c}$, show that $\vec{a}$ need not be equal to $\vec{c}$. When will $\vec{a}$ be equal to $\vec{c}$?
15. The frequency of vibration (n) of a stretched string is a function of tension (T), the length (l), and the mass per unit length (m). Prove that $n \propto \frac{I}{l}\sqrt{T/m}$ (dimensional analysis)

PROBLEMS

16. Express the following quantities in appropriate FPS units. 1.5 kg, 2 m², 3m³, 2 km/s. 2 m, 1 cm.
17. Express the following quantities in appropriate S.I. units. 1 inch, 2 lbs. 10 miles, 1 ft²,1 ft³, 1 yd,
18. A man wants to swim across a river 500 m wide. If he can swim at the rate of 4 km/h while the river flows at 3 km/h, then in what direction should he swim in order to reach a point exactly opposite his starting point? How much, time will he take to reach the other bank?

(*Ans.* 41°24′ with the bank,upstreom, $t = \frac{0 \cdot 5}{4 \sin 41^\circ 24'} = 11 \cdot 34$ min)

19. A picture weighing 2 kg is hung by a string fastened to the two upper corners of its frame and passing over a peg. If the two parts of the string are inclined at an angle of 60°, while the upper and the lower edges of the frame of picture remains horizontal, find the tension in the string.

(*Ans.* $T = \frac{2 \text{ kg}}{2 \cos 30^\circ} = 1.1547$ kg.)

2

LINEAR MOTION

A body is said to be in a state of *rest* when its position does not change with respect to its surrounding. On the other hand, if its position changes relative to the surrounding, it is said to be in a state of *motion*. The branch of physics that deals with the study of motion of objects is known as *mechanics*. Kinematics is that branch of mechanics that deals with the motion of a body without taking into account the forces or factors causing it. Generally there are three types of motion (*i*) linear or translatory motion in which the body moves along a straight line, (*ii*) circular or rotational motion in which the body revolves round a fixed point or rotates about a fixed axis, (*iii*) vibratory or oscillatory motion in which a body executes a to and fro periodic motion such as the swing of a pendulum.

The motion of a body is normally complicated; it may contain more than one of the three types of motion. This chapter will be devoted to the study of only linear motion of a body. The mass of the body will be treated as if concentrated in a point mass which will be referred to as a particle in discussions to follow to avoid other complications.

2.1. Displacement

A change of position of a body in a particular direction is called displacement. A body can move from *A* to *B* by infinite number of paths such as *ACB*, *ADB* etc. (Fig. 2.1). But its displacement is $\overrightarrow{AB}$ Displacement is a vector quantity. If the particle moves from *B* to *A*, its displacement is $\overrightarrow{BA}$. So though the magnitude is same in both cases, directon is different. Therefore displacement in the second case is opposed, to the displacement in the first case. We have seen in the previous chapter, in this case:

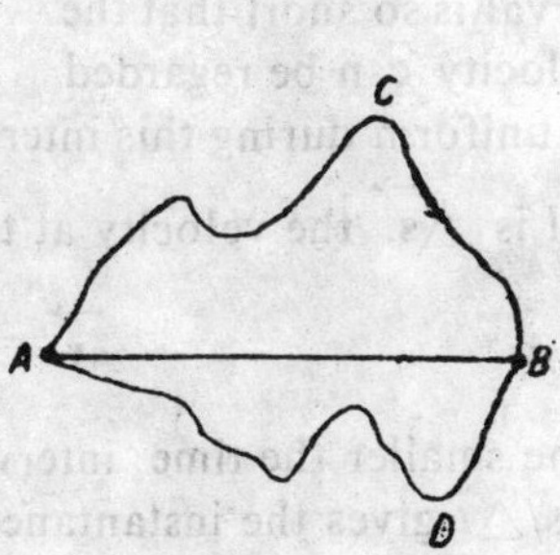

Fig. 2.1. Displacement.

$$\overrightarrow{AB} = -\overrightarrow{BA}$$

2.2. Speed and Velocity

The speed of a body in motion is its rate of change of distance *i.e.* the distance covered by the body in unit time. The velocity of a body is the rate of change of displacement *i.e.* the distance covered in a definite direction in unit time.

The speed is a scalar quantity whereas velocity is a vector quantity. The S.I. unit for both of them is meter per sec or m/s—and the dimensions are LT^{-1}.

If the displacements are equal for equal intervals of time, however, small these intervals may be, the body is said to be moving with a *uniform velocity*.

Suppose, a body is moving with uniform velocity of 60 km/h towards west. It means the direction is same throughout *i.e.* west and in every minute it covers a distance of 1 km; in every second it moves over 1/60 km and in every 1/1000th of a second it covers 1/60,000 km. Even for still smaller intervals of time distance covered will be equal. On the other hand, if velocity is different at different positions during displacement, the velocity is said to be non-uniform. In this case, two velocities of the body become relevant *viz. average velocity* and *instantaneous velocity.*

Consider a particle moving from *A* to *B* following the irregular path *ACB*, Fig. 2.2. To find the instantaneous velocity at any intstant of time *t*, (at *C*) we consider a very short time interval $\triangle t$ centring at *t* *i.e.* from $(t-\triangle t/2)$ to $(t+\triangle t/2)$. This time interval is so short that the velocity can be regarded as uniform during this interval. If the displacement during the interval is $\overrightarrow{\triangle s}$, the velocity at the instant *t*,

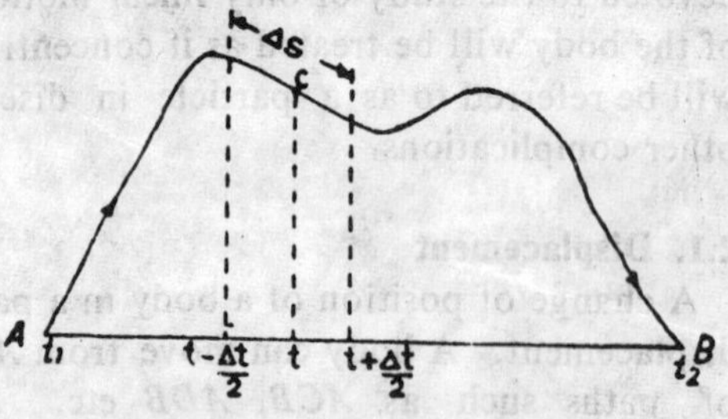

Fig. 2.2. Non-uniform velocity.

$$\vec{v}=\frac{\overrightarrow{\triangle s}}{\triangle t}.$$

The smaller the time interval $\triangle t$, the more accurately the ratio $\triangle s/\triangle t$ gives the instantaneous velocity of the particle. We may, therefore, write in the language of mathematics,

$$\text{Instantaneous velocity } \vec{v}=\lim_{\triangle t\to 0}\frac{\overrightarrow{\triangle s}}{\triangle t}=\frac{\overrightarrow{ds}}{dt} \qquad ...(2.1)$$

to the galactic centre and so on. Therefore, all objects supposed to be fixed on earth are actually moving with the speed of rotation of earth. Hence we cannot say a body to be in a state of absolute rest or in a state of absolute motion. Whatever we observe is only relative to some other object taken as reference. Let us take an example.

If two cars move with same speed in the same direction, their relative distance will remain unchanged. Taking one of the cars as reference, the velocity of the other car will be zero. Both the cars are having definite velocities but the velocity of one relative to the other is zero *i.e.* both of them are at rest with respect to each other. Similarly, if a car *A* going at a speed of 40 km/h overtakes another car *B* going at 30 km/h, the car *A* will appear to be moving at a velocity of 10 km/h to an observer in the car *B i.e.* the car *A* will have a relative velocity of 10 km/h with respect to the car *B*. On the other hand, the car *B* will have a relative velocity of −10 km/h with respect to the car *A*.

The relative velocity of one body with respect to another is the vector difference of the two velocities. Suppose a particle *A* and *B* are having velocities $\vec{v}_a$ and $\vec{v}_b$ respectively, (Fig. 2.6*a*). Then the relative velocity of *A* with respect to *B*.

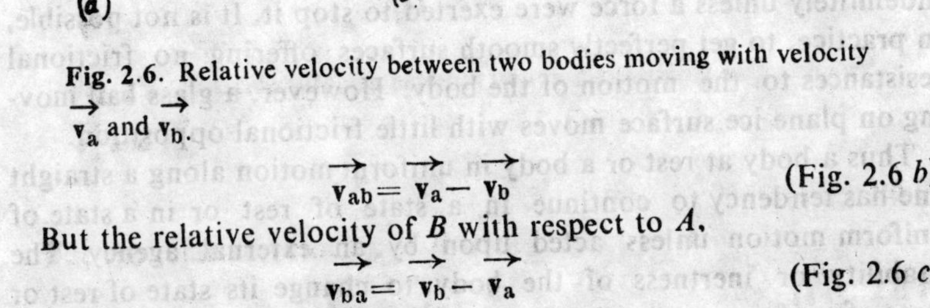

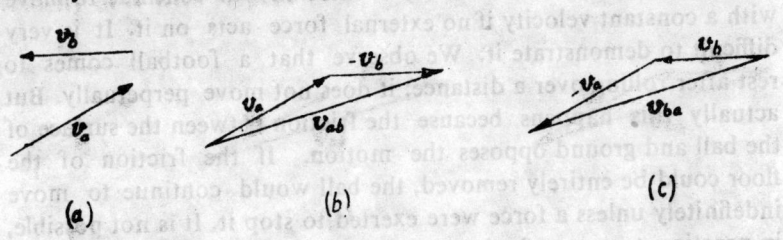

Fig. 2.6. Relative velocity between two bodies moving with velocity $\vec{v}_a$ and $\vec{v}_b$.

$$\vec{v}_{ab} = \vec{v}_a - \vec{v}_b \qquad \text{(Fig. 2.6 } b\text{)}$$

But the relative velocity of *B* with respect to *A*,

$$\vec{v}_{ba} = \vec{v}_b - \vec{v}_a \qquad \text{(Fig. 2.6 } c\text{)}$$

Special cases. If $\vec{v}_a$ and $\vec{v}_b$ are along the same straight line, simple algebraic addition or subtraction is necessary for deriving the relative velocity.

(*i*) If $\vec{v}_a$ and $\vec{v}_b$ are in the same direction and $v_a > v_b$.

$$v_{ab} = v_a - v_b \text{ and } v_{ba} = -(v_a - v_b)$$

(*ii*) If $\vec{v}_a$ and $\vec{v}_b$ are in opposite direction,

$$v_{ab}=v_a+v_b$$

and $$v_{ba}=-(v_a+v_b)$$

2.9. Newton's First Law of Motion—Galileo's Law of Inertia

We have studied the motion of bodies moving with uniform acceleration without considering external factors that produce such acceleration. We will now discuss the relationship between the acceleration and the external agencies that produce it.

A book lying on the table will remain at rest untill it is moved by some agency ; a car parked infront of the building will remain there untill it is moved by starting the engine; a heavy box on the floor will stay in place unless it is pushed or pulled. Thus a body at rest continues to remain at rest unless disturbed by some external agency.

What happens to a body in motion ? The Greek Philosopher Aristotle (384-322 B.C.) stated that a body will move with constant velocity so long as a constant force acts on it. After a period of nearly 2,000 years, Galileo (1564-1642 A.D.) contradicted this statement. From his study of motion on the surface of an inclined plane, he inferred that a body in motion will continue to move with a constant velocity if no external force acts on it. It is very difficult to demonstrate it. We observe that a football comes to rest after rolling over a distance; it does not move perpetually. But actually this happens because the friction between the surface of the ball and ground opposes the motion. If the friction of the floor could be entirely removed, the ball would continue to move indefinitely unless a force were exerted to stop it. It is not possible, in practice, to get perfectly smooth surfaces offering no frictional resistances to the motion of the body. However, a glass ball moving on plane ice surface moves with little frictional opposition.

Thus a body at rest or a body in uniform motion along a straight line has tendency to continue in a state of rest or in a state of uniform motion unless acted upon by an external agency. The inability or inertness of the body to change its state of rest or state of uniform motion by itself is called inertia— *inertia of rest* and *inertia of motion.*

Later Newton (1642-1727 A.D.) studied the problem of motion in detail and formulated them in three laws named after him. His first law is practically a re-statement of Galileo's law of inertia.

Statement of Newton's Ist Law of Motion. *Everybody continues* *in its state of rest or in state of uniform motion in a straight line unless it is compelled by an external impressed force to change that state of rest or uniform motion.*

This law gives a qualitative definition of force. The *force* is defined as that agency which acting on a body produces or tends to produce a change in the state of a body.

There are many illustrations of the 1st law of motion. When a bus in motion is suddenly stopped, the passengers experience a jerk in the forward direction. While the lower portion of the body in contact with the vehicle is stopped, the upper portion is still having the tendency or inertia of motion in the forward direction. On the other hand, when a vehicle is suddenly started, a passenger in it is likely to experience a jerk in the backward direction, obviously due to inertia of rest.

2.10. Newton's Second Law of Motion

Before we state the law, we will discuss the steps leading to it. Let us consider a simple experiment. We take a spring.

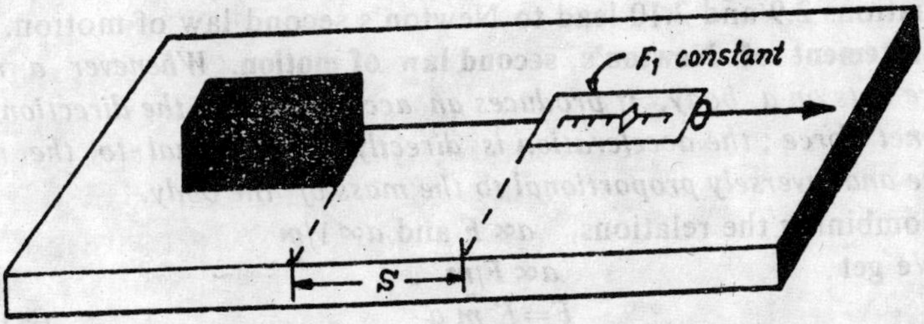

Fig. 2.7 Accelaration is proportional to the applied force.

balance and several identical block of metal. Let one metal block be placed on a horizontal smooth (frictionless) surface. The spring is attached to the metal block and the block is held while the spring is stretched for a known distance (Fig. 2.7). The reading in the spring balance is a measure of the force applied on the block. Let the force be F_1. The block is released but simultaneously the spring is also moved in the forward direction keeping the reading of the spring balance constant. Hence, a constant force is acting on the block. The distance, s, coverved by the body over a time interval, t, is noted.

The acceleration a_1 of the block$=2s/t^2$

$(\because\ s=v_1t+\frac{1}{2}a_1t^2$ and $v_1=0)$

The experiment is repeated with the spring stretched twice as much so that a force $F_2=2F_1$ is applied. The acceleration a_2 is determined. Further observations are taken with $F_3=3F_1$, $F_4=4F_1$ etc. and accordingly a_3, a_4 etc. are determined. From the experiment it is seen that,

$$\frac{F_1}{a_1}=\frac{F_2}{a_2}=\frac{F_3}{a_3}=\frac{F_4}{a_4}=\text{constant} \quad \cdots(2.8)$$

Therefore, $$a \propto F \quad \cdots(2.9)$$

This constant ratio of the force to the acceleration produced is a measure of the inertia of the body being accelerated and is known as *inertial mass* of the body.

In the second part of the experiment, let us take a number of blocks. A constant force F is applied while the spring is attached to one block. Let the acceleration produced be a_1. The same force F is applied when two blocks are attached to the spring; the corresponding acceleration produced in them is a_2. Similarly, for three blocks the accelaration is a_3 and so on.

If mass of a block is m_1, according to Equation 2.9 above,

$$F/a_1=m_1\ ;\ F/a_2=2m_1=m_2\ ;\ F/a_3=3m_1=m_3$$

or $$m_1a_1=m_2a_2=m_3a_3=\text{constant.}$$

Therefore $a \propto 1/m$

Equations 2.9 and 2.10 lead to Newton's second law of motion.

Statement of Newton's second law of motion. *Whenever a net force acts on a body, it produces an acceleration in the direction of the net force ; the acceleration is directly proportional to the net force and inversely proportional to the mass of the body.*

Combining the relations, $a \propto F$ and $a \propto 1/m$

We get $$a \propto F/m$$

or $$F=K\,m\,a \quad \cdots(2.11)$$

Where K is a constant of proportionality.

Newton's second law of motion is stated also in terms of change of momentum as follows.

The rate of change of momentum of a body is proportional to the net force and acts in the direction of the net force. Equation 2.11 can also be derived from this alternative statement. This is left as a problem for the student.

2.10 (a). Absolute Units of Force

In Equation 2.11, there is no restriction on units of force, mass and acceleration provided that the proper value is assigned to the constant K. For different systems the value of K is different. But for convenience, the unit of force is chosen in such a manner that K becomes equal to unity. In SI system the unit of force is chosen to be such a force which acting on a body of mass 1 kg produces an acceleration of 1 m/s^2 in it. This unit of force is named as 1 *Newton*. Putting these values in Equation 2.11,

$$1 \text{ Newton} = K \times 1 \text{ kg} \times 1 \text{ m/s}^2.$$

Thus $K = 1$.

Hence Equation 2.11, reduces to the form $\vec{F} = m\,\vec{a}$ ...(2.12)

We have restored the vector signs. This equation is used whenever any coherent set of units are used. From the definition of Newton, it is obvious that:

$$1 \text{ Newton (N)} = 1 \text{ kg m/s}^2.$$

In CGS system the unit of force is called *dyne*. A force of 1 dyne acting on a mass of 1 gm produces an acceleration of 1 cm/s^2 in it *i.e.* $1 \text{ dyne} = 1 \text{ gm. cm/s}^2$.

$$\begin{aligned} \text{Hence} \quad 1 \text{ N} &= 1 \text{ kg} \times 1 \text{ m/s}^2 \\ &= 1000 \text{ gm} \times 100 \text{ cm/s}^2 = (1{,}00{,}000)\text{gm} \times \text{cm/s}^2. \\ &= 10^5 \text{ dynes.} \end{aligned}$$

In FPS system the unit of force is *poundal*. A force of one poundal acting on a mass of 1 lb produces an acceleration of 1 ft/s^2 in it.

$$\begin{aligned} 1 \text{ poundal} &= 1 \text{ lb} \times 1 \text{ ft/s}^2 \\ &= (453.6 \text{ gms}) \times (30.54 \text{cm/s}^2) \\ &= 13{,}853 \text{ dynes} = 0.13833 \text{ Newtons.} \end{aligned}$$

The dimensions of force are $M\,L\,T^{-2}$.

It is evident that the first law gives a qualitative definition of force whereas the second law gives us a quantitative measure of the force (Equation 2.12).

2.10(b). Gravitational Systems of Units: Mass and Weight

The gravitational units of force is defined in terms of the pull of the earth upon a body. A body of mass m, is pulled towards the centre of the earth with a force of magnitude, mg, where g is the acceleration due to gravity (Equation 2.12). This force is termed as the *weight* of the body.

Thus weight $W = mg$

Hence if a force W acts on a body of mass m, it produces an acceleration g in it. This is utilised in defining the gravitational unit of force. The gravitational unit of force is that force which acting on a body of unit mass produces an acceleration equal to g in it. It is evident that the *gravitational unit of force is the weight of a unit mass.* In the SI system it is the weight of 1 kg, called 1 kg force or 1 kg. f. Normally this unit is not used. The gravitational unit of force in CGS system is the weight of 1 gm called 1 gm force (1 gm. f) In FPS system it is 1 lb weight.

It may be noted that gravitational unit of force

$$= g \times \text{absolute unit of force.}$$

$$1\ kgf = 1\ \text{Newton} \times g.$$

$$1\ gmf = 1\ \text{dyne} \times g.$$

$$1\ lb\ wt = 1\ \text{poundal} \times g.$$

The gravitational unit of force (also the weight of the body) is different at different locations on the surface of the earth as the value of g varies. But the absolute unit of force is same everywhere. Therefore, the absolute units of force will be used in this book.

It will be of interest to remember that weight of a body depends on the place of measurement though the mass is constant. The weight of a body is measured by a spring balance, but the mass of a body is measured by comparing its weight with that of a standard in a common balance

$$(\because\ w_1/w_2 = m_1g/m_2g)$$

Example 2.7. A net force of 6 N in applied to a body whose mass is 3 kg. Find the resulting acceleration.

Solution.

$$F = ma \quad \text{or} \quad a = F/m$$

$$= \frac{6\text{N}}{3\text{ kg}} = \frac{6\text{ kg. m. s}^{-2}}{3\text{ kg}} = 2\text{ m/s}^2.$$

Example 2.8. Two objects of mass 3 kg and 5 kg are attached to the ends of a cord which passes over a fixed frictionless pulley placed at 4.5 m above the floor. The objects are held at rest with 3 kg mass touching the floor and the 5 kg mass at 4 m above ground and then released.

(*a*) What is the acceleration of the system?

(*b*) What is the tension in the cord?

(*c*) How much time will elapse before the 5 kg object hits the floor ?

Given $g = 9.8\text{ m/s}^2$

Solution.

(*a*) The net force in the system

$$= (5-3) \times 9.8\text{ N}$$

$$= 19.6\text{ N}.$$

Total mass moving $= (5+3)$ kg

$$= 8\text{ kg}.$$

So acceleration of of the system

= Net force/Total mass moving

$$=\frac{19.6 \text{ N}}{8 \text{ kg}}=2.45 \text{ m/s}^2.$$

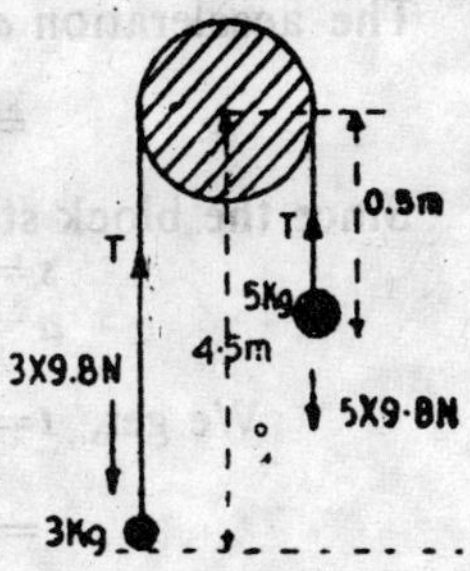

Fig. 2·8

Taking the downward direction as +ve, on 5 kg mass it will be $+2.45$ m/s² and -2.45 m/s² on 3 kg mass.

(*b*) Considering the force on 3 kg mass we have, the acceleration:

Net force/mass moving

$$=\frac{(3\times 9.8-T)\text{ N}}{3 \text{ kg}}=-2.45 \text{ m/s}^2$$

or $$T=36.75 \text{ N}.$$

If we start with 5 kg mass, the value of tension will be the same indicating tension is equal throughout the rope. However, the value of acceleration to be used in the equation will be $+2.45$ m/s².

(*c*) The 5 kg mass is at a height of 4 m initially. So to touch the ground, it will cover a distanse of 4 m starting from rest and moving with acceleration 2.45 m/s². Applying the relation

$$s=v_1 t+\tfrac{1}{2} a t^2$$

and on substitution $s=4$ m. $v_1=0$,

$$a=2.45 \text{ m/s}^2.$$

$$t=\sqrt{\frac{2\times 4}{2.45}} \text{ sec.}=1.81 \text{ sec.}$$

Example 2.9. A 25 kg block rests at the top of a smooth (frictionless) plane whose length is 2.0 m and whose height at elevated end is 0.50 m. How long will it take for the block to slide to the bottom of the plane when released?

Solution. The weight W of the block acts downwards as indicated in Fig. 2.9. The force W is resolved into two rectangular components, F acting along the inclined plane AC and R normal to the plane AC. The component F will produce motion of the block on the plane AC while R will have no effect in this direction.

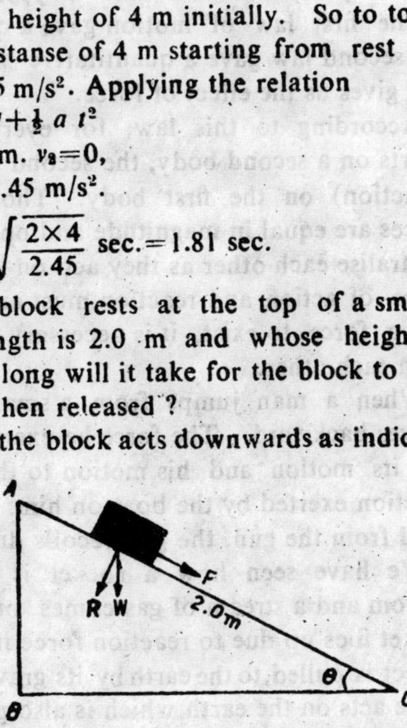

Fig. 2.9

$$F=W \sin\theta=m\, g \sin\theta$$

$$=(25 \text{ kg})(9.8 \text{ m/s}^2)\left(\frac{0.5\text{m}}{2.0\text{m}}\right)=61.25 \text{ N}$$

The acceleration $a = F/m$

$$= \frac{61.25 \text{ N}}{25 \text{ kg}} = 2.45 \text{ m/s}^2.$$

Since the block starts from rest, applying the relation

$s = v_1 t + \frac{1}{2} a t^2$ and putting $s = 2$ m, $v_1 = 0$

$a = 2.45$ m/s².

We get, $t = \sqrt{\frac{2 \times s}{a}} = \sqrt{\frac{2 \times 2.0 \text{ m}}{2.45 \text{m/s}^2}} = 1.2577$ sec.

$= 1.26$ seconds.

The block will slide down in 1.26 seconds.

2.11. Newtons Third Law of Motion

Statement. *To every action there is always an equal and but opposite reaction*; or, *mutual action of two bodies upon each other are always equal and directed in opposite directions.*

The first law of motion gave a qualitative definition of force ; the second law gave a quantitative measure of force and the third law gives us the effect of force.

According to this law, for every force (action) that one body exerts on a second body, the second body exerts an equal force (reaction) on the first body. Though the action and reaction forces are equal in magnitude and opposite in direction, they don't neutralise each other as they act on different objects. Since the force of action and reaction must exist together, it is obvious that for a force to exist, it is necessary that two bodies must interact with each other.

When a man jumps from a small boat to the shore, the boat moves backward. The force he exerts on the boat is responsible for its motion and his motion to the shore is due to the foree of reaction exerted by the boat on him. Similarly, when a bullet is fired from the gun, the gun recoils due to reaction force.

We have seen how a rocket is fired. The fuel burns at the bottom and a stream of gas comes out with a great force. The rocket flies up due to reaction force in the opposite direction. An object is pulled, to the earth by its gravitational force. The reaction force acts on the earth which is also pulled towards the object. But as the mass of the earth is large, the acceleration produced on it is too small to be observed. On the other hand, the object is accelerated by g. When a body is attached to the hook of a spring balance, the body pulls the spring downward with a force equal to its weight mg ; the spring also exerts an equal force, mg, on the body upwards.

These are few examples to illustrate the third law of motion.

2.12. Principle of Conservation of Linear Momentum

The linear momentum of a particle of mass m, moving with velocity $\vec{v}$ is a vector $m\vec{v}$. If no net force act on the body there is no acceleration and hence the body moves with a constant velocity $\vec{v}$. Hence its linear momentum $m\vec{v}$ remains constant.

Let us take the case of two objects in an isolated system (no external force acts on them). Suppose a small steel sphere A of mass

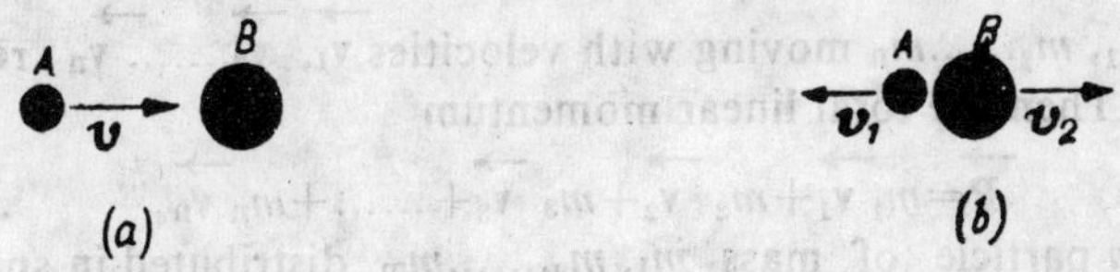

Fig. 2.10. Illustration of principle of conservation of linear momentum.

m_1, moves with a velocity $\vec{v}$ towards a big steel sphere B of mass m_2 and has a head on collision with it (Fig. 2.10 *a*). After the collision, A turns back with a velocity $\vec{v_1}$ and B moves forward with a velocity $\vec{v_2}$ (Fig. 2.10 *b*).

Since the velocity of the spheres are also in the same line we can take the magnitudes only. So momentum of the spheres before collision

$$=m_1v+0=p_1$$

Momentum of spheres after collision

$$=-m_1v_1+m_2v_2=p_2$$

(taking the direction from A to B as positive.)

If the velocities are measured, it will found

$$p_1=p_2$$

Similarly when a gun is fired, the bullet moves forward and the gun recoils. The momentum of the gun with bullet before firing is zero. After firing, the bullet of smaller mass moves forward with a very high velocity whereas the gun, whose mass is much larger, kicks backward with less velocity. The sum of the momentum of the system is still found to be zero.

Thus we can say that the vector sum of the linear momentum of two objects in an isolated system is constant. It is also found to be applicable to isolated systems containing more than two objects. It is stated as a general principle called 'Principle of Conservation of Linear Momentum'.

Statement. *If no net external force act on a system of several particles, the total linear momentum of the system is conserved (constant).*

Like the law of conservation of energy, the law of conservation of linear momentum is a fundamental law and of great importance in Physics. We explain below the physical significance of this principle.

The total linear momentum of the system is the vector sum of the linear momentum of each particle in the system.

Let us consider an isolated system containing n particles of mass $m_1, m_2......m_n$ moving with velocities $\vec{v}_1, \vec{v}_2,...... \vec{v}_n$ respectivelly. Then the total linear momentum

$$\vec{P}=m_1\vec{v}_1+m_2\vec{v}_2+m_3\vec{v}_3+......+m_n\vec{v}_n \qquad ...(2.13)$$

The n particle of mass $m_1, m_2,......m_m$ distributed in space can be replaced, for analytical purpose, by a single particle whose mass is the sum of the individual masses and the effect at a force on this single particle is the same as the effect on the system of particles. The point at which this imaginary particle is situated, called the centre of mass, depends on the mass, of the particles and their positions relative to one another.

So we replace the n particles by a single particle of mass M at the centre of mass of the system where

$$M=m_1+m_2+......m_n=\sum_{i=1}^{i=n} m_i$$

The velocity of this particle is written as $\vec{v}_{cm}$ which has the same effect as the agregate effect of velocities of individual particles of the system. Therefore, the momentum of this particle will be equal to the vector sum of the individual moments.

Hence, $$M\vec{v}_{cm}=m_1\vec{v}_1+m_2\vec{v}_2+......m_n\vec{v}_n$$

Therefore, from equation (2.13),

$$\vec{P}=M\vec{v}_{cm}$$

Differentiating with respect to time, we have

$$\frac{d\vec{P}}{dt}=M\frac{d\vec{v}_{cm}}{dt}=M\vec{a}_{cm}$$

where $\vec{a}_{cm}$ is the acceleration of the centre of the mass.

But $M\vec{a}_{cm}$=net force on the system (2nd law of motion).

Since the net force on the system = 0,

$$\frac{d\vec{\mathbf{P}}}{dt}=0.$$

Hence $\vec{\mathbf{P}}$ is contant *i.e.* the vector sum of the linear momentum is constant. Thus the principle of conservation of linear momentum is a direct consequence of Newton's second law of motion.

2.13. Third Law of Motion as a Consequence of Principle of Conservation of Linear Momentum

Suppose m_1 and m_2 be the mass ot two particles *A* and *B* constituting an isolated system. They are moving along the same line. Let *A* collide with *B*. The velocities of the particles change after collision. Let the change in velocity of *A* be $\Delta\vec{\mathbf{v}}_1$ and the change in velocity of *B* be $\Delta\vec{\mathbf{v}}_2$. Consequently, there is a change in momentum during a short interval Δt during which the particles are in contact. The change in momentum for *A* is

$$\Delta\vec{\mathbf{p}}_1=m_1\Delta\vec{\mathbf{v}}_1$$

and that for *B* is $\quad \Delta\vec{\mathbf{p}}_2=m_2\Delta\vec{\mathbf{v}}_2.$

According to the principle of conservation of linear momentum, the net change in the linear momentum in this isolated system is zero.

Thus $$\Delta\vec{\mathbf{p}}_1+\Delta\vec{\mathbf{p}}_2=0$$

or $$\Delta\vec{\mathbf{p}}_1=-\Delta\vec{\mathbf{p}}_2$$

or $$m_1\Delta\vec{\mathbf{v}}_1=-m_2\Delta\vec{\mathbf{v}}_2$$

Dividing both side by the time Δt and taking the limiting case when $\Delta t\to 0$.

$$m_1 \underset{\Delta t\to 0}{\text{Lt}}\frac{\Delta\vec{\mathbf{v}}_1}{\Delta t}=-m_2 \underset{\Delta t\to 0}{\text{Lt}}\frac{\Delta\vec{\mathbf{v}}_2}{\Delta t}$$

or $$\frac{m_1 d\vec{\mathbf{v}}_1}{dt}=-\frac{m_2 d\vec{\mathbf{v}}_2}{d}$$

or $$m_1\vec{\mathbf{a}}_1=-m_2\vec{\mathbf{a}}_2$$

Force on m_1 = −Force on m_2. Or Action = −Reaction

2.14. Collision—Elastic and Inelastic

If two objects in motion strike against each other or come close to each other such that the motion of one of them or both of them change abruptly, a collision is said to have takan place. The forces involved in collision may be large, but they act for only a very short time. The forces do not remain constant during this time. Their nature of variation is also very complicated. For this reason, it is usually convenient to study the problem of collision taking account the change in momentum. We come across many examples of collision daily. The coins in a carrom game colliding with one anothei or collision between two auotmobilies in a road accident etc. are examples of collision by physical contact. However, in case of two similar charged bodies they may not come in contact with each other but they will affect each other's motion when one is within the field of the other. This is also a type of collision but without physical contact.

In every collision or interaction, the linear momentum as well as the energy are conserved. But the type of energy usually changes because during collision, the transfer of kinetic energy takes place from one body to the other.

In any ordinary collision, some kinetic energy is lost in the form of heat and sound during collision. Hence the total kinetic energy usually decreases as the result of collision. Few cases where the total kinetic energy remains constant are known as *elastic collision*, and the rest of the cases where it decreases are known as *inelastic collission*. Certain collisions between atomic and nuclear particles are taken to be elastic. Two steel balls striking each other is an example of elastic collision but collision between two rubber balls is inelastic. If you throw a heavy metal ball to soft sand, the ball gets embedded in the sand and comes to rest. Here the kinetic energy of the ball is completely lost. This is an example of perfect inelastic collision.

One should note that when a dynamite cap is struck it releases lot of energy thus increasing the kinetic energy. Similarly, a neutron colliding with a nucleous releases nuclear energy with resulting high kinetic energy of the product particles. These examples should not be used to show that kinetic energy increases in collision. What happens actually is the releases of a new energy triggered by the collision.

Now we will analyse algebraically the problem of collision of two bodies. For convenience, a body is treated as a particle coinciding with the centre of mass of the body.

Head on collision. Suppose two bodies *A* and *B* of masses m_1 and m_2 respectively are moving along the same line with velocities $\vec{\mathbf{u}}_1$ and $\vec{\mathbf{u}}_2$ respectively. They collide and continue to move along the same line with velocities $\vec{v_1}$ and $\vec{v_2}$ respectively. This is an example of head on collision. Since the momentum remains constant:

$$m_1u_1+m_2v_2=m_1v_1+m_2u_2 \quad ...(2.14)$$

(Vector sum is represented as scalar sum since the velocities are in the same direction.)

From kinetic energy consideration:

$$\tfrac{1}{2}m_1v_1^2+\tfrac{1}{2}m_2v_2^2\leqslant\tfrac{1}{2}m_1u_1^2+\tfrac{1}{2}m_2u_2^2 \quad ...(2.15)$$

From equation (2.14), $m_1(v_1-u_1)=m_2(u_2-v_2)$ ...(2.16)

and from equation (2.15) $m_1(v_1^2-u_1^2)\leqslant m_2(u_2^2-v_2^2)$ ...(2.17)

Dividing Equation 2.17 by Equation 2.16,

$$\frac{v_1^2-u_1^2}{v_1-u_2}\leqslant\frac{u_2^2-v_2^2}{u_2-v_2}$$

or $(v_1+u_1)\leqslant(u_2+v_2)$

or $(v_1-v_2)\leqslant(u_2-u_1)$

Therefore, for an elastic collision $(v_1-v_2)=(u_2-u_1)=-(u_1-u_2)$ and in an inelastic collision $(v_1-v_2)<(u_2-u_1)$.

Thus in an elastic collision, the relative velocity after collision, (v_1-v_2), is equal to the negative of relative velocity, (u_1-u_2) before collisions *i.e.* the velocity of separation=–ve of velocity of approach. But in inelastic collision, the velocity of separation is less than the magnitude of velocity of approach.

Co-efficient of restitution. When one drops a small rubber ball from a height on a concrete floor, the ball rebounds. But it does not reach the same height from which it was dropped. This means the ball does not rebound or separate from the ground with the velocity with which it touched or approached the ground. If the velocity of separation were equal to the velocity of approach, the ball would have risen to the same height. If instead of a concrete floor, the ball is dropped on soft earth, the ball rises to a much smaller height. This obviously happens as the nature of collision between the rubber and the ground changes. This change is expressed by a term called co-efficient of restitution.

$$\text{Co-efficient of restitution, } e=-\frac{v_1-v_2}{u_1-u_2}=-\frac{\text{Velocity of separtion}}{\text{Velocity of approach}}$$

If the collision is perfectly elastic, $e=1$. If the collision is completely inelastic $e=0$; in this case, one body stick to the other *i.e.* velocity of separation is zero and they move as a single body after collision.

The velocities of the bodies *A* and *B* after an elastic collision can be calculated as follows.

Since the kinetic energy is conserved

$$\frac{1}{2} m_1 v_1^2 + \frac{1}{2} m_2 v_2^2 = \frac{1}{2} m_1 u_1^2 + \frac{1}{2} m_2 u_2^2$$

Simultaneous solution of this equation with Equation 2.14 gives:

$$v_1 = \frac{2m_2u_2 + u_1(m_1 - m_2)}{m_1 + m_2}$$

$$v_2 = [2m_1u_1 - u_2(m_1 - m_2)]/m_1 + m_2$$

For the special case in which the body B is at rest before collision,

$$u_2 = 0$$

$$v_1 = \frac{m_1 - m_2}{m_1 + m_2} u_1$$

and

$$v_2 = \frac{2m_1}{m_1 + m_2} u_1$$

Further, if $m_1 \ll m_2$, we have $v_1 \approx -u_1$ and $v_2 \approx 0$

This shows that a light body rebounds from a massive body with the same velocity as it approaches the massive body at rest.

Oblique incidence. We considered the case when the bodies after collision continues to move along the same line. But in actual practice, the directions of motion change after collision.

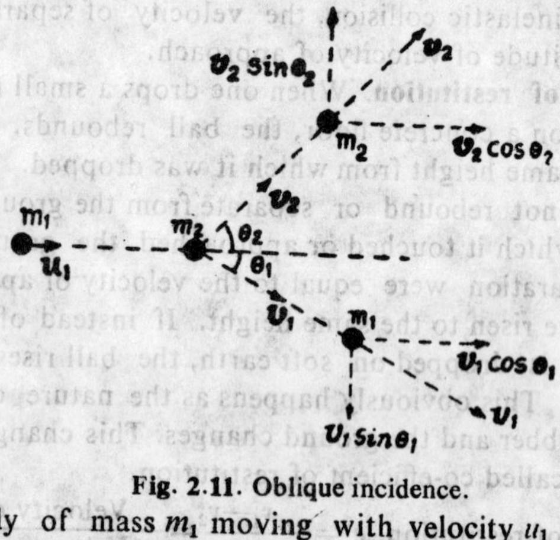

Fig. 2.11. Oblique incidence.

Let a body of mass m_1 moving with velocity u_1 collide with another particle of mass m_2 at rest, Fig. 2.11. After collision, let them move with velocities v_1 and v_2 respectively along directions inclined at angles θ_1 and θ_2 respectively with the initial direction of motion of m_1. Let the collision be perfectly elastic. To apply the relations for elastic collision equations, we have to resolve the velocities after collision into two mutually perpendicular components; x—component along the original direction of $\vec{u_1}$ and the y—component in a direction perpendicular to it.

Hence, applying the law of conservation of linear momentum to the x-component of velocities,

$$m_1 u_1 = m_1 v_1 \cos\theta_1 + m_2 v_2 \cos\theta_2 \qquad \ldots(i)$$

and similarly for y-component of velocities.

$$0 = m_1 v_1 \sin\theta_1 - m_2 v_2 \sin\theta_2 \qquad \ldots(ii)$$

Since the collision is perfectly elastic.

$$\frac{1}{2} m_1 v_1^2 = \frac{1}{2} m_1 v_1^2 + \frac{1}{2} m_2 v_2^2 \qquad \ldots(iii)$$

If m_1, m_2 and u_1 are known, the values of the rest four unknown quantities *viz.*, v_1, v_2, θ_1 and θ_2 can not be determined since only three equations are available. Hence the value of atleast one of these quantities should be known to predict the motion after collision.

Example 2.10. A body of mass 2.0 kg makes an elastic collision with another body at rest and turns back along the same line with a speed equal to one half of its original speed. Find the mass of the second body.

Solution. Applying law of conservation of linear momentum,

$$2 \times u = -2 \times \frac{u}{2} + mv$$

where u is the velocity of 2.0 kg body before collision; v is the velocity of the second body after collision and m is its mass.

So $$2u = -u + mv$$

or $$v = \frac{3u}{m} \qquad \ldots(i)$$

Since kinetic energy is conserved,

$$\frac{1}{2} \times 2u^2 = \frac{1}{2} \times 2\left(\frac{u}{2}\right)^2 + \frac{1}{2} mv^2$$

or $$u^2 = \frac{u^2}{4} + \frac{1}{2} mv^2$$

or $$v^2 = \frac{3u^2}{2m} \qquad \ldots(ii)$$

Substituting the value of v from (i),

$$\frac{9u^2}{m^2} = \frac{3u^2}{2m}$$

or $$m = 6 \text{ kg}.$$

2.15. Impulse

The concept of impulse is mostly used when a large force acts for a short duration such as in collision of two moving bodies, or in a sudden explosion or wind blowing at high speed during storm.

The force acts for a very small duration, but the magnitude

being high its effect is considerable. In collision, however, the magnitude of the force changes during collision though generally the direction is regarded constant. Further, the interval during which the force acts is two small to be measurable. Hence, the total effect of such a force is measured in terms of total change in momentum of the body due to the force.

According to Newton's second law of motion,

$$\vec{\mathbf{F}} = m\,\vec{\mathbf{a}}$$

where $\vec{\mathbf{F}}$ is the net force that produces an acceleration $\vec{\mathbf{a}}$ in the mass m. It can be written as

$$\vec{\mathbf{F}} = m\frac{d\vec{\mathbf{v}}}{dt}$$

or

$$\vec{\mathbf{F}}\,dt = md\,\vec{\mathbf{v}}$$

We can write

$$\int_{t_1}^{t_2}\vec{\mathbf{F}}\,dt = \int_{v_1}^{\mathbf{v}_2} md\,\vec{\mathbf{v}} = m(\vec{\mathbf{v}}_1 - \vec{\mathbf{v}}_2)$$

Where $\vec{\mathbf{v}}_1$ and $\vec{\mathbf{v}}_2$ are the velocities in the beginning and end of the time interval t $(t = t_2 - t_1)$

Substituting the values of respective momentums

$$\vec{\mathbf{p}}_1 = m\,\vec{\mathbf{v}}_1 \text{ and } \vec{\mathbf{p}}_2 = m\,\vec{\mathbf{v}}_2$$

$$\int_{t_1}^{t_2}\vec{\mathbf{F}}\,dt = \vec{\mathbf{p}}_2 - \vec{\mathbf{p}}_1$$

Thus, the impules of the force, is equal to the change in momentum of the body. Impulse is denoted by $\vec{\mathbf{J}}$.

$$\therefore \quad \vec{\mathbf{J}} = \int_{t_1}^{t_2}\vec{F}dt = \vec{\mathbf{p}}_2 - \vec{\mathbf{p}}_1 \qquad \ldots(2.18)$$

When F is either a constant force or an average force, then $\int Fdt$ can be replaced simply by Ft.

The unit of impulse is the impulse produced when one unit of force acts for 1 sec. In SI system, the unit is Newton. sec and in CGS system, it is dyne.sec. Its diamensions are MLT^{-1}.

Example 2.11. A ball of mass 0.1 kg is thrown in the horizontal direction against a vertical wall at a speed of 50 m/s. It rebounds with a velocity of 30 m/s in the opposite direction.

(a) What is the impulse of the collision? (b) If the ball remained in contact with the wall for 0.001 sec, what is the magnitude of the average force?

Solution. (a) The impulse $\vec{J}$ = change in momentum of the ball
$$= (0.1 \text{ kg}) (-30 \text{ m/s}) - (0.1 \text{ kg}) (50 \text{ m/s})$$
$$= -8 \text{ N·S}$$

The original direction of the ball is taken as positive.

(b) $\quad\quad J = Ft \text{ or } F = \dfrac{J}{t} = \dfrac{8 \text{ N·S}}{0.0015} = 8,000 \text{ N}$

2.16. Projectile Motion

When you throw a stone up the sky it travels through space for some distance before it falls to the ground. A bullet fired from the gun moves due to the initial velocity imparted to it but will gradually come down under the action of gravity. Neither the stone nor the bullet possess any driving power during its motion excepting the initial velocity imparted to them at the time of launching. An object launched into space, without driving power of its own, that travels freely under the action of gravity and air resistance alone, is called a *projectile*. The path described by a projectile is called a *trajectory*. Since the projectile has no other driving force, the only acceleration acting on it is due to earth's gravity acting downward. Hence the velocity changes continually both in direction and in magnitude. The motion remains on a vertical plane.

The analysis of projectile motion is made convenient by considering the velocity of the projectile to be the resultant of a horizontal and vertical component of the velocity. In otherwords, the velocity is resolved into a vertical and a horizontal component. Neglecting the air resistance, the horizontal component remains constant during its motion. But the vertical component continually decreases till it is zero, then the vertical component increases continually in the downward direction.

Equations of Motion of a projectile. Suppose a projectile is launched with an initial velocity $\vec{v_1}$ making an angle θ_0 with the horizontal OX. OX and OY are the co-ordinate axes, with O as the origin, in the vertical plane containing the velocity vector $\vec{v_1}$. The velocity $\vec{v_1}$ can be resolved into horizontal and vertical components of magnitudes $v_1 \cos \theta_0$ and $v_1 \sin \theta_0$ respectively.

The vertical acceleration of the projectile is in the negative (downward) direction along the Y-axis. Hence magnitude of the vertical component of the velocity at any instance,

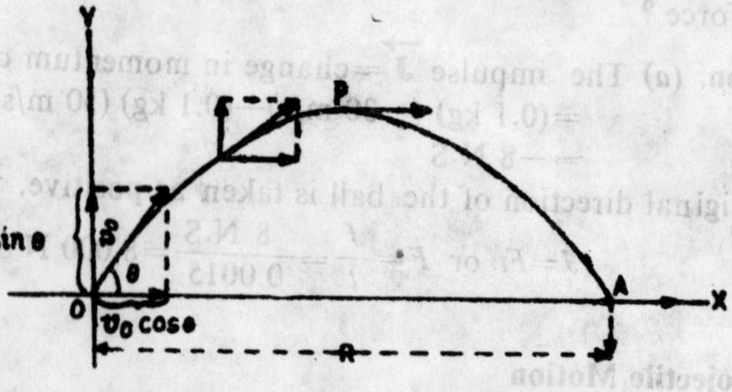

Fig. 2.12. Trajectory of a Projectile.

where t is the time interval from start upto that instant of time.

$$v_y = v_i \sin\theta_0 - gt$$

$$(\because \quad v = v_i + at) \qquad \ldots(i)$$

Neglecting air resistance, the horizontal acceleration is zero. *i.e.* the magnitude of velocity remains constant at any time t.

So $\qquad v_x = v_i \cos\theta_0 \qquad \ldots(ii)$

As the value of t increases the vertical component of the velocity v_y decreases in magnitude till at reaches the highest point P where the projectile will be momentarily horizontal. At P the Y-component of the velocity will be zero and the X-component will have the constant value given by (ii).

Thus at $\qquad P, \; v_x = v_i \cos\theta_0$

and $\qquad v_y = v_i \sin\theta_0 - gt = 0$

or $\qquad t = \dfrac{v_i \sin\theta_0}{g} \qquad \ldots(2.19)$

i.e. the projectile will reach the highest point after a time t.

The magnitude of the velocity of the projectile at any instant is

$$v = \sqrt{v_x^2 + v_y^2}$$

and the angle this resultant velocity makes with the horizontal is given by

$$\tan\theta = \frac{v_y}{v_x}$$

The distance covered in the horizontal direction at any instant t,

$$x = v_x t = (v_i \cos\theta_0)\, t \qquad \ldots(2.20)$$

Similarly, the distance covered in the vertical direction at any instant t, $\qquad y = (v_i \sin\theta_0)\, t - \frac{1}{2} gt^2 \qquad \ldots(2.21)$

The projectile rises to the highest point P after a time t given by Equation 2.19. If the height of P is h, then,

Putting the value of t in Equation 2.21

$$h=(v_i \sin \theta_o)\,\frac{v_i \sin \theta_o}{g}-\frac{1}{2}\,g\left(\frac{v_i \sin \theta_o}{g}\right)^2$$

$$=\frac{v_i^2 \sin^2 \theta}{2g} \qquad ...(2.22)$$

Equation of the trajectory of a projectile. From Equation 2.20,

$$t=\frac{x}{v_i \cos \theta^o}$$

Putting the value of t in equation 2.21,

$$y=(v_i \sin \theta_o)\,\frac{x}{v_i \cos \theta_o}-\frac{1}{2}\,g\left(\frac{x}{v_i \cos \theta_o}\right)^2$$

$$=\tan \theta_o \,.\, x-\frac{gx^2}{2v_i^2 \cos^2 \theta_o}=px-qx^2 \qquad ...(2.23)$$

where p and q are constant replacing $\tan \theta_o$ and

$$\frac{q}{2v_i^2 \cos^2 \theta_o}$$

respectively. This is the equation of a parabola. Thus, the trajectory of a projectile is a parabola.

The Range of a projectile. In Fig. 2.12, O is the point of projection and A is the point where the trajectory intersects the horizontal axis. The distance OA is called '*horizotal range*' or simply range of the projectile. It is denoted by R.

At the point A, the Y-coordinate is zero. If the projectile reaches A after a time interval T, from equation (2.21),

$$0=(v_i \sin \theta_o)\, T-\tfrac{1}{2}\, gT^2$$

or

$$T=\frac{2v_i \sin \theta_o}{g} \qquad ...(2.23)$$

(Note that T is known as *time of flight*, this time is twice the time taken by the projectile to reach the highest point of the trajectory, Equation 2.19.)

The range R is the horizontal distance covered by projectile in in this time interval, T.

Hence

$$R=v_x T=(v_i \sin \theta_o)\left(\frac{2v_i \sin \theta_o}{g}\right)$$

$$=v_i^2\,\frac{\sin 2\,\theta}{g} \qquad ...(2.25)$$

From this expression, it is evident that for a given velocity of projection, the range R is maximum when $\sin 2\theta$ is maximum *i.e.* when $2\theta=90^\circ$ or $\theta=45^\circ$.

Example 2.12. A projectile is thrown with a speed of 98 m/s in a direction 30° above the horizontal. Find the time of flight, range and the height to which it rises.

($g=9.8$ m/s²)

Solution. Time of flight

$$T=\frac{2v_1 \sin \theta_0}{g}$$

Here $v_1=98$ m/s $\theta_0=30°$ $g=9.8$ m/s².

Hence $T=\frac{2\times 98 \text{ m/s} \times \sin 30°}{9.8 \text{ m/s}^2}=10$ seconds.

Range $R=v_x\, T=v_1 \cos \theta_0 \,.\, T$

$=98 \text{ m/s}\times(\cos 30°)\times 10$ s

$=848.7$ sec. $=14$ min. 8.7 sec.

Max height $h=\frac{v_1^2 \sin^2 \theta_0}{2g}$

$=\frac{(98 \text{ m/s})(\sin 30°)^2}{2\times 9.8 \text{ m/s}^2}$

$=122.5$ m.

Example 2.13. A shell is fired from the top of a building 78.4 meter high in the horizontal direction with a velocity of 300 m/s. (*a*) How long does the shell take to reach the ground? (*b*) What is the range? (*c*) With what vertical velocity does the shell strike the ground? ($g=9.8$ m/s²).

Solution. (*a*) At the time of projection, $v_y=0$, $v_x=300$ m/s. If t is the time taken by the shell to reach the ground travelling a distance of 78.4 meters in the downward direction, we get from Equation 2.21,

$78.4 \text{ m}=0+\frac{1}{2}\times(9.8 \text{ m/s}^2)(T)^2$

or $T=4$ s

(*b*) Range $R=v_x T$

$=300 \text{ m/s}\times 4$ s

$=1200$ m.

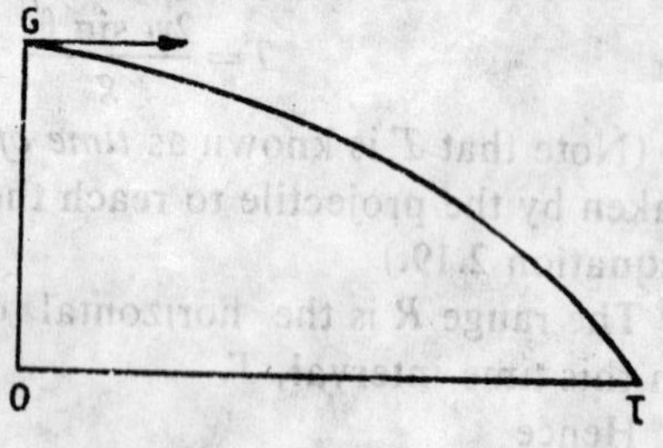

Fig. 2.13

(*c*) The projectile will hit the ground with a vertical velocity v_y, when $v_y=0+9.8 \text{ m/s}^2\times 4 \text{ s}=39.2$ m/s.

WORK, ENERGY AND POWER

2.17. Work.

In our day to day life we do many types of work. But in physics, the term work is restricted to cases in which the point of application of a force undergoes displacement. Thus a man with a load on his head does no work, though he may experience sufficient physical strain. But in lifting the load to his head, he does work as the force represnted by the weight of the load is displaced.

The amount of work done by a force or against a force is given by the product of the force and the distance through which the point of application of the force moves in or against the direction of the force.

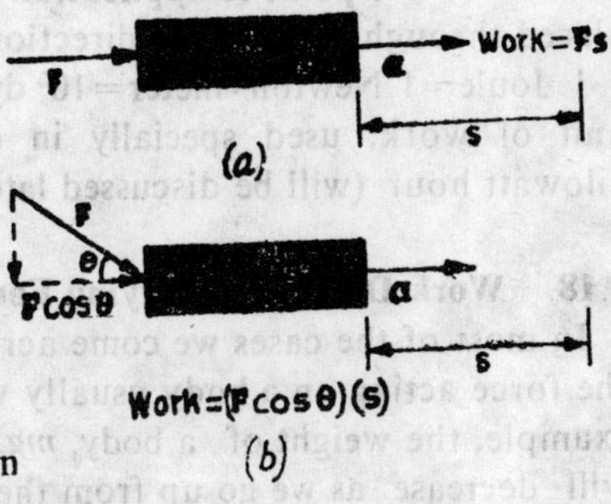

Fig. 2.14. Work done by a force. $W = \vec{F} \cdot \vec{S}$

$$W = F \, . \, S \text{ (Fig. 2.14 } a) \qquad \ldots(2.26)$$

where W is work done, F is the magnitude of the force and S is the distance moved by the point of application of the force in the direction of the force.

But the force and displacement may not be in the same direction always. Let θ be the angle between the direction of the force $\vec{\mathbf{F}}$ and the displacement $\vec{\mathbf{S}}$. Then the magnitude of component of $\vec{\mathbf{F}}$ parallel to $\vec{\mathbf{S}}$ is $F \cos \theta$ (Fig. 2.14 *b*).

$$\text{Hence work done, } W = (F \cos \theta)\,(S) = \vec{\mathbf{F}} \cdot \vec{\mathbf{S}} \qquad \ldots(2.27)$$

i.e. the work done is the scalar product of force vector and displacement vector. Though force and displacement are vector quantities work is a scalar quantity.

When $\theta = 0$, Equation 2.27 reduces to Equation 2.26. When $\theta = 90°$, $W = 0$ as $\cos \theta = 0$. Hence no work is done by the force even though there might be a displacement of the point of application of the force. For example, a body rotating in a circular orbit is subjected to a centripetal force (*Chapter 3*) directed towards the centre of the orbit. Since the motion is always perpendicular to this force, no work is done by this force.

Unit of work. A unit of work is done when the point of application of a unit force is displaced through unit distance in the direction of the force. In SI system it is Newton-meter or Joule

(J). In CGS system, the unit of work is erg which is the work done when the point of application of a force of 1 dyne is displaced through 1 cm in the direction of the force.

1 Joule = 1 Newton·meter = 10^5 dynes. 10^2 cm = 10^7 ergs. Another unit of work, used specially in electrical measurement is the kilowatt hour (will be discussed later).

2.18. Work Done by a Varying Force

In most of the cases we come across in nature, the magnitude of the force acting on a body usually vary during displacement. For example, the weight of a body, *mg*, will decrease as we go up from the surface of the earth. Hence the work done in raising body over a great height will involve displacement of the point of application of a varying force. Of course, when the height is small compared to the radius of the earth, *g* and hence weight is regarded constant in all calculation.

To calculate the work done by a varying force we follow the following method.

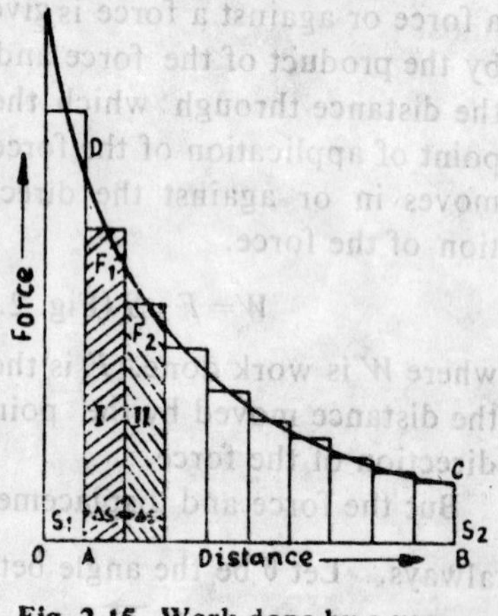

Fig. 2.15. Work done by a varying force.

Let us consider a force whose magnitude is changing as the point of application of the force is displaced through a distance. The nature of the variation is presented in Fig. 2.15. Divide the total displacement into very small equal parts $\Delta \vec{s}$ each.

Over this small displacement, the force represented by the ordinate drawn at the middle of the interval may be regarded as constant.

We want to calculate the work done when the point of application of the force is displaced from *A* to *B*. The ordinate in the middle of the first interval after *A* represent a force $\vec{F}_1$ which is regarded as constant over the interval Δs. So the work done when the point of application of the force moves from *A* over this small interval Δs is

$W_1 = \vec{F}_1 \cdot \Delta \vec{s}$ = Area of the shaded rectangle (I)

Similarly the work done for next interval,

$W_2 = \vec{F_2} \cdot \vec{\Delta s}$ = Area of the rectangle (II)

Therefore, the total work done for displacement S from A to B, is

$$W = W_1 + W_2 + \ldots\ldots = \vec{F_1} \cdot \vec{\Delta s} + \vec{F_2} \vec{\Delta s} + \ldots\ldots$$

which is given by the sum of the areas of all the rectangles Corresponding to all small displacements. Since the figure is magnified, the upperside of the rectangles and the force-displacement curve appear to be different. But in the limit when $\Delta s \to 0$, the number of rectangles tend to infinity and the upperside of the rectangles and the force-displacement curve very nearly coincide with each other. So the sum of the areas of all rectangles will be almost equal to the area under force-displacement curve $ABCD$ *i.e.* the area bounded by the curve and both displacement and force axes,

$$\text{Thus} \qquad W = \lim_{\Delta s \to 0} \sum_{s_1}^{s_2} \vec{F} \; \vec{\Delta s} = \int_{s_1}^{s_2} \vec{F} \cdot \vec{ds} \qquad \ldots(2.28)$$

2.19. Energy: The Capacity to do Work

The capacity of doing work by a body or system of bodies is called energy. This means a body possessing more energy is capable of doing more work. When work is done on a body in the absence of frictional force, energy of the body is increased by the amount of work put in. On the otherhand, when work is done by the body, energy of the body decreases by the same amount. The units in which energy is expressed are the same as the units for work.

Energy can exist in many forms such as mechanical, heat, light, sound, electrical, magnetic, chemical, nuclear etc. Energy from one form can be transformed into another. Electric energy is converted to mechanical energy in a fan: to heat energy in an electric stove, to light energy in an electric bulb. On burning coal, chemical energy of the coal changes into heat energy. The heat energy may produce steam which can move a turbine (machanical energy). If the turbine is coupled with the rotor of a dynamo, electricity is generated. When you rub your palms vigorously, they get heated up—mechaical energy is converted into heat energy. Electric current flowing in a coil gives magnetic property to it.

The generation of hydro-electricity is another example of conversion of mechanical energy to electrical energy. In India we have many hydro-electric stations such as in Hirakund, Machkund, Bhakra-Nangal, Sileru etc. A hydro-electric station is based at a place when a river has a steep gradient such as in hilly region. A dam is constructed just before the location where the river moves in a greater incline. Pipes are laid from the reservoir to the bottom and regulated flow of water through them hits the blades of the turbines at a great speed. The turbine moves and generates electricity.

Einstein showed (1905) that even mass and energy are equivalent to each other and he established a relation connecting mass and energy. The relation is $E=mc^2$ ...(2.29)
Where E is energy, m is mass c is the speed of light.
As $c=3\times 10^8$ m/s, the energy equivalent of even a small mass of 0.1 gm is $(1\times 10^{-4}$ kg$)\times(3\times 10^8$ m/s$)^2$
or 9×10^{12} joules.

In this universe, most of the energy exist in the form of mass. We know vast amount of nuclear energy is released when small part of the mass is converted to energy in nuclear fission (atom bomb) and fusion (hydrogen bomb) reactions. The sun and stars derive their energy from the conversion of mass into energy during the fusion of hydrogen inside them. However, the mechanism of complete conversion of any mass into energy is not yet known.

2.20. Potential and Kinetic Energy

Potential energy. Energy can also be classified as potential energy and kinetic energy. The energy possessed by a body by virtue of its position or configuration or internal mechanism is known as potential energy. Electrical energy, elastic energy, magnetic energy, nuclear energy and gravitational energy are most important forms of potential energy. When we wind a clock, the workdone by us is stored in the spring and the spring runs the clock by unwinding itself. The spring possess potential energy because of its deformed condition. The most common form of potential energy is the gravitational energy. In raising a body to a height from the surface of the earth work is done against the force of attraction due to earth. The amount of work done is stored as potential energy of the body and is recovered when the body is released to come down. Thus, by virtue of its position at a higher level, the body possesses more capacity to do work than it had when it was at ground level.

Let a body of mass m be raised to a height h, above the surface of the earth. The weight of the body, $W=m\,g$ is the force acting towards the centre of the earth. Hence, increase in potential energy *i.e.* work done to raise the body

$\triangle E_p = \text{Force} \times \text{distance}$

$= W\,h = mgh$ ...(2.30)

Conventionally ground level is taken as the reference level where the body is regarded to be possessing zero potential energy. Thus instead of referring as increase in potential energy, $\triangle E_p$ is simply written as E_p and referred as Gravitational Potential Energy.

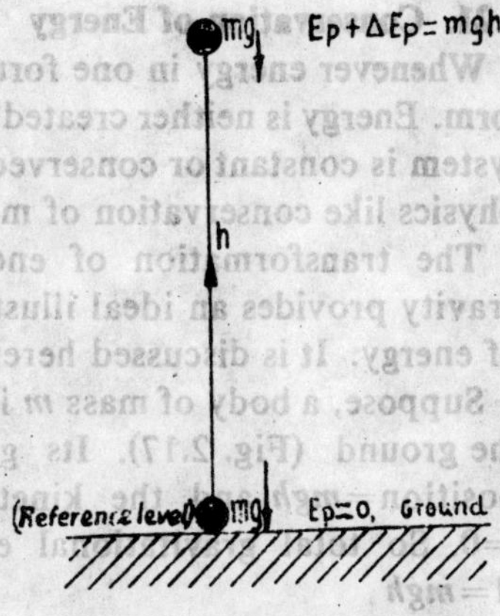

Fig. 2.16. Gravitational potential energy.

Kinetic energy. Energy possessed by a body due to motion is known as kinetic energy. An automobile in motion has kinetic energy whereas when it is stationary it has no such energy.

The kinetic energy of a moving body is equal to the amount of work it will perform before it is brought to rest. Alternatively, it is also equal to the amount of work originally needed to impart the velocity to the body.

Let a body of mass m, moving with an initial speed v_1 be acted upon by a steady force F over a distance S in the direction of the force. Let the speed at the end be v_f.

The work done W is given by $W = \int F ds$

We have $F = ma = m \dfrac{dv}{dt}$ But $\dfrac{dv}{dt}$ can be written as $\dfrac{dv}{ds} \cdot \dfrac{ds}{dt}$

so that $$W = \int m \left(\frac{dv}{ds} \cdot \frac{ds}{dt}\right) ds$$

$$= \int m \frac{ds}{dt} dv$$

on replacing $\dfrac{ds}{dt}$ by v, $W = \displaystyle\int_{v_1}^{v_f} mv\, dv = \tfrac{1}{2}mv_f^2 - \tfrac{1}{2}mv_1^2$

The work done on the body by the force appears as a change in kinetic energy *i.e.* $\triangle E_k = \frac{1}{2}\, m\, (v_f^2 - v_1^2)$.

If a body is initially at rest, $v_1 = 0$ and the gain in kinetic energy is simply termed as kinetic energy.

Thus the kinetic energy of a body at any instant,

$$E_k = \tfrac{1}{2} mv^2 \qquad ...(2.31)$$

2.21. Conservation of Energy

Whenever energy in one form disappears it reappears in another form. Energy is neither created nor destroyed. The total energy in a system is constant or conserved. This is a fundamental principles in physics like conservation of momentum.

The transformation of energy in a body falling freely under gravity provides an ideal illustration of principle of conservation of energy. It is discussed here:

Suppose, a body of mass m is held at rest at A at a height h above the ground (Fig. 2.17). Its gravitational potential energy in this position $= mgh$ and the kinetic energy $= 0$. So total gravitational energy at

$A = mgh$ ...(*i*)

Let the body be released and come down to the position B after covering a distance s. Its velocity at

$B = \sqrt{2gs}$ $(\because v^2 = 0 + 2gs)$

So Potential energy at $B = mg\ (h - s)$

Kinetic energy at B

$= \frac{1}{2}mv^2 = \frac{1}{2}m.2gs = mgs$

So the total gravitation energy at B

$= mg\ (h - s) + mgs = mgh$ ...(*ii*)

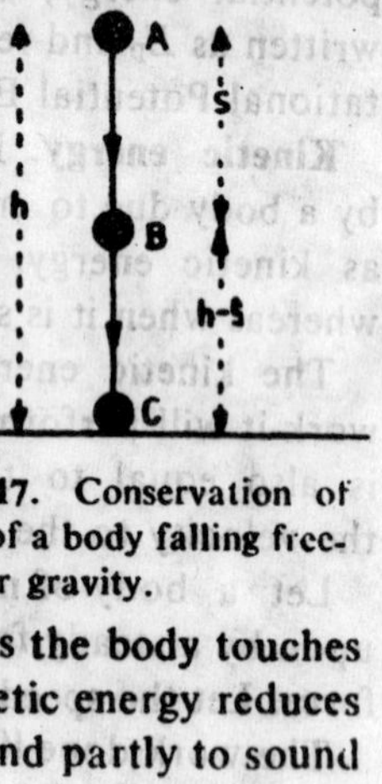

Fig. 2.17. Conservation of energy of a body falling freely under gravity.

Thus total energy at A = Total energy at B. As the body touches the ground, both gravitational potential and kinetic energy reduces to zero. They are partly transformed into heat and partly to sound energy.

2.22. Power

Power is the time rate at which work is done. If the rate at which work is done is constant and in a time interval t, the total amount of work done is W, then,

$$P = W/t$$

Hence work done in time t = Power $\times\ t$.

The units of power in the MKS system is Joule per second *i.e.* the work is done at the rate of 1 Joule per second. This is also called *Watt* named after James Watt, the inventor of steam engine. The *kilowatt* (1000 watts) is more frequently used. The CGS unit, erg per second is very small and usually not used. There is another unit of power in use called the *horse-power*. One horse power is equal to 746 watts.

Since work is the product of power and time, any power unit murtiplied by a time unit may be used as unit of work. We have seen the ratings of electric bulbs say 100 watts written on it. This means that the bulb consumes electrical energy at the rate of 100 watts *i.e.* 100 Joules per sec. So if this bulb is lighted for an hour, the energy consumed is $(100\times 60\times 60)$ watt-sec. or 100 watts $\times 1$ hour. This is expressed as 100 watt-hour. Electrical energy consumed is measured in terms of kilowatt hour or 1000 watt hours. One kilowatt hour of energy is referred as 'one unit' in the electric bills given to us by Board of Electricity. Thus we pay for energy, not for power.

$$1 \text{ kwh} = 1000\times 60\times 60 \text{ Joules} = 36\times 10^5 \text{ Joules.}$$

Example 2.13. A box is pushed along a horizontal surface with a constant velocity for a distance of 3.0 m against a frictional force of 500 N. How much work is done?

Solution. Since there is no acceleration, there is no net force acting on the box *i.e.* the force of friction is equal to the force with which the box is pushed.

So, work done $\quad W = F.s. = (500 \text{ N})(3.0 \text{ m})$

$= 1500 \text{ N.m.} = 1500$ Joule.

Example 2.14. A 10 kg stone is raised to the top of a building 30 m high. What is the increase in potential energy of the stone, (*a*) if it is raised vertically? (*b*) if it is moved along a ramp, inclined at an angle of 30° with the horizontal?

$(g = 9.8 \text{ m/s}^2)$

Solution. (*a*) $\quad E = F.s. = (mg)(s)$

$= (10 \text{ kg})(9.8 \text{ m/s}^2)(30 \text{ m})$

$= 2940 \text{ N.m.} = 2940$ Joule

(*b*) Neglecting friction,

$E = F.s. \cos\theta$ where θ is the angle between the direction of the force and displacement.

Value of *s* in this case is *CA*,

From Fig. 2.17,

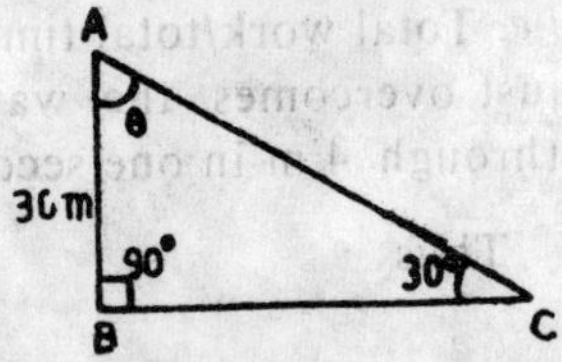

Fig. 2.18

$$\sin 30° = \frac{30\text{m}}{CA}$$

or $\quad CA = 60$ m.

and, $\quad \theta = 60$.

So $\quad E_p = (10 \text{ kg})(9.8 \text{ m/s}^2)(60 \text{ m}) \cos 60° = 2940 \text{ N.m.}$

$= 2940$ Joule.

The increase in potential energy is the same whatever the path may be provided the height is same.

Example 2.15. A bullet of mass 10 g moving with a speed of 500 m/s gets embedded in a tree after penetrating 5 cms into it.

Calculate the average resisting force exerted by the wood on the bullet?

Solution. Kinetic energy of the bullet just before striking the wooden block is:

$$E_K = \frac{1}{2}mv^2 = \frac{1}{2}(0.01 \text{ kg})(500 \text{ m/s})^2$$
$$= 1250 \text{ N.m.}$$

This kinetic energy of the bullet is uesd in doing work against the average resisting force $\overline{\mathbf{F}}$ of the wood.

Hence $(\overline{\mathbf{F}})(0.05 \text{ m}) = 1250 \text{ N.m.}$

or $\overline{\mathbf{F}} = 5{,}000 \text{ N.}$

Example 2.16. A force of 200 N acts on a body of mass 20.0 kg so that it accelerates from rest untill it attains the velocity of 50 m/s. Through what distance was force exerted?

Solution. The amount of work done on the body=Kinetic energy gained by it. Let the distance through which the force is applied be S;

Then; $(200 \text{ N})(S) = \frac{1}{2}(20 \text{ kg})(50 \text{ m/s})^2.$

or $$S = \frac{(10 \text{ kg})(50 \text{ m/s})^2}{200 \text{ N}} = 125 \text{ m}$$

Example 2.17. What is the energy consumed by a 200 watt electric bulb in 20 hours?

Solution. Energy consumed $= 200 \text{ watt} \times 20 \text{ hours.}$

$= 4{,}000$ watt hour $= 4$ kwh.

Example 2.18. A motor boat moves at a steady speed of 4 m/s. If the water resistance to the motion of the boat is 4,000 N, calculate the power of the engine?

Solution. Power of the engine is the rate at which work is done *i.e.* Total work/total time. Since there is no acceleration the engine just overcomes the water-resistance. So a 4,000 N force moves through 4 m in one second.

Thus $$P = \frac{(4000 \text{ N} \times 4\text{m})}{1s} = 16000 \text{ N m/s}$$
$$= 16 \times 10^3 \text{J/s} = 16 \text{ kw.}$$

2.23. Friction

If we want to slide a block of wood on a horizontal table top, we must continue to apply a steady horizontal force for the box to slide uniformly over the surface. Since there is no acceleration, we know from Newton's second law of motion, no resultant force is acting on the body. We, therefore, conclude that a force acting in opposite direction cancels the force applied. Hence it is obvious that there is a force parallel to the surfaces in contact, acting in a direction to oppose motion. This opposing force is called force of *friction.*

If it were possible to have a perfectly frictionless surface, according to Newton's First Law of motion, a body would continue its non-stop motion. But in nature we donot come across a perfectly frictionless surface. We try to reduce frictional force by making the surfaces as smooth as possible. To reduce the friction between carom dice and the surface of the board, we spray powder on the body. Yet it is not possible to completely eliminate friction.

The magnitude of the force of friction depends upon the nature of surfaces in contact. The frictional force when a block of wood moves over an wooden surface will be different from the frictional force when the same block moves over a steel surface.

The friction produces many undesirable effects. It increases the work necessary to operate machinery. Work done, in overcoming the friction is converted to heat and does not produce any useful work. Heat causes further damages. Steps are taken in designing machineries to reduce friction. Lubrication is applied in moving parts of machinery to reduce friction. Electric fans, motors, bicycles etc. are required to be oiled and greased regularly so as to reduce friction. In some big machines, continuous supply of some lubricant is provided for reducing friction in moving parts. Air is also used as a lubricant. Purified air compressed between the two surfaces forms an elastic cushion between the running parts and eliminates the friction between the moving parts.

The friction, however, is desirable in many respects. We are able to walk because of the friction between the feet and ground. On a smooth or slippery surface it is difficult to walk due to the lack of friction. Nails and screws hold to the surfaces due to friction. In cycling, driving a car, striking a match and sewing a piece of cloth and even a writing on paper is made possible due to friction.

2.24. Origin of Friction

The frictional forces are caused due to adhesion of one surface to the other and by interlocking of irregularities of surfaces in contact. Even a highly polished surface when examined under a very powerful microscope reveals dents and irregularities, Fig. 2.19. Due to interlocking of the sufraces, the adhesive forces between

Fig. 2.19. The surfaces of two bodies in contact. The irregularrities on the surfaces have been magnified.

dissimilar molecules in contact, will have components parallel to the surface. When one body moves over the other, the relative motion is opposed by this force. This opposing force is the force of friction.

2.25. Static and Kinetic Friction

We have seen earlier that force of friction comes into play whenever there is relative motion between the surfaces in contact. But the friction is present even for stationary surfaces in contact. This is called *static friction.* Consider the following example, Fig. 2.20.

A rectangular block of wood A is kept on a horizontal table. A string attached to one end of the block passes over a pully P and a scale pan, S is suspended from the free end of the string. The weight of the scale pan pulls the block towards the pulley, but the block does not move as the force of friction just balance the force of gravity. Let a small weight W_1 be added to the scale pan. The pull on the block increases by W_1, yet the block remains static. Thus the force of friction has increased to balance the increased pull. But this does not continue indefinitely. The static friction, F_s has a limiting value beyond which it cannot increase further. So when the weight on the scale pan provides a pull just in excess of this limiting value of the static frictional force called *limiting friction*, F_L, the block starts moving. After the block starts moving the frictional force is called *kinetic friction.*

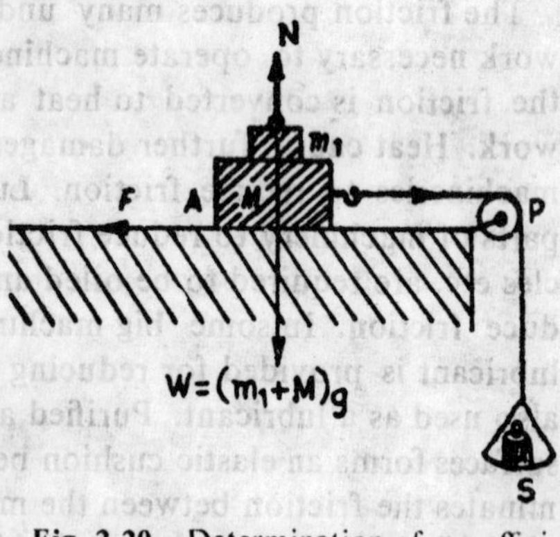

Fig. 2.20. Determination of co-efficient of static friction.

But as the block slides the magnitude of the frictional force decreases. The new value of the force of friction is called the force of kinetic friction, F_k, or force of sliding friction. F_k is always less than F_L For the block to start sliding, the applied force should be atleast F_L. Therefore, as the block moves, a unbalanced force $F_L - F_k$ acting on the body accelerates it. If the pull on the block is decreased so that it equals F_k, the block will move with uniform speed. The work done in overcoming the friction, $\int F_k ds$ is converted to heat and does not cobtribute to the kinetic energy of the body.

Let us refer to the Fig. 2.20 again. The weight W of the block acts vertically downward and the normal reaction N, exerted by the table on the block is equal to the weight of the block. These two forces acting in opposite directions press the two surfaces together, Suppose, we put a weight say a 20 gf on the block; the normal reaction is increased by this amount. Then we determine the magnitude of F_L. This will found to have increased compared to the previous case *i.e.* when no weights are put on the block. Thus varying the normal reaction, the corresponding limiting friction can be determined. It is found that the *limiting friction is proportional to the normal reaction.*

Suppose the block is replaced by another block of larger surface area in contact with the table. The nature of the surface of this new block and its weight are kept identical to the previous block. It will be found that the magnitude of the limiting friction is very close to the previous value.

2.26. Law of Static Friction

From the foregoing discussions, the following empirical relations emerge for limiting friction for dry and unlubricated surfaces.

(*i*) The limiting friction is almost independent of the area of surface of contact.

(*ii*) The limiting friction is proportional to the normal reaction.

Co-efficient of friction. The ratio of the limiting friction to the normal reaction is called co-efficient of static friction for the two surfaces in contact.

Co-efficient of static friction,

$$\mu_s = \frac{F_L}{N}$$

or $F_L = \mu_s N$ (The value of μ_s can be determined from the experiment described in 2.25).

Hence when the applied force is insufficient to move the body, the force of static friction is less than $\mu_s N$ (*i.e.* $< F_L$). μ_s is constant for the pair of surfaces in contact. It depends upon the nature of surfaces in contact and the condition of the surfaces *viz.* polish, roughness, grains etc. A list of μ_s for some pair of surfaces are given in Table 2.1.

2.27. Laws of Kinetic Friction

Similar to the static friction, there are two empirical laws for kinetic friction between dry and unlubricated surfaces.

(*i*) The frictional force is almost independent of the area of surfaces in contact.

(*ii*) The frictional force is proportional to the normal reaction.

Co-efficient of kinetic friction. The ratio of the force of kinetic friction to the normal reaction is called co-efficient of kinetic or sliding friction. It is denoted by μ_K.

$$\mu_K = F_K/N \quad \text{or} \quad F_K = \mu_K N$$

Since $\quad F_L > F_K$ and $\mu_S > \mu_K$

Like μ_S, μ_K is constant for all pairs of surfaces in contact and is different for different types of surfaces. It is also roughly independent of the speed of sliding provided that the resulting heat does not alter the condition of two surfaces.

TABLE 2.1. Co-Efficient of Friction (Approximate).

Material	*Condition*	μ_S	μ_K
Steel on steel	Dry	0.15	0.09
Steel on steel	Greased	0.03	0.03
Rubber tyre on dry concrete road	Dry	01.0	0.70
Rubber tyre on Wet concrete road	Wet	0.70	0.50
Steel on wood	Dry	0.30	0.18
Wood on wood	Dry	0.45	0.25
Leather on wood	Dry	0.50	0.35
Stone on concrete	Dry	0.75	0.45

2.28. Limiting Angle (Angle of Repose)

There is a simple and alternative method for determing co-efficient of static friction without measurement of forces. AC is a plane surface whose end A is hinged to a horizontal surface AB, (Fig. 2.21). The C end can be lifted and secured to a vertical plane. Thus the inclination of the plane AC can be varied. A body of weight W is kept on the plane AC and the plane AC is gradually lifted upwards untill the body just begins to slide down the incline.

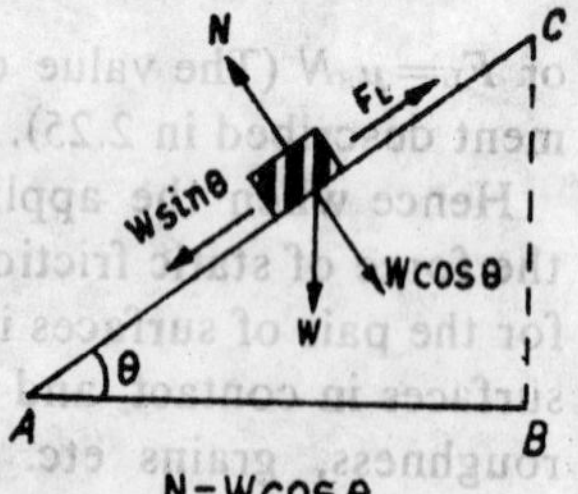

Fig. 2.21. Limiting angle.

Consider the forces acting in the process. Resolve the weight W of the body into components $W\cos\theta$ acting perpendicular to the AC and $W\sin\theta$ acting along the plane AC. The component $W\cos\theta = N$ presses the two surfaces together. The component

$W \sin\theta$ acting down the slope is equal to the limiting friction, F_L.

Thus $$\mu_s = \frac{F_L}{N} = \frac{W \sin\theta}{W \cos\theta} = \tan\theta.$$

The angle θ is called the limiting angle or angle of repose.

2.29. Rolling Friction

When one body rolls over another such as a wheel or cylinder or a spherical ball rolling on the ground, the frictional resistance called rolling friction develop between the two surfaces. This resutls mostly due to deformation produced on the surfaces. When an automoblie tyre is in contact with ground you will notice slight

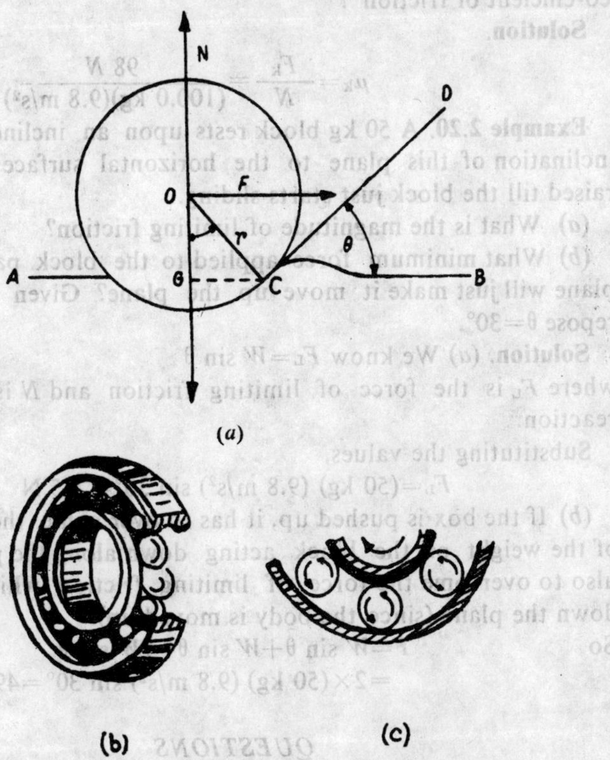

Fig. 2.22. (*a*)Rolling friction is due to deformation of surfaces in contact, (*b*) Ball bearing is used to reduce friction, (*c*) A section of (*b*), showing the relative motion of the parts in a ball-bearing.

flattening of the portion in contact with the ground. The surface is continually distorted as the wheel rolls. The deformation of the two surfaces produce internal friction in the two bodies. But when a metal ball rolls on a rigid surface, less deformation is produced. So force of rolling friction is less the more rigid the surface. It also varies inversely as the radius of the rolling surface. The co-efficient of rolling friction $\mu_r = F_r/N$.

μ_r is ordinarily much smaller than μ_s for the same two surfaces. Since the rolling friction reduces frictional opposition, it is much easier to pull a load supported on wheels rather than to pull the load sliding on the ground. Use of ball bearing in the axle of a wheel replaces the sliding friction between the axle and wheel by rolling friction.

Example. 2.19. A 98 N force is sufficient to pull a 100 kg block on a horizontal table at uniform speed. What is the value of the co-efficient of friction ?

Solution.

$$\mu_k = \frac{F_k}{N} = \frac{98\ N}{(100.0\ \text{kg})(9.8\ \text{m/s}^2)} = 0.1$$

Example 2.20. A 50 kg block rests upon an inclined plane. The inclination of this plane to the horizontal surface is gradually raised till the block just starts sliding.

(*a*) What is the magnitude of limiting friction?

(*b*) What minimum force applied to the block parallel to the plane will just make it move up the plane? Given the angle of repose $\theta = 30°$.

Solution. (*a*) We know $F_L = W \sin\theta$

where F_L is the force of limiting friction and N is the normal reaction.

Substituting the values,

$$F_L = (50\ \text{kg})\ (9.8\ \text{m/s}^2) \sin 30° = 245\ \text{N}$$

(*b*) If the box is pushed up, it has to over come the component of the weight of the block acting down along the plane and has also to overcome the force of limiting friction which acts now down the plane (since the body is moved up).

So
$$F = W \sin\theta + W \sin\theta = 2W \sin\theta$$
$$= 2 \times (50\ \text{kg})\ (9.8\ \text{m/s}^2) \sin 30° = 490\ N$$

QUESTIONS

1. Can the average velocity of a body be a null vector when the average speed has a definite non-zero magnitude?

2. Prove graphically the relation $v = v_1 + at$.

3. A man sitting in the compartment of a moving train throws a ball vertically upwards. Will the ball come back to his hand if (*i*) the train moves with constant speed (*ii*) the train is taking up speed (*iii*) the train is slowing down? Explain your answer.

4. Can a body have an acceleration with zero velocity ? Give example.

5. Give an example where the velocity changes though the speed remains constant.

6. Draw the velocity time curve for a body falling freely under gravity. Neglect air resistance.

7. A man is trying to cross a river in a direction perpendicular to the current. Draw vector diagram to show the velocity of the man relative to the bank.

8. Which starts quickly—an empty vehicle or a loaded vehicle? Give reasons.

9. Do the force of action and reaction cancel each other? Justify your answer with examples.

10. Give two examples, not mentioned in this book, of physical events or devices depending upon the conservation of momentum principle.

11. Is the weight of a body constant at all places on the earth?

12. A stone tied to the end of a rope is lowered inside a deep dry well and then brought out. When is the tension in the string higher, while lowering the stone or while lifting it? Explain your answer.

13. Draw a curve to illustrate the variation of range of a projectile with angle of projection.

14. A man rowing a boat against current in a river remains stationary with respect to the bank. Is he doing any work ? Explain your answer.

15. What is the work done by tension in the string when a simple pendulum swings?

16. If you throw a ball in a moving train does its kinetic energy depend upon the speed to the train?

17. Explain how the ultimate source of energy available in a steam engine is the sun.

18. Show that the sum of potential and kinetic energy of a body falling freely under gravity remains constant throughout its journey.

19. When one hits a nail with a hammer, whether the nail or the hammer performs work?

20. A tram car moving on the steel track and an automobile on concrete road are moving side by side with the same speed. Which vehicle can stop in a shorter distance? (Co-efficient of friction for rubber on concerte is 0.6 and for steel is on steel is 0.2).

21. Which is higher, work done in going to the top of a hill by directly up the slope or by following a winding road? Why is it preferred to drive the automobile along the winding path?

PROBLEMS

1. A rifle bullet acquire a speed of 500 m/s in traversing inside the barrel. If the mass of the bullet is 10 gms and length of the barrel is 50 cms. find out the average acceleration and the accelerating force.

[*Ans.* $250\ km/s^2$; *2500N*]

2. A ball of mass 50 gms, is dropped from a height of 200 meters above the ground. How long will it take to reach the ground? What is the distance covered in the last second of its journey? What is the accelerating force on

the mass? (g=9.8 m/s²). [*Ans.* 6.45 s 57.7 m 0.49 N]

3. A boy on a cliff 19.8 m high drops an iron ball. One second later he throws a second ball after the first. Both the balls reach the ground at the same time. With what speed did the boy throw the second ball? (g=9.8 m/s²). [*Ans.* 15 m/s]

4. A ball thrown up is caught by the thrower 4 sec after it was thrown. How high did it go and with what speed was it thrown? How far below its highest point was it 3 sec after start (g=9.8. m/s²)

[*Ans.* 19.6 m, 19.6 m/s, 4. 9m]

5. A parachutist jumps out from an aeroplane flying at a height of 2875 m. He falls freely for 10 *s* before the parachute opens out after which he descends with a net retardation of 2 *m/s²*. What is his velocity when he reaches the ground? What is the time taken after the parchute opens?] (g=9.8 m/s²)

[*Ans.* 8 m/s, 45 s]

6. A train travelling at a speed of 72 km/h is brought to rest in 15 s by applying brakes. How far did it travel before coming to rest? What was the retardation produced? [*Ans.* 100 m,—2 m/s²]

7. A ballon is ascending at a speed of 28 m/s at a height of 64 m above the ground when an iron ball is dropped from the balloon. How long will the ball take to reach the ground ? What will be the velocity with which it strikes the ground ? [*Ans.* 7.464 s, 2038.4m/s]

8. A bullet striking a wall at a velocity of 500 m/s gets embeded in it after penetrating into it a distance of 25 cm. If the thickness of the wall were 15 cms. with what velocity the bullet would have emerged out of the wall ?

[*Ans.* 316.23 m/s]

9. In a rocket, fuel is consumed at the rate of 50 kg/s. If the exhaust gases are ejected at a speed of 3.5×10^4 *m/s* what is the thrust experienced by the rocket ? [*Ans.* 1.75×10^6 N]

10. A constant force of 20 N acts for 2 seconds on a body of mass 2 kg placed on a frictionless, horizontal surface and intially at rest. Calculate,

(*i*) The velocity of the body at the end of 1 sec, 2 sec, and 3 sec.

(*ii*) The distance covered in 1 sec, 2 sec, 3 sec.

11. An automobile is moving at a speed of 40 km/h due east. A second automobile starting simultaneously moves due north at a speed of 60 km/h. What is the velocity of the first automobile with respect to the second?

12. A mass of 1 kg is suspended by a string from an elevator. If the elevator is ascending with an acceleration of 2 m/s², what is the tension of the string? What will be the tension if the elevator descends with the same acceleration? [*Ans.* 11.8 N, 7.8. N].

13. In problem 10, calculate the momentum of the body at the end of 1 sec, 2 sec, and 3 sec.

14. A man weighing 45 kg in sitting in a lift which is moving vertically upwards with an acceleration of 2 m/s². Compare the pressure on the base of the lift while ascending with that while descending. (g=9.8 m/s²)

[*Ans.* 531 N, 351 N]

15. Two objects of mass 3 kg and 6 kg are connected by a weightless string which passes over a frictionless pulley. The objects are initially held at a height of 15 m. from the ground with equal lengths of string on either side of the pulley. Find,

(*i*) The acceleration of the system,

(*ii*) The velocity of the objects at the instant the heavier mass goes down a distance of 13.08 *m*. (g=9.8 m/s^2) [*Ans* 3.27 m/s^2, 6.54 $\sqrt{2}$ m/s]

16. A ball moving with a speed of 12 m/s strikes a similar stationary ball. The direction of each ball after collision makes an angle of 45° with the original line of motion. Find the speed of two balls after the collision. Is the collision elastic?

[*Ans*. $6\sqrt{2}$ *m/s* ; *Yes, since kinetic energy is conserved*]

17. A packet is dropped from an aeroplane flying horizontally at a height of 7840 m with a speed of 245 m/s. Find the time taken by the bomb to reach the ground and also the magnitude and direction of the velocity with which it strikes the ground. (g=9.8 m/s^2) [*Ans. 40 s*]

18. A man draws water from a well 10 meter deep. If the mass of the bucket with water in it is 10 kg what is the amount of work done in drawing a bucket of water. g=9.8 m/s^2 [*Ans*. 9.8×10^2 *J*]

19. A two kilo-watt motor is used to pump water from a well 20 m deep. Calculate the quantity of water pumped out per second. [*Ans. 10.204 kg*]

20. How much work will be done in dragging through 100 m. a 200 kg mass on a level road if co-efficient of friction is 0.20? [*Ans*. 3.92×10^4 *J*]

21. Three 200 watt incandescent bulbs, five 40 watt fluorescent tubes, five 60 watt fans and one 1 K watt immersion heater are used in a building. Assuming all of them are put to simultanous use, how many units of electricity are consumed in 6 hours? [*Ans. 12.6*]

22. A pump can lift 10,000 kg of water per hour from underground source 120 m deep. Calculate the power of the pump in watts, assuming that its efficiency is 70%. [*Ans 5.600 Watt*]

23. A vehicle with a mass 1500 kg is travelling at a speed of 60 km/h. Suddenly the brakes are applied causing all tyres to skid. How far will the car travel before coming to stop? [*Ans*. μ=0.04]

3

CIRCULAR MOTION

In the previous chapter we considered the motion along a straight line. A body moves along a straight line with uniform speed when no net force act on the body. If a net force acts along the direction of motion, the body accelerates in the same direction. But, the direction of motion changes when a net force acts on a body in a direction different from the original direction of motion. Thus a body can have acceleration in a direction other than the original direction of motion. We come across such accelerations when we move along a curved path on a bicycle or when a car turns a corner. The motion of a planet in its orbit around the Sun or the motion on a motorbike in the globe in a circle or motion of electron in the orbit around the atomic nucleus are also examples of such acceleration. The departure from rectilinear motion is a common feature in nature. Therefore, the study of motion along a curved path assumes importance. However, we shall confine our study to the uniform circular motion, the simplest type of motion in which there is no change in speed but only change in direction.

3.1. Angular Velocity

Consider a particle A moving on a circular path of radius r, Fig. 3.1. The particle has moved from A to B in a time interval t. During this interval, the line OA has rotated to the position OB. The angle $\angle AOB=\theta$ is the angular displacement of the particle during this time. The *average angular speed* $\bar{\omega}$ (omega) is defined as the rario of angular displacemet θ to the time interval t,

Thus $$\bar{\omega}=\frac{\theta}{t}$$

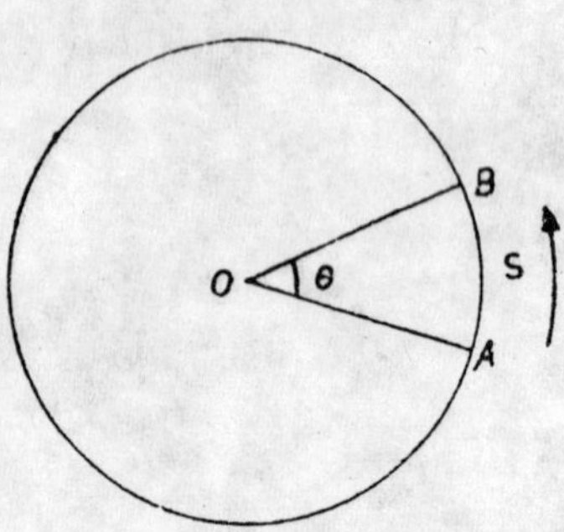

Fig. 3.1. Angular velocity $\omega=\frac{\theta}{t}$.

Let us consider the particle moving with constant speed along the circumference. In such a case, the angular speed of the particle will be constant and same as the average angular speed: *i.e.*

$$\omega=\frac{\theta}{t} \qquad ...(3.1)$$

The angle θ is usually expressed in radians. For θ in radians and t in seconds, ω will be in radian/sec. or rad/s. When the angle θ is measured in radians, the arc length S is related to the radius of the circle r by $S=r\theta$

or $$\theta=\frac{S}{r}$$

$$\omega=\frac{\theta}{t}=\frac{\frac{S}{r}}{t}=\frac{1}{r}\left(\frac{S}{t}\right).$$

But $\frac{S}{t}=v$, the constant linear speed of the particle.

Thus $$\omega=\frac{1}{r}v \text{ or } v=r\omega \qquad ...(3.2)$$

Linear speed = Radius of the circular path × angular speed.

The distinction between speed and velocity is also applicable in case of angular motion. An angular velocity is represented by a vector along the axis of rotation. Its direction is indicated, by the thumb of the right hand grasping the axis while other Fingers encircle axis in the direction of rotation. The general vector equations are quite complicated and we shall deal with its scalar properties only, though we may refer to the terms velocity and acceleration. The dimension of the angular velocity are:

$$\frac{\frac{L}{T}}{T}=T^{-1}.$$

3.2. Uniform Circular Motion

Centripetal force. When a particle moves in a circular path with constant speed, its velocity continually changes since the direction of motion always changes though the speed remains constant. This is a case of change of velocity due to change in direction. Here particle experiences acceleration in a direction at right angle to the direction of motion. The direction of motion of the particle at any time point is given by a tangent to the circle at that point. Thus the direction

2π radians=360°; 1 radian=$360/2\times\frac{22}{7}$ degree=57.3 degree.

[2]The instantaneous angular speed (for non-uniform circular motion)

$$\omega=\lim_{\Delta t\to 0}\frac{\Delta\theta}{\Delta t}=\frac{d\theta}{dt}.$$

of the acceleration is always radial. We will derive an expression for this acceleration and force associated with it.

Suppose a particle of mass m move along the circular path of radius r with constant angular velocity ω. The linear speed of the particle $v=r\omega$ (Fig. 3.2).

Let A be the position of the particle at any instant of time and B its position after a short time interval $\triangle t$. The velocity vectors at A is $\vec{v}_1$ and the velocity after time $\triangle t$ *i.e.* at B is $\vec{v}_2$. It should be noted that the magnitude of velocity v is same in both cases. To calculate the acceleration, we draw the vector diagram of the velocities.

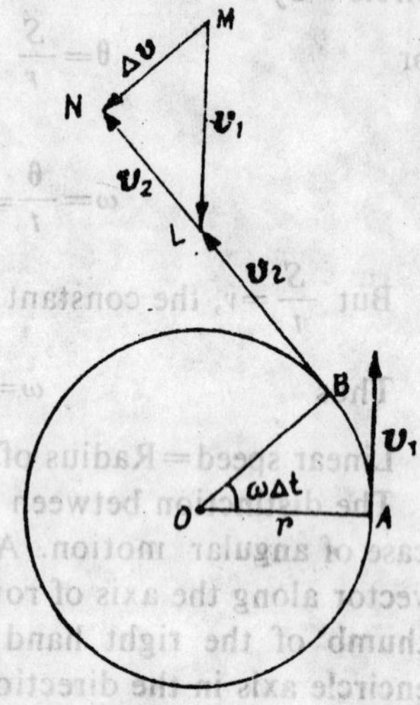

Fig. 3.2. Acceleration in uniform circular motion.

Let LM and LN represent the two velocity $\vec{v}_1$ and $\vec{v}_2$ respectively. Here $LM=LN$; only the directions are different. The change in velocity *i.e.* vector difference of final velocity and initial velocity,

$$\triangle \vec{v} = \vec{v}_2 - \vec{v}_1 = (-\vec{v}_1) + (\vec{v}_2)$$
$$= \vec{ML} + \vec{LN} = \vec{MN}.$$

Thus the vector $\vec{MN}$ represents the change in velocity in a time interval $\triangle t$. It is directed inwards.

In the $\triangle$s LMN and OAB,

$$\angle BOA = \angle NLM \text{ and } \frac{OA}{OB} = \frac{LM}{LN} = 1.$$

Thus the two triangles are similar.

Therefore, $$\frac{MN}{AB} = \frac{LM}{OA} \quad \ldots(i)$$

As $\triangle t \to 0$, the angle θ also tends to zero and the arc AB approximates to the chord AB.

So $$AB = r\theta = r\omega\triangle t = v\triangle t.$$

On replacing AB by $v\triangle t$, MN by $\triangle v$, and LM by v in relation (*i*) above,

$$\frac{\triangle v}{v\triangle t} = \frac{v}{r} \quad \left(\text{Note } v = |\vec{v}_1| = |\vec{v}_2|\right)$$

Hence acceleration

$$\left|\vec{\mathbf{a}}\right|=\frac{\Delta v}{\Delta t}=\frac{v^2}{r} \qquad \ldots(ii)$$

This expression also gives the average acceleration for non-uniform speed. The instantaneous acceleration in that case is given by

$$\underset{\Delta t \to 0}{Lt} \frac{\Delta v}{\Delta t}.$$

Replacing v by $r\omega$ in (*ii*)

$$\left|\vec{\mathbf{a}}\right|=\frac{v^2}{r}=r\omega^2 \qquad \ldots(3.3)$$

In the limit when $\Delta t \to 0$, the angle θ approaches zero, the sides *LM* and *LN* become almost parallel to one another. Then *MN* joining the ends of two porallel lines will be perpendicular to *LM* or *LN*. Thus $\Delta \vec{\mathbf{v}}$ and hence $\vec{\mathbf{a}}$ will be perpendicular to the velocity vector. Since the direction of the velocity at any point on the circular path is along the tangent to the circle, the acceleration at that point will be along the radius through that point directed towards the centre 0. This acceleration directed towards the centre is called *Centripetal Acceleration*. Since acceleration is always caused by a net force, there must be a force acting on the body and directed towards the centre of the circular path. This force is called *Centripetal* or 'centre seeking force.'

The magnitude of this force,

$$F=ma=mr\omega^2$$

is constant for a body moving with constant angular speed along a fixed circular orbit. Since the angular speed is expressed rad/s, the units of a or F obviously will depend upon the units in which m and r are expressed.

Let us take an example to show how the centripetal force acts. Suppose you tie a stone at the end of a string and whirl it in a circular path. You will have to exercise a pull inwards to keep the stone moving in the circular orbit. Thus your finger provides the centripetal force. If the string breaks, the centripetal force will disappears and the stone will fly tangentially instead of moving in the circular path. The paths taken by sparks from a grinding wheel in a workshop are illustrations of this force. The centripetal force is utilised in centrifuge which is a device to separate materials of different densities in a mixture into its individual components.

No work is done by the centripetal force as the direction of the centripetal force is perpendicular to the direction of motion.

Neglecting friction therefore no energy is spent in moving an object in a horizontal circular orbit. Hence the moon revolves around the earth continually without slowing down.

3.3. Centrifugal Reaction

The centripetal force produces a centrifugal reaction. So when you whirl the stone tied to a string, the stone will exert an outward force on your finger. This is the centrifugal reaction force which is equal in magnitude but opposite in direction to the centripetal force (Newton's third law of motion). It must be noted that the centripetal force acts on the rotating body and the centrifugal reaction acts on the body at the centre. Therefore, the two forces do not cancel each other.

Very often, we come across a wrong use of the term 'centrifugal force'. A person travelling in a car at high speed around a curve experiences outward movement of the body. Generally it is said to be due to 'centrifugal force'. But actually this is a consequence of Newton's first law of motion—inertia of motion. The body tries to continue its motion in the straight line while the car has already taken a turn along the curved path.

3.4. Banking of Curves

When a man runs around a curve, he usually leans inwards to provide centripetal force. On a level ground if the person does not lean the horizont centripetal force is provided by friction between his feet and ground. The upward reaction force exerted by the ground due to his pressing the ground just balances his weight.

But when he leans, the upward reaction force of the ground is no more perpendicular to the ground. The vertical component of the reaction force balances the weight of the man and the horizontal component provides centripetal force (Fig. 3.3).

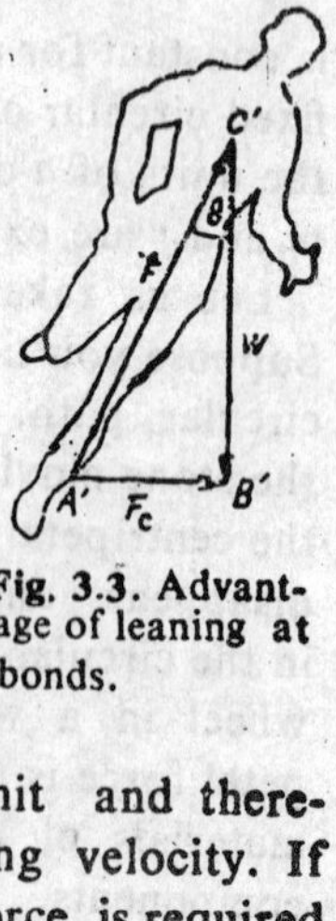

Fig. 3.3. Advantage of leaning at bonds.

When a car moves on a curved path on a level road, the centripetal force is provided by the frictional force acting between the wheel and the road. The frictional force has a maximum limit and therefore it can provide centripetal force up to a limiting velocity. If the velocity exceeds that limit, higher centripetal force is required but the frictional forces cannot provide that. So the car will skid off the road. Hence the driver has to reduces his speed. The speed has to be further reduced on a slippery track as the friction is less. This difficulty is overcome if the road is made sloping downwards towards the inner circumference at the bends. In this case, a part of the horizontal centripetal force is still supplied by friction, but a greater part is provided by the horizontal component of the normal reaction of the surface. This system of providing an inclination for the road at bends is called banking of curves. In railway tracks, banking is done at the curves also. The outer rail is elevated with respect to the inner rail in the curved part of the track.

Suppose a car is moving on a 'banked' curved road (Fig. 3.4). AC is width of the road and it is inclined to the horizontal at angle θ called 'angle of banking'. The normal reaction of the ground N, which is perpendicular to the surface AC is resolved into two components. Vertical component $N \cos \theta$ balances the weight of the car and the horizontal component $N \sin \theta$ provides centripetal force.

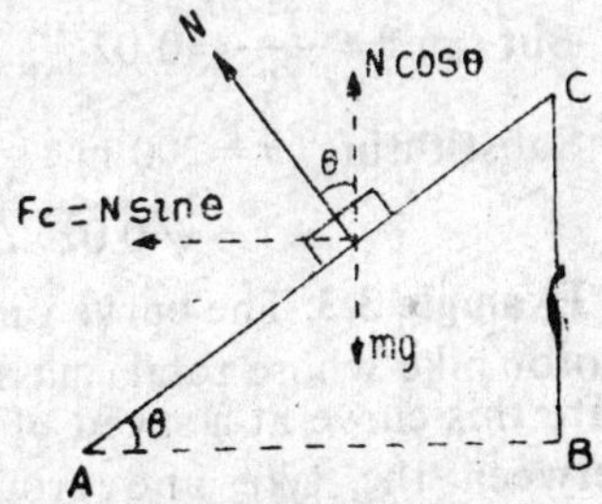

Fig. 3.4. Banking of curves (vertical section).

Thus $N \cos \theta = m g$ ...(*i*)

and $N \sin \theta = mv^2/r$ ...(*ii*)

where v is the speed of the car, r is the radius of the curved track and m is the mass of the car. Here we have neglected the contribution of friction to the centripetal force.

From (*i*) and (*ii*) above,

$$\tan \theta = v^2/rg \qquad ...(3.4)$$

Since the angle of banking depends upon the speed, the curve can be ideally banked for only one speed. For higher speeds, the force of friction must be depended upon to prevent slipping. In addition to contributing to road safety, the banking of curve reduces wear and tear of the tyres and wheels by reducing the lateral force of friction on them. A person negotiating a turn on a level road, a cyclist going round a curved path, should lean himself or his cycle from the vertical at an angle θ given by Equation 3.4 to avoid slipping.

Example 3.1. A vehicle of mass 1000 kg moves round a curve whose radius of curvature is 200 m at a constant speed of 108 km/h. What is the centripetal force on the car ?

Solution. The centripetal force $F = \frac{mv^2}{r}$.

Here, $m = 1000$ kg, $v = 108$ km/h $= 30$ m/s and $r = 200$ m

$$F = \frac{(1000 \text{ kg})(30 \text{ m/s})^2}{200 \text{ m}} = 4,500 \text{ N.}$$

Example 3.2. A curve on a road forms an arc whose radius of curvature is 200 m. If the width of the road is 30 m and its outer edge is 0.6 m higher than the inner edge, for what speed it is ideally banked?

Solution. If θ is the angle of the bank,

$$\tan \theta = \frac{\text{difference in the level of two edges of the road}}{\text{width of the road}}$$

$$= \frac{0.6 \text{ m}}{30 \text{ m}} = 0.02$$

But $\tan \theta = \frac{v^2}{rg} = 0.02.$

Substituting $r = 200$ m., $g = 9.8 \text{ m/s}^2$

$$v = \sqrt{0.02 \times 200 \text{ m} \times 9.8 \text{ m/s}^2} = 6.26 \text{ m/s.}$$

Example 3.3. The curve on a level road has a radius of 75.0 m. A motor bike whose total mass with its rider is 120 kg has to negotiate this curve at a speed of 54 km/h. If co-efficient of friction between the tyre and ground is 0.4, will the rider be able to go round the curve without slipping? At what angle should he lean over to avoid slipping?

Solution. The required centripetal force $= mv^2/r$.

Here $m = 120$ kg, $v = 54$ km/h $= 15$ m/s, and $r = 750$ m.

$$F_{cp} = \frac{(120 \text{ kg})(15 \text{ m/s})^2}{75.0 \text{ m}}$$

$$= 360 \text{ N.}$$

The total weight of the rider and the motor bike is

$$(120 \text{ kg})(9.8 \text{ m/s}^2) = 1176 \text{ N.}$$

We know the maximum force of friction,

$$F_L = \mu_s N$$

$$= 0.4 \times 1176 \text{ N} = 470.4 \text{ N.}$$

Since $F_L > F_{cp}$, the rider can negotiate the curve. Now, if θ is the angle at which the rider should lean from the vertical,

$$\tan \theta = \frac{mv^2/r}{mg}$$

$$= \frac{360 \text{ N}}{1176 \text{ N}} = 0.3060$$

$$= 17$$

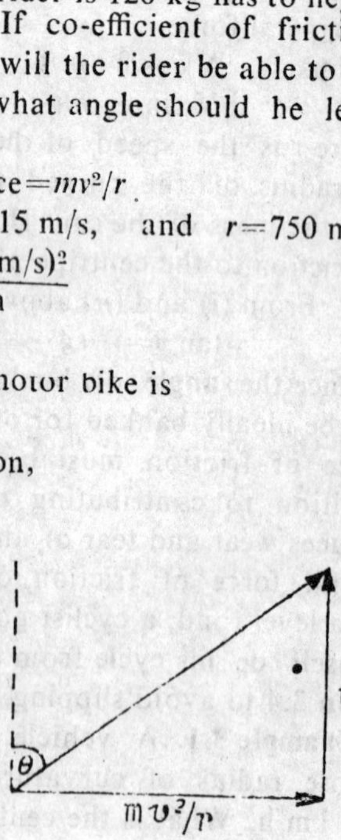

Fig. 3.5

Example 3.4. In a circus, the motor cycle feat is being shown in a spherical cage of radius 6 m. What should be minimum speed of the motor cycle for the performance ?

Solution. Since the motor cycle also moves in vertical loops during the performance, it is necessary that the cycle does not fall under its own weight when it is on the upperside of the cage. The requirements is that the centripetal force should be greater than the weight.

$$i.e. = \frac{mv^2}{r} \geqslant mg$$

The minimum speed is given by the condition

$$\frac{mv^2_{min}}{r} = mg$$

$$\therefore \quad v_{min} = \sqrt{rg}$$

Putting $r = 6$ m and $g = 9.8$ m/s²

$$v_{min} = \sqrt{(6\text{m})(9.8\text{ m/s}^2)}$$
$$= 7.668 \text{ m/s}$$

3.5. Planetary Motion and Kepler's Laws

Ptolemy, a Greek astronomer of the second century A.D., was perhaps the first to make a detailed study of the motion of planets. The theory he proposed is called the '*Geocentric Theory*' according to which the earth is the centre of the universe around which the moon, other planets, sun and even other stars revolve. The theory was very complicated as different types of complex orbits were proposed for different celestial bodies to fit into the theory. In the sixteenth century A.D., copernicus propounded the '*Heliocentric Theory*' to explain the motion of earth and other celestial bodies. This theory puts the sun at the centre while the planets revolve around the sun in their respective orbits. Ptolemy's view putting the earth as the centre of the universe had been accepted for centuries by the Christian Church which controlled every sphere of activity including education and administration. So Copernicus found it very difficult to put forward his theory though he was convinced of its soundness. The Controversy continued for a long time, but ultimately the 'heliocentric theory' of Copernicus was accepted.

It will be of interest to note here that three distinguished Indian astronomers, Aryabhatta, Bhaskaracharya and Samant Chandrasekhar made remarkable contributions in the field. Aryabhatta had stated from his calculations that earth moves about its axis. Samant

Chandrasekhar had proposed that neither the sun nor the earth is the centre of the universe, they rotate about a common point on the line joining them. Due to lack of communication between east and west and the treatise of Indian workers being in Sanskrit, their achievements did not receive worldwide recognition.

Johannes Kepler (1571-1630) was assistant to Tycho Brache (1546-1601), a great astronomer, for sometime. After the death of Tycho Brache, Kepler carefully analysed his observations and calculations and observed certain regularities with regards to the motion of the planets. He proposed three laws of planetary motion to fit into the data. These laws are known as Kepler's Laws.

Kepler's Laws. (1) *The law of orbit.* The planets revolve around the sun in elliptical orbits with the sun at a common focus of the ellipse.

(2) *The law of area.* The radius vector from the sun to the planet (the line which joins a planet to the sun) sweeps out equal area in equal times.

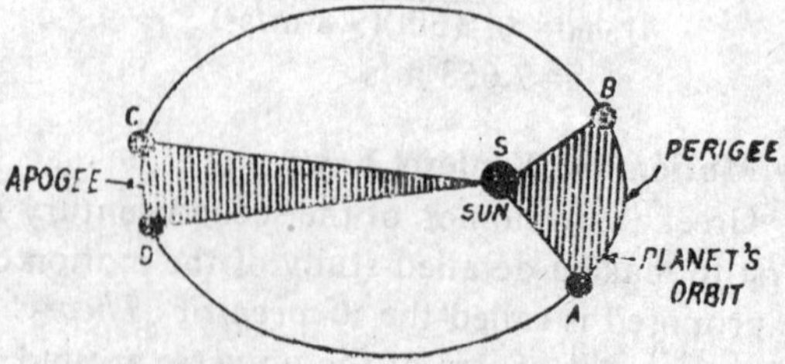

Fig. 3.6 Planetary orbit around the sun. In equal time the area swept out when the planet moves from *A* to *B* equals the area swept out when it moves from *C* to *D*. Thus its orbital velocity is greatest between *A* and *B*.

(3) *The law of periods.* The square of the periods of planets are proportional to the cubes of their mean distances from the sun.

In otherwords, T^2/r^3 is the same for all planets where T is time taken by the planet for one complete revolution around the sun (period) and r is the mean distance of the planet from the sun. The mean distance refers to the semi-major axis of the ellipse.

Kepler's laws were empirical and based on observations only. Newton in his '*Principia Mathematica*' (1687) showed that the kind of planetary motion described by Kepler's laws could be deduced from the universal law of gravitation.

3.6. Newton's Universal Law of Gravitation

A material object falls to the earth when released from above. Everybody saw thus, accepted the phenomenon as inevitable, but none thought about its cause till Newton (1665 A.D.) saw the famous apple falling to the ground. He thought an object falls to the earth due to the force of gravitation. The same force of gravitation, he reasoned, also provides centripetal force necessary to keep moon moving round the earth. This, of course, was contrary to the view of Aristotle, that the motion of the celestial bodies is governed by laws different from those governing the motion of bodies on the surface of the earth.

Newton suggested that while the earth exerts attraction on moon, moon also exerts attraction on the earth. The force acting on a body depends on its mass ($\because \vec{F} = m\vec{a}$). So the gravitational force of the earth on the moon should depend upon the mass of the moon. Similarly, the gravitational force of the moon on the earth should depend upon the mass of the earth. Further, he estimated the gravitational acceleration on a body near the surface of the earth. He also calculated the cetripetal acceleration of the moon (provided by the gravitational force) on the basis of its period of revolution and the radius of orbit ($a=v^2/r$). It was found to be about 0.00267 m/s², a value much lower than the value of acceleration due to gravity on earth. To explain this difference, he suggested that the force of attraction is inversely proportional to the square of the distance of the body from the earth. Considering the mass of the earth to be concentrated at its centre, he was able to justify this assumption.

He generalised his ideas in the form known as Newton's Universal Law of Gravitation as stated below.

Every material object in the universe attracts every other material object towards itself. The force of attraction between them varies directly as the product of their masses and inversely as the square of the distance between their centers of mass.

In symbolic notation, it may be expressed as

$$F \propto \frac{m_1 . m_2}{d^2}$$

where m_1 and m_2 are the mass of the two objects, d is the distance between the centres of mass and F is the magnitude of the force that one object exerts on the other. Introducing a constant of proportionality factor G, we can write

$$F = G \frac{m_1 . m_2}{d^2} \quad ...(3.5)$$

G is called the universal gravitational constant. The value of G depends upon the system of units used in Equation 3.5. If the force is given in Newtons, the mass in kg and distance in meters,

$$G = 6.670 \times 10^{-11} \text{ N.m}^2/\text{kg}^2$$

The dimensions of G are $L^3M^{-1}T^{-2}$.

Newton applied laws of gravitation to analyse the data based upon the motion of moon in its orbit and found fair agreement between his calculations and observations. He could also derive Kepler's law of periods on the basis of his idea about gravitation. Thus Newton could explain the motion of celestial bodies as well as motion of objects near earth by the same laws.

The work of Kepler and Newton also confirmed the heliocentric theory of the universe.

Since the force of attraction between two small objects is very small, the measurement of force of that order was considered impossible. Henry Cavendish (1731-1810) measured for the first time in 1798 the force of attraction between two small golden spheres and established the validity of the law of gravitation after more than a century of its formulation. The value of G was also determined by him.

Example 3.5. The radius of the earth is 6.4×10^3 km. Assuming acceleration due to gravity on the surface of the earth is 9.8 m/s², calculate the value of acceleration due to gravity at a distance 6.4×10^3 km from the surface of earth.

Solution. A body on the surface of the earth is situated at a distance of 6.4×10^3 km and in the second position at a distance of 12.8×10^3 km from the centre of the earth. Suppose M_e and m are the mass of the earth and the body respectively.

Earth's force of attraction on the body on the surface of earth is :

$$\frac{G M_e m}{(6.4\times10^6 m)^2}=mg \qquad ...(i)$$

Similarly, for the second position :

$$\frac{G M_e m}{(12.8\times10^6 m)^2}=mg_1 \qquad ...(ii)$$

When g_1, is the acceleration due to gravity in this place.

Dividing equation (*ii*) by (*i*) and putting the value of g,

$$g_1=\frac{(6.4\times10^6\ \text{m})^2}{(12.8\times10^6\ \text{m})^2}\times9.8\ \text{m/s}^2$$

$$=2.45\ \text{m/s}^2.$$

Example 3.6. Calculate the mass of the earth, given that

$$G=6.67\times10^{-11}\ \text{N m}^2/\text{kg}^2,$$

the radius of the earth $=6.4\times10^3$ km and $g=9.8$ m/s².

Solution. Taking (*i*) of the previous example,

$$mg=G M_e m/(6.4\times10^6\ \text{m})^2$$

Putting the value of G and g

$$M_e = \frac{(9.8 \text{ m/s}^2)(6.4\times 10^6 \text{ m})^2}{(6.670\times 10^{-11} \text{ N.m}^2/\text{kg}^2)}$$
$$= 6.0181\times 10^{24} \text{ kg.}$$

Example 3.7. Two lead balls whose mass are 2.5 kg and 6.5 kg respectively are placed with their centres 50 cms apart. With what force do they attract each other ? What is the acceleration produced on the bigger mass ?

Solution.

$$F = \frac{G\, m_1 m_2}{d^2}$$
$$= \frac{(6.67\times 10^{-11} \text{ N.m}^2/\text{kg}^2)(2.5 \text{ kg})(6.5 \text{ kg})}{(0.5 \text{ m})^2}$$
$$= 4.335\times 10^{-9} \text{ N}$$

acceleration, $a = F/m$

$$= 0.667\times 10^{-9} \text{ m/s}^2$$

It may be noted that acceleration is so small that the body is practically unaffected.

N.B. Derivation of Kepler's law of period from law of gravitation :

Suppose, mass of a planet is m_p, mass of the sun is m_s, mean radius of the orbit is r and the time taken to complete one revolution of the planet around the sun be T. Then the average centripetal force on the planet

$$F_{cp} = m_{cp}\, \omega^2 r$$

where ω is the angular velocity of the planet.

Substituting $\omega = \frac{2\pi}{T}$,

$$F_{cp} = \frac{4\pi^2 m_p r}{T^2}$$

According to Newton's universal law of gravitation, the gravitational force of attraction on the planet

$$F_g = G\, \frac{m_p\, m_s}{r^2}$$

Since the centripetal force is provided by the gravitational force, equating then, we get,

$$G\, \frac{m_p m_s}{r^2} = \frac{4\pi^2\, m_p r}{T^2}$$

or $$\frac{T^2}{r^3} = \frac{4\pi^2}{G\, m_p} = \text{Constant.}$$

This proves law of periods.

3.7. Inertial Mass and Gravitational Mass

While dealing with Newton's second law of motion, we have seen that a net impressed force is necessary to produce acceleration in body. But same force produces different acceleration in different bodies. Thus each body responds to the same force in different degrees. The measure of the difficulty of accelerating a body by application of force is referred to as to inertial mass of the body. *The inertial mass of a body is the ratio of force applied to the acceleration producd in it i.e.*

$$m = F/a.$$

But according to law of gravitation, the force of attraction between two bodies is proportional to the product of their masses. Thus the mass can be considered as that property of matter by virtue of which everybody exerts a force of attraction on every other body. This property is called *gravitational mass.*

Let us examine the relationship between the inertial and gravitational mass.

Suppose two masses m_1 and m_2 are to be compared by observing acceleration produced due to impressed net force. Let two forces F_1 and F_2 acting on m_1 and m_2 respectively produce equal acceleration a in them.

Then $$F_1/F_2 = m_1 a/m_2 a = m_1/m_1 \quad \text{...}(i)$$

Here m_1 and m_2 are inertial mass of the bodies.

Let us consider the gravitational effect.

The force of attraction between a body and the earth is defind as its weight.

Thus $$W_1 = G\, M_e\, m_1/R^2$$

and $$W_2 = G\, M_e\, m_2/R^2$$

Where W_1 and W_2 are the weight of the two bodies. M_e is the mass of earth and R is mean radius of earth. The centre of mass of both bodies are assumed to be on the surface of the earth. Thus

$$W_1/W_2 = m_1/m_2 \quad \text{...}(ii)$$

Here m_1 and m_2 are gravitational mass of the two bodies. But the force W_1 and W_2 on the bodies with inertial mass m_1 and m_2 produce same acceleration g.

So according to (i),

$$W_1/W_2 = m_1/m_2 \quad \text{(inertial)}$$

Hence gravitational force is proportional to the mass measured by the inertial property. Thus mass measured by observing either inertial effect or gravitational effect are same, though they appear to be different by definition. In fact, Newton used the proportionality between gravitational force and inertial mass in accounting for planetary motion.

3.8. Variation of g on the Surface of Earth

We know that a body is attracted towards the earth due to earth's gravitational attraction. The force due to earth's gravitation varies inversely as the square of the distance of the body from the centre of the earth. Hence acceleration due to gravity also changes in the same manner ($\because a=F/m$). Therefore, as we move away from the surface of earth into space, the value of g decreases rapidly. In the outerspace, therefore, value of g is practically zero. That is why astronauts have to face weightlessness in the outerspace.

All places on the surface of earth are not situated at equal distances from its centre. The distance between the top of a mountain peak say Everest and the centre of the earth is more than the distance between a place at sea level and the centre of earth. Therefore the value of g at the top of Everest is lower than the value of g at the sea level.

Earth is not a perfect sphere. Its equitorial diameter is more than the polar diameter. Hence the value of g at poles is more than the value of g at equator. There is another reason also for g at equator being lower than g at poles. Since the earth is rotating about its axis each particle on it also completes one rotation during a day. But a particle at the equator moves round in a greater orbit (orbital diameter is equal to the equitorial diameter) and other particles move round in smaller orbits as the latitude of the place increases from 0 to 90°. The diameter of the orbit for a particle at the poles is practically zero. So the linear speed, of particles at the equator is highest and that at the poles is the lowest. A particle moving with higher speed in a circular orbit experiences a greater centripetal force. Since the centripetal force is provided by a component of the gravitational force the net gravitational force on the body is minimum at the equator as the centripetal force is highest and maximum at poles. Therefore the value of g also is lowest at the equator and highest at the poles. The standard value of g is taken at sea level at 45 latitude.

On the other hand, as we go into the earth say inside a mine, the value of g also decreases. We can explain this in the following manner. Suppose a body is placed inside the earth at a depth h. If the radius of earth is R_e, then the mass of the portion of earth with radius (R_e-h) exerts gravitational attraction on it in the direction towards the centre of earth. At the same time, the portion of the earth in the annular sphere of thickness h surrounding the body exerts gravitational pull away from the centre of earth. So the gravitational pull that the body experiences is less than the pull on the surface of earth. Hence values of g inside the earth is less than the value of g on the surface of the earth. If the earth were a uniform solid sphere, the value of g at its centre would have been zero.

3.9. Gravitational Field

A compass needle brought near a magnet shows deviation from its usual north-south orientation. Similarly, an electric charge brought near another charge experiences a force of attraction or repulsion. These effects we explain by saying that there exists a magnetic field in the space surrounding the magnet or an electric field near the charge.

A body is attracted towards the centre of earth by gravitational force. On the analogy of electric and magnetic field, it can be said there exists a gravitational field surrounding the earth and any mass brought into this field experiences a force of attraction towards the centre of the earth.

We proceed to define the Gravitational Field strength as follows.

According to Newton's law of gravitation, the gravitational force F due to earth on a body of mass m situated at distance x from the centre of earth ($x > R_e$, the radius of earth) is given by

$$F = G\frac{M_e\, m}{x^2}$$ where M_e is the mass of the earth.

When $m = 1$, the force $I = G\, M_e/x^2$ ...(3.6)

is known as the gravitational field strength due to the earth at that point.

Hence, *the force of attraction on a unit mass at a point in the earth's gravitational field is called the gravitational field strength at that point*. The gravitational field strength I varies inversely as the square of distance from the centre of the earth. Thus as the distance increases it decreases very rapidly. It may be noted that the value of I at a point is nothing but the weight of a body of unit mass. Therefore on the surface of the earth, it is numerically equal to g (acceleration due to gravity on the surface of earth) and decreases as we move into space. In outer space the value of I is practically zero.

The gravitational field is not unique to the earth. Each and every material body has its own gravitational field. But the field strength due to a small body is too small in comparision to the earth's gravitational field strength. In general, we discuss the earth's gravitational field only in the vicinity of the earth. But when we move deep into outer space, the earth's field decreases considerably and we will come in the range of the gravitational field due to other celestial bodies like moon when we go near them.

3.10. Gravitational Potential Energy

Since a gravitational force of attraction exists between two material objects, work has to be done against the force of attraction to separate them. Similarly, work is to be done by the force if the bodies are to be brought closer. The work done against the force increases the energy of the system whereas the work done by the force decreases the energy of the system. Since the work done is gravitational in nature, the change in energy of the system is known as change in gravitational potential energy of the system.

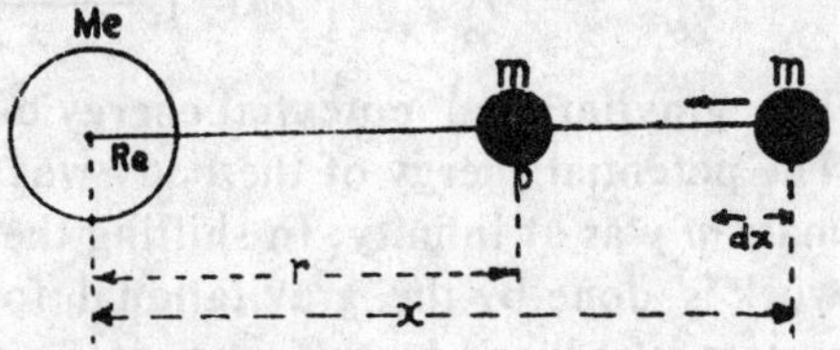

Fig. 3.7. Gravitational energy of a mass in the earth's field.

The gravitational force of attraction of a body on another body placed at infinity is zero. Thus when a second body is displaced at infinity no work is done. Hence energy of the system remains constant. This is taken as reference or zero level for measuring gravitation potential energy. Hence the gravitational potential energy E_p of one body with respect to another is zero when the distance between them is infinity.

Therefore, the potential energy of a small body at a distance x from a very large body may be defined as the work done by an external agent in bringing the small body from infinity to a distance x from the centre of mass of the large body. If the large body is the earth, the potential energy is named as gravitational potential energy due to earth. We derive an expression for the gravitational potential energy of a body in the earth's gravitational field in the following manner.

Let a body of mass m be situated at a distance x from the centre of the earth, (x is greater than R_e, the radius of earth) whose mass is M_e.

The force of attraction on the body,

$$F = G\frac{M_e m}{x^2}.$$

Suppose the mass m is shifted towards the earth through an inifinitesimal distance dx. The work done in the process Fdx, is the change in gravitational potential energy of the body. The gravitational potential energy of the body at P, situated at a distance r from the centre of the earth ($r \geqslant Re$), *i.e.* the work done in bringing the body from infinity upto the P is obtained by integrating

$F\,dx$ within limits $x=\infty$ and $x=r$

So the potential energy of mass m at

$$P=\int_{\infty}^{r} Fdx=\int_{\infty}^{r}\frac{G\,M^{e}m}{x^{2}}\,dx=\left[\frac{-GM_{e}m}{x}\right]_{\infty}^{r}=-\frac{GM_{e}m}{r} \quad ...(3.7)$$

The gravitational potential energy of a body is always negative. The potential energy of the body was taken to be zero. When the mass m was at infinity. In shifting the body from infinity to P, the work is done by the gravitational force; hence the energy of the system is reduced by this amount from original zero value. This is how the gravitational potential energy is always negative.

The potential energy of the body close to the surface of the earth

$$E_p=-\frac{G\,M_e\,m}{R_e} \quad (\because\ R\approx R_e)$$

Usually, the term gravitational potential is used instead of gravitational potential energy.

3.11. Earth Satellites

Some planets of the solar system have natural satellites orbiting them. The moon is the natural satellite of earth. It revolves around the earth in nearly circular orbit of mean radius 3.85×10^5 km once in 27.3 days.

On October 4, 1957, Sputnik-I, the first artificial satellite was launched into orbit around the earth by Russian scientists. Since then many countries have joined the race for space exploration. Man has already set foot on the moon. Vehicles have been launched to probe distance planets and space schuttles are also in use now. Artificial satellites launched by many countries are presently revolving around the earth in orbits at a height of few hundred kilometers. They are being put to a variety of uses such as scientific investigation of the conditions in the upper atmosphere, weather studies etc. and also for taking photographs of vital installations in other countries for spying activities. Another type of satellites, synchronised with the rotation of the earth (geo-stationary) have made possible world-wide microwave communication. Recently television broadcast in India during Asiad '82 was possible due to one such sattellite.

India's first artificial satellite, *Aryabhatta*, named after the famous 5th century astronomer, was launched into orbit on *April, 19, 1975*. This was followed by Bhaskar.

India had also put into orbit the communication satellite INSAT-1 (A) which unfortunately did not function as expected.

Others satellites in series are to be launched soon.

Escape velocity. Artifical satellites are always put in to orbit at a height of a few hundred kilometers above the surface of the earth because at a lower attitude, the speed will be quickly reduced due to resistance of air. The satellite instead of remaining in the orbit will be pulled to earth by gravitational force. Therefore, it is necessary to put the satellite into orbit at much higher attitude where the air is in a highly rarefied state. A satellite released horizontally at such a height remains moving in a nearly circular orbit. As the air resistance is negligible, no work is done to keep the satellite revolving in the orbit.

Rockets are used for putting a satellite into orbit. We have seen a stone thrown upwards gets returned and come down due to earth's gravity. So the rocket must be fired with sufficiently high velocity in order to overcome the earth's gravitational attraction. The minimum velocity with which the rocket must be fired in order to escape completely from the earth's gravitational attraction is called *Escape Velocity*. Let us estimate the escape velocity.

Suppose, a rocket of mass m (including the mass of satellite) is fired upwards. We donot take into account the mass of the fuel which is to be consumed ultimately. Since the height at which the fuel is burnt out is small compared to the radius of the earth, the distance of this point from the centre of the earth can be taken equal to the radius of the earth, R_e. Let the velocity of the rocket at this stage be v.

At this position,

The kinetic Energy of the rocket, $E_k = \frac{1}{2} mv^2$

The Potential Energy of the rocket $E_p = -G\, M_e m / R_e$

Where M_e is the mass of the earth.

As the rocket moves up the velocity reduces. Hence the kinetic energy reduces and at the same time the potential increases by the same amount, neglecting other losses. At infinite distance, the potential increases to its maximum possible value *i.e.*$=0$. Thereafter, there can be no further change in either potential energy or in kinetic energy of the rocket. So, if some amount of kinetic energy is still available, the rocket will continue to move with constant speed.

Therefore, $E_k + E_p$

i.e. $$\left(\frac{1}{2} mv^2 - G\,\frac{M_e M}{R_e}\right)$$

should be positive for the rocket to escape the gravitational [illegible] the earth. If it is $-$ve, the rocket may either go into orbit [illegible] back to the earth.

Thus the rocket to escape from the earth's field.

$$\frac{1}{2}mv^2 > G\, M_e \frac{m}{R_e}$$

or
$$v > \sqrt{2G\frac{M_e}{R_e}}$$

Substituting, $G = 6.67 \times 10^{-11}$ N . m²/kg²

$M_e = 5.98 \times 10^{24}$ kg

and $R_e = 6.4 \times 10^6$ m

$v > 11.16$ km/s

or $v > 4 \times 10^4$ km/h

So a rocket must be fired upwards with a velocity greater than 4×10^4 km/h to escape the earth's pull. This velocity is called escape velocity.

It should be noted that a rocket has to be fired with volocity very close to escape velocity but not equal to escape velocity in order to lift the satellite upto the orbit. For a rocket to proceed to a distant planet it must be fired at a velocity higher than the escape velocity. A multistage rocket (Fig. 3.8) is usually used for lanuching the satellite for two reasons. The high speed necessary for take off can not be attained by a single rocket. Even if such a velocity is attained, the rocket and the satellite can not withstand the tremendus amount of heat generated due to friction in the lower atmosphere. Therefore, the rocket contains number of independent stages; the second stage to be fired after the fuel in the

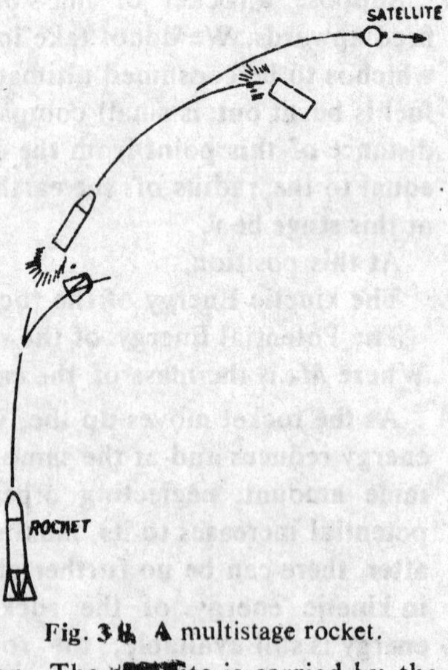

Fig. 3.8 A multistage rocket.

1st stage is burnt out and so on. The satellite is carried by the last stage. The first stage (booster) moves with less velocity in the dense atmosphere and carry the other stages of the rocket with its load. After the first stage has lifted the entire load and has burnt out, it gets detached from the rocket. The second stage, then fired, continues to accelerate and move with higher speed but in apparently rarer atmosphere. The last stage of the rocket discharges the satellite horizontally to the orbit at required velocity called orbital velocity.

Orbital velocity. The orbital velocity is fixed for a particular height or for the type of orbit. If the horizontal velocity at which the satellite is launched into the orbit is less than the orbital velocity required at that height, the satellite will spiral down to earth. On the other hand, if the launching velocity is more than the orbital velocity, the satellite will move outward to larger orbits or will completely escape to the space.

Suppose m is the mass of a satellite and h is the distance of the orbit from the ground. The gravitational force F on the satellite towards the earth is:

$$F=G\frac{M_e m}{(R_e+h)^2}$$

where M_e is the mass of the earth; R_e is the mean radius of the earth and G is universal gravitational constant.

If this force produces on acceleration, a, of the satellite towards earth,

$$a=\frac{F}{m}=G\frac{M_e}{(R_e+h)^2}$$

If the satellite is moving with a constant velocity v in the circular orbit (radius $=R_e+h$), the centripetal acceleration necessary to keep it moving in the orbit.

$$a_c=\frac{v^2}{(R_e+h)}$$

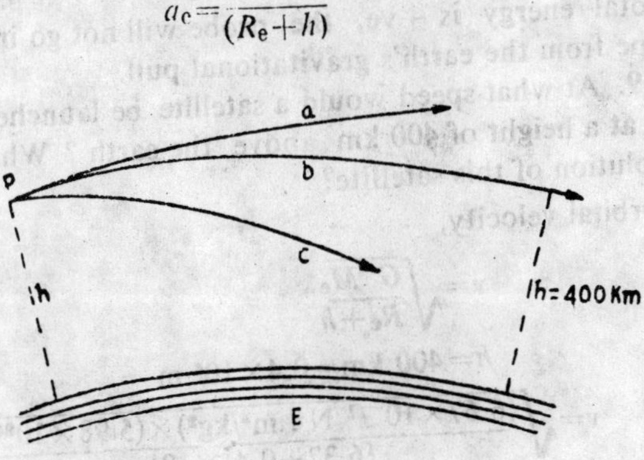

Fig. 3.9. The lanuching of a satellite. If the satellite is discharged horizontally with sufficient velocity, it moves in a circular orbit.

a = 15 km/s, satellite escapes to space.

b = 7.5 km/s ; satellite moves in circular orbit at a height of 400 km.

c = 4 km/s ; satellite cannot overcome earth's gravitational pull.

The acceleration a_c is provided by the gravitational force *i.e.*

$$a = a_c$$

$$\frac{v^2}{R_e+h} = \frac{G\,M_e}{(R_e+h)^2}$$

or

$$v = \sqrt{\frac{G\,M_e}{R_e+h}} \quad \ldots(3.8)$$

This is the horizontal velocity at which the satellite should be launched at a height h in order to move in a circular orbit. If the satellite is not launched horizontally, instead of moving in circular orbit, it moves in an elliptical orbit.

Example 3.8. A 100 kg space probe is shot from the earth with an initial speed of 12 km/s. Can the probe move in a circular orbit around the earth?

Given $M_e = 5.98 \times 10^{24}$ kg ; $G = 6.67 \times 10^{-11}$ N m²/kg².
$R_e = 6.37 \times 10^6$ m.

Solution. At take off, if the total energy is +ve, the probe will escape from the earth's attraction and if it is −ve, it may go into orbit.

At take off,

$$E_p = -G\frac{M_e\,m}{R_e}$$

$$= (-6.67 \times 10^{-11}\ \text{N m}^2/\text{kg}^2)\left(\frac{5.98 \times 10^{24}\ \text{kg})(100\ \text{kg})}{6.37 \times 10^6\ \text{m}}\right)$$

$$= -6.262 \times 10^7\ \text{J}$$

$$E_k = \tfrac{1}{2}mv^2 = \tfrac{1}{2}(100\ \text{kg})(12 \times 10^3\ \text{m/s})^2 = 720 \times 10^7\ \text{J}$$

Total energy $E = E_p + E_k = 713.738$ J.

Since the total energy is +ve, the probe will not go into orbit and will escape from the earth's gravitational pull.

Example 3.9. At what speed would a satellite be launched into a circular orbit at a height of 400 km above the earth ? What is the period of revolution of this satellite?

Solution. Orbital velocity,

$$v = \sqrt{\frac{G\,M_e}{R_e+h}}$$

Here

$$h = 400\ \text{km} = 0.4 \times 10^6\ \text{m}$$

∴

$$v = \sqrt{\frac{(6.67 \times 10^{-11}\ \text{N . m}^2/\text{kg}^2) \times (5.98 \times 10^{24}\ \text{kg})}{(6.37 + 0.4) \times 10^4\ \text{m}}}$$

$$= 7.852 \times 10^3\ \text{m/s.}$$

The circumference of the orbit

$$= 2\pi\,(R_e + h)$$

$$=2\times 3.14\times 6.77\times 10^{6}\text{ m}$$
$$=42.5156\times 10^{6}\text{ m.}$$

$\therefore$ Period of revolution of the satellite around the earth

$$=\frac{\text{circumference}}{\text{velocity}}$$
$$=\frac{42.5156\times 10^{6}\text{ m}}{7.852\times 10^{3}\text{ m/s}}=5.415\times 10^{3}\text{ s}$$
$$=1\text{ hour }30\text{ min }15\text{ sec}$$

3.12. Rotation of Rigid Bodies

Rigid body. We have considered the mass of a body concentrated at a point or point mass for application of forces and consequent motion. Now we will consider the rotatory motion of rigid body taking into consideration its actual distribution of mass. In an ideal rigid body the distance between any two particles on it remains always constant and hence there is no change in shape.

If a rigid body moves such that any straight line in it makes a constant angle with a fixed direction in space, the body is said to have translatory motion. In such a motion all particles constituting the body move along parallel and equal paths and hence the motion of any particle on it represents the motion of the entire body. That is why a rigid body is represented by a particle at its centre of gravity so far its motion and position are concerned.

When the motion of a rigid body is such that each particle in it remains at a constant distance from a fixed straight line and moves in a plane perpendicular to the fixed line (axis), the body is said to have rotatory motion. Evidently each particle of the body not on the axis, moves in a circle in a plane perpendicular to the axis with centre of the circle on the axis. The angular displacement of each particle during a time interval is the same.

For example, P and Q are two points on a disc rotating about an axis OA. As the disc rotates P has moved to P' and Q to Q'. In this case :

$$\angle POP'=\angle QOQ'$$

Distance PQ = Distance $P'Q'$.

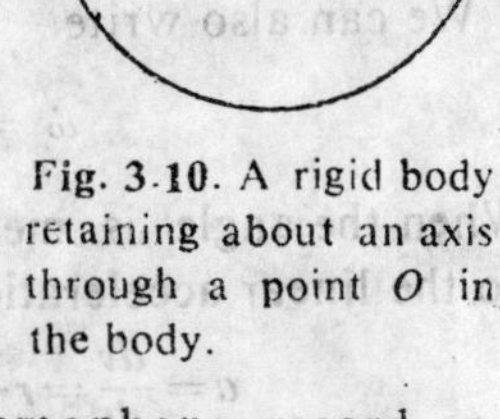

Fig. 3.10. A rigid body retaining about an axis through a point O in the body.

At any time the motion of a body may consist of translation or rotation or a combination of translation and rotation. A grinding wheel, pulley, the blades of a fan, a gramophone record, a turbine, a flywheel etc., rotating about an axis are examples of rigid bodies in rotational motion.

The wheel of a train rotates about its axle and at the same time moves ahead *i.e.* it has translational motion besides rotational motion. Earth rotates about its own axis and simultaneously moves along the orbit. A spinning top has rotational and translational motion; its axis also changes with time. We shall discuss only the case of rotatory motion of rigid bodies about a fixed axis.

3.13. Rotatory Motion

As a disc rotates about an axis through its centre all particles on it have same angular speed ω though the linear speed of the particles will be different.

Since $v=r\omega$, a particle at the edge must move with higher linear speed than one near the axis.

In Article 3.1, we have seen that the average angular velocity of a particle is :

$$\bar{\omega}=\frac{\theta}{t}$$

and the instantaneous angular velocity

$$\omega=\underset{\Delta t\to 0}{Lt}\frac{\Delta\theta}{\Delta t}=\frac{d\theta}{dt}=\dot{\theta}$$

As in the case of linear motion, angular motion may be uniform or accelerated. *Angular acceleration* α *(alpha) is the time rate of change of angular velocity.*

The average angular acceleration,

$$\bar{\alpha}=\frac{\omega-\omega_i}{t}$$

where ω_i and ω are the initial and final angular velocities at the beginning and end of the time interval t. But if the angular acceleration is variable, the angular acceleration at any instant.

$$\alpha=\underset{\Delta t\to 0}{Lt}\frac{\Delta\omega}{\Delta t}=\frac{d\omega}{dt}=\dot{\omega}$$

We can also write

$$\dot{\omega}=\frac{d\omega}{dt}=\frac{d}{dt}\left(\frac{d\theta}{dt}\right)=\frac{d^2\theta}{dt^2}=\ddot{\theta}$$

When the angle is measured in radiana, $v=r\omega$ (Ref. Eq. 3.2). So the linear acceleration

$$a=\frac{dv}{dt}=r\frac{d\omega}{dt}=r\dot{\omega} \qquad ...(3.9)$$

i.e. linear acceleration = angular acceleration × distance of the particle from the axis of rotation.

The unit of angular acceleration is rad/sec^2. When the angular acceleration is zero, linear acceleration is also zero. Though the angular acceleration for all particles on a rigid body may be equal at any instant, their linear acceleration will be different as in case of angular and linear velocity.

For a body undergoing uniformly accelerated rotatory motion we can state the following relations in analogy with similar relations for linear motion.

$$\left.\begin{aligned} \omega &= \omega_1 + dt \\ \theta &= \omega_1 + \tfrac{1}{2}\alpha t^2 \\ \omega^2 &= \omega_1^2 + 2\alpha\theta \end{aligned}\right\} \qquad \ldots(3.10)$$

where ω_1 and ω are the initial and final angular velocities, θ is the angular displacement during the time interval t and α is the uniform angular acceleration.

3.14. Rotational Kinetic Energy—Moment of Inertia

Let a rigid body (Fig. 3.11) rotate about an axis, perpendicular to the plane of the diagram and passing through O, with an angular velocity ω.

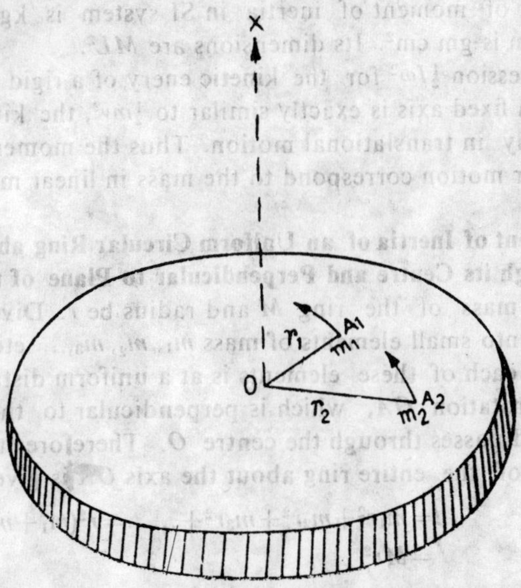

Fig. 3.11. Rotational kinetic energy of a rigid body.

Let A_1 be a small element of the rigid body. Its mass is m_1 and it is situated at a distance r_1 from O.

So the linear velocity of the element A_1 is given by $v_1 = r_1\omega$.

The kinetic energy of this element is

$$\frac{1}{2} m_1 v_1^2 = \frac{1}{2} m_1 r_1^2 \omega^2$$

The body is made of large number of such elements. The total kinetic energy of the body is equal to the kinetic energy of all such elements. Let the mass of the elements of which the body is made of be m_1, m_2, m_3,......etc and their respective distance from the axis be r_1, r_2, r_3,......etc. Since all elements have the same angular velocity, the kinetic energy of the body = The sum of the kinetic energy of all elements.

Thus K.E. of the body

$$= \frac{1}{2} m_1 r_1 \omega^2 + \frac{1}{2} m_2 r_2 \omega_2 \omega^2 + \ldots\ldots$$

$$= \frac{1}{2}(\Sigma_1 m_1 r_1^2)\omega^2 = \frac{1}{2} I\omega^2 \qquad \ldots(3.11)$$

where $I = \Sigma_1 m_1 r_1^2$

The quantity I is called the moment of inertia of the body about the axis through O. Since it depends upon the mass of the elements and their distances from the axis, its value will change if either the shape of a body is altered (even without changing the mass) or the position of the axis is changed. Otherwise, it has the same value.

The unit of moment of inertia in SI system is kg m² and in CGS system is gm cm². Its dimensions are ML^2.

The expression $\frac{1}{2} I\omega^2$ for the kinetic enery of a rigid body rotating about a fixed axis is exactly similar to $\frac{1}{2} mv^2$, the kinetic energy of the body in translational motion. Thus the moment of inertia I in angular motion correspond to the mass in linear motion.

3.15. Moment of Inertia of an Uniform Circular Ring about an Axis through its Centre and Perpendicular to Plane of the Ring

Let the mass of the ring M and radius be r. Divide the ring (Fig. 3.12) into small elements of mass m_1, m_2, m_3,.....etc. Since the ring is thin each of these elements is at a uniform distance r from the axis of rotation OA, which is perpendicular to the plane of the ring and passes through the centre O. Therefore the moment of inertia of the entire ring about the axis OA is given by:

$$I = m_1 r^2 + m_2 r^2 + m_3 r^2 + \ldots\ldots = r^2(m_1 + m_2 + \ldots\ldots)$$

$$= Mr^2 \qquad \ldots(3.12)$$

It may be of interest to note that the moment of inertia of the ring about one of the diameters as the axis is $\dfrac{Mr^2}{2}$.

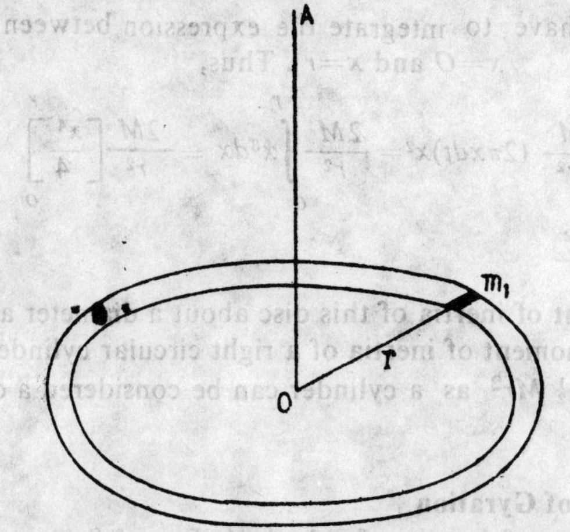

Fig. 3.12. The moment of inertia of a uniform thin ring.

3.16. The Moment of Inertia of a Thin Uniform Circular Disc about an Axis through its Centre and Perpendicular to the Plane of the Disc

Let the mass of disc be M and radius be r. Divide the disc into a large number of thin circular and concentric rings (Fig. 3.13). Let us consider one such typical ring of radius x, and width dx, as shown in Fig. 3.13. Since the ring is very thin, its area $=$circumference$\times$width

$$=2\pi x.dx$$

The mass of unit area of the disc

$$=\frac{M}{\pi r^2}$$

So the mass of the ring

$$=\frac{M}{\pi r^2}\,2\pi x\,dx$$

The moment of inertia of the ring about the axis OA,

$$dI=\frac{M}{\pi r^2}(2\pi x dx)x^2$$

The entire disc can be considered to be made of such rings with redius ranging from O to r.

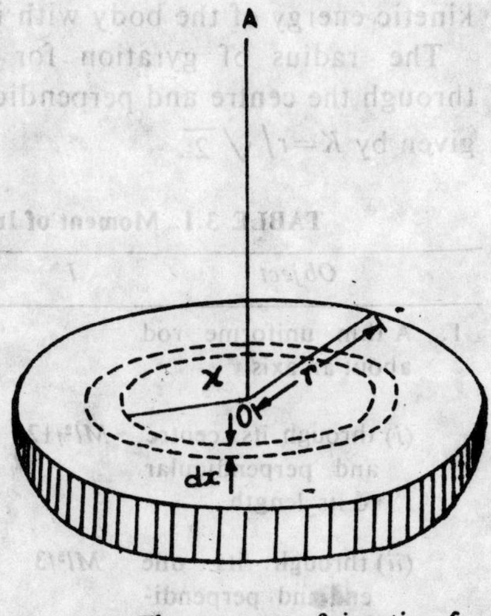

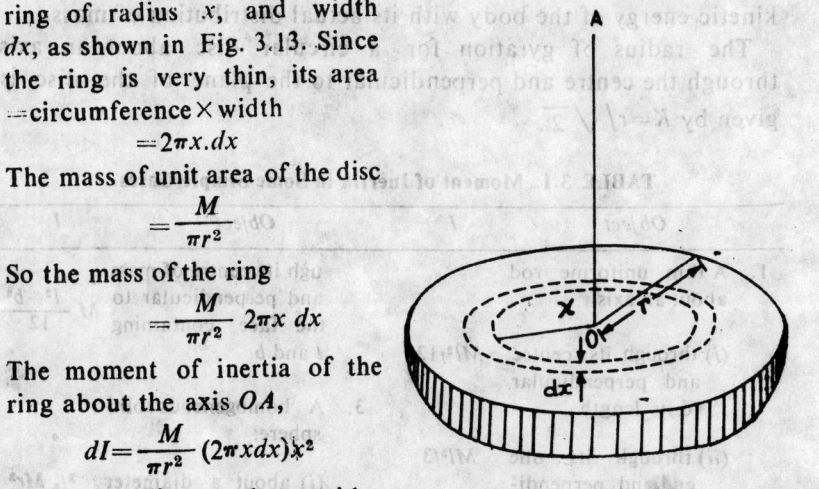

Fig. 3.13. The moment of inertia of a uniform circular disc.

Hence to get the moment of inertia of the whole disc about the axis OA, we have to integrate the expression between the limits $x=O$ and $x=r$. Thus,

$$I=\int_{O}^{r}\frac{M}{\pi r^2}(2\pi x dt)x^2=\frac{2M}{r^2}\int_{O}^{r}x^3dx=\frac{2M}{r^2}\left[\frac{x^4}{4}\right]_{O}^{r}$$

or $$I=\frac{Mr^2}{2} \qquad ...(3.13)$$

The moment of inertia of this disc about a diameter as the axis is $Mr^2/4$. The moment of inertia of a right circular cylinder about its axis is also $\frac{1}{2}Mr^2$ as a cylinder can be considered a disc of large thickness.

3.17. Radius of Gyration

The moment of inertia of a body $I=\Sigma_i m_i r_i^2$. Suppose the whole mass of the body is concentrated at a point at a distance K from the axis so that the particle possesses the same moment of inertia as the body possesses about the same axis. Thus $I=MK^2$ ...(3.14) where $M=\Sigma_i m_i$, the total mass of the body. K is called the radius of gyration of the body about the given axis.

Hence if the whole mass of the body is concentrated in a particle at a distance equal to radius of gyration from the axis, the kinetic energy of rotation of the particle, will be the same as the kinetic energy of the body with its actual distribution of mass.

The radius of gyration for a circular disc about an axis through the centre and perpendicular to the plane of the disc is given by $K=r/\sqrt{2}$.

TABLE 3.1. Moment of Inertia in Some Simple Cases.

Object	*I*
1. A thin uniforme rod about an axis	
(*i*) through its centre and perpendicular to its length.	$Ml^2/12$
(*ii*) through its one end and perpendicular to its length.	$Ml^2/3$
2. A rectangular Bar of length l and breadth b about an axis through its centre of mass and perpendicular to the face containing l and b.	$M\frac{l^2+b^2}{12}$
3. A homogeneous solid sphere:	
(*i*) about a diameter as axis.	$^2/_5\ Mr^2$
(*ii*) about a tangent as axis.	$^7/_5\ Mr^2$

Example 3.11. A uniform circular disc of mass 1.0 kg has a radius of 20 cm. What is its moment of inertia about an axis through the centre perpendicular to the plane of the disc? What is the radius of gyration about the axis?

Solution.

$$I=\tfrac{1}{2}Mr^2=\tfrac{1}{2}(1.00\ \text{kg})\times(0.2\ \text{m})^2$$
$$=2\times 10^{-4}\ \text{kg.m}^2$$

$$K=\frac{r}{\sqrt{2}}=\frac{0.2}{\sqrt{2}}=0.141\ \text{m}$$

3.18. Torque

Moment of force. Consider a heavy wheel mounted on an axis at 0. It is our common experience that if we apply a force along the direction of the spoke at *A*, the wheel does not turn. But when the force is applied at *B* in a direction perpendicular to the spoke it rotates. The rotation becomes easier if the force is applied farther from the axis, say at *C*. Similarly when a wrench is used to turn a nut, the rotating effect is greatest when the wrench is held farthest from the nut.

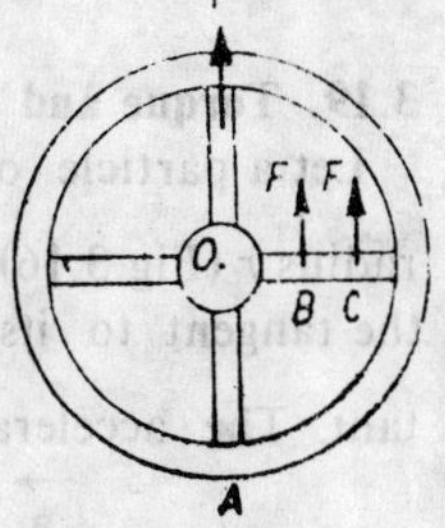

Fig. 3.14. Torque due to a force.

Thus it is obvious that the factor that determines the effect of a given force upon the rotational motion is the perpendicular distance from the axis of rotation to the line of action of the force. This distance is called moment arm.

But for a given moment arm, a larger force produces greater effect on rotational motion. The two quantities, force and moment arm, are of equal importance in rotatory motion. They are combined into a single quantity, *torque* which is a measure of the effect of the force in changing the rotation about the chosen axis.

The two vectors involved in the torque are the force $\vec{\mathbf{F}}$ and the length vector $\vec{\mathbf{r}}$ (Fig. 3.15). The torque $\vec{\tau}$ (tau) is defined as the vector moment of the force about *O*

$$\vec{\tau}=\vec{\mathbf{r}}\times\vec{\mathbf{F}}$$

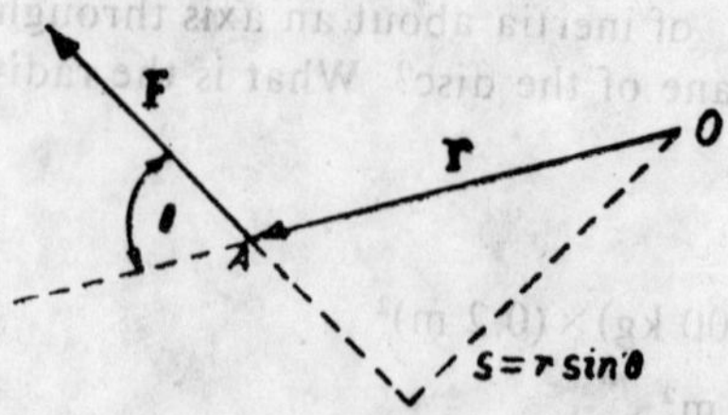

Fig. 3.15. Torque is the vector product of the force vector $\vec{F}$ and length vector $\vec{r}$. A clockwise torque $\vec{\tau} = \vec{r} \times \vec{F}$

The torque is a vector quantity and has the magnitude $rF \sin\theta$ and its direction is perpendicular to the plane of the paper (given by the righthanded screw rule). From Fig. 3.15, we see that $r\sin\theta = S$ *i.e.* the perpendicular distance between O and line of action of the force.

Thus $\tau = FS$ ...(3.15)

So τ is the moment of the force about O *i.e.* the product of the force and perpendicular distance between O and the line of action of the force. The unit of torque in SI system is Newton-metre and in CGS system is dyne-cm. Its dimensions are ML^2T^{-2}.

3.19. Torque and Moment of Inertia

Let a particle of mass m be kept moving in a circular path of radius r (Fig 3.16). Suppose a force $\vec{F}$ acts on the particle along the tangent to its path at every instant. The acceleration $\vec{a}$ produced is

$$\vec{a} = \frac{\vec{F}}{m}$$

If the angular acceleration of the particle is ω, we have

$$a = r\dot{\omega}$$

$$\therefore \quad F = ma = mr\,\dot{\omega}$$

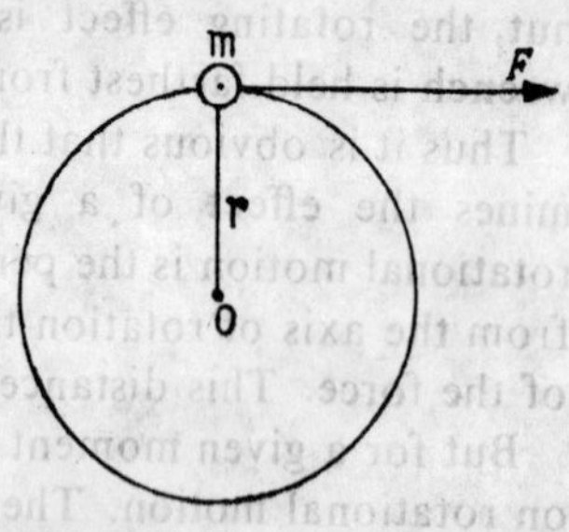

Fig. 3.16. Torque and moment of inertia.

Thus the magnitude of torque due to force F about the axis of rotation through O,

$$\tau = Fr = mr^2\,\dot{\omega} = I\,\dot{\omega} \quad ...(3.16)$$

where $I = mr^2$ is the moment of inertia of the particle about the axis. For any rotating body, this relation for τ is applicable provided I and τ refer to the same axis.

3.20. Torque Due to Number of Forces

The effect of torque due to a number of forces is determined b the vector sum of the torques due to individual forces. If the ve tor sum of the torques acting on a body is zero, the body will have no angular acceleration. Thus a body at rest or moving with uniform angular velocity will maintain that condition.

We know in linear motion, a body is in linear equilibrium when vector sum of the forces acting on the body is zero. In rotational motion, the body remains in rotatational equilibrium when the vector sum of torques acting on the body is zero. Therefore, torque plays the same role in rotational motion as force in translational motion. We can put these conditions for equilibrium in symbols as follows :

1st condition : $\Sigma \vec{F}=0$ —condition for linear equilibrium.

2nd condition : $\Sigma \vec{\tau}=0$ —condition for rotational equilibrium.

3.21. Torque Produced by a Couple

Whenever one or more forces are applied on a body, the body will have both rotational and translational acceleration unless one of the equilibrium conditions given in previous article are satisfied.

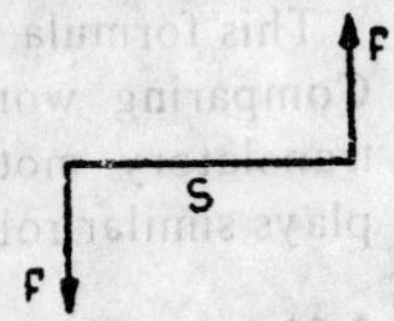

Fig. 3.17. Torque due to a couple of forces $\tau=FS$

There are two exceptions however. When a single force is applied along a line passing through the centre of gravity of the body, the body has translational acceleration only. Similarly, when a pair of forces equal in magnitude opposite in direction and not in the same line (Fig. 3.17) are applied to an object, the object will have no translational acceleration, but only rotational acceleration. This pair of forces are called a *couple*. The magnitude of the torque due to the couple is equal to the product of one of the forces and perpendicular distance between them and is independent of the position of the axis. For a couple

$$\tau=FS \qquad ...(3.17)$$

3.22. Work Done by Torque

Let two equal forces F constituting the couple act tangentially on the rim of a disc at the end of a diameter (Fig. 3.18). Thus the torque $\tau=2Fr$, where r is the radius of the disc.

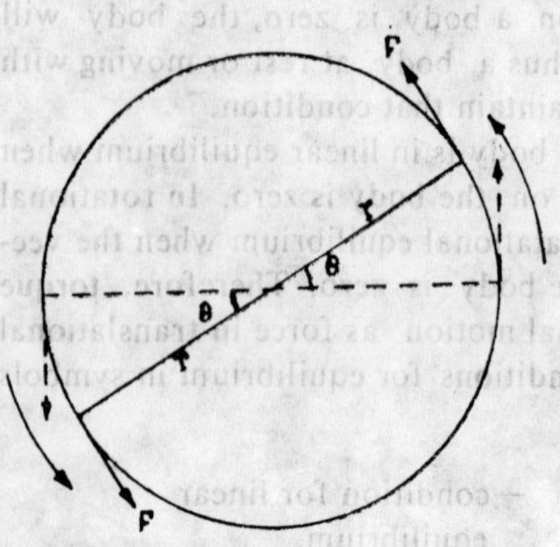

Fig. 3.18. Work done by the torque.

Let, the disc be free to rotate about an axis passing through its centre 0, perpendicular to the plane of the disc. Suppose, the couple rotates the body through a very small angle θ. The displacement l of each force is the lenght of the arc,

$$l=r\theta$$

when θ is in radians. Since θ is very small the displacement l and the direction of force F are regarded as the same. Hence, work done by a single force is

$$Fl=Fr\theta.$$

Tho total workdone by both the forces constituting the couple

$$=2Fr\theta.$$

But $\tau=2Fr$.

So work done by the torque$=\tau\theta=$torque$\times$angular displacement. ...(3.18)

This formula also applies even when the value of θ is large. Comparing workdone by the torque with workdone by a force in translatory motion, we find again the torque in rotatory motion plays similar role as force in translatory motion.

3.23. Angular Momentum

Let us consider a particle of mass m moving with a velocity $\vec{\mathbf{v}}$ along the circumference of a circle. Its linear momentum,

$$\vec{\mathbf{p}}=m\,\vec{\mathbf{v}}.$$

The particle has an angular momentum about an axis through 0 (perpendicular to the plane containing $\vec{\mathbf{r}}$ and $\vec{\mathbf{v}}$) given by

$$\vec{\mathbf{L}}=\vec{\mathbf{r}}\times\vec{\mathbf{p}}=\vec{\mathbf{r}}\times m\,\vec{\mathbf{v}}$$

As angular momentum is the product of linear momentum and a moment arm, it is also called moment of momentum.

The magnitude of $\vec{\mathbf{L}}$ is given by:

$$L=rp\sin\theta=mrv\sin\theta.$$

where θ is the angle between $\vec{\mathbf{r}}$ and $\vec{\mathbf{p}}$ *i.e.*

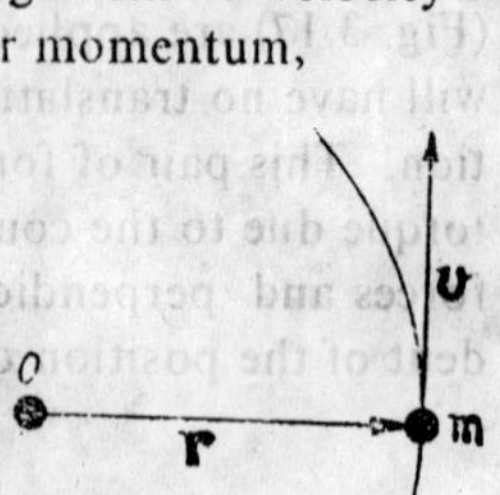

Fig. 3.19. Angular momentum

$$\vec{\mathbf{L}}=\vec{\mathbf{r}}\times m\,\vec{\mathbf{v}}$$

$\vec{v}$ and $\vec{r}$ (Fig. 3.19). Since the velocity at any instant in circular motion is perpendicular to r, $\theta = 90°$.

So $$L = mrv. \qquad ...(i)$$

The direction of $\vec{L}$ is perpendicular to the plane containing $\vec{r}$ and $\vec{v}$ and in a sense determined by the right handed screw rule.

Since $v = r\omega$ where ω is the angular velocity,

$$L = mr^2\omega \qquad ...(ii)$$

In the rotation of rigid bodies angular momentum about the axis of rotation is an important property. All particles on the rigid body has same angular velocity. Hence the total angular momentum of a rigid body is

$$L = \Sigma_i\, mr^2\omega = (\Sigma mr^2)\omega = I\omega \qquad ...(3.19)$$

$$= \text{moment of inertia} \times \text{angular velocity}$$

where I is the moment of inertia of the body.

The unit of angular momentum in SI units is kg m²/s and in CGS units is gm cm²/s.

Its dimensions are ML^2T^{-1}

(Compare Equation 3.19 with $p = mv$ for linear motion).

3.24. Conservation of Angular Momentum

If a net force acts on a body in linear motion, the body is accelerated and its momentum changes. The rate of change of linear momentum is a measure of the net force (Chapter 2). Similarly a torque acting on a body produces rotational acceleration in it and its angular momentum ($L = I\omega$) changes.

We know,

$$\tau = I\dot{\omega} = \frac{d}{dt}(I\omega) = \frac{dL}{dt} \quad (\because\ L = I\omega)$$

On analogy with linear momentum, it is seen that the rate of change of angular momentum is a measure of the torque applied. In the absence of a net torque, a rigid body continues to rotate with its initial angular velocity, in both direction and magnitude. Hence the rotating body tends to maintain the same plane and rotation and its angular momentum $\vec{L}$, remains unchanged.

On analogy with the principle of conservation of linear momentum we can restate the law for rotatory motion in terms of angular momentum *viz.* if there is no net external torque acting upon a body, the angular momentum of the body remains unchanged. This is law of conservation of angular momentum.

If the distribution of mass of a rotating body is changed during its rotation, the moment of inertia I changes. Hence the angular velocity must change to maintain the same angular momentum. Take for example a man standing on a turn table and holding a pair of weights in his outstretched hands. The turn table is free to rotate with negligible friction about a vertical axis (Fig. 3.20). When the turn table is rotating at constant speed, if he brings down his hands suddenly his moment of inertia is decreased. The speed of rotation will be found to increase considerably. If there is a way of measuring the value of I and ω, it will be found that $I\omega$ is same is both cases.

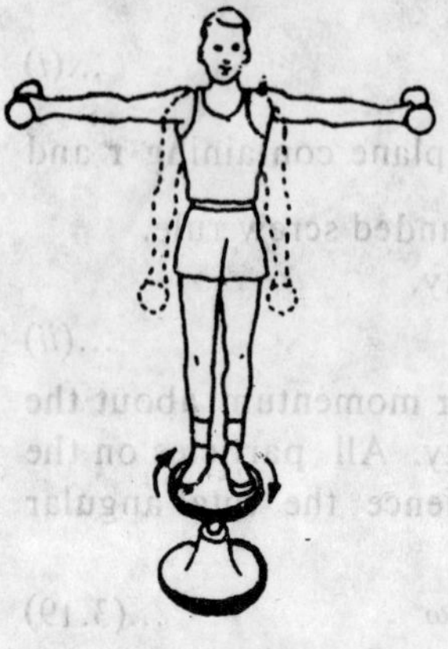

Fig. 3.20. Angular momentum is conserved.

3.25. Analogy between Linear and Rotatory Motion

We summarise below the analogy between linear motion and rotational motion. The quantities relating to linear motion are given in the first column and the corresponding quantities in rotational motion are given in the second column.

TABLE 3.1

Linear motion		*Rotational motion*
Linear displaceement	S	Angular dispiacment θ
Linear velocity	$v=ds/dt$	Angular velocity $\omega=d\theta/dt$
Linear acceleration	$a=dv/dt$	Angular acceleration $\alpha=\dot{\omega}=d\omega/dt$
Mass	m	Moment of Inertia I
Force	$F=m\,a$	Torque $\tau=I\,\alpha$
Work done	$=F\,S$	Work done$=\tau\,\theta$
Kinetic energy	$E_K=\frac{1}{2}\,mv^2$	Kinetic energy $E_K=\frac{1}{2}\,I\,\omega^2$
Power	$=F\,v$	Power$=L.\omega$.
Linear momentum	$P=m\,v$	Angular momentum $L=I\,\omega$
Condition for linear equilibrium	$\Sigma\,\vec{F}=0$	Condition for rotational equilibrium $\Sigma\,\vec{\tau}=0$.

Example 3.10. A wheel and its axle have a total mass of 1000 kg. The axle has a diameter of 10 cm. A 5 kg ball is fixed to the end of a cord and the cord is coiled round the axle. When released, the ball moves downward a distance of 2.5 m in 5.0 seconds.

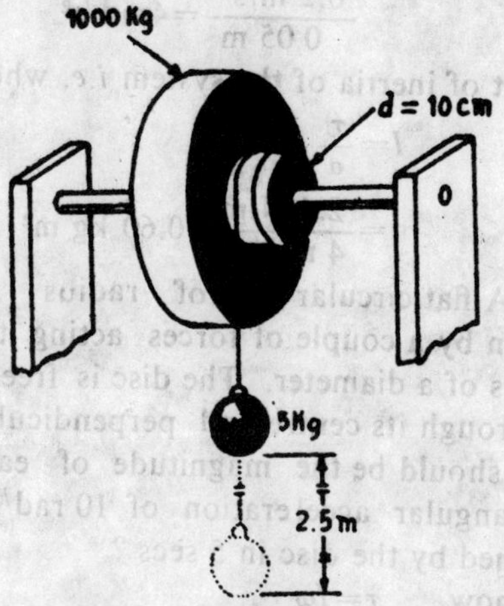

Fig. 3.21

Determine (*i*) the torque acting on the system.

(*ii*) the angular acceleration of the system.

(*iii*) the moment of inertia of the wheel and axle about the horizontal axis through the centre of the wheel.

Given $g = 9.8$ m/s² and friction is negligible.

Solution. (*i*) The linear acceleration of the ball is,

$$a = \frac{2s}{t^2}$$

$$= \frac{2 \times 2.5 \text{ m}}{25 \ s^2} = 0.2 \text{ m/s}^2$$

$$[\because s = 0 + \tfrac{1}{2}at^2]$$

The net force acting downward

$$= (5.0 \text{ kg})\ (0.2 \text{ m/s}^2) = 1 \text{ N}.$$

But the net downward force on the ball is also $W - T$ where W is the weight of the ball and T is the tension acting upwards.

$$W - T = (5.0 \text{ kg})\ (9.8 \text{ m/s}^2) - T$$

$$= 49 \text{ N} - T$$

So $\quad 49 \text{ N} - T = 1 \text{ N}$

or $\quad T = 48 \text{ N}$

The tension provides the torque,

$$\tau = r\ T$$

$$= (0.05 \text{ m})\ (48.0 \text{ N}) = 2.4 \text{ N.m}.$$

(*ii*) The angular acceleration of the system

$$\alpha = \frac{a}{r}$$

$$\frac{0.2\ \text{m/s}^2}{0.05\ \text{m}}=4\text{rad/s}^2 \qquad [\because a=\alpha r]$$

(*iii*) The moment of inertia of the system *i.e.* wheel and axle.

$$I=\frac{\tau}{\alpha}$$

$$=\frac{2.4\ \text{N.m}}{4\ \text{rad/s}^2}=0.60\ \text{kg m}^2$$

Example. 3.11. A flat circular disc of radius 0.2 m and mass 10 kg is acted upon by a couple of forces acting tangential to the rim at the two ends of a diameter. The disc is free to rotate about an axis passing through its centre and perpendicular to the plane of the disc. What should be the magnitude of each force so that the disc has an angular acceleration of 10 rad/s² ? What is the kinetic energy gained by the disc in 5 secs ?

Solution. We know $\tau=I\omega$

In this example, I for disc$=Mr^2/2$

$\therefore$ Torque $$\tau=\frac{Mr^2}{2}\omega$$

$$=\frac{(10.0\ \text{kg})\ (0.2\ \text{m})^2}{2}\times 10\ \text{rad/s}^2$$

If each force of the couple is F, the moment of couple *i.e.* torque

$$=F\times 2r=F\times(0.4\ \text{m})$$

So, $$F\times 0.4\ \text{m}=(5.0\ \text{kg})\ (0.2\ \text{m})^2\ (10\ \text{rad/s}^2)$$

or $$F=5\ \text{kgm/s}^2=5\ \text{N}.$$

Angular velocity at the end of 5 secs $=10\ \text{rad/s}^2\times 5\ s$

$$=50\ \text{rad/s} \qquad [\because \omega=0+\alpha t]$$

$\therefore$ Kinetic energy gained, $\frac{1}{2}I\omega^2$

$$=\frac{\frac{1}{2}(10.0\ \text{kg})\ (0.2\ \text{m})^2\ (50\ \text{rad/s})^2}{2}$$

$$=250\ \text{J}.$$

QUESTIONS

1. Centripetal force and centrifugal reaction are equal in magnitude, but opposite in direction. Do they balance each other?
2. Is any work done by centripetal force ? Explain.
3. Why does a cyclist find it easier to go round a curve of larger radius than in a curve of smaller radius?
4. A car takes a sudden turn to the right while travelling on a straight road. A passenger in the front seat finds himself sliding towards right door. Explain the reason?
5. A person is sitting on a merry-go-round rotating with high speed. Explain, different forces coming into play to keep the person in balance.
6. The term 'centrifugal force' is used mostly in the wrong context Explain.

7. Every object in the universe attracts every other object towards itself according to Newton's universal law of gravitation. Do we experience it while moving on the surface of earth?

8. An astronaut is speeding from earth to the moon. Draw a rough sketch to show how his weight changes between the earth and the moon.

9. Assuming the earth to be a perfect sphere with level ground, where would g be minimum?

10. How does g change when you go (*i*) to the top of Everest (*ii*) inside Kolar mine?

11. The escape velocity on earth is calculated to be 11.4 km/s. If a rocket is fired at this speed from the surface of Jupiter, is it likely to go round the Jupiter into orbit?

12. If the polar ice cap increases suddenly how would the period of rotation of the earth be affected?

13. A body at the end of a string moves in a vertical circle while you pull the string always towards the centre of the circle. Can the speed of the body be constant?

14. Give two examples, not mentioned in this book, to illustrate conservation of angular momentum?

15. Two equal and opposite forces act on a body, yet the body is not in equilibrium, explain the reason?

16. There are two hollow spheres of same mass, volume and colour. One is made with wood and the other with iron. How will one distinguish between the two ?

PROBLEMS

1. Assuming earth to be a perfect sphere of radius 6.4×10^3 km, calculate the reduction in the value of g at the equator due to earth's rotation.

[*Ans.* 3.4×10^{-2} *m/s*^{2}]

2. What is the minimum speed at which an aeroplane can execute a loop of 150 m radius so that there will be no tendency for the pilot to fall out at the highest point ? [*Ans. 38.34 m/s*]

3. Calculate the angle which the bicycle and rider must make with the vertical while going round a curve of 8 m. radius at 18 km/h. [*Ans.* $17°36'$]

4. The designer of a highway wish to have automobiles round a curve at 100 km/h. If the width of the road is 60 m, and its outer edge is 6 m higher than the inner edge, what should be the ideal value for radius of curvature at the curve ? [*Ans. 7.886 km*]

5. Assuming the distance between two rails to be 1.2 m determine the difference in the level of the outer rail and the inner rail if a train has to negotiate safely a curve of 1 km radius of curvature at a speed of 48 km/h.

[*Ans. 2.18 cm*]

6. Calculate the period of Jupiter if its mean distance from the sun is 7.78×10^8 km. Given, distance of earth from the sun is 1· 9 $\times 10^7$ km (11.93 years).

7. Determine the height above the surface of earth where value of g reduces to half its value on the surface (2.66×10^3 km). Radius of earth $= 6.4 \times 10^3$ km. [Hints :- $g_1/g = (R_e/R_e+h)^2$]

8. What would the necessary circular orbital velocity be for a satellite at an attitude of 800 km ? *[Ans 26.720 km/h]*

9. Calculate the velocity of escape from the moon's gravitational field given that its radius is $1 \cdot 7 \times 10^3$ km and the value of g at its surface is 1.65 m/s² *[Ans. 2.368 km/s]*

10. If the earth suddenly contracted to quarter of its present radius, what would be the duration of a day? *[Ans. 1.5 h]*

11. Calculate the mass of the sun from the following data: Radius of the earth's orbit taken to be circular$=1.5\times10^8$ km ; $G=6.67\times10^{-11}$ N.m.^{2} kg^{-2} and the period of revolution is 365 days. *[Ans. 2.01×10^{30} kg]*

12. A flywheel rotates about a horizontal axle of radius 2 cm passing through its centre. A mass of 1 kg hanging from the end of a light string wound round the axle is held at rest. When released the wheel rotates and the hanging mass descends 50 cm. in 10 sec. Calculate the moment of inertia of the flywheel. *[Ans. 0.3916 kg. m²]*

13. A grind stone has a moment of inertia of 6.0 kg. m² about its axis. A constant torque is applied so that in 10 seconds after starting from rest the grind stone atains a speed of 75 r.p.m. What is the value of the torque ? *[Ans. 1.713 kg. m². rad. s^{-2}]*

4

OSCILLATORY MOTION AND WAVE MOTION

We have discussed earlier about two types of motions, namely, longitudinal and circular motion. In this chapter we will deal with a third type of motion called oscillatory motion. Oscillatory motions are responsible for generation of all types of wave motions which will be dealt in this chapter also.

When a body moves back and forth repeatedly about a mean position, its motion is called oscillatory. We find a very familiar example of oscillatory motion in the motion of the pendulum of a wall clock. The pendulum moves in one direction and then turns back, crosses the middle position or mean position and then goes to the otherside, turns back again. This motion goes on repeating. If a mass suspended by a spring is pulled in downward direction and then released, it will be found to execute oscillatory motion in vertical direction. Many such examples can be cited. The vibration of the drum of a *tabla* or the wire of a violin are also oscillatory motions. One should not however think that the oscillatory motion is limited only to the to and fro motion along a straight line. If a body suspended by a string is given a twist and then released, the body will swing back and forth about a mean position. This is an example of angular oscillation.

4.1. Simple Harmonic Oscillation

The motion of the planets around the sun is periodic motion *i.e.* it takes equal time to go once round the sun. But the motion is not oscillatory as the planets move in one direction only.

The oscillatory motion may also be periodic if it takes equal time to come back to its original position after completing the to and fro motion once. Such a periodic motion is called harmonic motion. Thus a harmonic motion is periodic but not all periodic motion are harmonic.

If a system is disturbed from its position of equilibrium and left to itself, in the absence of friction the harmonic oscillation it executes about the mean position of equilibrium, is called simple harmonic oscillation. This simple phenomenon was observed by

Galileo in case of a simple pendulum and ultimately the clocks and watches developed due to this discovery. The term simple harmonic oscillation is used because such an oscillation can be expressed in terms of sine and cosine functions (single harmonic function). Simple harmonic motion is most fundamental type of oscillatory motion. All other oscillations can be analysed in terms of combination of number of simple harmonic oscillations. Hence the study of simple harmonic oscillation assumes importance in all branches of Physics. In mechanics disturbances from a state of equilibrium are studies by analysing in terms of combinations of number of SHM. Vibration of a material object produces sound waves which is also a combination of number of SHM. In every type of wave motion, the particles of the medium in which the wave travels execute SHM or combination of SHM. Even for propagation of light and radio waves (electromagnetic waves) this is true except that instead of material particles it is the electric and magnetic field intensities that oscillate. Thus simple harmonic oscillation or motion is responsible for all types of wave propagation. Hence study of SHM becomes still more important.

4.2. Characteristics of Simple Harmonic Motion

The motion of a simple pendulum is the most familiar example of a simple harmonic motion.

A small body of mass *m* suspended by a light inextensible thread from a rigid support makes a simple pendulum, Fig. 4.1. Let the mass be pulled aside to the position indicated in the diagram and then released. Let us examine the forces acting in this position.

The force due to weight *mg* acts in the downward direction. This force can be resolved into two rectangular components, $mg\cos\theta$ and $mg\sin\theta$. The component $mg\cos\theta$ acts along the string to provide tension. But the component $mg\sin\theta$ acts perpendicular to the direction of the string and is directed towards the equilibrium position. So the pendulum when left to itself moves towards the equilibrium position and ultimately crosses over to the otherside of equilibrium position due to inertia of motion.

As it moves over to the otherside, the direction of the force $mg\sin\theta$ is reversed. So the velocity of the pendulum decreases till it becomes zero and after that it turns back towards the equilibrium position. Thus, whatever may be the position of the pendulum in its oscillatory motion the force $mg\sin\theta$ is always directed towards the equilibrium position. In other words, this force tends to restore the pendulum from its displaced position to its equilibrium position. Therefore, this force is called *restoring force*. The restoring force for sipmle harmonic motion of a loaded spring, of prongs of a tuning fork etc. are provided by elasticity.

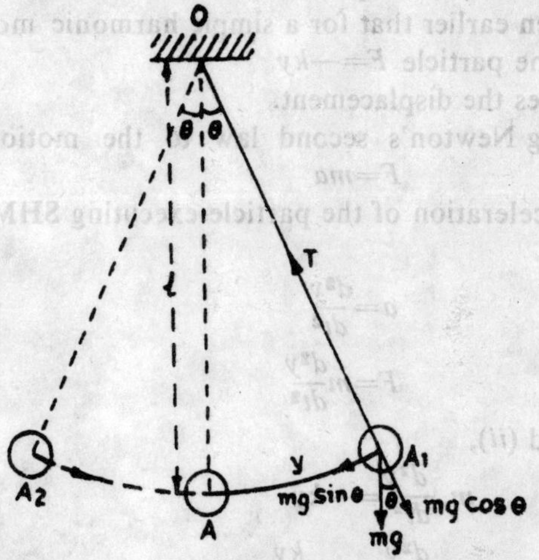

Fig.4.1 Forces acting on a simple pendulum.

The maximum angular displacement of a simple pendulum, is always limited to 4°. Under this condition, $\sin\theta=\theta$ (in radians). So, the restoring force $mg \sin\theta$ can be taken as $mg\,\theta$. Putting $\theta=y/l$ in radian measure, where l is the length of the pendulum and y is the length of the arc (taken equivalent to displacement) the restoring force,

$$F=mg\theta=mg\ y/l$$

or $$F_r \propto y$$

As the restoring force is always directed opposite to the displacement, $F_r=-ky$, where k is a constant. Thus the magnitude of the restoring force is proportional to displacement. This is true for all types of simple harmonic motion. k is constant for that system and is known as *force constant* or *spring factor*. It is the magnitude of the force per unit displacement. Its unit for linear oscillation is N/m. However, in case of angular oscillation, the restoring force is replaced by restoring torque.

In the light of above discussion, we can formulate SHM as follows.

If a body makes a to and fro periodic motion about a mean position under the action of a force (or a torque) always directed towards the mean position and magnitude of the force is proportional to the displacement of the body from the mean position, then the body is said to be executing simple harmonic motion or oscillation.

4.3. Representation of Simple Harmonic Oscillation

We have seen earlier that for a simple harmonic motion, restoring force on the particle $F=-ky$ ...(*i*)
where y denotes the displacement.

But applying Newton's second law to the motion of the body

$$F=ma$$

where a is acceleration of the particle executing SHM.

Since

$$a=\frac{d^2y}{dt^2}$$

$$\therefore \quad F=m\frac{d^2y}{dt^2} \quad \text{...(ii)}$$

From (*i*) and (*ii*),

$$m\frac{d^2y}{dt^2}=-ky$$

or

$$\frac{d^2y}{dt^2}=-\frac{ky}{m}$$

Putting

$$\frac{k}{m}=\omega^2$$

$$\frac{d^2y}{dt^2}=-\omega^2y$$

or

$$\frac{d^2y}{dt^2}+\omega^2y=0 \quad \text{...(4.1)}$$

Equation 4.1 is the differential equation of a simple harmonic motion. On solving this equation we can find the expression for

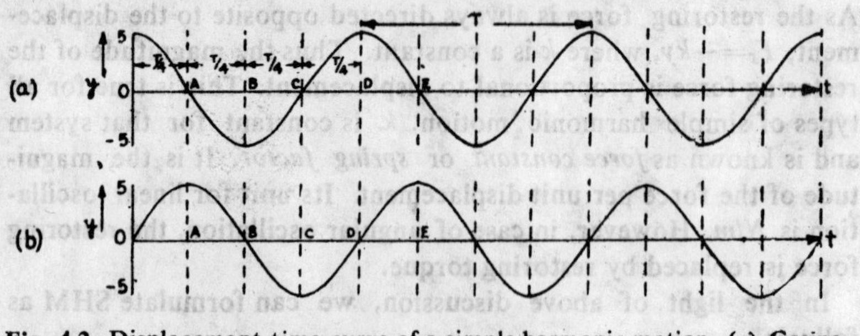

Fig. 4.2. Displacement time curve of a simple harmonic motion. (*a*) Cosine function (*b*) Sine function.

displacement of the particle at any instant of time. The solution of this equation is beyond the scope of this book. But as we have stated earlier, a simple harmonic motion can be expressed in terms of sine or cosine function. So the displacement of a body executing a simple harmonic motion can be represented either by a cosine curve or a sine curve as shown in Fig. 4.2.

For Fig. 4.2 (*a*), we can write a simple relation for y and t.

$$y=a\cos 2\pi\frac{t}{T} \quad \ldots(4.2)$$

and for Fig. 4.2 (*b*), we can write similar expression,

$$y=a\sin 2\pi\frac{t}{T} \quad \ldots(4.3)$$

The equation (4.2) can also be written as:

$$y=a\sin\left(\frac{2\pi t}{T}+\frac{\pi}{2}\right) \quad \ldots(4.4)$$

Referring to Fig. 4.2 we see that we may shift the curve 4.2 (*a*) by $T/4$ forward in the time-scale to make it fit with the curve 4.2 (*b*). It will be explained later that $T/4$ in the time scale is related to a quantity called phase $\pi/2$, that is, $\pi/2$ is the phase change corresponding to time change of T/4. So a harmonic equation may be represented in general by

$$y=a\sin\left(\frac{2\pi t}{T}+\alpha\right) \quad \ldots(4.5)$$

Solution of equation 4.1 comes in the form,

$$y=a\sin(\omega t+\alpha) \quad \ldots(4.6)$$

That equation 4.6 is a solution of equation (4.1) can be checked as follows.

Differentiating equation 4.6 twice, we get

$$\frac{d^2y}{dt^2}=-\omega^2 y$$

which is same as the equation 4.1. Thus it is confirmed that equation 4.6 is a solution of equation 4.1. As we have discussed earlier this displacement could be a linear displacement (loaded spring) or an angular displacement (as in the balance wheel of a watch) or an electrical charge, (as in electrical circuits) or even a temperature (as in thermal oscillations). However, we will confine ourselves to mostly linear mechanical oscillations. In linear mechanical oscillations the unit of y is meter and in angular oscillation, it is radian.

In all types of simple harmonic motion, the displacement-time curve gives a wave pattern as in Fig. 4.1 and the displacement is described by equations 4.5 or 4.6. This wave pattern has certain distinctive characteristics defining the harmonic oscillation. We will now discuss them below

Amplitude. The maximum magnitude of the displacement of the oscillating body from its mean position is known as the amplitude of oscillation. In Fig. 4.2(*a*) it is shown that the displacement varies between the limits $+5$ and -5 So this quantity 5 is the amplitude of harmonic oscillation. Referring to equations 4.5 and 4.6., we fiind that $y_{max}=\pm a$ as the value of a sine function varies between $+1$ and -1. Thus the amplitude is a.

Period. The minimum interval of time at which the motion repeats itself is called the periodic time or simply period. It is the time taken for completing one oscillation. Actually, for a given curve T may be found out by locating two consecutive points with similar displacement and then finding the time difference between them.

In Fig. 4.2 (*b*), the distance between O and D on the time axis is the period T. Substituting the value of t in equation 4.3 in terms of T one can see how the displacement indicated in the curve 4.1 (*b*) fits with the magnitude displacement arrived at from this equation.

$$\left.\begin{aligned}
&\text{At } 0, \quad t=0.\\
&\text{From equation 4.2,}\\
&\qquad y=a\sin 0=0.\\
&\text{At } A, \quad t=\frac{T}{4}\\
&\qquad y=a\sin 2\pi\frac{\frac{T}{4}}{T}=+a\\
&\text{At } B, \quad t=\frac{T}{4}+\frac{T}{4}=\frac{T}{2}\\
&\qquad y=a\sin 2\pi\frac{\frac{T}{2}}{T}=0\\
&\text{At } C, \quad t=\frac{T}{4}+\frac{T}{4}+\frac{T}{4}=\frac{3T}{4}\\
&\qquad y=a\cos 2\pi\frac{3T}{4}=-a\\
&\text{At } D, \quad t=\frac{T}{4}+\frac{T}{4}+\frac{T}{4}+\frac{T}{4}=T\\
&\qquad y=a\cos 0=0.
\end{aligned}\right\}\quad (i)$$

Hence it is seen that O and D are nearest points where the displacement is similar. The displacement at O and B are not similar since the direction is different *i.e.* at O it is going from

$-$ve to $+$ve whereas at B it is going from positive side to negative.

Frequency. The number of oscillations completed per unit time is called frequency, denoted by ν.

Obviously $$\nu=\frac{1}{T}.$$

The unit of ν is (second) $^{-1}$, written as S^{-1}. But usually it is expressed as 'cycles per second' or cps. Presently cps has been replaceed by the term Hertz (Hz),

i.e. 1 cps$=$1 Hz.

Phase : In Equation 4.6, the angle $(wt+\alpha)$ is called the phase angle or simply phase of the vibrating particle at the instant t. It is denoted by ϕ.

$$\phi=(wt+\alpha).$$

From equation 4.6, we find

$$\left.\begin{array}{ll} \text{if} \quad \phi=0 & y=0. \\ \phi=\frac{\pi}{2} & y=a. \\ \phi=\pi & y=0. \\ \phi=\frac{3\pi}{2} & y=-a. \\ \phi=2\pi & y=0. \end{array}\right\} \quad (ii)$$

So, for a change in phase angle by 2π, the motion repeate itself. Comparing the expressions with expressions (i), we find a phase change of 2π corresponds to change in time scale by T.

Further at $t=0$, $\omega t+\alpha=\alpha$.

Therefore, the quantity α is the initial phase of the vibrating particle and is called the phase constant or the *epoch.* This phase constant gives the clue about the initial state of motion. Referring to Fig. 4.2 (b), $\alpha=0$, and in Fig. 4.2 (a), $\alpha=\pi/2$. This means, if we discribe a simple harmonic oscillation using a cosine function, the phase differs by $\pi/2$. For this reason, it is necessary to use equation in one form only. In this book we will use the equation in sine form. The initial phase α, depends upon where we choose to fix the origin or zero for time measurement.

Angular Frequency. It is obvious from the foregoing discussion, that the phase change of 2π corresponds to one cycle or time T. Therefore, the quantity $2\pi/T=2\pi\nu$, called the angular frequency is often used for convenience. Comparing equations (4.5 and 4.6) we see,

$$\omega=\frac{2\pi}{T}=2\pi\nu \quad \ldots(4.7)$$

The unit of angular frequency is rad/s.

4.4. Time Period in Terms of Constants of the System

We know, from article 4.3,

$$\omega^2=\frac{K}{m}$$

or

$$\omega=\sqrt{\frac{K}{m}} \qquad ...(i)$$

From equation 4.7

$$\omega=2\pi\nu \qquad ...(ii)$$

So

$$\nu=\frac{1}{2\pi}\sqrt{\frac{K}{m}} \qquad ...(4.8)$$

$$T=\frac{1}{\nu}\sqrt{\frac{m}{K}} \qquad ...(4.9)$$

In angular harmonic oscillation, restoring force is to be replaced by restoring torque, and the mass by moment of inertia. So equation 4.8 and 4.9 will take the forms,

$$\nu=\frac{1}{2\pi}\sqrt{\frac{C}{I}} \qquad ...(4.8\ a)$$

$$T=2\pi\sqrt{\frac{I}{C}} \qquad ...(4.9\ a)$$

where C is the restoring couple for unit twist and I is the moment of inertia of the body about the axis of oscillation.

As a general nomenclature, K is called the *spring factor* and m the *inertia factor*. So we can say, in general,

$$\nu=\frac{1}{2\pi}\sqrt{\frac{\text{Spring factor}}{\text{Inertia factor}}}$$

4.5. Particle Velocity and Acceleration

The displacement of a particle executing simple harmonic oscillation is given by equation 4.6, *viz.*

$$y= a\sin(\omega t+\alpha) \qquad ...(4.10)$$

So the velocity of the particle at any instant is given by

$$\dot{y}=v=\frac{dy}{dt}=a\omega\cos(\omega t+\alpha)$$

$$=a\omega\sin\left(\omega t+\alpha+\frac{\pi}{2}\right) \qquad ...(4.11)$$

$$[\because \cos\theta=\sin(90+\theta)]$$

and the instantaneous acceleration is given by

$$\ddot{y}=\frac{dv}{dt}=\frac{d}{dt}\left(\frac{dy}{dt}\right)=-\omega^2 a\sin(\omega t+\alpha)$$

$$=\omega^2 a\sin(\omega t+\alpha+\pi) \qquad ...(4.12)$$

Comparing these equations, we arrive at following conclusions, (*i*) The particle velocity amplitude is ωa *i.e.* ω times the displacement amplitude, so

$$V_0=\omega a=\frac{2\pi}{T}a=2\pi\nu a \quad \text{...(4.13)}$$

where V_0 is the maximum particle velocity.

(*ii*) The particle acceleration amplitudes is $\omega^2 a$ *i.e.* ω^2 times the displacement amplitude.

$$g_0=\omega^2 a=\left(\frac{2\pi}{T}\right)^2 a=\frac{4\pi^2}{T^2}a=4\pi^2\nu^2 a \quad \text{...(4.14)}$$

where g_0 is the maximum particle acceleration.

(*iii*) The velocity is $\frac{\pi}{2}$ ahead of the .phase relative to the displacements and the acceleration is π ahead of the phase relative to the displacement or $\frac{\pi}{2}$ ahead of the phase relative to the velocity.

We have seen earlier, a phase change of 2π corresponds to time change T.. So, ahead of phase by $\frac{\pi}{2}$ is equivalent to advancement in time by $\frac{T}{4}$ This means, the 'maximum displacement' and maximum velocity' do not occur simultaneously. The difference, in the time scale, between them is $\frac{T}{4}$. Taking the example of a simple pendulum, therefore, the pendulum will have maximum velocity when it is at the mean position which differs by $\frac{T}{4}$ from the extreme positions where the displacement is maximum. Therefore, in graphical representation, the velocity-time graph is shifted $\frac{T}{4}$ towards the earlier, time compared to the veloeity displacement

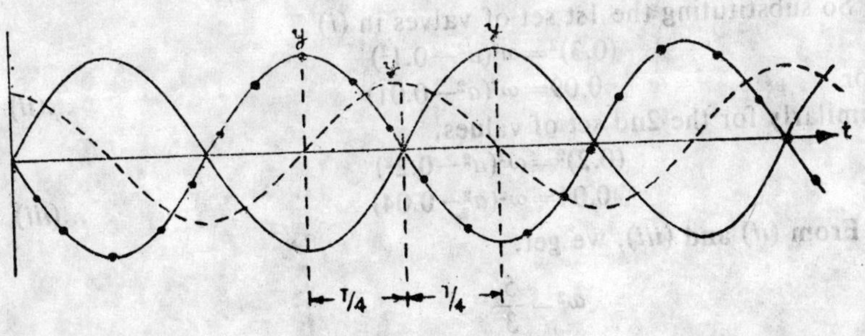

Fig. 4,3. y, y, y for the same oscillation plutted against time

placement graph. Similarly the acceleration-time graph is shifted further $\frac{7}{4}$ towards the earlier time relative to the $y-t$ graph. These three graphs have been shown in Fig. 4.3.

Example 4.1. An oscillating simple pendulum whose time period is 2.05 has the amplitude of oscillation as 0.05 m.

Calculate the velocity and acceleration amplitude during oscillation ?

Solution. We have,

$$a=0.05 \text{ m}$$
$$T=2s$$

So
$$v_0=\frac{2\pi}{T}a=2\times\frac{22}{7}\times\frac{1}{2s}\times 0.05 \text{ m}$$
$$=0.157 \text{ m/s}.$$
$$g_0=\frac{4\pi^2}{T^2}a=4\times\left(\frac{22}{7}\right)^2\times\frac{1}{(25)^2}\times 0.05 \text{ m}$$
$$=0.493 \text{ m/s}^2.$$

Example 4.2. A particle is executing SHM along a straight line. Its velocity has the value 0.3 m/s and 0.2 m/s when its distances from the mean positions are 0.1m and 0.2m respectively. Calculate the period and amplitude of SHM.

Solution. We know

$$y=a\sin(\omega t+\alpha)$$

and
$$v=\frac{dy}{dt}=a\omega\cos(\omega t+\alpha)$$

so
$$v^2=a^2\omega^2\cos^2(\omega t+\alpha)$$
$$=\omega^2\{a^2-a^2\sin^2(\omega t+\alpha)\}$$
$$=\omega^2(a^2-y^2) \quad \ldots(i)$$

Here when $y=0.1$ m, $v=0.3$ m/s.

and $y=0.2$ m, $v=0.2$ m/s.

So substituting the 1st set of valves in (*i*)

$$(0.3)^2=\omega^2(a^2-0.1^2)$$

or
$$0.09=\omega^2(a^2-0.01) \quad \ldots(ii)$$

Similarly for the 2nd set of values,

$$(0.2)^2=\omega^2(a^2-0.2^2)$$

or
$$0.04=\omega^2(a^2-0.04) \quad \ldots(iii)$$

From (*ii*) and (*iii*), we get,

$$\omega^2=\frac{5}{3}$$

or
$$\omega=\sqrt{\frac{5}{3}}\ S^{-1}$$

Hence the period

$$T=\frac{2\pi}{\omega}=2\pi\sqrt{\frac{3}{5}}\ S=4.86\ S$$

Putting the value of ω in (*ii*), we get,

$$a^2=\frac{0.32}{5}\ \text{m}^2$$

or $$a=0.0253\ \text{m}.$$

4.6. Energy in Simple Harmonic Oscillation

In a simple harmonic oscillation, the energy of the oscillating body is partly Potential, E_p and partly Kinetic, E_k.

We have seen earlier the velocity maximum occurs when displacement is minimum. A simple pendulum while oscillating has maximum velocity in the mean or equilibrium position. So its kinetic energy is maximum at this position. But the potential energy is zero fot this position where displacement is zero because the equilibrium position is taken as the reference state for measuring potential energy. But as the pendulum oscillates from the mean position its velocity decreases. So gradually its PE increases at the expense of KE. At the extreme positions ($+a$ or $-a$) the velocity becomes movementarily zero. At this extreme position, therefore, the total energy becomes potential energy and kinetic energy is zero. By the principle of conservation of energy, the sum af the potential and kinetic energy is constant if we neglect frictional losses. Hence, the total energy of oscillation E equals the maximum value of either the potential energy or kinetic energy,

$$E=E_{k(max)}=E_{p(max)}$$

Taking the expression kinetic energy, therefore,

$$E=E_{k(max)}=\tfrac{1}{2}mv_0^2=\tfrac{1}{2}m(2\pi\nu a)^2$$
$$=2\pi^2m\nu^2a^2=\tfrac{1}{2}ma^2\omega^2 \qquad \ldots(4.15)$$

(Substituting from equation 4.13)

where v_0 is the velocity amplitude.

Also $$E_{p(max)}=2\pi^2m\nu^2a^2=\tfrac{1}{2}m\omega^2a^2$$

If the oscillations are angular,

$$E=2\pi^2I\nu^2\theta_0^2=\tfrac{1}{2}I\omega^2\theta_0^2 \qquad \ldots(4.16)$$

as m is replaced by moment of inertia I and linear amplitude a by angular amplitude θ_0.

For any intermediate position between the mean and extreme position potential energy,

$$E_p=E-E_k=E_{(max)}-E_k$$
$$=\tfrac{1}{2}m(v_0^2-v^2)$$

Substituting the value of v and v_0 from equation 4.11 and 4.13,

$$E_p = \tfrac{1}{2}m\{a^2\omega^2 - a^2\omega^2 \cos^2(\omega t + \alpha)\}$$

$$= \tfrac{1}{2}ma^2\omega^2 \sin^2(\omega t + \alpha) = \tfrac{1}{2}m\omega^2 y^2 \qquad ...(4.17)$$

From equation 4.17 also we see $E = E_{p(max)} = \tfrac{1}{2}m\omega^2 a^2$ ...(4.18)

This agrees with the value of $E_{k(max)}$ given in equation (4.15).

4.7. Some Examples of Simple Harmonic Oscillation

Some typical cases in which simple harmonic oscillations occur are discussed below.

(*a*) **Simple pendulum.** We have referred to the motion of a simple pendulum earlier in art. 4.2. The value of force constant or spring factor,

$$K = \frac{mg}{l}$$

where m and l are the mass of the bob and length of the pendulum respectively.

So from equation 4.8, frequency $\nu = \dfrac{1}{2\pi}\sqrt{\dfrac{K}{m}}$

$$= \frac{1}{2\pi}\sqrt{\frac{mg/l}{m}} = \frac{1}{2\pi}\sqrt{\frac{g}{l}} \qquad ...(4.19)$$

So time period of a simple pendulum,

$$T = \frac{1}{\nu} = 2\pi\sqrt{\frac{l}{g}} \qquad ...(4.19a)$$

(*b*) **A loaded spring.** When a spring is stretched or compressed restoring forces set in it due to its elastic property. If the deformation is small, say about 10 per cent of the original length of the spring, the restoring force $F = -k\,y$, where y is the displacement of the spring relative to its undisturbed position.

In Fig 4.4, a spring is shown in (*a*). A load P attached to its lower end stretches the spring by an amount d (*b*). If P is pulled down and released, the spring oscillates about the equilibrium position O (*c*).

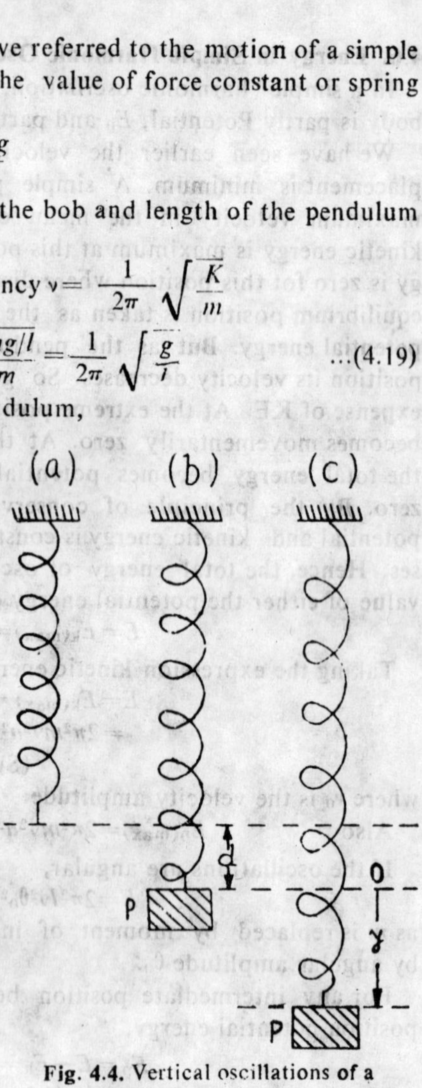

Fig. 4.4. Vertical oscillations of a loaded spring

The same thing will also haopen in case of a stretched rubber band.

Frequency $\nu = \frac{1}{2\pi}\sqrt{\frac{K}{M}}$

If the mass of spring is negligible comparea to the mass m of the load P, the value of K will depend upon the quality of the spring alone. The acceleration due to gravity at the place will have no ffect upon the frequency.

(*c*) **A torsional pendulum.** A heavy body suspended by a wire when given a twist executes angular oscillations. For a small angular displacement θ from its equilibrium position, the wire develops a restoring torque $C\,\theta$ and if the moment of inertia of the body about the axis of oscillation is I, we get,

$$I\,\frac{d^2\theta}{dt^2} = -C\,\theta \text{ or } \frac{d^2\theta}{dt^2} = -\frac{C}{I}\,\theta$$

So $$\nu = \frac{1}{2\pi}\sqrt{\frac{C}{I}} \text{ or } T = 2\pi\sqrt{\frac{I}{C}} \qquad \ldots(4.20)$$

In a watch, the balance wheel also executes harmonic oscillation and its time period is given by the equation 4.20.

(*d*) **Oscillation of a Small Bar Magnet Suspended in a Uniform Magnetic Field.** Suppose a bar magnet of pole strength m is suspended freely by means of a torsionless thread in earth's magnetic field H. If it is displaced from the equilibrium position by a small angle and then left to itself, it will execute simple harmonic oscillation. Let us consider, the forces acting on the magnet in the displaced position.

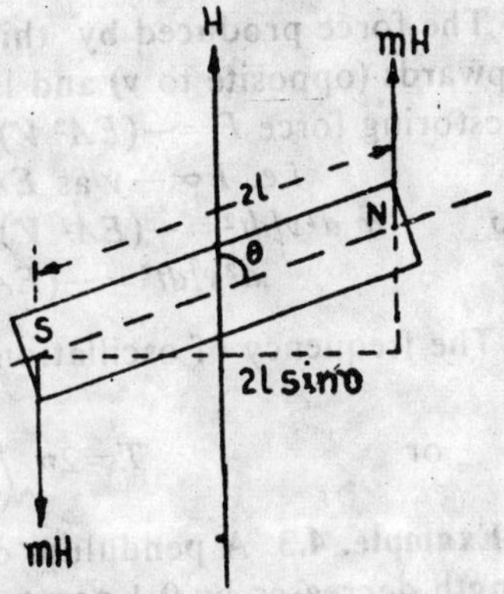

Fig. 4.5. Simple harmonic motion of a bar magnet in an uniform magnetic field.

Two equal and opposite forces of magnitude mH act on the magnet as shown in fig. 4.5. These two forces constitute a couple that provides the restoring torque to bring back the magnet to the equilibrium position.

The magnitude of this restoring couple,

$$\tau = -mH.\,2\,l\sin\theta = -MH\sin\theta$$

where $M = 2\,m\,l$, the magnetic moment of the magnet ; $2\,l$, the length of the magnet.

Let I be the moment of inertia of the magnet about the suspension thread as axis. If θ is the angular displacement then the angular acceleration is $d^2\theta/dt^2$.

So $$I\frac{d^2\theta}{dt^2}=-MH\sin\theta=-MH\theta \text{ as } \sin\theta=\theta \text{ when } \theta \text{ is smal.}$$

$$\frac{d^2\theta}{dt^2}=-\frac{MH}{I}\theta=\omega^2\theta$$

So $$\nu=\frac{1}{2\pi}\sqrt{\frac{MH}{I}} \text{ or } T=2\pi\sqrt{\frac{I}{MH}} \quad \ldots(4.21)$$

(*e*) **The Case of an air chamber with a neck.** An air chamber of volume V has a narrow and long neck of cross-sectional area A as shown in Fig. 4.6. A ball of mass m fits smoothly in the neck. If the ball is pressed down by a small distance of y and then released, it will execute SHM as shown below.

If the ball is pressed down a distance y, the volume of air in the chamber decreases by yA and, therefore an excess pressure P is produced. The excess pressure, $P=Ey\,A/V$;

where E is the elasticity of air. (Students should refer to the properties of solids chapter to derive this expression).

The force produced by this excess pressure is upwards (opposite to y) and is PA. So we get,

Restoring force $F=-(EA^2/V)y$

i.e. $F\propto -y$ as $EA^2/V=K$ is a constant.

So $m\,d^2y/dt^2=-(EA^2/V)y$

or $d^2y/dt^2=-(EA^2/Vm)y$

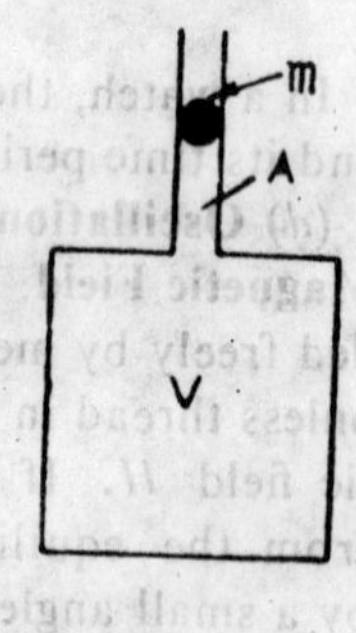

Fig. 4.6. A ball in the neck of an air chamber.

The frequency of oscillation, $\nu=\dfrac{1}{2\pi}\sqrt{\dfrac{EA^2}{Vm}}$

or $$T=2\pi\sqrt{\frac{Vm}{EA^2}}$$

Example. 4.3. A pendulum clock shows accurate time. If the length decreases by 0.1 per cent, find the error in time per day.

Solution. The correct number of seconds per day is 86,400. Suppose, the change in the number of seconds per day is x then from equation 4.19 *i.e.*

$$\nu=\frac{1}{2\pi}\sqrt{\frac{g}{l}}$$

$$86{,}400=\frac{1}{2\pi}\sqrt{\frac{g}{l}} \quad \ldots(i)$$

$$86{,}400+x=\frac{1}{2\pi}\sqrt{\frac{g}{(l-0{\cdot}001\,l)}} \quad ...(ii)$$

where l is the original length of the pendulum. Dividing (*ii*) by (*i*),

$$\frac{86{,}400+x}{86{,}400}=\sqrt{\frac{l}{l-0{\cdot}001\,l}}=(1-0.001)^{-1/2}=1+\tfrac{1}{2}\times 0.001$$

So, $\frac{x}{86{,}400}=\frac{1}{2}\times 0{\cdot}001$ or $x=43$ seconds

The clock gains 43 seconds each day.

Example. 4.4. A rectangular block of hollow aluminium chamber of cross-section $10\times 10^{-4} m^3$ and mass 0.13 kg is floated over water with an extra weight of 0.7 kg attached to its bottom. The block floats vertically. If it is slightly depressed and then released, calculate the frequency of oscillation of the block taking $g=10$ m/s². (Neglect water resistance).

Solution. Cross sectional area of the block, $A=10\times 10^{-4}$ m². Total mass of the block, $m=0.13$ kg$+0.7$ kg$=0.2$ kg. Density of water, $\rho=10^3$ kg/m³.

Let the block be depressed through a small distance of y metres and then released. The restoring force F is equal to the weight of the water displaced (Buoyancy).

$$F=Ay\,\rho g$$
$$=(10\times 10^{-4}\text{m}^2)\,(y\ m)\,(10^3\text{kg/m}^3)(10\text{m/s}^2)=10\ yN.$$

So, force constant *i.e.*, Restoring force per unit displacement,

$$K=F/y=10\ N/m.$$

So, the frequency of oscillation,

$$\nu=\frac{1}{2\pi}\sqrt{\frac{K}{m}}=\frac{1}{2\pi}\sqrt{\frac{10\ N/m}{0.2\ \text{kg}}}=1.126\ \text{HZ}.$$

4.8. Free and Forced Vibrations

If a body capable of oscillating is disturbed from its equilibrium position and then left to itself, the body oscillates with a frequency determined by the physical dimension and the elastic properties of the body. We have already discussed number of such systems and their frequencies of oscillation. This oscillation is called *free oscillation* because the system is left free to itself after feeding some energy initially. The frequency of free oscillation is called the natural frequency of the system.

According to Equation 4.8, the natural frequency

$$\nu_0=\frac{1}{2\pi}\sqrt{\frac{K}{m}}$$

The subscript ν_0 has been added to indicate that it is the frequency of free oscillation

In a free oscillation, it is assumed that there is no loss of energy due to air resistance or friction in the medium. Hence

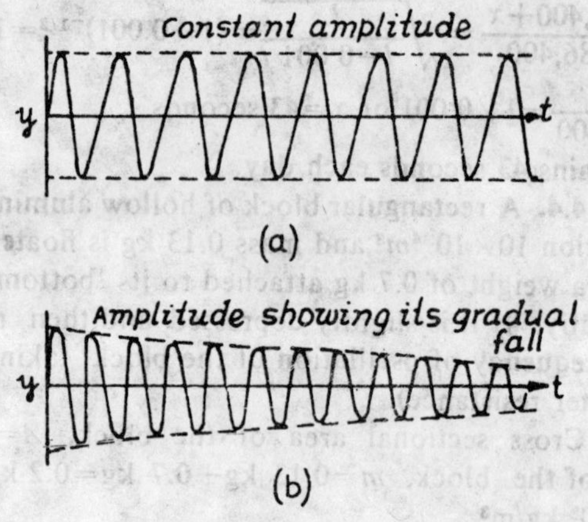

Fig. 4.7. (*a*) Free oscillation, (*b*) Damped oscillation.

the amplitude of oscillation a is taken to be constant. But in practice, it is found that the medium in which the body oscillates always offer some amount of resistance. So the energy of the oscillating system reduces, hence the amplitude gradually falls or the oscillation gets damped. Such oscillations are called damped harmonic oscillation, Fig. 4.7(*b*). Left to itself, the system will stop oscillating after some time. But we can feed energy to the oscillating system to keep it oscillating. If this energy is fed continuously, so that loss of oscillating energy is just balanced by the external supply of energy, the frequency of oscillation will still remain close to ν_0. These are called *maintained oscillations.* We can take the example of a swing. The swing oscillates after being given an initial push, but will normally come to a stop after some time. But when we give a gentle push to the swing in the direction of the motion the amplitude is unaffected if the magnitude and frequency of the push is suitably adjusted.

Forced oscillation. The energy from an external source may also be fed with a definite frequency ν. Let us name this external agent or source as driver and the oscillating system to which the energy is supplied as driven. So ν is the frequency of the driver and ν_0 is the frequency of the driver. Under such conditions, after sometime, the natural vibration of frequency ν_0 dies out and the system starts oscillating with the frequency ν of the driver. The amplitude of this oscillation of frequency ν increases gradually till a stage is reached where the power lost due to friction etc.

equals the power fed by the driver. The amplitude then remains constant. These are called forced oscillations. Note that the frequency of the forced oscillations is ν not ν_0.

Resonance. If ν and ν_0 have a large difference, the power fed in forced oscillation is very small. But if $\nu = \nu_0$ the supply of power by the driver to the oscillating system is in step with the oscillation. So the amplitude of oscillation increases considerably. This particular case of forced oscillation where the driver frequency is equal to the natural frequency of the driven system is called *resonances i.e.*, the two systems are in resonance. If there are drivers of many frequencies, the driven system accepts maximum energy from just that driver whose frequency agrees with its frequency. This is just what happens when we tune our radio or television to a particular station. Waves from many stations are present around the antenna, but we tune only to the desired station. This is just done by making the natural frequency of a circuit in the receiver equal to the frequency of the desired station. The musical instruments like flutes work on resonance principle. A device called Helmhotz resonator working on this principle is used to analyse complex musical sounds. Tachometer, an instrument for measuring the speed of various motors is also based on this principle.

Resonance can also be disastrous. A passing vehicle or an aeroplane causes rattling of doors and windows of a house if their natural frequency of vibration happens to be equal to that of sound waves coming from the plane or the vehicle. When an army marches in steps on a bridge, it is possible that the frequency of footsteps may be equal to one of the natural frequencies of the bridge. In that case the bridge may pick up a larger amplitude of vibration causing damage to a bridge. Therefore, while crossing the bridge the army is asked to break steps.

4.9. Waves and Wave Propagation

If we drop a stone in a calm pond small waves or ripples are generated and they spread over the entire surface of the pond. Even in a small water basin, we could generate such ripples by dipping a finger. All these are waves on the surface of water. Whenever there is a disturbances on the surface of water, water waves are generated that spread in all directions. The big waves we observe in the sea are also caused by disturbance in the water surface.

If we observe the spreading of disturbance in the pond, we will find a series of alternate crests (elevation) and troughs (depressions) on the water surface. These crests and troughs are not stationary ; they will be found travelling along the surface of water. This succession of movement of crests and troughs are

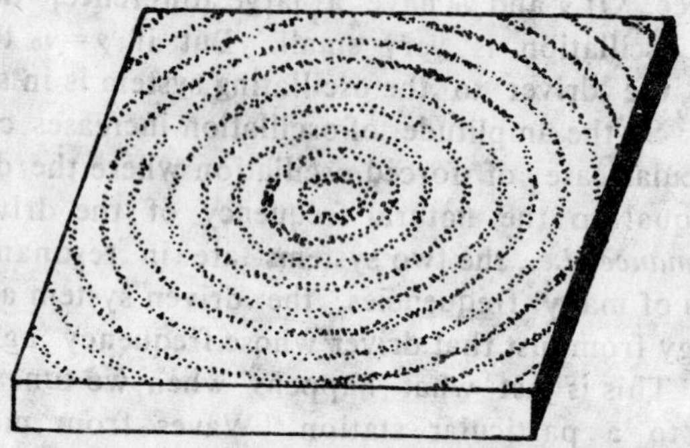

Fig. 4.8. Waves on the water surface.

called *waves* and the material in which the wave travel is called the *medium*. In this case, the medium is water surface. Suppose, we put a small bit of straw or any floating object on the surface of water. The object will move up and down along with the passage of a crest and trough respectively, but its position will remain unchanged. This shows water particles do not move from one end of the pond to the other end when wave travels over the surface of water. It is only the disturbance caused at one end that spreads by causing all particles in the medium to undergo same cycle of disturbance in succession.

Let us consider another example. Suppose a string is fixed horizontally to two rigid supports so that the string is under tension. If one end of it is suddenly pulled aside and then left to itself a kink or pulse starts off there and proceed along the string to the other end. You will observe that each part of the string undergoes a disturbance about its own mean position and the disturbance occurs for the different parts in regular succession. Fig. 4.9 shows the travelling of a pulse at different stages with passage of time, The stages *a*, *b*, *c*, *d*, *e*, *f*, corresponds to time $t=0$, t_1, $2t_1 \ldots 5t_1$. The distance has been divided into number of equal intervals marked x_1, $2x_1$, $3x_1$..etc. The pulse which is at the left end at the time $t=0$ travels by a distance x_1 in each time interval t_1. You will observe that the pulse has travelled a distance $3x_1$ over a time interval $3t_1$.

We observe that in this process no part of the string travels along with the pulse *i.e.* no matter is transferred. Only the disturbance caused at one end (it is also called the information) is transferred along the string. This property we also observed in case of water waves, and is true for all types of waves. So, one important characteristic of wave motion is that *no matter is transferred along the direction of wave transmission.*

The velocity with which the wave travels in the medium is called wave velocity. Let us denote it by C. If we fix our attention on a particular crest, its velocity of travel will give us the wave velocity. For example, in Fig. 4.9, the pulse starting at $t=0$ travels a distance of x_1 in each time interval t_1. So its speed $C=x_1/t_1$. In the same figure we have shown another pulse starting when $t=3t_1$. The pulse is shown a small one to indicate that it is a weak pulse. But this pulse also covers the distance x_1 in the time interval t_1. This means the velocity of this weak pulse is also x_1/t_1 *i.e.* same as that of the strong pulse. A third pulse of a different shape starts when $t=4t_1$. But this also travels with the same velocity x_1/t_1. So the wave velocity does not depend on the size and shape of the pulse. This is also true for water waves and all other types of waves. Of course, if the medium is moving the velocity will increase in the direction of movement of the medium. But relative to the medium, it will remain constant.

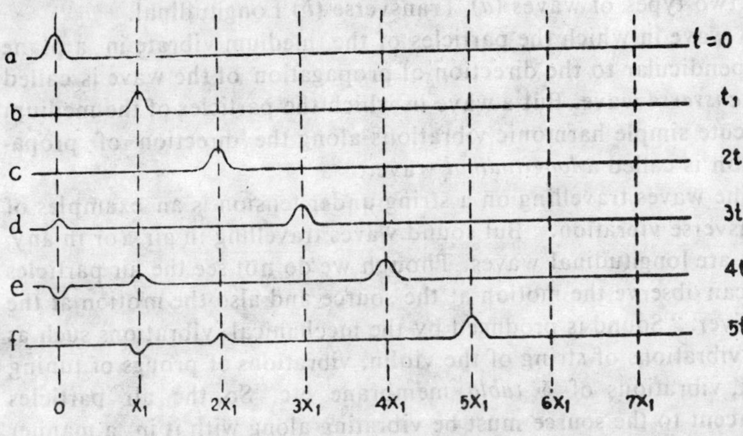

Fig. 4.9. The travelling pulses over a string. *a, b, c*...refer to the stages at successive intervals.

But wave velocity will be different in different medium. So the second important characteristic of wave motion is that the wave *velocity relative to the medium depends only on the nature of the medium.* The nature of the source of waves, such as putting a big stone in the pond or putting the finger in a water basin, has practically no effect on the wave velocity. Of course, there is some variation due to the nature of sources, but we will neglect them for our consideration.

If the source producing the waves, is moving relative to the medium like a speeding boat in water then also the velocity of waves relative to the medium remains-unchanged. Suppose the boat velocity is V, then the wave velocity relative to the boat is $(C-V)$ in the forward direction and $(C+V)$ is the direction behind the source. On the otherhand, if the medium itself is moving with a velocity V while the source is stationary, the waves will move with velocity $(C+V)$ in the forward direction and $(C-V)$ in the reverse direction. But in any case, the wave velocity relative to the medium is unchanged.

4.10. Longitudinal and Transverse Waves

In the wave propagation, particles of the medium do not move along with the wave. It is the disturbance which moves forward. The particles simply execute vibrations about their mean positions. All waves are not necessarily periodic. But we will consider the simple case of periodic waves in which the particles of the medium execute simple harmonic oscillations about their mean positions as the wave travels. Such a wave is called a harmonic wave. There are two types of waves (*a*) Transverse (*b*) Longitudinal.

A wave in which the particles of the medium vibrate in a plane perpendicular to the direction of propagation of the wave is called a *transverse* wave. But a wave in which the particles of the medium execute simple harmonic vibrations along the direction of propagation is called a *longitudinal* wave.

The waves travelling on a string under tension is an examples of transverse vibration. But sound waves travelling in air (or in any, gas) are longitudinal waves. Though we do not see the air particles we can observe the motion at the source and also the motion at the receiver. Sound is produced by the mechanical vibrations such as the vibrations of string of the violin, vibrations of prongs of tuning fork, vibrations of a *tabla* membrane etc. So the air particles adjacent to the source must be vibrating along with it in a manner to execute to and fro motion along the direction of propagation of sound wave. If you observe the diphragm of a microphone carefully, it will be found to be vibrating. This means that air particles from the source to the receiver vibrate along the direction of propagation. A sudden compression at one end of a long spring sets longitudinal waves (Fig. 4.10). In Fig. 4.10 (*a*) the left end has been compressed and is then suddenly released. The portion *A* expands as a result the turns at the right of *A* gets compressed so that there is a new compression at *B*. In the mean time in the process of release from the compressed condition the portion *A* even crosses the original (equilibrium) condition of the spring and becomes rarefied (Fig. 4.10 *b*). Then likewise the state of compression is handed over from *B* to *C* and state of rarefaction from *A* to *B*. So, it is alternate compressions and rarefactions that travel forward. The material particles of the spring, therefore, executes to and fro periodic motion about their mean position position of the turns in original condition).

Sound waves are also combination of similar compressions and rarefactions. Take the example of the vibration of a tuning fork (Fig. 4.10 *c*). In normal position, the prong is at *A*. As it starts moving towards *B*, air particles close to *B* will be forced to move towards right till the prong reaches *B*. But other particles beyond them will continue to remain in their normal positions. So the affected air particles close to the prong will be compressed.

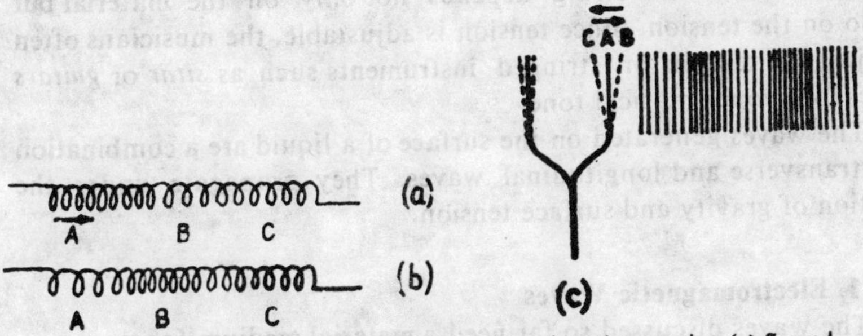

Fig. 4.10. (*a* and *b*) Longitudinal vibrations in a spring (*c*) Sound waves are alternate compressions and rarefactions.

The prong after reaching *B* will turn back towards *A* leaving a partial vacuum behind it. So the air particles which were moving towards right will now have freedom to move towards left. They will continue to move to the left till the prong reaches *C*. So there will be rarefaction of air particles. But before these particles turned back, they had hit the particles to the right of them. So the state of compression proceeds to the right followed by state of rarefaction. As the prong turns again from *C* to *B* another compression starts to be followed by rarefaction for movement from *B* to *C*. So for one cycle of oscillation, there will be one compression and one rarefaction. So all air particles in the medium will alternately get compressed and rarefied along the direction of propagation *i.e.* will execute to and fro periodic motion along the direction of propagation. Since a compression and rarefacton means variation of pressure, the longitudinal waves are also called *pressure waves*. A bomb explosion produces pressure waves of much larger magnitude called *shock waves*. The glass panes of the buildings quite away from the explosion site gets damaged due to sudden variation of pressure due to rapid compression and rarefaction in air.

Transverse waves can travel only in mediums having volume elasticity and rigidity. Therefore, transverse waves can travel only in solids. But longitudinal waves can travel in solids, liquids and gases. Though both transverse waves (caused by shape deformation) and longitudinal waves (caused by compression) can be transmitted in solids the speed of transmission of two kinds of waves are usually unequal.

It is of interest to note that a string may be too soft and possess negligible rigidity. So normally it is not expected to transmit a transverse wave. It is mainly due to tension applied on the string that wave transmission is possible. Therefore, the velocity of wave transmission in the string depends not only on the material but also on the tension. Since tension is adjustable, the musicians often adjust the tension on stringed instruments such as *sitar* or *guitars* etc. for proper musical tone.

The waves generated on the surface of a liquid are a combination of transverse and longitudinal waves. They propagate under the action of gravity and surface tension.

4.11. Electromagnetic Waves

The waves discussed so far need a material medium for transmission. Sound can not travel in vacuum. Therefore, all these waves are grouped together as mechanical waves or elastic waves. But there are a different type of waves that require no material medium for its transmission,it can travel in vacuum or free-space. One such example is of light waves which reach us from the sun after travelling through free-space. Radiowaves, microwaves, X-rays and gammarays etc. also belong to this category. They are all called *electromagnetic waves*. The transmission of electromagnetic waves actually means transmisssion of two mutually perpendicular oscillating electric and magnetic fields in the direction of transmission. The direction of both these fields are perpendicular to the direction of propagation. Therefore, electromagnetic waves are transverse waves.

All electromagnetic waves travel in free-space or vacuum with the speed of light *i.e.* 3×10^8 m/s. Mechanical waves travel much slower. For normal air, sound waves travel with a velocity nearly 350 m/s or nearly 1200 km/hr.

4.12. Graphical Representation of Wave Motion

As a wave is transmitted in a medium, each particle of the medium in its path undergoes a cycle of displacement *i.e.* simple harmonic motion. All particles along the path of transmission passes through similar displacements with passage of time. Thus three parameters are involved in wave motion along a line : distance x of the particle from a fixed position, time t and the displacement y, of the particle from the mean position. But in a two dimensional plot such as in a graph paper, we can either plot y against t (when x is fixed) or y against x (when t is fixed). Let us consider how a transverse wave can be represented graphically.

In transverse vibration, each particle in the medium executes simple harmonic motion in a direction perpendicular to the direction of propagation. But all particles will not be in same state or

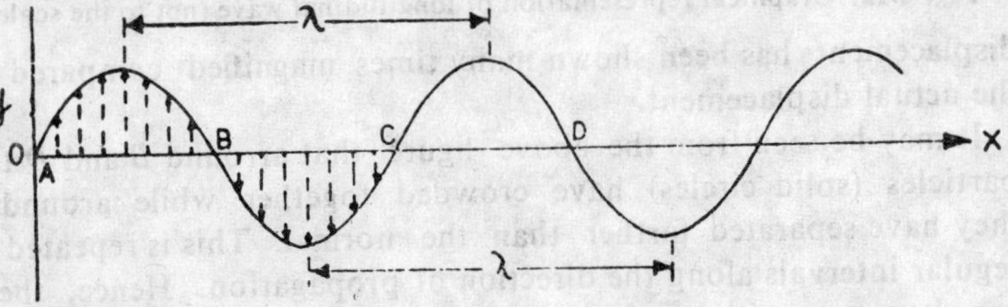

Fig. 4.11. Graphical representation of transervse wave (not to the scale).

phase of vibration at the same time. As the wave moves with time the particies will be in various phases of displacement at any fixed time (Fig. 4.11). The dotted lines represent the magnitude of displacement and the arrow indicates its direction. All particles from A to B have positive displacement and those from B to C have negative displacement. If we join the tips of the displacement vectors at any time it comes in the shape of a wave as indicated as in Fig. 4.11. The distance between any two consecutive crasts or troughs is called *wavelength* λ (lambda). It may be noted that different scales have been used for displacement y and distance x in this graph. For normal vibrations, the displacement is much smaller than the displacement indicated in the diagram.

In the longitudinal wave, the displacement is along the direction of propagation, yet it can be represented graphically like transverse wave (Fig. 4.12). All particles in the normal position have been shown by hollow circles. Particles from A to C are in various phases of displacement indicated by the arrow heads. The new positions of the particles are given by solid circles. Displacement of particles from A to B is to the right of the mean position and from

B to *C* is to the left of the mean position. Lines are drawn along *y* direction proportional to the displacement. The displacements from *A* to *B* are shown as +ve displacement and those from *B* to *C* are shown as−ve displacement in Fig. 4.12. Here also, the

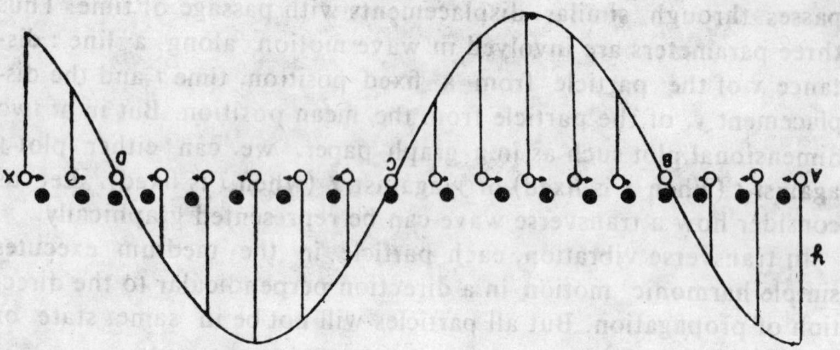

Fig. 4.12. Graphical representation of longitudinal wave (not to the scale).

displacement has been shown many times magnified compared to the actual displacement.

It may be seen from the above figure that arround *B* and *D* the particles (solid circles) have crowded together while around *C* they have separated farther than the normal. This is repeated at regular intervals along the direction of propagation. Hence, there are alternate positions of maximum density(compression) and mini-

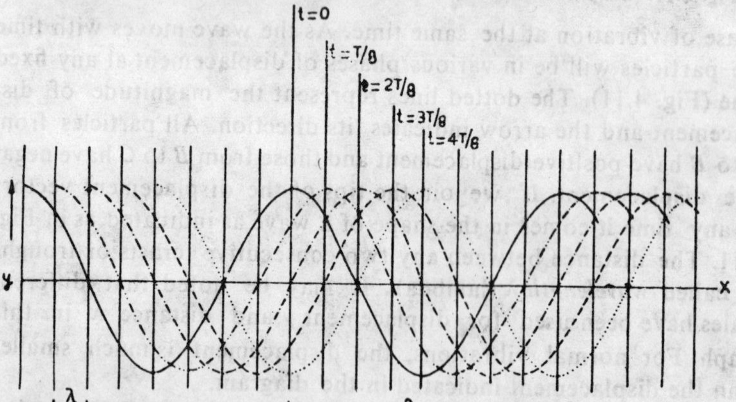

Fig. 4.13. A travelling wave. The *y*—*x* graph shown at time intervals T/8 each.

mum density (rarefaction). Thus in longitudinal waves, alternate compressions (high pressure) and rarefactions (low pressure) are transmitted with the waves.

Let us now discuss how the motion of the wave can be represented graphically. In Fig. 4.13, a series of $y-x$ curves have been drawn at successive equal time intervals of $T/4$ such as for $t=0$, $t=T/8$, $t=2T/8, \ldots t=T$ etc. If we look at a crest or a trough, we will find it moving ahead with time. In each time interval $T/8$, they move the distance $\lambda/8$ in the x-direction. In other words, the waves covers a distance λ in a time T.

4.13. Analytical Representation of Progressive Wave

We have considered the graphical representation of the propagation of a harmonic wave. The harmonic wave which moves in a medium is also called progressive wave. We will consider below the analytical representation of a progressive wave.

As a wave is transmitted in a medium, the displacement of a particle in its path at any instant can be described by either equation 4.2 or 4.3. But all particles along the path of transmission of the wave will have similar displacement with passage of time *i.e.* the state of displacement will be transmitted. Suppose a wave is transmitted along the X-axis with velocity C. This means that a particular displacement advances with a velocity C. Hence, whatever displacement a particle is having at the origin, $(X=0)$ at the time t, the same displacement a particle will have at $X=x$ after a time x/c *i.e.* at the time $(t+x/c)$. Conversely, we can say that the displacement at $X=x$ at the time t must have occurred at the origin at the time $(t-x/c)$.

Let us represent the displacement at $X=x$ at a time t by Equation 4.3.

$$y=a \sin 2\pi\, t/T \qquad \ldots(4.23)$$

But a particle at the origin must have gone through this displacement before a time interval x/c. Therefore, Equation 4.23 can be modified by putting $(t-x/c)$ in place of t:

$$y=a \sin \frac{2\pi}{T}(t-x/c) \qquad \ldots(4.24)$$

This is the equation for a harmonic wave travelling along the positive direction of the X-axis. It gives the displacement at any position x at any time t. If the wave is moving in the negative direction of X, then Equation 4.24 is changed as,

$$y=a \sin \frac{2\pi}{T}\left(t+\frac{x}{c}\right) \qquad \ldots(4.24a)$$

It can be seen that at a given position (x fixed), the value of y repeats itself at time interval of T. Equation 4.24 can be expanded to the form.

$$y=a\left\{\sin 2\pi \frac{t}{T}\cos\frac{2\pi x}{CT}-\cos\frac{2\pi t}{T}\sin\frac{2\pi x}{CT}\right.$$

$$\left.\begin{array}{lll} \text{So at } t=0, & y=-a\sin 2\pi x/CT \\ t=T/4 & y=+a\cos 2\pi x/CT \\ t=T/2, & y=+a\sin 2\pi x/CT \\ t=3T/4 & y=-a\cos 2\pi x/CT \\ t=T & y=-a\sin 2\pi x/CT \end{array}\right\} \quad (i)$$

Thus displacements at $t=0$ and at $t=T$ are similar. At no other time between these two instants the displacements are similar *i.e.* T is the minimum time interval when the displacements are similar. T is the *time period* of the harmonic wave. A particle located at x, passes through all phases of displacements of harmonic motion during this period T. Thus the time period of the harmonic wave is the same as the time period of oscillating body that produces the harmonic wave.

Further, at a given time t, there will be number of points at equal distances from each other where the displacements will be similar. Displacement at a distance x from the orgin at a time t

$$y_{(t,\,x)}=a\sin\frac{2\pi}{T}(t-x/C)$$

$$=a\sin\left(\frac{2\pi t}{T}-\frac{2\pi x}{CT}\right)$$

At the same instant, displacement at a distance CT apart,

$$y_{(t,\,x+CT)}=a\sin\left[\frac{2\pi t}{T}-\frac{2\pi(x+CT)}{CT}\right]$$

$$=a\sin\left[\frac{2\pi t}{T}-\frac{2\pi x}{CT}-2\pi\right]=a\sin\left(\frac{2\pi t}{T}-\frac{2\pi x}{CT}\right)$$

Thus we see that the displacement at a space interval of CT is same. Since C is the velocity of the wave transmission, CT is the distance covered during time interval T. T is the time period of harmonic oscillation. Referring to Fig. 4.13., we see that the distance covered by the displacement during T is λ or *wave-length.*

Thus $\lambda=CT$ ...(4.25.)

viz: The wavelength is the distance covered by the wave during one period or one cycle of oscillation of the source.

Thus we see that the displacement repeats itself after a time interval T and a space interval λ.

4.14. Other Terms Connected with Harmonic Waves

(*a*) **Amplitude.** Referring to set of expressions (*i*) in the previous article, we see that for a given value of x, the value of displacement varies between $+a$ and $-a$.

The magnitude of the maximum displacement, a, is known as the *amplitude* of the harmonic wave.

Frequency. If a source is continuously generating waves, the waves will spread and during unit time the space surrounding the source will be filled by these waves whose number will be equal to the number of oscillation of the source during unit time.

The number of waves generated by the source per unit time is called the frequency of the hormonic wave,

Obviously, $$\nu=\frac{1}{T} \qquad ...(4.25)$$

Wave Number. Number of waves present in unit distance is called wave number. It is denoted by σ.

$$\sigma=\frac{1}{\lambda} \qquad ...(4.26)$$

Comparing equations 4.25 and 4.26, one can seen that wave number is a quantity for oscillations (source for wave) in space analogous to the frequence ν for oscillations in time. The frequency ν is measured as cycles per second or in hertz (Hz). The wave number is expressed in cycle per metre.

Phase and phase difference. For a harmonic wave, at a given λ each particle undergoes a cycle of displacement which repeats at regular intervals of T. So the state of the particle at the time t, $t+T$, $t+2T$, $t+3T$......$t+nT$ is the same. This is also expressed as saying that 'phase' is the same. So T is the minimum time interval for which phase of the vibration is the same.

Similarly, at a given time t, the state of the particles at x, $x+\lambda$, $x+2\lambda$......is the same. Hence λ is the minimum space interval for which phase of the vibrations is the same.

Though we mention here the 'the phase is same', actually it means as phase change of 2π at these intervals of λ or T. We explain that below.

Equation 4.24. can be written as,

$$y=a\sin\left[2\pi\left(\frac{t}{T}-\frac{x}{CT}\right)\right]$$

$$=a\sin\left[2\pi\left(\frac{t}{T}-\frac{x}{\lambda}\right)\right] \qquad ...(4.27)$$

The argument of the sine is called the 'phase angle' or simply 'phase' ϕ. But a more general expression is to write,

$$y=a\sin\left[2\pi\left(\frac{t}{T}-\frac{x}{\lambda}\right)+\alpha\right] \qquad ...(4.27a)$$

where α is called the 'initial phase'. Thus the phase at x at the time t in a wave represented by the equation 4.27 is given by,

$$\phi=2\pi\left(\frac{t}{T}-\frac{x}{\lambda}\right)+a \qquad \text{...(4.28)}$$

So the phase changes both with the time t and the distance x. If $\triangle\phi$ is the change of phase for a change of time $\triangle t$,

$$\phi+\triangle\phi=2\pi\left(\frac{t+\triangle t}{T}-\frac{x}{\lambda}\right)+a$$

Substracting equation 4.28 from this expression,

$$\triangle\phi=\frac{2\pi}{T}\triangle t=2\pi\nu\triangle t \qquad \text{...(4.29)}$$

Similarly, the phase changes with position x as,

$$\triangle\phi=-\frac{2\pi}{\lambda}\triangle x \qquad \text{...(4.30)}$$

The negative sign in this equation indicates that the wave is moving towards the positive X direction and the forward points are behind the phase *i.e.*, they have to reach the successive stages of vibration later.

From equation 4.29, for $\triangle t=T$, $\triangle\phi=2\pi$ and from equation 4.30., for $\triangle x=\lambda$, $\triangle\phi=2\pi$ *i.e.*, a path difference of λ is equivalent to a phase change of 2π.

Thus, the phase changes by 2π during one cycle (time variation T or space variation λ) *i.e.* for each cycle it changes by 2π. Thus vibrations having a phase difference of integral multiple of 2π *i.e.*, $\triangle\phi=\pm 2n\pi$, are said to be in same phase ($n=0, 1, 2...$).

During the cycle, the phase change lies between 0 and 2π. For example, for a quarter cycle *i.e.*, $T/4$ or $\lambda/4$,

$$\triangle\phi=\frac{\pi}{2}$$

At half cycle, *i.e.*, $T/2$ or $\lambda/2$.

$$\triangle\phi=\pi$$

At $3T/4$ or $3\lambda/4$, $\triangle\phi=3\pi/2$

and at T or λ $\triangle\phi=2\pi$.

The phase at $T/2$ (or $x=\lambda/2$) is said to be opposite to the phase at $t=0$ or T (or $x=0$ or λ) and *vice-versa*. Thus the vibrations which have a phase difference of old multiple of π *i.e.*, $\triangle\phi=\pm(2n+1)\pi$, are said to be in opposite phase, ($n=0, 2, 3,...$).

Angular wave number. The phase change per unit distance or path difference is called the angular wave number. It is denoted by K.

$$K=\frac{2\pi}{\lambda} \qquad \text{...(4.31)}$$

It is measured in radians per meter. We have defined the term angular frequency earlier.

$$\omega = \frac{2\pi}{T}.$$

Comparing K and ω, we find the angular wave number is a quantity for oscillations in space analogous to the angular frequency ω for oscillations in time.

Examples 4.5. For a plane wave $y = 2.0 \times 10^{-.01}$ sin $(1000t - 0.9x + \pi/2)$. Write down (*i*) the general expression for phase angle ϕ (*ii*) the phase at $x=0$, $t=0$ (*iii*) the angular frequency ω (*iv*) the phase difference between two points separated by 10 cm along the X-axis (*v*) the phase change in a millisecond (*vi*) the amplitude at $x=100$ metre. Units of y, t and x are taken as 10^{-6} m, S and m respectively.

Solution. Compair with general expression for ϕ

$$\phi = 2\pi\left(\frac{t}{T} - \frac{x}{\lambda}\right) + a$$

$$= \left(\omega t - \frac{2\pi x}{\lambda}\right) + a$$

(*i*) $\phi = (1000\,t - 0.9\,x + \pi/2)$.

(*ii*) $a = \pi/2$

(*iii*) $\omega = 1000$ rad/s.

(*iv*) $\triangle\phi = -0.9\,x = -0.9 \times 0.1$

$= -0.09$ radian (*see* Equation 4.30)

(*v*) $\triangle\phi = 1000\,t = 1000 \times 10^{-3} = 1$ radian.

(*vi*) Amplitude at x is $2.0 \times 10^{-.01x} \times 10^{-6}$ m where x is in metre.

At $x = 100$ m, the amplitude

$a = 2.0 \times 10^{-0.01 \times 100} \times 10^{-6}$m

$= 2.0 \times 10^{-1} \times 10^{-6}$ m $= 2.0 \times 10^{-7}$m.

4.15. Wave Velocity and Particle Velocity

Particle velocity. The wave moves in a medium with a constant velocity c, but the particle velocity undergoes a cyclic change. The particle velocity at any time t and in any position x can be found by differentiating Equation 4.27 for displacement y,

$$y = a \sin 2\pi\left(\frac{t}{T} - \frac{x}{\lambda}\right)$$

$$v = \frac{dy}{dt} = \frac{2\pi a}{T} \cos 2\pi\left(\frac{t}{T} - \frac{x}{\lambda}\right)$$

$$= V_0 \cos 2\pi\left(\frac{t}{T} - \frac{x}{\lambda}\right)$$

where $V_0 = \frac{2\pi}{T} a = \omega a$ ...(4.32)

Thus we note that the a particle velocity $\dot{y}$ is a varying quantity. The variation is harmonic in nature and the velocity-amplitude V_0 is $2\pi/T$ times the displacement amplitude a.

Comparing the expressions for y and $\dot{y}$, one can see that the particle velocity is described by the cosine function whereas displacement is described by a sine function of the same quantity. Thus there is a phase difference of $\pi/2$ between y and $\dot{y}$. (see Art 4.5). It can also be shown that the particle acceleration $\ddot{y}$ differs in phase by $\pi/2$ with $\dot{y}$.

The wave velocity is different for different medium and is given by $C=\lambda T=\nu\lambda$ where ν is the frequency of the source. Therefore, for the same source wave length λ will vary if medium is changed.

Example 4.6. A body vibrating with certain frequency sends waves 20 cm long in a medium A and 30 cm long in another medium B. The velocity of wave in A is 10 m/s. Find the wave velocity in B.

Solution. Let ν be the frequency of the vibrating body for the medium A, $\lambda_A=0.2$ m; $C_A=10$ m/s for medium B, $\lambda_B=0.3$ m; $C_B=?$

We know $\lambda=CT=C/\nu$

Since the frequency is unchanged,

or $\nu=C/\lambda$

$$\frac{C_A}{\lambda_A}=\frac{C_B}{\lambda_B}$$

or $$C_B=C_A\times\frac{\lambda_B}{\lambda_A}=10 \text{ m/s}\times\frac{0\cdot3 \text{ m}}{0\cdot2 \text{ m}}=15 \text{ m/s}.$$

4.16. Energy Transmission in a Progressive Wave

In any oscillatory motion, there is a continuons change of enery from potential to kinetic and vice-versa. In a progressive wave, the oscillations are transferred with a velocity C and hence the energy is also transferred with this velocity.

A particle executing SHM has maximum velocity or velocity amplitude when it crosses the mean position of rest. In this position, the total energy is kinetic. So the energy possessed by a particle of mass m and velocity amplitude V_0,

$$E=\tfrac{1}{2}\, m\, V_0^2 \qquad \ldots(i)$$

$$=\tfrac{1}{2}\, m\, a^2\, \omega^2 \qquad \ldots(ii)$$

$(\because \; V_0=a\omega)$

where a is the amplitude and ω is the angular frequency.

Suppose, n is the number of particles per unit volume of the medium in which the wave is moving. Then the total energy of particles in unit volume.

$$u = \frac{1}{2}\, m\, n\, a^2\, \omega^2 = \frac{1}{2}\, \rho\, a^2\, \omega^2 \qquad ...(iii)$$

where ρ is the density of the medium.

Let C be the wave velocity. If the wave moves across a unit area of the medium, then the volume of the medium covered by the wave in unit time is C. Therefore, the energy that flows across a unit area or the intensity of the wave is

$$I = uc = \frac{1}{2}\, \rho\, c\, a^2\, \omega^2$$

So, $$= 2\pi^2 \rho\, c\, a^2 \nu^2 \qquad ...(4.33)$$

$I \propto a^2$ when ρ, C and ω are constant.

$I \propto \omega^2$ *i.e.* $I \propto \nu^2$ when ρ, C and a are constant *i.e.* the intensity of the wave is proportional to the square of the amplitude and also to the square of the frequency of vibration.

Example. 4.7. A 10 watt source sends out waves in air at a frequency 500 Hz. Deduce the intensity at 50 m distance, assuming spherical distribution. If $C = 350$ m/s and $\rho = 1.3$ kg/m, deduce the displacement.

Solution. Intensity at 50 m distance;

$$I = \frac{\text{power}}{\text{area}} = \frac{10 \text{ watt}}{4\pi\,(50 \text{ m})^2}$$

$$= 3.1847 \times 10^{-4} \text{ watt. m}^{-2}$$

From Equation 4.33,

$$I = 2\pi^2 \rho\, c\, a^2 v^2 = 2\pi^2\,(1.3 \text{ kg/m}^3)\,(350 \text{ m/s})(500\, s^{-1})^2\, a^2$$

$$= 224.3 \times 10^7 \times a^2 \text{ watt. m}^{-2}$$

Equating the two values of I

$$a = 0.3768 \times 10^{-6} \text{ m.}$$

4.17. Wave Fronts

So far we have considered waves moving in one direction only. But when a source generates waves, it spreads in the space around the source. Suppose we repeatedly dip a finger in water in a calm pond. A series of crests and troughs will move out. At any instant, each crest has a circular form with the finger at its centre. Similarly each trough will also be circular in shape. The particles along the crest have upward displacement equal to the amplitude and particles along the trough have downward displacement also equal to the amplitude. So all particles in a given crest are in the same phase. Similarly, all particles

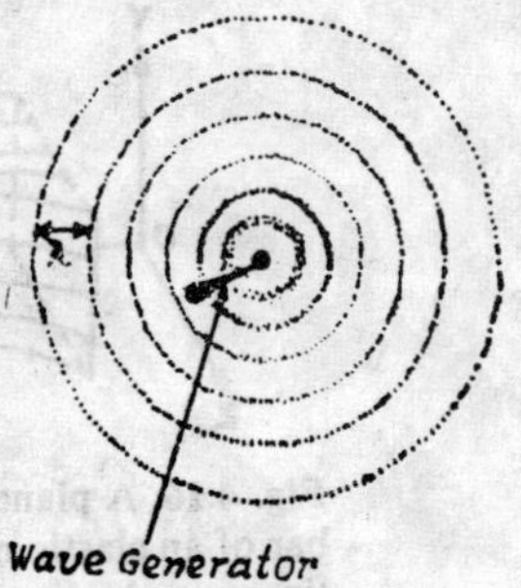

Fig. 4.14. Circular wave front.

along a trough are also in same phase. Around the source, we can locate other points at same phase and join them, they will also be circular in shape.

The locus of points in the same phase of vibration is called a *wave front*. Obviously, the wave front in this case is a circle (Fig. 4.14). The wave-front moves with the wave velocity C in the medium. But if the wave travels in 3-dimension such as sound

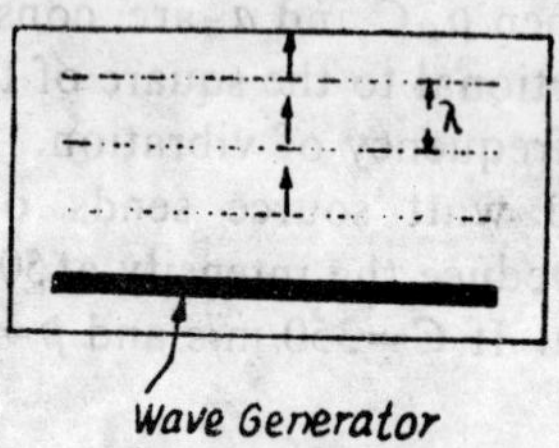

Fig. 4.15. Parallel wave fronts in water.

being produced in air, the sound travels with the same speed in all directions. The wave front in this case is spherical. Similarly if a flat plate is regularly dipped in water, it produces parallel wave fronts *i.e.* the succession of crests and troughs are parallel lines on the water surface (Fig. 4.15) As the wave spreads on water surface, it will be observed that the heights of crest go on decreasing with increase of distance from the source. This means, greater the radius r smaller is the height of the crest. This happens because same energy distributes over bigger circles as r increases. As the circumference is

Fig. 4.16. A plane transverse wave propagating in a rectangular bar of an elastic material. The wave fronts shown by dotted lines are planes perpendicular to the direction of propagation.

$2\pi r$, the amount of energy associated with unit area *i.e.* the intensity $I \propto 1/r$ and hence the amplitude $a \propto 1/\sqrt{r}$ ($\because I \propto a^2$). But in case of spherical wave front in 3-dimensional space, the energy spreads over a sphere of surface area $4\pi r^2$.

So $I \propto 1/r^2$ and hence the amplitude $a \propto 1/r$.

The Equations 4.24, 4.24(*a*), 4.27 given previously are, therefore, to be modified by altering the value of a (by introducing the influence of r) for application into circular and spherical wave fronts.

The Equations 4.24, etc. are for plane harmonic waves only. In plane waves, the wave fronts are planes perpendicular to the direction of propagation. In these waves since the phase at a given time depends only on x, the wave fronts are parallel to the yz plane. Since the amount of energy flowing across a unit area of each wave front is the same therefore, the amplitude a remains nearly constant. Of course the amplitude might decrease due to the damping of the medium. However, we are not considering that aspect here.

Plane transverse waves can be produced in a long bar of uniform cross-section say a rectangular bar by giving periodic vibration along the plane of the face at one end. The wave fronts travel along the length and are always parallel to the face. (**Fig. 4.16.**)

4.18. Velocity of Transverse Wave along a Stretched String

Transverse waves are propagated along a string under tension. Let us consider the forces acting on a small section of the pulse moving in the string. We know that the string is stationary along the direction of movement of the pulse, but the pulse is moving with a velocity C relative to the string. Since we are interested in the relative speed between the string and pulse for analytical purposes, we can consider the string moving over the top portion of a *stationary* pulse with the relative speed C. Consider the top portion as a circular arc (Fig. 4.17) whose radius of curvature is r, the length of the arc,

$$l = r\Delta\theta \qquad \ldots(i)$$

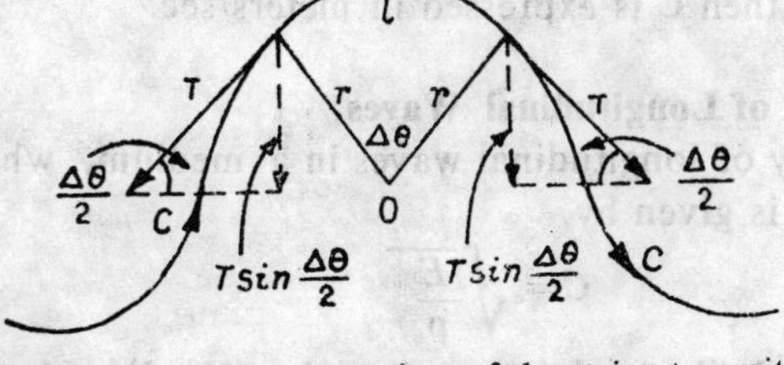

Fig 4.17. Forces acting on a section of the string trasmitting a pulse.

The tension T acting on this portion can be resolved into rectangular components. The two horizontal components cancel each other, but the downward components, $T \sin \Delta\theta/2$ act in the same direction. So the resultant force acting downward.

$$F = T \sin \frac{\Delta\theta}{2} + T \sin \frac{\Delta\theta}{2}$$

$$= 2T \sin \frac{\Delta\theta}{2}$$

As $\Delta\theta$ is small, $\sin \Delta\theta/2 = \Delta\theta/2$

So $$F = T\Delta\theta \qquad \ldots(ii)$$

This force provides the centripetal acceleration. But the centripetal acceleration in the circular arc portion of the string is,

$$a_{cp} = C^2/r \qquad \ldots(iii)$$

So, the centripetal force acting on this portion of the string of mass M.

$$Ma_{cp} = \frac{MC^2}{r}$$

$$= \frac{mlC^2}{r} = \frac{m\,(r\Delta\theta)\,C^2}{r}$$

$$= mc^2\,\Delta\theta \qquad \ldots(iv)$$

where m is the mass per unit length of the string.

Equating RHS of (ii) and (iv),

$$C^2 = \frac{T}{m} \quad \text{or} \quad C = \sqrt{\frac{T}{m}} \qquad \ldots(4\ 34)$$

So a transverse wave will travel along the string under tension with constant speed untill it reaches the end of the string. The assumption that the angle $\Delta\theta$ is small means that Equation 4.34 holds strictly only for transverse waves of small amplitude, but is independent of the shape of the pulse. The speed increases with increase of T but decreases with increase of m. Thus we can regulate the speed of transverse virbration along a string by altering the value of T or m.

In Equation 4.34 if the tension on the string is in Newtons, and m is in kg/m, then C is expressed in meters/sec.

4.19. Velocity of Longitudinal Waves

The velocity of longitudinal waves in a medium, whether solid, liquid or gas, is given by

$$C = \sqrt{\frac{E}{\rho}} \qquad \ldots(4.35)$$

where E is the bulk modulus of elasticity (N/m^2) and ρ is the density (kg/m^3) of the medium and C is the wave velocity (m/s). The density of a solid is much larger than that of a gas, but the bulk modulus of elasticity is larger by a greater factor. Therefore, waves in a solid travel faster than those in a gas.

The bulk modulus of elasticity in a gas has different values depending on whether the conditions are adiabatic or isothermal. Newton assumed that for a gas, longitudinal waves such as sound waves are propagated under isothermal condition. This means that there is no temperature variation during compression and rarefaction. According to Boyle's Law, at constant temperature,

$$PV = \text{constant}.$$

where P and V are the pressure and volume of a given mass of the gas. Differentiating we get,

$$Vdp + PdV = 0$$

or
$$dp = -P\frac{dV}{V}$$

or
$$\frac{dp}{dV/V} = -P \quad \text{...(4.36)}$$

The expression on the left hand side is the volume elasticity E. The negative sign indicates that the volume decreases with increase of pressure. So neglecting the negative sign and substituting P for E, from Equation 4.35 we get,

$$C = \sqrt{\frac{P}{\rho}} \quad \text{...(4.37)}$$

This is Newton's formula for the velocity of longitudinal waves in a gas.

But it was found that according to this relation, the velocity of sound waves in air at NTP is 280 m/s whereas experimentally it is found to be 332 m/s. To account for this difference, Laplace pointed out the vibartions causing compressions for sound waves take place so rapidly that there is no time for heat either to leave from or enter into the gas medium. Thus the change takes place under adiabatic condition. For an adiabatic change,

$$PV^{\gamma} = \text{constant}$$

where
$$\gamma = \frac{\text{Specific heat of the gas at constant pressure}}{\text{Specific heat of the gas at constant volume}}$$

Differentiating, we get

$$V^{\gamma}\, dp + P\gamma . V^{\gamma-1}\, dV = 0$$

Divinding both sides by $V^{\gamma-1}$

$$Vdp + \gamma PdV = 0$$

or
$$dP/\frac{dV}{V} = -\gamma P$$

Hence
$$E = \gamma P$$

Substituting the value of E in Equation 4.36, we get

$$C = \sqrt{\frac{\gamma P}{\rho}} \quad \text{...(4.38)}$$

Since for air $\gamma=1.4$, C at NTP$=\sqrt{1.4\times280}$ m/s $333=$m/s. This value is close to the experimental value.

Since the condition of a gas changes rapidly with other factors, it is necessary to know how the velocity of longitudinal waves is influenced by them.

Effect of pressure. At a given temperature P/d is constant for a gas as shown below.

Let V_1 be the volume of a mass m of a gas at pressure P_1 and let V_2 be the volume of the same mass of the gas at pressure P_2. Since the temperature remains constant, according to Boyle's Law,

$$P_1V_1=P_2V_2$$

or
$$P_1\frac{m}{\rho_1}=P_2\frac{m}{\rho_2}$$

where ρ_1 and ρ_2 are the values of the density of the gas in the two cases.

So
$$\frac{P_1}{\rho_1}=\frac{P_2}{\rho_2}$$

or
$$\frac{P}{\rho}=\text{constant}$$

Hence, for a gas at a given temperature, since γ is fixed the velocity C is independent of pressure according to Equation 4.38.

Effect of density. From Equation 4.38 it is evident that for gases, having the same value of γ, the velocity of longitudinal waves in them is inversely proportional to the square root of their densities. Thus the velocity of sound in hydrogen is four times its velocity in oxygen at the same pressure since the density of oxygen is 16 times that of hydrogen.

Effect of temperature. Let C_t be the velocity at t°C, C_0 be the velocity at 0°C, ρ_t be the density at t°C and ρ_0 be the density at 0°C.

Then
$$C_0=\sqrt{\frac{\gamma P}{\rho}} \quad \text{and} \quad C_t=\sqrt{\frac{\gamma P}{\rho_t}}$$

$$\therefore \quad \frac{C_t}{C_0}=\sqrt{\frac{\gamma P}{\rho_t}\Big/\frac{\gamma P}{\rho_0}}=\sqrt{\frac{\rho_0}{\rho_t}} \qquad \ldots(i)$$

But
$$\rho_0=\rho_t\left(1+\frac{t}{273}\right)=\rho_t\left(\frac{273+t}{273}\right)$$

$$=\rho_t\frac{T}{T_0}.$$

Therefore, from (*i*)

$$\frac{C_t}{C_0}=\sqrt{\frac{T}{T_0}} \qquad ...(4.39)$$

Hence the velocity of sound in a gas is directly proportional to the square root of the absolute temperature of the gas.

From Equation 4.39 we can write,

$$C_t=C_0\left(1+\frac{t}{273}\right)^{\frac{1}{2}}$$

$$=C_0\left(1+\frac{1}{2}\times\frac{t}{273}+\text{terms with higher powers of }\frac{t}{273}\right)$$

In the normal temperature range, $t/273$ is a small quantity. Hence terms containing its higher powers are so small that they are neglected in comparision.

So $$C_t=C_0\left(1+\frac{1}{2}\times\frac{t}{273}\right)$$

$$=C_0+C_0\times\frac{t}{546}$$

$$\therefore \quad C_t-C_0=C_0\times\frac{t}{546}$$

Taking the velocity of sound in air,

$$C_0=332 \text{ m/s, for } t=1°\text{C},$$

$$C_t-C_0=332\times\frac{1}{546}\text{ m/s}$$

$$=0.61 \text{ m/s}.$$

Hence we find that the change in velocity of sound in air for 1°C change in temperature (from 0°C) is 0.61 m/s.

Effect of moisture. The density of water vapour is less than the density of dry air for the same pressure and temperature. Therefore, the density of moist air is always less and hence the velocity of a longitudinal waves in moist air is always greater than its velocity in dry air.

Effect of wind. We have discussed earlier that the velocity of the medium affects the velocity of wave propagation. If the wind blows with a velocity W, the wave is propagated in the direction of the wind with a velocity $(C+W)$ and with a velocity $(C-W)$ against it.

Example 4.8. Find the temperature at which the velocity of sound in air is 2 times its velocity at 10°C.

Solution. 10°C=283°K.

Suppose at T °K the velocity is two times of its value at 283°K.

$$\frac{C_T}{C_{283}} = \sqrt{\frac{T}{283}}$$

But $\frac{C_T}{C_{283}} = 2$

$\therefore$ $\sqrt{\frac{T}{283}} = 2$

or $T = 1132°K$ or $859°C$.

4.20(a). Reflection of Waves

In the case of lightwaves reflection from smooth surfaces occur according to well known laws of reflection. Radio waves are also reflected by obstacles and this principle is utilised in the *radar*.

Mechanical waves are also reflected just like the electromagnetic waves. The most common example of sound (mechanical waves) reflection is observed in echoes heard in big halls or even in open fields close to some buildings or hills etc. Bats have no eyes but they utilise sound reflection to locate the position and direction of their destination. They send out high frequency sound waves (ultrasonic) and from the reflections received from obstacles they can estimate the nature, size, distance and direction of the obstacle. A device called *sonar* which is used for measuring depth of sea and to locate under water rocks, icebergs etc, also utilises the principle of sound reflection. Suppose a sound is sent out by a source and then its echo is received there. If the interval of time between the production of sound and receiving echo is t, the velocity of sound be C then distance between the sound source and the reflecting obstacle, $d = Ct/2$. ...(4.40)

This is the principle used in sonar and also in radar.

The sensation of sound lasts in our brain for 1/10th of a second. Assuming the velocity of sound under normal conditions to be about 350 m/s, an obstacle should be situated at least 17.5 m away so that the reflected sound is received after 1/10th of a second. Otherwise, the impression of the echo will not be distinguished from the original sound.

The sound waves obeys same laws of reflection as the light waves. But the wave length of sound waves being larger, we require bigger reflecting surfaces. Two simple experiments are described below to demonstrate this

Experiment 1. T_1 and T_2, are two hollow tubes. A wooden board B is clamped vertically. The two tubes T_1 and T_2 are placed in front of the board in the same horizontal plane with their axes inclined to the board and interesecting at a point on the surface of the board. A stop watch W is held near the end of the tube T_1. Place your ear near the end of the tube T_2. You will hear the ticking of the watch. A screen S of asbestos or felt is kept in between the two tubes so that the sound of the watch does not reach the ear directly. Keeping the tube T_1 fixed, if the inclination of T_2 is gradually altered, in one position the sound heard will be maximum. Hence this position of T_2 gives the dircetion of reflected sound. In this position it is observed,

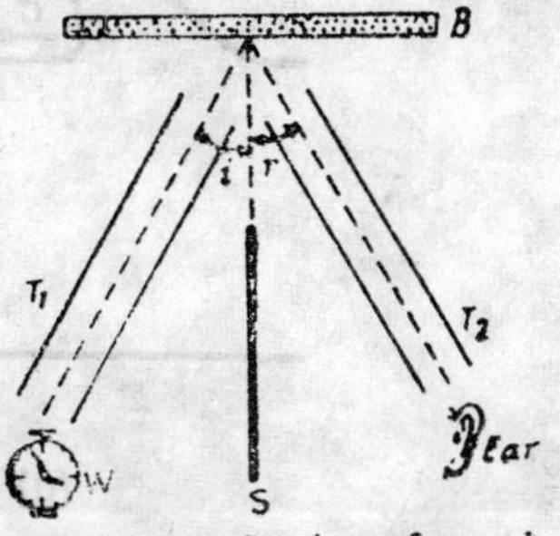

Fig. 4.18. Reflection of sound.

1. That the axes of the tubes T_1 and T_2 make equal angle with the board. Hence angle of incidence=angle of reflection.

2. That the axes of the tubes T_1, T_2 and the normal to the board at the point of their intersection lie in the same plane. This statement is analogus to first law of reflection for light.

Experiment 2. Two large concave parabolic mirros M_1 ann M_2, made of any polished metal are placed co-axially facing each other at some distance Fig. 4.19. A watch is suspended at the principal focus of M_1. A small funnel with a rubber tube attached to its end is kept close to M_2. Then the funnel is moved about while the open end of the rubber tube is held near the ear. It will be observed that when the funnel is placed at the principal focus of M_2, ticking of watch will be heard clearly and thus the intensity of sound is maximum.

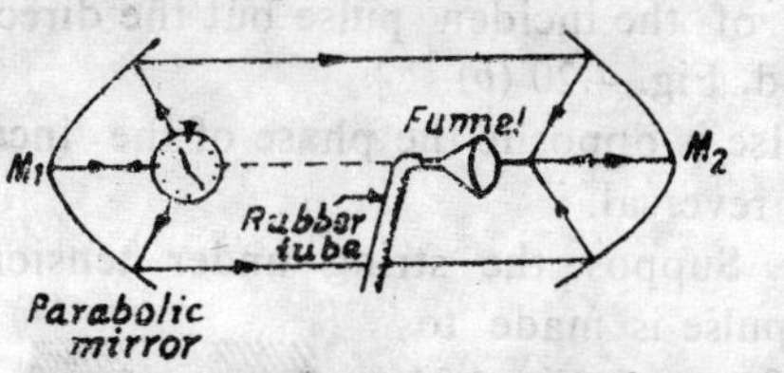

Fig. 4.19. Reflection of sound from curved surface.

This is also expected to happen when two concave mirrors face each other. An object placed at the principal focus of one of them produces its image at the principal focus of the other mirror.

It is possible also to demonstrate that laws of reflection are also obeyed by water waves.

4.20(*b*). Nature of Reflected Waves

The nature of reflected waves depends on the reflecting surface. We will consider two such cases.

(*a*) **Reflection on a stretched string.** Suppose one end of a string

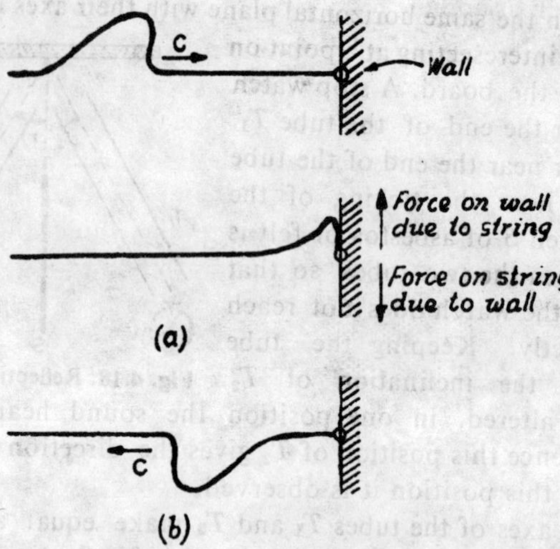

Fig 4.20. Reflection of a pulse at a fixed end.

under tension is attached to a rigid wall. Let a wave pulse advance towards the fixed end, Fig. 4.20 (*a*). When the pulse arrives at the fixed end, it exerts a force on the support. The reaction of this force generates a reflected pulse in the string. The shape of the reflected pulse is same as that of the inciden pulse but the direction of displacement is reversed, Fig. 4.20 (*b*).

So the phase of reflected pulse is opposite the phase of the incident pulse *i.e.* there is a phase reversal.

(*b*) **Consider a different case.** Suppose the string under tension is held vertically and a wave pulse is made to move from the rigid end to the free end, Fig. 4.21 (*a*). As the incident wave reaches the free end, the free end also sets into vibration. A reflected wave pulse is set up upwards Fig. 4.21 (*b*). But in this case the shape of the reflected pulse and direction of displacement are same as those of the incident wave *i.e.* the reflected and incident pulse are in same phase.

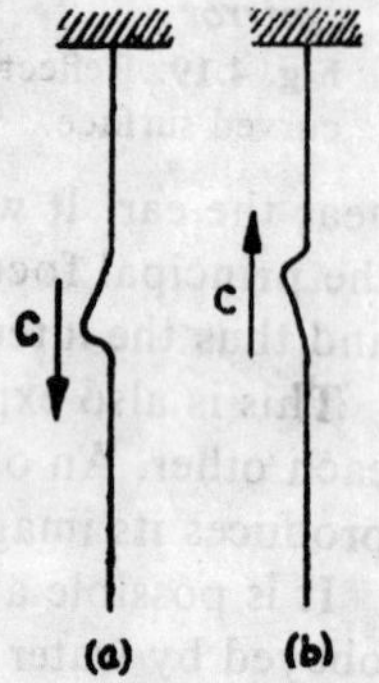

Fig 4.21. Reflection of a wave pulse at a free end.

Thus a transverse wave is reflected from a fixed end with reversal of phase whereas from a free end it is reflected in the same phase. Longitudinal waves are also reflected at a boundary. Like transverse waves, reflection occurs with a phase reversal at a fixed end (closed organ pipe) and without reversal of phase at a free end (open organ pipe).

4.21. Nature of Reflection and Refraction of Waves

In case of light, we have seen that when light waves travelling in one medium reaches the boundary of another medium, part of the light is reflected back to the original medium and another part is transmitted or refracted into the second medium. This is also applicable in case of elastic material waves.

Suppose two strings *AO* and *OB* made of different materials are joined together at *O* and kept under tension, Fig. 4.22. The mass per unit length for string *OB* is greater than that for the string *AO*. So wave velocity c_1 in *AO* is greater than the wave velocity c_2 in

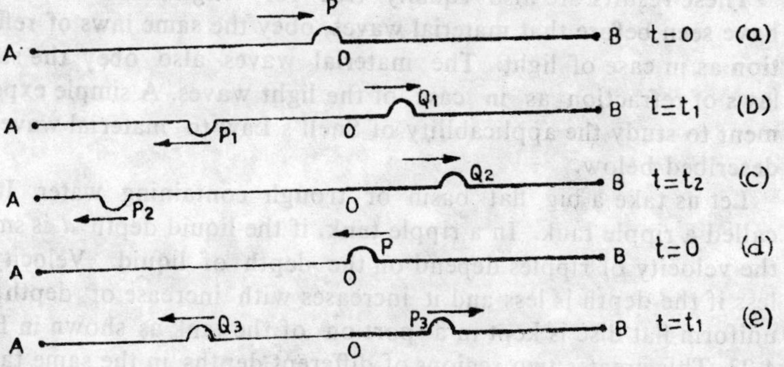

Fig.4.22. Reflection and refraction of a pulse at junction of two strings.

OB *i.e.* $c_1 > c_2$ (Refer Equation 4.34). Let a pulse be generated in *AO*. At time $t=0$, let the pulse *P* just reach the junction. We will observe, after this, that a pulse proceeds forward to the string *OB* while simultaneously a pulse proceeds backward in the string *OA*. At time $t=t_1$, Q_1 and P_1 show the position of these two pulses. They further move forward to Q_2 and P_2 at time $t=t_2$. The following features can be noticed.

(*a*) Reflected and transmitted pulses are created simultaneously at the junction of the two dissimilar materials.

(*b*) Since the pulse moves faster in the string *AO* *i.e.* $c_1 > c_2$, the distance P_1P_2 is greater than the distace Q_1Q_2.

(*c*) The direction of the transmitted pulse Q_1 is in the same as the incident pulse *P* (upwards) but the reflected pulse P_1 is in the opposite direction (upside-down). Thus there is no phase change of the transmitted pulse whereas the reflected pulse is in opposite phase *i.e.* there is phase reversal.

Suppose, we initiate the pulse in the string *OB* Fig. 4.22(*d*). The incident pulse *P* generates a reflected pulse P_3 and transmitted pulse Q_3. It will, however, be noticed.

(*d*) The reflected pulse is in the same direction as the incident (pulse upwards). Experiments with several pair of strings made of different materials always lead to these common results.

1. The wave coming from a medium where the wave velocity is greater to a medium where the wave velocity is smaller is reflected upside-down *i.e.* with reversal of phase.

2. The wave coming from a medium where the wave velocity is smaller, is reflected with no change of phase.

3. The wave transmitted or refracted into the second medium always goes without change of phase.

These results are also equally true for longitudinal waves. We have seen before that material waves, obey the same laws of reflection as in case of light. The material waves also obey the same laws of refraction as in case of the light waves. A simple experiment to study the applicability of Snell's Law to material waves is described below.

Let us take a big flat basin or trough containing water. It is called a ripple tank. In a ripple tank, if the liquid depth d is small the velocity of ripples depend on the depth of liquid. Velocity is less if the depth is less and it increases with increase of depth. A uniform flat disc is kept in a portion of the tank as shown in Fig. 4.23. This creates two regions of different depths in the same tank.

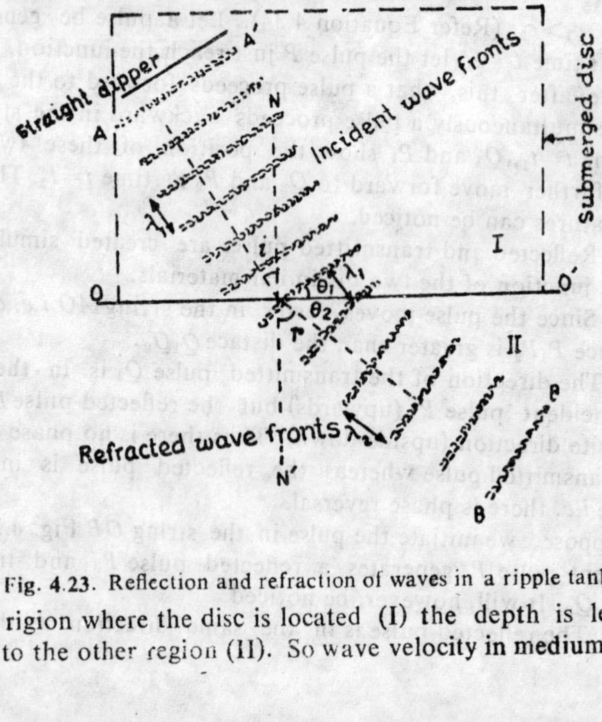

Fig. 4.23. Reflection and refraction of waves in a ripple tank.

In the rigion where the disc is located (I) the depth is less compared to the other region (II). So wave velocity in medium *I* where water depth is less is smaller than the wave velocity in the medium II (Fig. 4.23.).

Dip a flat rod—called straight dipper in the tank as shown in the tank. As the dipper is moved in and out of water slowly, ripples are generated in the medium I. The wave fronts spreading out are parallel lines with respect to the dipper surface. That is why these waves are also called straight waves.

Fig. 4.23 shows the wave fronts as AA' in medium I and wave fronts BB' emerging into the medium II. The incident wave front makes an angle θ_1 and the refracted wave front makes an angle θ_2 with the surface OO' separating the two mediums. It will be seen that,

$$\frac{\sin \theta_1}{\sin \theta_2} = \frac{C_1}{C_2} = \text{constant}$$

where C_1 and C_2 are the respective velocities of the ripples in the two mediums.

Looking at the wave fronts, it will be found that distance between consecutive crests in the two regions are not the same. This shows that wavelengthe λ_1 end λ_2 are also unequal. If λ_1 and λ_2 are measured accurately, it will be seen that,

$$\frac{\sin \theta_1}{\sin \theta_2} = \frac{C_1}{C_2} = \frac{\lambda_1}{\lambda_2} = \text{constant} \qquad \ldots(4.41)$$

Thus Snell's law of refraction is also obeyed for water waves. It can also be shown that laws of refraction are also obeyed by sound waves.

A lens for sound can be made to show that lens equation obeyed for light is also obeyed by sound waves. This means sound waves, which are longitudinal in nature obeys the same laws of refraction. Two circular plastic or rubber seats are sealed around the rim and then it is filled with a gas other than air. It will have a convex shape with a bulge at the centre. Such a device can be used as a 'sound lens'. It should be noted that the size of this lens should be quite large as sound waves, have large wave lengths. The diameter of this lens should be about 1 m; thickness at the centre about 30 cm and the sheets should be thin. The lens is kept vertically and a stop watch is fixed at some distance along the axis on one of its sides. A funnel, to one end of which a rubber tube is attached, is moved along the axis on the other side. Holding the end of the rubber tube to the ear one hears clear and loud ticking of the watch at a particular position. The watch is the source (like object in light) and the tunnel is the receiver (image). If we take the distance between the lens and the source as 'p' and the distance between the lens and the receiver as 'q', then it can be shown approximately that,

$$\frac{1}{p}+\frac{1}{q}=\text{constant for a given lens as in light.}$$

4.22. Dispersion of Waves

Lights of different colours have different wavelengths and frequencies. The velocity of light waves in a medium changes with frequency (hence wave length). Different coloured lights will, therefore, have different angles of refraction. White light passing through a prism thus separate into coloured bands. This is called *dispersion* and the band of coloured light is called *spectrum*.

The refractive index

$$\mu=\frac{\text{Wave velocity } C \text{ in vacuum}}{\text{Wave velocity } C' \text{ in the medium}} \quad ...(4.42)$$

Experiments show that

$$\lambda \text{ violet} < \lambda \text{ red and } \mu \text{ violet} > \mu \text{ red.}$$

Hence from 4.42,

$$C' \text{ red} > C' \text{ violet.}$$

Therefore, in the case of light dispersion (*i*) μ decreases with increase of λ *i.e.* $d\mu/d\lambda$ is negative and (*ii*) the velocity of light in a medium increases with wave length *i.e.* $dC'/d\lambda$ is positive.

In mechanical waves, however, the wave velocity does not change appreciably with frequency except at very high frequencies. Therefore, it is not simple to show dispersion. Hence, the wave-velocity C' of a mechanical wave in a given medium is treated independent of frequency for all practical purposes. For example, velocity of sound waves of all frequencies. in a given medium is taken to be equal.

4.23. Polarisation in Waves

In longitudinal waves, the vibration or displacement occurs along the direction of propagation of the wave. Therefore, only one mode or type of vibration is possible in case of longitudinal waves. But the situation is different for transverse waves. For example, consider the transverse waves over a string. If the length of the string is along the X-axis, the transverse vibrations will be in the Y—Z plane. The vibrations could be along Y-axis or Z-axis or at any angle to the Y or Z direction. In principle, the vibrations have infinite possiblities. Of course, in the string usually we generate vibrations in one particular direction. But if we have a wave in which vibrations in all possible directions in the transverse plane are present, the wave is called unpolarised. If however the vibrations take place in one direction only, the wave is said to be *polarised*. (fixed direction). It is obvious, the problem does not arise for longitudinal waves.

A simple experiment may be performed to show the distinction between polarised and unpolarised waves. A narrow rectangular

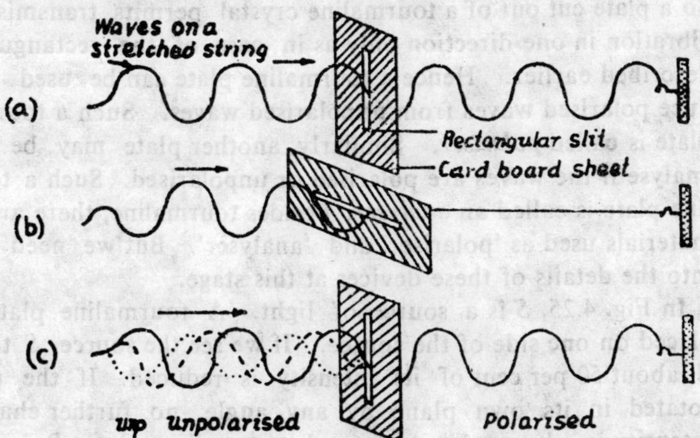

Fig. 4.24. (*a*) Transverse waves pass through the slit when the plane of vibration is along the length of the slit. (*b*) It is stopped when the slit is rotated through 90°. (*c*) If the wave is having transverse vibrations in number of planes, only the vibration along the length of the slit passes beyond the slit.

slit is made in a card board. A string under tension is stretched horizontally and the string passes through the opening in the slit, Fig. 4.24. Let transverse waves be generated in the string. Now rotate the card board. It will be observed that the vibration will pass through the slit only in one position of the slit *i.e.* when the direction of vibration is along the lenth of the slit, Fig. 4.24 (*a*). But when the length of the slit and the direction of vibrations do not coincide, the vibrations are stopped by the slit. Conversely, if it is possible for the string to vibrate simultaneously in a number of directions in the transverse plane then all such vibrations cannot pass through the slit. Only one mode of vibrations *i.e.* parallel to the length of the slit, is transmitted. In such a case, the waves before the slit are unpolarised and those beyond the slit are polarised. Longitudinal virbrations, however, are not affected by the rotation of the slit. In mechanical waves, the polarisation does not have much importance. Sound waves are longitudinal waves. On strings, normally we see polarised waves.

The phenomenon of polarisation, however, is important for light waves. Experiments on polarisation for light provides evidence of its wave nature instead of being corpuscles or particles. They also conclusively establish that vibrations for light waves are transverse in nature.

Experiment to produce plane-polarised waves. A tourmaline crystal allows vibrations to pass through it in one direction only. So a plate cut out of a tourmaline crystal permits transmission of vibration in one direction only as in case of the rectangular slit described earlier. Hence a tourmaline plate can be used to produce polarised waves from unpolarised waves. Such a tourmaline plate is called *polariser*. Similarly, another plate may be used to analyse if the waves are polarised or unpolarised. Such a tourmaline plate is called an *analyser*. Besides tourmaline, there are other materials used as 'polariser' and 'analyser'. But we need not go into the details of these devices at this stage.

In Fig. 4.25, *S* is a source of light. A tourmaline plate *A* is placed on one side of the source. If we see the source *S* through *A*, about 50 per cent of its intensity is reduced. If the plate is rotated in its own plane by any angle, no further change in intensity is observed. A second tourmaline plate *B* is placed beyond *A* along the path of light from *S*. Suppose *B* is rotated in its own plane, it will be observed that in particular position the

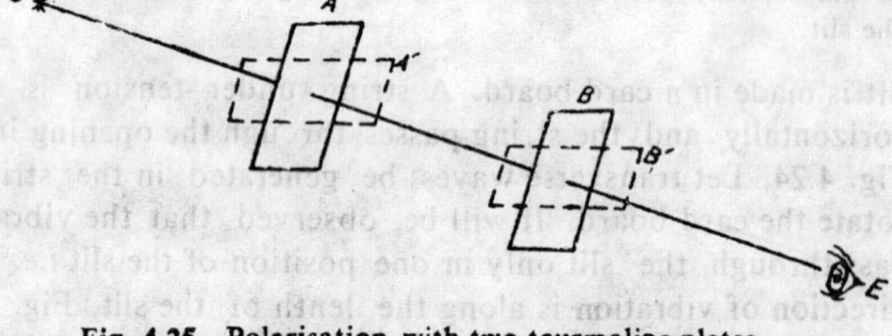

Fig. 4.25. Polarisation with two tourmaline plates.
Position *A-B*: light from *A* pass through *B* undiminished
Pos ition *A-B* : light from *A* is stopped by *B*.
Pos tion *A-B* : effect is same as in case of *A-B*.

light passed by *A* will also pass through *B* undiminished in intensity. If *B* is rotated from this position, the light emerging from *B* becomes weaker and weaker. When *B* is rotated through 90° (to the position *B′*) light is completely cut off. With fur[illegible] n, the intensity of transmitted light increases and reach the maximum value when the angle of rotation is 180°. Continuing further, at 270°, again transmission is cut off and when we come back to the original position, transmission is maximum again. Here we call *A* as the 'polariser' and *B* as the 'analyser'. Alternatively, if *B* is kept fixed and *A* is rotated, same cycle of changes in intensity will also occur.

If you consider light as a stream of corpuscles or particles coming from the source, this phenomenon cannot be explained. This can happen only if light has wave nature. Also if light waves are longitudinal waves, this type of variation in the intensity of transmitted light will also not occur. The only possible reason to explain these facts is to consider light as transverse wave motion. Following is the explanation of these experimental results.

In Fig. 4.26, the experimental set up given in Fig. 4.25 is repeated with few additions. The axes *X*, *Y*, *Z* are introduced and few circles such as *P*, *Q*, *R*, *S* and *T* are also included. *X*-axis is the direction of transmission of waves and the vibrations in *Y*-*Z*

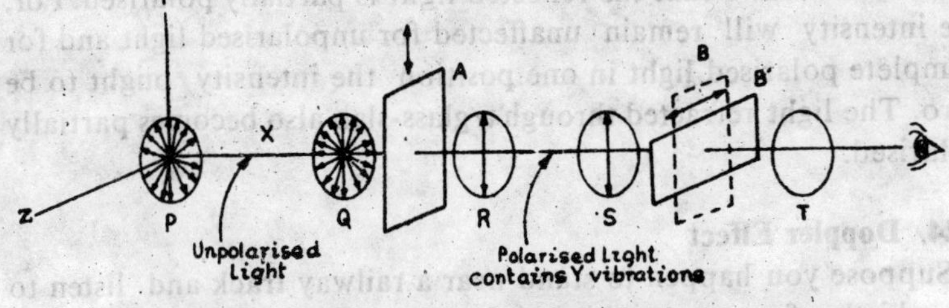

Fig 4.26. Explanation of the function of polariser and analyser, *PQ*-unpolarised light. *R*, *S*-vibration along *Y*-direction *T*-No vibration.

plane are shown in these circles. The circle *P* is situated at the source. The vibrations are uniformly distributed in the *Y*-*Z* plane. Though only a limited number of directions of vibration are shown, in principle, there is no limit to such vibrations. All vibrations are still uniformly present (circle *Q*) before the waves reach the polariser *A*. But the polariser *A* allows vibrations in one direction only say *Y*-direction to pass and stops other vibrations. The displacements in the *Y*-*Z* plane (transverse vibration) are vector quantities.

Each vector can be resolved into two rectangular components *i.e.* one along the *Y*-axis and the other along *Z*-axis. All *Y*-components pass through *A* and all *Z*-components are stopped. Hence after passing through the polariser *A*, the light has vibrations only along the Y-direction. It is indicated in the circle R. Now the Y-vibrations travel upto the 'analyser' B (circle S). If the analyser is parallel to the polariser, then the analyser allows only Y-vibrations. Since the light incident on B contains only Y-vibrations, light passes through B undiminished. On the otherhand, if B is rotated through 90°, (B' position) it would permit only Z-vibrations to pass through it. Since the incident light has no Z-vibrations, light is completely cut off by B. Circle T shows vibrations completely absent. The analyser and polariser are said to be in 'crossed-position' in this case.

The 'polariser' and 'analyser' are called polaroids. Since they have same characteristics, either polaroid can be used as a polariser or analyser. The polarised light considered here is also known as 'plane polarised' light.

The light reflected from a plane glass surface or even water surface is found to be partially polarised. If light is incident at an angle of nearly 55° on a glass-surface and the reflected light is seen through a polaroid, the intensity will vary widely but not become zero. This shows that the reflected light is partially polarised. For, the intensity will remain unaffected for unpolarised light and for complete polarised light in one position the intensity ought to be zero. The light refracted through a glass-slab also becomes partially polarised.

4.24. Doppler Effect

Suppose you happen to stand near a railway track and listen to the whistle of an approaching train engine. You will find that the sound of the whistle when the train approaches will appear to be different from the sound of the whistle when the train moves away from you. This difference is caused because the frequency of sound reaching your (observer) ear is changing when the engine is in motion. But the frequency of the whistle-sound originating from the engine (source) has remained the same throughout. Hence this apparent change of frequency is associated with the relative motion between the 'source' and the 'observer'. This effect was first observed by Doppler and therefore is known as Doppler effect.

Doppler observed that when either the source or the medium or the observer are in motion, the frequency of sound as received by the observer is different from the frequency of sound emitted by the source. The specific examples are given below.

(*a*) **Source is in motion—observer stationary**. Suppose, the frequency of the sound emitted by the source is ν. Then the wave length,

$$\lambda=\frac{C}{\nu} \quad ...(4.42)$$

where C is the velocity of sound when the source is stationary and λ is the corresponding wavelength,

Let the source move with a velocity V_s towards the observer *i.e.* position of S changes to S_1 (Fig. 4.27) So the waves generated by the source in one second will now occupy the distance $(C-V_s)$ (for stationary source it is simple C). But the number of waves in this space are ν.

Therefore, the wavelength λ_1 under this condition is given by

$$\lambda_1=\frac{C-V_s}{\nu}$$

If the medium is also in motion such as blowing of wind, the relation needs a further correction. Suppose the medium has a

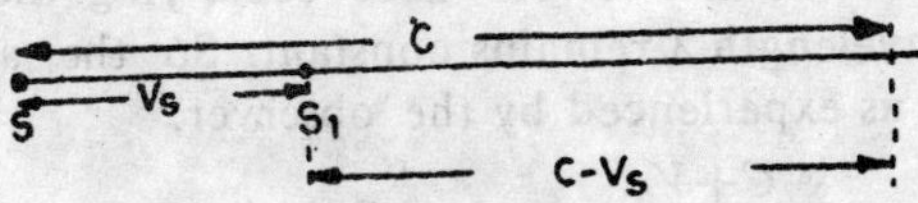

Fig. 4.27. Doppler effect. Apparant change in frequency when the source is in motion. S—Source. O—Observer. C—Sound velocity. V_s—Source velocity.

speed V_m towards the observer. Then the velocity of sound towards the observer should be $(C+V_m)$ instead of simply C.

So $$\lambda_1=\frac{C+V_m-V_s}{\nu} \quad ...(4.43)$$

If should be remembered that the change of wavelength from λ to λ_1, is a real change. The apparent frequency of sound as experienced by the observer, is given as

$$\nu_1=\frac{\text{Velocity of sound relative to the observer}}{\lambda_1}$$

$$=\frac{C+V_m}{\frac{C+V_m-V_s}{\nu}}=\frac{C+V_m}{C+V_m-V_s}\,\nu \quad ...(4.44)$$

So the apparent frequency is higher than the true frequency when the source is approaching the observer.

It can be shown that if the source is moving away from the observer, the apparent frequency,

$$\nu_1=\frac{C+V_m}{C+V_m+V_s}\,\nu$$

i.e. the apparent frequency is lower than the true frequency when the source is moving away from the observer.

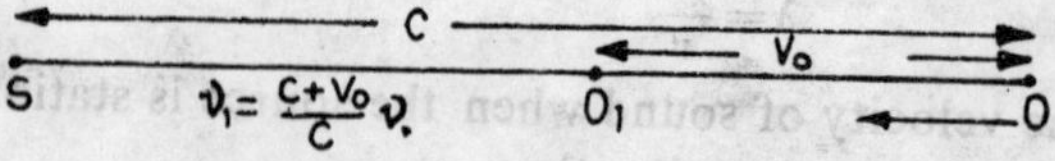

Fig. 4.28. Doppler effect. Apparent change in frequency when the observer is in motion. *S*—Source *O*—Observer.

If the medium velocity V_m is in opposite direction, $-V_m$ is to be substituted is place of $+V_m$ in Equations 4.44 and 4.45.

(*b*) **Observer in motion—source stationary.** Suppose the source *S* is stationary and the observer *O* move with a velocity V_o towards *S*. Let the observer move from *O* to O_1 in 1 second *i.e.*, $OO_1 = V_0$.

For a stationary observer at 0, the number of sound waves reaching in one second$=\nu$.

But in this case, number of waves reaching the observer during one second is equal to the number of waves occupying the length $(C+V_o)$. But the wavelength λ remains constant. So the apparent frequency of sound as experienced by the observer,

$$\nu_1 = \frac{C+V_0}{\lambda}$$

But $$\lambda = \frac{C}{\nu} \text{ (Equation 4.42)}$$

So $$\nu_1 = \frac{C+V_0}{C}\,\nu \quad \text{...(4.46)}$$

Hence the apparent frequency of the sound increases if the observer moves towards the source.

It can be similarly shown that when the observer moves away from the source, the apparent frequency

$$\nu_1 = \frac{C-V_0}{C}\,\nu \quad \text{...(4.47)}$$

i.e., the apparent frequency is lower than the true frequency when the observer moves away from the source.

In addition, if the medium is also moving, sound velocity should be either $C+V_m$ or $C-V_m$ depending on whether medium is moving from the source to the observer or opposite to that. So Equation 4.46 or 4.47 may be modified accordingly.

(*c*) **Both source and observer in motion.** Suppose both the source and the observer are moving from left to right.

The apparent frequency of sound experienced by the observer, (without considering the condition of the source), is

$$\nu_2 = \frac{C - V_0}{C} \nu_1 \quad \text{...(from 4.47)}$$

where ν_1 is the frequency of the source.

But since the source is also moving ν_1 is different from the frequency of a stationary source.

$$\nu_1 = \frac{C}{C - V_s} \nu \quad \text{...(from 4.44)}$$

where ν is the actual frequency of the source.

Substituting the value of ν_1,

$$\nu_2 = \frac{C - V_0}{C - V_s} \nu \quad \text{...(4.48)}$$

If the medium is also moving towards the observer,

$$\nu_2 = \frac{C + V_m - V_0}{C + V_m - V_s} \nu \quad \text{...(4.48}a\text{)}$$

It can be shown that Equation 4.44, is a special case of Equation 4.48 when $V_0 = 0$. Similarly Equation 4.47 is a special case of Equation 4.48 when $V_s = 0$ and $V_m = 0$. It should be borne in mind that Equation 4.48 can be used to determine the apparent frequency in all cases if we apply the following procedure.

1. The velocity of sound C should always be taken as positive.
2. V_m, V_s and V_0 should be considered as positive when their directions are from the source to the observer.
3. V_m, V_s and V_0 should be considered as negative when their directions are from the observer to the source.
4. After observing the proper sign conventions, the numerical values of V_m, V_s and V_0 may be substituted to determine the value of apparent frequency.

The principle of Doppler effect is applied in the sonar to determine the speed of mobile underwater objects such as a submarine. Ultrasonic waves sent by the sonar are reflected from the submarine and received back at the sonar stationed in a ship. So if a submarine approaches the ship at a speed of 10 m/s, it is equivalent to the 'source' approaching at 20 m/s. Taking the sound velocity in water,

$C = 1500$ m/s, the apparent frequency

$$\nu_1 = \frac{1500}{1480} \nu = \left(1 + \frac{20}{1480}\right) \nu$$

For $\nu = 30{,}000$ HZ, the increase in frequency *i.e.*, $(\nu_1 - \nu)$, therefore is about 400 HZ. This difference in frequency is measured by the sonar. In practice, the sonar is calibrated directly to give the approach velocity of the mobile target.

A bat determines the location and nature of objects by sending ultrasonic waves. It is likely that bats have also developed sensory devices to find the speed of the objects in their surrounding by sensing the Doppler frequency shift.

The radar uses electromagnetic waves to locate the position of objects in the sky. The speed of approaching planes and streams can also be determined by noting the frequency shift. Since for electromagnetic wave propagation, a medium is not necessary,

$$\nu_1=\frac{C-V_0}{C-V_s}\,\nu$$

gives the apparent frequency.

Here C is the velocity of electromagnetic waves. Since $V_s/C \ll 1$ and $V_0=0$.

$$\nu_1=\nu\,\frac{C}{C-V_s}=\nu\left(\frac{1}{1-V_s/C}\right)=\nu\left(1-\frac{V_s}{C}\right)^{-1}=\nu\left(1+\frac{V_s}{C}\right) \quad \text{...(4.49(}$$

As in case of sonar, if the aeroplane is approaching at a speed of 1000 km/h, source velocity is to be treated as 2000 km/h.

The Doppler effect has an important application in astronomy. Light from an incandescent source when examined through a spectrograph gives some spectral lines of different colours. Thus each line is associated with a given frequency. These spectral lines for a given element has a fixed pattern. The light from distant stars also gives a number of spectral lines corresponding to the elements present in them. But compared with the corresponding spectral lines from a source on the earth, these lines show a slight frequency change or shift. The shift generally occurs towards the red end of the spectrum and hence is referred to as 'red shift'. Red light has greater wave length or lower frequency. So 'red shift' means still lowering of the frequency. So this shift is said to be caused due to the stars in distant gallaxies moving away from us. This evidence also leads us to believe that the universe is expanding.

Example 4.9. A man standing near a railway track hears the whistle of an engine approaching him at a speed of 30 m/s. If the true frequency of the whistle is 3200 HZ what frequency does the man hear? What will be the frequency heard by him when the train goes past him?

Take velocity of sound=350 m/s.

Solution. Given, $V_0=0$ and we assume $V_m=0$, $\nu=3200$ HZ,
$C=350$ m/s.

Hence $$\nu_1=\nu\frac{C}{C-V_s}$$

$$=3200\times\frac{350}{350-30}\text{HZ}=3500\text{ HZ}.$$

When the engine is going away,

$$\nu_1=3200\frac{350}{350+30}\text{HZ}=2947.4\text{ HZ}.$$

Example 4.10. It is observed in a sonar placed in a ship that there is a frequency increase of 200 HZ in the signal reflected from a submarine. Compute the velocity of approach of the submarine. (velocity of sound in water=1500 m/s. Frequency of the sonar source=20 KHZ).

Solution. Here $V_0=0$, $C=1500$ m/s, $\nu=20{,}000$ HZ.

The apparent frequency, $\nu_1=20{,}000$ HZ$+200$ HZ.

$=20{,}200$ HZ.

Here, $\nu_1=\nu\frac{C}{C-V_s}$ or $=\frac{\nu}{\nu_1}=\frac{C-V_s}{C}=\left(1-\frac{V_s}{C}\right)$

or $V_s=\left(1-\frac{\nu}{\nu_1}\right)C=\left(\frac{\nu_1-\nu}{\nu_1}\right)C$

$$=\frac{200}{20{,}200}\times1500\text{ m/s}=14.85\text{ m/s}$$

The velocity of approach of the submarine$=\frac{V_s}{2}=7.425$ m/s.

4.25. Superposition of Waves

When number of waves reach simultaneously in any part of the space, their 'effects' add together according to the principle of superposition. The principle of super position is stated as: If $\vec{y}_1$, $\vec{y}_2$, $\vec{y}_3$,...are displacement vectors due to the waves 1 2, 3...at a point when they act separately then the net displacement at the point when all the waves act together is given by the vector sum of the individual displacements *i.e.*

$$\vec{y}=(\vec{y}_1+\vec{y}_2+\vec{y}_3+\ldots) \qquad \ldots(4.50)$$

For simplicity, we shall take the displacement components to be along the same line. Hence, the vector sum given in Equation 4.50 will be replaced simply by algebraic sum of the individual displacements. Further, we will consider only two waves superposing on each other. Then

$$y=y_1+y_2$$

It should be remembered that since y_1 and y_2 are functions of time and space, y will also vary with time and distance. Three important cases of super position will be considered here.

1. Two waves of the same frequency both moving either in the

forward or backward direction (interference of waves).

2. Two waves of slightly different frequencies, both moving in the forward (or backward direction) (Beats).

3. Two waves of the same frequency moving in the opposite direction (stationary waves)

4.26. Interference of Waves

Earlier in Art. 4.14. while dealing with progressive waves, we have seen that for particles separated by a path of λ, the phase difference is 2π. This means a path difference of λ corresponds to a phase difference of 2π and a path difference of $\lambda/2$ corresponds to a phase difference of π. In general, a path difference of x, is associated with a phase difference of $2\pi/\lambda\ x$. Suppose two waves start simultaneously from the same source and cover equal distances travelling over different paths and then they superpose on each other. Since they started from the same source simultaneously their frequency and initial phase (or epoch) are the same. As they have travelled equal distances, the phase difference introduced in the path is also zero. Thus the displacement due to the two waves when they superpose will be in the same phase. Hence the total displacement will be the sum of individual displacements. So the intensity due to the waves will be reinforced. If the path traversed by the two waves is such that the displacement due to individual wave is maximum *i.e.* the displacement amplitudes are a_1, a_2 or $-a_1$, $-a_2$ then the total displacement after super position will be $\pm(a_1+a_2)$ (Here a_1 and a_2 are individual amplitudes). Since the maximum intensity due to a wave,

$$I_{max} \propto a^2$$

the maximum intensity due to super position therefore will be,

$$I_{max} \propto (a_1+a_2)^2$$

Thus the intensity can be increased considerably. If the path difference is λ, then also the displacements will be in phase. This principle is utilised in the stethoscope used by a doctor to hear the sound of heart-beats of a patient. The sound waves are transmitted to the ears through two tubes of equal length. Thus sound waves of the hearth-beat arrive in the same phase.

Let us consider the opposite case. Suppose, two waves starting simultaneously from the same source cover different distances before they superpose on each other. Let the difference in the path traversed be $\lambda/2$. So the displacements due to the two waves when they superpose will have a phase difference of π *i.e.* the displacements will be in opposite phase. Thus the total displacement will be the difference of the individual displacements. So the intensity due to the waves will be reduced. If both the waves arrive when the displacements are maximum, (one $+a_1$ and the other $-a_2$) the total displacement after super position will be (a_1-a_2). Thus the intensity also will be small *i.e.* $I \propto (a_1-a_2)^2$.

In general, when the path difference is 0, λ, 2λ,.........$n\lambda$, the intensity after super position will be maximum. On the other hand, for path difference $\lambda/2$, $3\lambda/2$, $5\lambda/2$ .....$(n+1/2\ \lambda)$, the intensity will be minimum.

Thus, path difference corresponding to I_{max}

$$=0, \lambda, 2\lambda.........n\lambda \qquad ...(4.51)$$

Path difference for I_{min}

$$=\lambda/2, 3\lambda/2,......(n+\tfrac{1}{2})\lambda \qquad ...(4.51a)$$

Hence the phase difference for I_{max}

$$=2\,n\pi \qquad ...(4.52)$$

The phase difference for I_{min}

$$=(2\,n+1)\,\pi \qquad ...(4.52a)$$

where n is any integer (0, 1, 2......etc.)

Between the two extreme cases, for other path differences the displacements will be at various phases. So the algebraic sum of the displacements and hence the intensity will have intermediate values.

Thus the interference between two waves will produce a resultant wave with displacements different from the individual waves.

We can observe the effect of interference in a ripple tank. S_1 and S_2 are two dippers in a shallow ripple tank containing water. Both the dippers are made to vibrate electrically with same frequency. At any instant of time, the crests and troughs produced by S_1 and S_2 are shown by full-line and dotted line curves, respectively (Fig. 4.29) waves from both the sources S_1 and S_2 will reach over the entire surface of water after some time. So interference is observed on the entire surface. Wherever a crest of the waves coming from S_1 meets with the crest of the waves coming from S_2, there will be a stronger crest. Similarly, when a trough meets another trough, there will be a stronger trough and when a crest meets a trough they will cancel out each other's effect. As a result of interference, therefore, waves of larger amplitude will be observed in the region of interference. In Fig. 4.29, the points of first category, amplitude (a_1+a_2) are marked by ●, the points of second category, amplitude $-(a_1+a_2)$, are marked by ○ and the points of third category, amplitude (a_1-a_2) are marked by □. With the passage of time, the new crests and troughs will also change their position. So a point like P_1 will alternately have high crests (a_1+a_2) and high troughs $-(a_1+a_2)$ *i.e.* the disturbance is maximum. But a point like P_2, will always be in the same condition *i.e.* the disturbance will be zero or very weak.

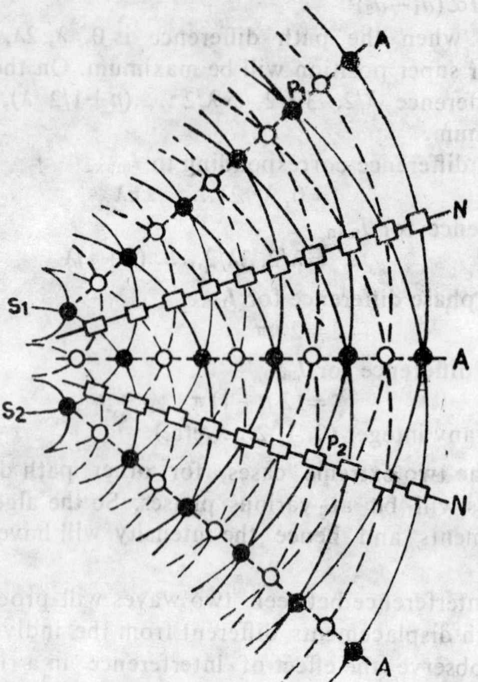

Fig. 4.29. Interference due to two wave trains from S_1 and S_2. A-antinodes. N-Nodes.

Referring to Equations 4.51 and 4.51 (*a*), we can get,

Path difference $= S_2 P_1 - S_1 P_1$

$= n\lambda$ — for maximum intensity.

and Path difference $= S_2 P_2 - S_1 P_2$

$= (n+\frac{1}{2})\lambda$ — for minimum intensity.

Here, of course, we have assumed that the initial phase of the waves starting from S_1 and S_2 are same. The interference pattern will still be observed even if there is initial phase difference between the waves from both sources. But the locations of P_1 and P_2 will be changed. If the initial phase difference between the waves from the two sources is π, P_1 becomes a point of minimum displacement and P_2 becomes a point of maximum displacement. In general, if S_2 is ahead of the phase relative to S_1 by ϕ_0, the conditions for maxima and minima to be observed at a point P

become

$$\phi_0+\frac{2\pi}{\lambda}(S_2P-S_1P))=2n\pi\text{—for maxima}$$

$$\phi_0+\frac{2\pi}{\lambda}(S_2P-S_1P)=(2n+1)\pi-\text{for minima} \quad ...(4.53)$$

Compare the Equations 4.53 with Equations 4.52 and 4.52 *a*. Interference is also observed in radio waves, light waves *i.e.* electromagnatic waves. The same conditions for maxima and minima are followed in these cases also.

The occurrence of interference for sound waves can be demonstrated by a simple experiment described below:

Fig. 4.30 shows a quinckes tube *APB* and *COD* are two *U*-shaped tubes. The tube *CQD* can slide into the other tube. In the tube *APB*, there are openings at *A* and *B*.

A vibrating tuning fork is held close to *A* and listner's ear is kept close to *B*. Sound from the opening *A* can reach *B* over two

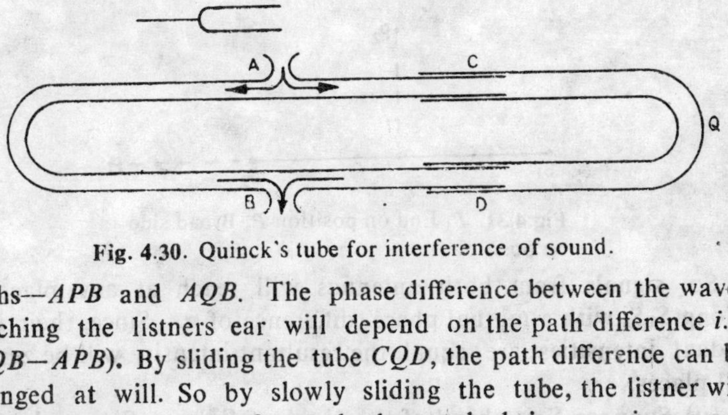

Fig. 4.30. Quinck's tube for interference of sound.

paths—*APB* and *AQB*. The phase difference between the waves reaching the listners ear will depend on the path difference *i.e.* (*AQB*—*APB*). By sliding the tube *CQD*, the path difference can be changed at will. So by slowly sliding the tube, the listner will observe the intensity of sound alternately being maximum and minimum according as the conditions are satisfied for Equations 4.52 or 4.52*a*.

4.27. Coherent Sources

While discussing the interference of waves due to two sources, we have taken the initial phase difference between the waves, coming from both of them to be either zero or constant. Even if the sources send out waves at random phases, interference between the waves will be observed. But the pattern of interference will not be predictable. Moreover, the points of maximum and minimum displacements such as P_1 and P_2 in Fig. 4.29 will change their positions at random. Therefore, it is required to have two such sources which will send out waves either at the same phase or with a constant phase difference. Any change in phase for S_1 should be accompanied by a similar change in the phase of signals sent by S_2. The sources satisfying these conditions are known as coherent sources. Further details about the coherent sources will be discussed in optics.

Example. 4.11. Two vertical radio antennas are sending out waves of $\lambda=300$m and of equal intensity. If the separation between them is 150 m, discuss the intensity of resultant waves for maxima and minima along the line joining the sources (end on position) and along the perpendicular bisector (broad-side on position) when (*i*) the two sources are in the same phase (*ii*) sources are in the opposite phase.

Solution. *For end on—position.* (*i*) In this case the only phase difference is due to the difference of the path, $S_1P_1-S_2P_1=S_1S_2$

But $S_1S_2=150 \text{ m}=\lambda/2$.

So phase difference $\triangle\phi=\delta=\pi$.

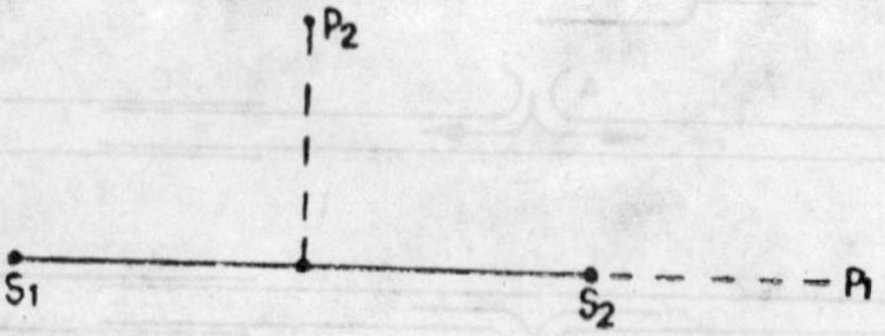

Fig.4.31. P_1 End on position P_2 Broad side on position.

So, signals from both antennas will reach at any place P_1 along S_1S_2 with a mutual phase difference of π. Since, their individual intensities are equal, the resultant intensity will be zero at all places.

(*ii*) Suppose S_2 is ahead of the phase of S_1 by π. Since, the path difference of S_1S_2 corresponds to a phase difference of π, the total phase advance of $S_2=\pi+\pi=2\pi$

So, signals from both sources will reach in the same phase at all points along S_1S_2 or S_2S_1. Since the resultant displacement will be $2a$, the intensity $I=4I_1$ where I_1 is the intensity due to individual source.

Broad-side-on position. (*i*) In this case, all points along the perpendicular bisector are situated at equal distances from S_1 and S_2. So, the signals will reach at the same phase at all points.

Hence $I=4I_1$

(*ii*) Here, the signal from both the sources reach in opposite phase as the sources are having a phase difference of π and the signals travel equal distances. So the intensity along the perpendicular bisector will be zero.

Example 4.12. If sound from two sources of intensity ratio 25:1 interfere, calculate the ratio of intensities at the maxima and minima of interference pattern.

Solution. $$\frac{I_1}{I_2}=\frac{a_1^2}{a_2^2} \quad ...(i)$$

where I_1 and I_2 are the intensities and a_1 and a_2 are the amplitudes respectively of the two sources.

So $$\frac{a_1}{a_2}=\sqrt{\frac{I_1}{I_2}}=\sqrt{25/1}=5$$

But $$\frac{I_{max}}{I_{min}}=\left(\frac{a_1+a_2}{a_1-a_2}\right)^2=\left(\frac{a_1/a_2+1}{a_1/a_2-1}\right)^2 \quad ...(ii)$$

So $$\frac{I_{max}}{I_{min}}=\left(\frac{5+1}{5-1}\right)^2=9/4$$

4.28. Beats

When waves from two sources of slightly different frequencies superpose on each other, the resultant wave alternately varies in intensity with passage of time. If the sources are sound sources at a given place alternate loud and low sounds will be heard. This phenomenon is called *beats*. The time interval between successive loud sound or low sound is called *one beat-period*. The number of beat-periods occurring in one second is called the *beat-frequency*. It will be shown below that the beat-frequency is equal to the difference of frequency between the two sources.

Suppose, at the observation point the displacements due to the two sources are written as,

$$y_1=a_1 \sin 2\pi\frac{t}{T_1}=a_1 \sin 2\pi\nu_1 t \quad ...(4.54)$$

$$y_2=a_2 \sin 2\pi\frac{t}{T_2}=a_2 \sin 2\pi\nu_2 t \quad ...(4.55)$$

Here, the frequencies of the two sources are ν_1 and ν_2 and we take $\nu_1 \approx \nu_2$, *i.e.* $(\nu_2-\nu_1)$ is very small.

We take that both the sources are emitting at the same phase. So, the phase difference between the two waves at a given place is given by

$$\delta=2\pi\nu_2 t-2\pi\nu_1 t$$
$$=2\pi(\nu_2-\nu_1)t \quad ...(4.56)$$

Thus, the phase difference changes with time. The resultant displacement will be maximum when both the displacements are in the same phase or the phase difference is any multiple of 2π,

i.e., $\delta=0, 2\pi, 4\pi......2n\pi$

Similarly minimum will take place when the two waves reach in the opposite phase.

i.e., $\delta=\pi, 3\pi, 5\pi......(2n+1)\pi$

From Equation (4.56), we can see that the value of δ passes through 0, π, 2π, 3π etc. as the value of t increases thus producing a maximum and minimum intensity alternately with passage of time. Between two consecutive maxima or minima the phase difference is 2π. So, for a phase difference of 2π, we get one *beat*. From Equation 4.56, we see that the phase change during one second is $2\pi(\nu_2-\nu_1)$. So during one second we get $2\pi(\nu_2-\nu_1)/2\pi=(\nu_2-\nu_1)$ beats. *i.e.*, the beat frequency is equal to the frequency difference of the two sources.

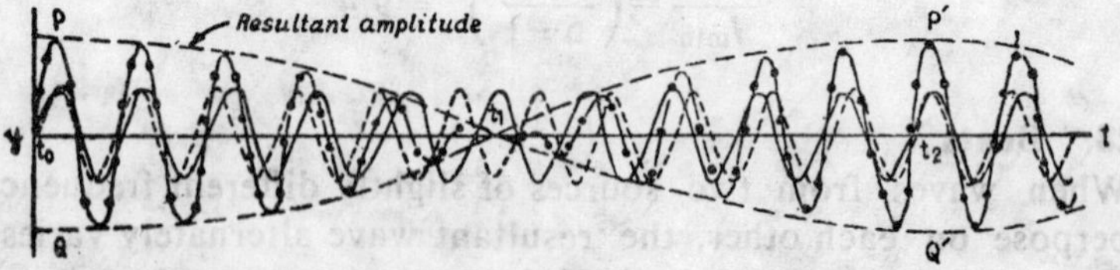

Fig. 4.32. Production of beats.

In Fig. 4.32 the formation of beats is illustrated graphically. The $y-t$ curve for two waves are represented one by the full line curve and the other of slightly higher frequency (lower period) by dotted line curve. In the beginning at t_0 the two oscillations are in the same phase. But as time advances, they gradually differ in phase and at t_1, they are in opposite phase. Again at t_2, they arrive in the same phase. So at t_0, the resultant amplitude is maximum and at t_1, it is minimum to become maximum again for t_2 and so on. The resultant amplitude is shown by the beaded curve (—•—•—). The dashed line touching the amplitude of the resultant curve shows the periodic variation in the amplitude of the resultant wave. The time period of this resultant fluctuation is (t_2-t_1) or one beat period. Referring to the Fig. 4.32 one can see the during this period, one wave completes x oscillations, the other wave completes $(x+1)$ oscillations. Thus the beat frequency is equal to the frequency difference of the sources.

It is of interest to note that both in interference and beats we come across periodic variation in the intensity due to superposition of two trains of waves. But in case of interference, the periodicity is observed in *space*, *i.e.*, at a given time, alternate maxima and minima are uniformly formed in space. But in case of beats the periodicity is observed in *time, i.e.*, at a given place, alternate maxima and minima are formed at regular time intervals.

In laboratory, beats can be produced by two tuning forks of the same frequency. If some amount of bee's—wax is attached to the prongs of one a tuning fork and then they are sounded, bath the forks will not vibrate any more with the same frequency. There will be a few cycles difference in frequency of vibration. While vibrating, if the stems of the tuning forks are touched to a table or any bigger surface, one can distinctly hear alternate *waxing* and *waning* of sound.

In matching the frequencies of two musical instruments, musicians utilise the principle of beats. A small difference in the frequencies may not be detected by sounding them separately, but when sounded together, beats will be produced. So adjustments are made till both the instruments are in unison *i.e.* till beats disappears. So this principle can also be used to determine an unknown frequency by comparison. In electronics, beat-frequency is used in many cases. In super-heterodyne radio receivers, this principle is utilised in producing intermediate-frequency (IF). A very low frequency oscillator can be made by mixing the output of two high frequency oscillators of slightly different frequencies. Such an oscillator is called a beat-frequency oscillator. Utilising this method, an unknown frequency can also be determined if the output of a standard oscillator of known frequency is mixed with it till the beats disappear.

Example 4.13. A tuning fork *A* produces 4 beats/sec. with another tuning fork *B*. It is found that by loading *B* with some bee's wax, the beat-frequency increases to 6 beats/sec. If the frequency of *A* is 320 HZ, determine the frequency of *B*, when loaded.

Solution. Since the beat-frequency is 4, the frequency of *B*, either $(320+4)=324$ or $(320-4)=316$.

By leading *B*, its frequency will decrease. Thus if 324 is the original frequency, the beat frequency will reduce. On the other hand, if it is 316, the beat frequency will increase which is the case here so the original frequency of the tuning fork $B=316$ HZ and when loaded, it is $316-2=314$ HZ.

4.29. Stationary Waves

So far we have discussed the effect of super position of two waves travelling either in the forward or backward direction. Now we will consider the superposition of two waves of equal frequencies and equal amplitudes travelling through the same medium in opposite directions. The resultant effect is the production of a wave which does not travel in either direction with passage of time. Therefore, such a wave is called *stationary wave*. As we have discussed earlier, the waves superposing to produce a stationary wave are *progressive waves* or travelling waves as they travel with a definite velocity in a medium.

The production of stationary waves can be shown both analytically and graphically. We will, however, keep ourselves confined to the graphical method only.

The graphical method. In this method, the displacement distance (y, x) graphs for the two progressive waves travelling in opposite

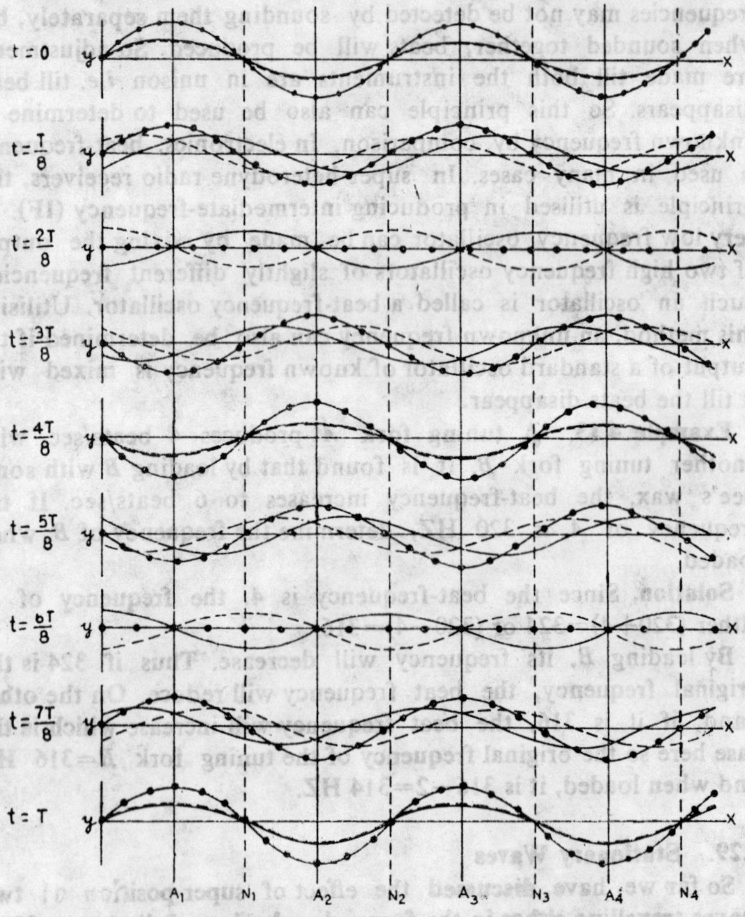

Fig. 4.33. Formation of stationary waves — Curve moves to the right — Curve moves to the left.

directions are drawn at several equal time interval and then the resultant is deduced by adding (y_1+y_2) at each location x). In Fig. 4.33 one wave is shown by the full-line curve and the other by the dotted line curve.

The solid line wave is moving to the right and the other one is moving to the left. The conditions at time intervals of $T/8$ each in eight steps are shown. At the start, $t=0$, the peaks of both waves are coinciding. In this next interval, $t=T/8$, the full line $y-x$ curve has moved to the left by $\lambda/8$ while the dashed—line curve has moved to the left by $\lambda/8$ during the same interval. The resultant of the two waves *i.e.*, $y=y_1+y_2$ are shown by the beaded line. Mark the resultant amplitudes at positions marked A_1, N_1, A_2...etc. At $t=0$, the peak of one coincides with the peak of the other. Therefore, the amplitude of the resultant is maximum $2a$ *i.e.*, the sum of the individual amplitudes. Looking at the $y-x$ curve at $t=0$, one can see that the resultant displacement are either $+2a$ or $-2a$ at points marked A_1, A_2, A_3 etc. whereas at points N_1, N_2...it is zero.

But at $t=T/8$, the individual displacements at A_1, A_2...etc are less, but both their phases are either +ve or −ve. So the resultant also reduces.

But at N_1, N_2...etc, the curves are out of phase (phase difference $=\pi$). One has +ve displacement whereas the other has −ve displacement of the same magnitude. So the positions N_1, N_2...etc, continue to have zero displacement.

Proceeding further, at $t=2T/8$, one curve has further advanced by $\lambda/8$, to the right and the second curve by $\lambda/8$ to the left. At all points A_1, A_2...the positive peaks of one wave coincide with the negative peak of the other. Thus at these points the resultant displacement is zero. At the same time, the resultant displacements at N_1, N_2...etc, are also zero as the individual displacements are zero at these points.

At $t=3T/8$, the waves move further by $\lambda/8$ in opposite directions. The displacements at A_1, A_3, A_5...etc, add to give a resultant negative value whereas at A_2, A_4, A_6 etc, the resultant is positive. But at N_1, N_2, N_3, etc the resultant displacements are zero.

At $t=4T/8=T/2$, the resultant displacements at A_1 A_3, A_5...is $-2a$ and those A_2, A_4, A_6...is $+2a$. But at N_1, N_2... They are still zero.

As time advances, the resultant displacements go on altering till for $t=T$ they become identical to those at $t=0$ It can be shown that this phenomenon is repeated for each time interval of

T. Observing the Fig. 4.33 one can see, that the displacements of particles at points marked A_1, A_2, A_3...etc. undergo a periodic variation from $-2a$ to $+2a$ in a periodic time T. In contrast, particles at N_1, N_2... remain at rest at all times. The particles at other points also undergo periodic vibration in the same interval T but their amplitudes are less than $2a$.

The points where particles undergo maximum displacement are called *antinodes* (A_1, A_2, A_3...) and the points where particles have no displacement are called *nodes* (N_1, N_2...). Though the displacment is zero at the nodes for all times, the particle velocity is not always zero.

From Fig. 4.34 one can see that all the particles reach their maximum displacement simultaneously (the amplitude ranging from

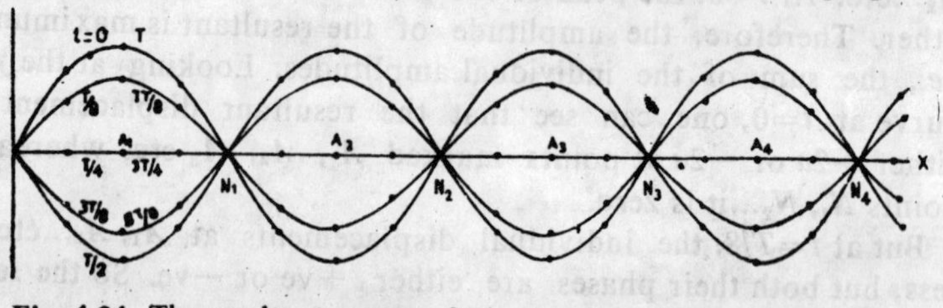

Fig. 4.34. The resultant curves of Fig. 4.33 in sight stage from $t=0$ to $t=T$.

$2a$ to zero) at $t=0$, or $t=T/2$ or $t=T$ and so on. Similarly, all particles also have zero displacement simultaneously *i.e.*, at $t=T/4$, $3T/4$ etc. This means the phase of the resultant wave does not change with distances x. It changes with time simultaneously for all positions. Therefore, the displacement does not proceed with time as in case of a progressive wave. Hence it is called a stationary wave. The position of the vibrating particles in a stationary wave at different time intervals are shown in Fig. 4.34. This is actually taken from Fig. 4.33. The distance between two successive nodes or antinodes is $\lambda/2$ and the distance between a node and its nearest antinodes is $\lambda/4$. It will be shown later that it is possible to produce stationary waves in a stretched string where the nodes and antinodes are a set of points. In stationary waves produced on water surface or in a stretched membrane, they are a set of lines. Stationary waves can also be produced in space such as in organ pipes. The nodes and antinodes form a set of planes—nodal planes and antinodal planes in those cases.

The two oppositely travelling waves are practically created by reflection at a boundary. Incident wave travels in the forward direction and the reflected wave travels in the opposite direction.

In Article 4.19, we have seen that reflection can occur with or without change of phase at the boundary. If there is no phase change at the boundary, both the incident and reflected waves will always be in the same phase at the boundary. Thus the signs of y_1 and y_2 are always same. So this position will always be an antinode. On the other hand, if there is phase change at the boundary, y_1 and y_2 will always have opposite signs. Thus the boundary will be the seat of a 'node'. In dealing with stationary waves any node or antinode may be taken as the origin. Normally. the boundary at which reflection takes place and hence the opposite moving wave originates is taken as the origin ($x=0$).

4.30. Differences between Progressive and Stationary Waves

1. In progressive waves, the displacement is transmitted from particle to particle with the passage of time. So the maximum displacement (also minimum displacement) is achieved at different points at different times. But in stationary waves, all particles reach the maximum displacement (also minimum displacement) simultaneously.

2. In progressive waves, the amplitude is the same at all position. But in stationary waves, the amplitudes vary from place to place, being maximum at the antinodes and zero at the nodes.

3. The phase of displacement of particles vary with distance in progressive waves whereas at any instant all the particles are in the same phase in stationary waves.

4. In a stationary wave, all the particles in the medium have maximum displacement twice in each cycle. In that position, all the particles are momentarily at rest. Therefore, the kinetic energy is zero and the entire energy appears as potential energy only. Similarly, twice in each cycle all the particles in the medium have zero displacement so that the potential energy is zero. So the total energy is kinetic. Hence, in stationary waves, the total energy alternately becomes wholly potential or wholly kinetic. In progressive waves, since the particles are in various phases of displacement, all particles do not come to rest simultaneously. Hence the total energy cannot be wholly potential or wholly kinetic. If we consider the energy of all the particles in a distance of one wavelength, it is distributed equally between potential and kinetic.

5. In longitudinal progressive waves, there are alternate compressions (hence increase in pressure) and rarefaction (decrease in pressure). So the variation in pressure is transmitted with time. Graphically, the excess pressure is proportional to the slope of

$y-x$ curve for stationary waves. We see in Fig. 4.34 that this slope is always zero at the antinodes and maximum at the nodes. Hence there is no variation of pressure at antinodes and the displacement antinodes, therefore, can be called pressure nodes. Similarly, the displacement nodes have maximum variation of pressure and may be called pressure antinodes. Between a pressure node and pressure antinode, the pressure variation has intermediate values.

Thus, the pressure amplitude is same for all positions in a progressive wave and it moves with time. In the stationary waves, it is different at different positions and is fixed for that position.

4.31. Stationary Waves in Strings

In open air or in an infinitely long string or an infinite expansion of water surface, the waves travel in any direction with a velocity determined by the properties of the medium and the nature of the wave transmitted. We have seen in Article 4.21 if two different strings are joined together, there is reflection at the boundary. So if a string is held under tension and one end is connected to a vibrating source say a tuning fork, waves will travel to the other end. On reaching the other end, they will be reflected. Hence at anytime, along the length of the string, there will be incident and reflected waves moving with the same velocity. So they will produce stationary waves on the string. As the velocity of the waves along the string depends on the applied tension and mass per unit length of the string (see Equation 4.34.) the pattern of stationary waves will also be determined by these factors besides the frequency of the source. The formation of stationary waves along a stretched string can be studied by sonometer and in Melde's experiments. We will, however, not go into the details of these devices.

4.32. Stationary Waves in Air Columns

If vibrations are set at one end of a straight long pipe, progressive waves proceed inside the pipe. On reaching the other end, reflected waves will be set up at the boundary. Thus stationary waves will be generated inside the tube. Boundary conditions may be of two types (1) Both ends are open (2) one end open and the other end closed. The first category is known as open organ pipe and the second type is known as closed organ pipe. We consider both case separately.

(*a*) **Open organ pipe.** At both ends, there will be antinodes because open atmosphere is a free or loose boundary relative to the confined air in the pipe. So reflection occurs without change of

phase. The simplest case is of two antinodes at the end and a node in the middle Fig. 4.35 (*a*). Therefore, if the length of the pipe is l,
$$\lambda/2=l, \text{ or } \lambda=2l$$
and the frequency,
$$\nu=\frac{C}{\lambda}=\frac{C}{2l} \qquad ...(4.57)$$
where C is the velocity of sound in air ν therefore is the frequency of the fundamental note. The next possibility is two nodes and one antinode in the middle in addition to the antinodes at the ends, Fig. 4.35(*b*).

So $$2\frac{\lambda}{2}=l \text{ or } \lambda=l$$

So, frequency of the second harmonic,
$$\nu_2=\frac{C}{\lambda}=\frac{C}{l}=2\nu$$

Similarly the third harmonic, $\nu_3=\dfrac{3C}{2l}=3\nu$

and the pth harmonic, $\nu_p=\dfrac{pC}{2l}$

where $p=1, 2, 3$ ....(4.58)

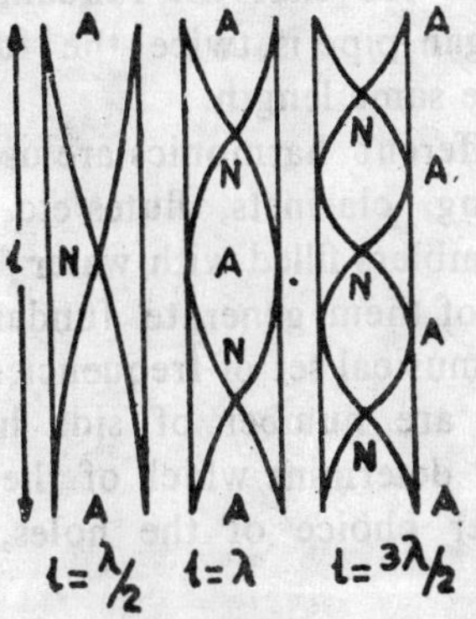

Fig 4.35. Stationary waves in open organ pipes. (*a*) Fundamental or 1st harmonic, (*b*) Second harmonic, (*c*) Third harmonic.

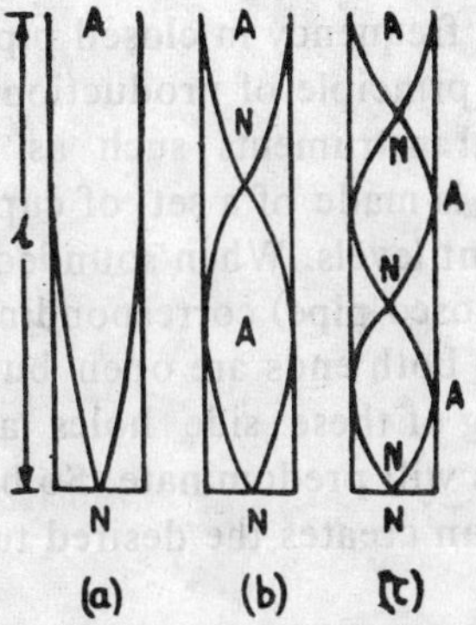

Fig. 4.36. Stationary waves in closed organ pipe. (*a*) Fundamental, (*b*) Third harmonic, (*c*) Fifth harmonic.

(*b*) **Closed organ pipe.** There will be always a node N, at the closed end and an antinode, A, at the free-end. Simplest case is shown in Fig. 4.36 (*a*).

If the length of the pipe is l, then, $\dfrac{\lambda}{4}=l$ or $\lambda=4l$.

So the frequency of the fundamental,

$$\nu = \frac{C}{\lambda} = \frac{C}{4l} \qquad \ldots(4.59)$$

The next possibility is one antinode and one node in the middle, Fig. 4.36(*b*). In this case

$$3\frac{\lambda}{4} = l \qquad \text{or} \qquad \lambda = \frac{4l}{3}$$

So the frequency is

$$\frac{C}{\lambda} = \frac{3C}{4l} = 3\nu$$

Thus it is third harmonic.

Similarly, the next harmonic will be fifth harmonic, $\nu_5 = \frac{5C}{4l} = 5\nu$

In general, the *p*th frequency will be of $(2p+1)^{th}$ harmonic.

$$\nu_p = \frac{(2p+1)C}{4l} \qquad \ldots(4.60)$$

where $p = 0, 1, 2 \ldots\ldots$

Thus we find that it is possible to generate different harmonics in organ pipes. In open organ pipe, all harmonics are generated whereas in the closed pipe only odd harmonics are generated. Comparing Equations 4.57 and 4.59 we can see that the fundamental frequency of vibration in an open organ pipe is twice the fundamental frequency in closed pipe of the same length.

The principle of production of different harmonics are used in musical instruments such as Jaltarang, clarinets, flutes etc. Jaltarang is made of a set of cups or tumblers filled with water upto different levels. When sounded, each of them generate fundamentals (closed pipe) corresponding to a musical set of frequencies. In a flute, both ends are open but there are number of side holes. Closing of these side holes actually determine which of the harmonics will predominate. So by proper choice of the holes, the musician creates the desired tune.

4.33. Determination of Velocity of Sound by Resonance Air Column

The arrangement consists of vertical glass tube *PQ* (Fig. 4.37) connected to a water reservoir *R* by a rubber tube *T*. The water level *S* in the tube can be altered by changing the height of *R*. So the tube portion *PS* is a closed organ pipe and is adjustable for fundamental frequency as given in Equation 4.59.

A vibrating tuning fork is held near *P* while the reservoir is lowered to increase the length *PS*. The air inside is set into vibration, but its amplitude remains small till the length corresponds to the length for fundamental frequency equal to the frequency of

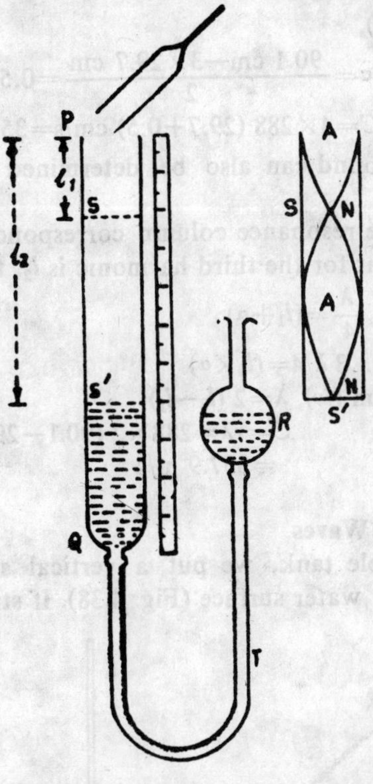

Fig. 4.37. The resonance air column.

the tuning fork when resonance occurs and loud sound is heard. Since under this condition, the fundamental frequency is equal to the frequency of the tuning fork.

$$\nu = C/4l \quad \text{or } C = 4\,l\nu$$

where ν is the frequency of the tuning fork and l is the length of the air-column. However, the antinode at P is actually forced little outside the tube. Therefore an end correction factor 0.6 r where r is the radius of the tube is added to the length of the tube. Thus,

$$C = 4\,(l + 0.6\,r)\,\nu \qquad \ldots(4.61)$$

Example. 4.14. A resonance air column produces resonance with a tuning fork of $\nu = 288$ HZ at column length 29.7 cm and 89.1 cm. Deduce (*i*) the end correction and (*ii*) the speed of sound in air.

Solution. Let a be the end correction. The first resonance occurs when the length of air column corresponds to the fundamental and the second the 3rd harmonic.

So, $C=4\nu\,(29.7+a)$ ...(*i*)

and $C=\frac{4\nu}{3}(90.1+a)$ ...(*ii*)

Equating (*i*) and (*ii*),

$$a=\frac{90.1\text{ cm}-3\times 29.7\text{ cm}}{2}=0.5\text{ cm}$$

So $C=4\times 288\,(29.7+0.5)\text{ cm/s}=357.9\text{ m/s}$

The velocity of sound can also be determined without calculating a.

If the length of the resonance column corresponding to the funmental is l_1, and that for the third harmonic is l_2, then we know,

$$\frac{\lambda}{4}=(l_1+a) \qquad ...(iii)$$

and $3\,\lambda/4=(l_2\times a)$ ...(*iv*)

Subtracting (*iii*) from (*iv*), $\lambda=2\,(l_2-l_1)$

$$C=\nu\lambda=288\times 2\,(90.1-29.7)\text{ cm/s}$$
$$=357.9\text{ m/s}$$

4.34. Diffraction of Waves

Suppose in a ripple tank, we put a vertical sheet *KL* with an opening slit *S* on water surface (Fig. 4.38). If straight waves are

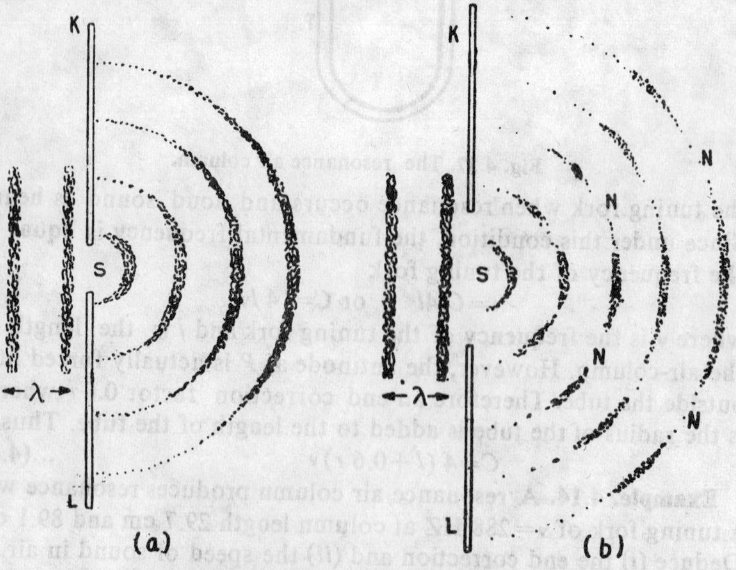

Fig. 4.38. Ripples bending round the corner, (*a*) Slit width<λ, (*b*) Slit width>λ.

generated on the left of the sheet by repeated dippings of a plane sheet portion of the waves will be obstructed by *KL*, but a portion will pass through the slit. If the width of the slit is smaller than the wavelength λ, the ripples will spread in all directions on the right side (Fig. 4.38 *a*). But if the width is larger than λ. then the ripples spread upto some angle from the normal direction, becomes absent in regions *NN* and then again appear with reduced amplitude. The main point in both cases is that the waves spread to the regions not directly in its line of propagation *i.e.* they bend or spread around obstacles. This phenomenon of spreading of waves around obstacle is called *diffraction*.

Diffraction is a characteristic property of all waves. It is readily observed in case of sound waves. Standing on one side of a low wall one can hear someone talking from the other side due to diffraction of sound waves. Diffraction is observed in light also. But as the wave length is small, the obstacles or slits should naturally be of much smaller dimensions. Details about diffraction of light will be covered in the chapter on optics.

QUESTIONS

1. True simple harmonic motions are extremely rare in nature. Why ?

2. Simple harmonic motion is periodic, but all periodic motions are not simple harmonic . Discuss.

3. Are the following motions simple harmonic? Explain.

(*a*) Motion of the balance wheel of a watch.

(*b*) Motion of the piston in the cylinder of a steam locomotive engine.

(*c*) Vibration of an unloaded spring.

(*d*) Motion of a rubber ball on rebounding from a concrete floor.

4. A vertically oscillating spring is taken to the top of a hill. What will happen to its time period ?

5. The load suspended by a vertically vibrating spring is increased. How would the time period of oscillation be affected?

6. A simple pendulum might be used to assist in locating oil fields. Explain.

7. What would be the effect on the period of a simple pendulum when taken inside a deep mine?

8. Could two simple harmonic motions be added to produce a resultant simple harmonic motion? Illustrate your answer?

9. Soldiers are asked to break steps while marching on a bridge. Why?

10. Give some examples of resonance in nature.

11. Give two examples of transverse waves and longitudinal waves not mentioned in the book. To which category does the waves on water surface belong?

12. Show that every part of the string through which a pulse passes is governed by Hook's law of elasticity?

13. Is it possible for a transverse or longitudinal wave to exist if the vibratory motion were not simple harmonic? Explain your answer.

14. Verify, by dimensional analysis, the correctness of Equations 4.34, 4.35, 4.36 and 4.37.

15. Is there any relation between the velocity of the wave proceeding along a string and velocity of the particles of the string vibrating in transvers direction?

16. The velocity of transverse vibration along a string can be altered by the experimentor, but he cannot change the velocity of sound in air. Why?

17. Arrange the following in increasing order.

(*a*) Velocity of sound in dry air. (*b*) Velocity of television signals. (*c*) Sound velocity in water. (*d*) Sound velocity in steel. (*e*) Velocity of light. (*f*) Velocity of sound in moist air.

18. The energy in an oscillating body is partly potential and partly kinetic. Are they equal to each other in any position?

19. When two waves interfere, does one affect the speed of the other?

20. When a ripple is reflected from the edge of a ripple tank, what happens to its phase?

21. Is there a transfer of energy through the medium when a stationary wave is produced in it? Explain.

22. When a number of knots are tied at irregular intervals along the length of a string, stationary waves cannot be set up in it. Explain the reason.

23. In longitudinal stationary waves, displacement nodes are pressure antinodes. Explain?

24. Which is richer in harmonics, notes from a closed organ pipe or from an open organ pipe?

25. Can you hear beats when the sounding sources have a frequency difference of 200 HZ?

26. Why echo is not heard in a small room?

27. Can a stationary wave be set up when the two progressive waves have different amplitudes?

28. There are two simple pendulums of the same length and two others with different lengths. All of them are suspended from a horizontal string. One of the pendulums is moved aside and left to oscilate. What would you observe now?

PROBLEMS

1. A particle of mass 10 g executing SHM has an amplitude of 4 cm. If the frequency of vibration is 32 HZ, find its (*i*) maximum velocity and (*ii*) energy when at the mean position. [*8.038 m/s 0.323*J]

2. A particle describes SHM along a line 0.4 cm long. Its velocity at the mean position is 1 m/s. Find the frequency? [*79.618 HZ*]

3. If a second's pendulum is increased in length by 1%, how many seconds will it lose in a day? [*7 min. 12 sec.*]

4. Two bodies of mass 2, 4, 8, 16 kg are oscillating in turn on a spring of force constant 200 N/kg. Deduce the angular frequencies? [*10, 7.07, 5, 3.535 HZ*]

5. A harmonic oscillation is represented by $y=0.40$, sin $(2000\ t+0.72)$ where y and t are in mm and S, respecitively. Deduce (*i*) the amplitude, (*ii*) the frequency, (*iii*) the angular frequency, (*iv*) the period and, (*v*) the initial phase. [*0.40 mm, 2000 HZ, 1000 HZ* $\pi/$, $\pi/1000$ *S, 0.72* rad]

6. A wave of amplitude 2 cm and frequency 100 HZ is travelling in the positive direction of the x-axis with a velocity of 0.3 m/s. Calculate the displacement, velocity and acceleration of a particle of the medium situated at 1m from the origin at $t=3S$. [$-1.732\times10^{-2}m$, -6.28 m/s, 6.825×10^{3} m/s^2]

7. Plane sound waves of frequency 480 HZ in a gas have an amplitude of 0.0020 mm. Assuming the density of the gas to be 0.001426 g/cm and the velocity of sound to be 350 m/s, calculate the energy density of sound.

[51.83×10^{-3} J/m^3]

8. An addition of 2.0 kg-wt to the tension of a stretched string changed the frequency of the string to two times the original frequency. What was the original tension? [*2/3 kg-wt*]

9. On a road parallel to the railway track a man is moving in a car at 118 km/h towards north while the train is moving to the south with the same speed. If the engine is blowing a whistle at 1000 HZ, what will be the apparent frequency of the whistle to the man in the car when the engine approaches him? What will be the apparent frequency when the train goes past him ? Velocity of sound 350 m/s. [1208.2, 827.7 HZ]

10. An organ pipe has a length of 50 cm. Neglecting the end-correction, find the frequency of its fundamental and next harmonic when it is (*a*) open and (*b*) closed at one end, taking the velocity of sound in air as 350 m/s,

[(*a*) *350 HZ, 700 HZ,* (*b*) *175 HZ, 525 HZ*]

11. A string having mass per unit length m, and under tension T is stretched between two bridges. If the string betweeen the bridge is made to gener rate transverse vibrations, calculate its fundamental frequency when the distance between the bridges is l?

$$\left[\nu=\frac{1}{2l}\sqrt{T/m}\right]$$

5

LIQUIDS

5.1. Intermolecular Attraction

Matter is composed of molecules that are in perpetual motion at any temperature above absolute zero. That is how the smell of a gas spreads in a room. While trying to account for the behaviour of matter on the basis of molecular structure and motion, it was assumed that there is no interaction between molecules in the matter and the molecules themselves occupy negligible space.

The molecules in a solid are relatively close to each other. In liquids, on the average, molecules are farther apart. In a gas, of course, the distance between molecules is the largest. Therefore, the approximation that interaction is negligible could be best applied to the case of a gas. Taking into account intermolecular attraction and also introducing a correction term for volume occupied by the molecule, Vanderwall modified the equation of state of an ideal gas to account for the behaviour of a real gas. The intermolecular interaction is therefore known as Vanderwall interaction and the forces involved in the interactions are known as Vanderwall forces. The Vanderwall interaction exists in gases and liquids and in some solids also (moleculer solids, Chapter 6.) In general, in solids, the molecules are closely packed. Hence the molecular interaction is the strongest. The molecular interaction in liquids comes between those of solids and gases.

The molecular interaction forces are basically electric in nature. The atoms constituting the molecule contain equal amount of positive and negative charge and as such the net charge on the molecule is zero. But the positive and negative charges on the molecule are not evenly or symmetrically distributed. The atoms of a molecule are so arranged that the centre of mass of positive charge may not coincide with that of the negative charge. As a result, it forms an electric dipole, in which equal amount of positive and negative charges are separated by a small distance. Such dipole is associated with a dipole moment, $m=er$ where e is the the change and r is the distance between them.

The molecules like water, alcohols, aniline, acetone, aldehydes, esters, different acids, polymers etc have permanent dipole moment and are known as *polar* molecules. In molecules like O_2, N_2, CO_2, CS_2, CCl_4, benzene etc the charge distribution is symmetrical; the centre of mass of the positive and negative charges coincide and their dipole moment is zero. They are called *non-polar* molecules.

An electric dipole behaves in an electric field in the same way as a magnet behaves in a magnetic field. A dippole interact with the neighbouring dipoles in the same manner a magnet interacts with neighbouring magnets. Since the dipole-dipole interaction is stronger most of the polar molecules, remain as liquid at ordinary temperature and pressure.

Diple-dipole force is the most important cause for Vanderwalls interaction. There are two more categories of Vanderwall's interaction namely, interaction due to induced dipole forces (also called induction forces) and interaction due to dispersion forces. These two are also of dipole-dipole type, but less strong. The dispersion force, the weakest, is present in all cases and even in monoatomic gases like argon etc. However the nature of these forces is beyond the scope of our present discussion. In addition to these Vanderwall forces, there is also gravitational attraction. But it being smaller than other forces by a factor of about 10^{29}, is usually neglected.

The exact characteristic of the intermolecular forces is quite complicated. The molecular forces are attractive at large distances,

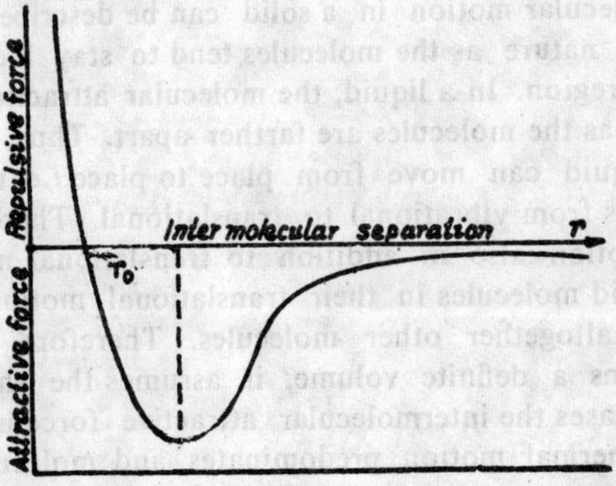

Fig. 5.1. Variation of intermolecular forces with separation.

but have to be repulsive at very close distances. In the absence of repulsive forces all matter would have collapsed to a point. When the molecules are very close to each other, the similar point charges of the atoms constituting the molecule repel each other. Thus the repulsive force is electric in origin.

The natue of intermolecular forces has been shown in Fig. 5.1. The attractive force increases rapidly with decrease in distance, passes through a maximum and then decreases to zero at a particular separation r_0. When the separation is less than r_0 the force becomes repulsive. The distance r_0, called the equilibrium distance, is of the order of 3×10^{-10} m. In this position the potential energy due to repulsion and attraction is minimum. The repulsive force when $r<r_0$, varies more sharply with decreasing intermolecular distance approximately as $1/r^9$

$$F r \propto 1/r^9$$

For larger separation, the attractive force varies inversely as the seventh power of r *i.e.* $F r \propto 1/r^7$. It decreases so rapidly that usually its effect is considered non-existant beyond $r=10^{-9}$ m. This distance is called molecular range. A sphere of 10^{-9} m radius around the molecule is called the sphere of influence of the molecule. The molecule will exert attractive force on other molecules lying within this sphere and practically have no interaction with molecules lying outside this sphere.

Besides the intermolecular forces, the molecules are also subject to thermal agitation. Their combination results in three states of matter, namely solid, liquid, and gas. In solid the strong attractive forces can overcome the thermal agitation to hold the molecules in a regular pattern and thus maintain definite volume and shape. In fact the molecular motion in a solid can be described as being vibrational in nature as the molecules tend to stay localised and remain in one region. In a liquid, the molecular attraction is comparatively less as the molecules are farther apart. Thus, the molecules in a liquid can move from place to place *i.e.* the type of motion changes from vibrational to translational. There may be vibrational motion also in addition to translational motion. Of course the liquid molecules in their translational motion are not free to leave altogether other molecules. Therefore, while the liquid maintains a definite volume, it assumes the shape of the container. In gases the intermolecular attractive force is so weak that random thermal motion predominates and molecules are almost free to move anywhere. A gas therefore has neither shape nor volume of its own but assumes those of its container when enclosed. The liquids and gases are frequently grouped as fluids, since they flow readily and do not resist shearing stress. For the sake of simplicity a liquid is sometimes treated as dense gas.

5.2. Cohesive and Adhesive Force

The attractive forces due to interaction between similar molecules are called cohesive forces. Similarly the forces between dissimilar molecules are called adhesive forces. When a glass rod is dipped into water, the molecule of glass and those of water experiences adhesion and the water molecules between themselves experience cohesion. If the rod is lifted out of water, a layer of water sticks to the surface of the glass rod. This shows that adhesion between glass and water is greater than cohesion between water molecules. On the other hand, mercury does not stick to the surface of the rod; the cohesion between mercury molecules is greater than adhesion between mercury and glass molecules.

The cohesive force in a liquid restricts the movements of molecules in a liquid so as to keep its volume constant. Therefore, to separate two molecules, energy supplied to them should be at least equal to intermolecular interaction energy. That is why, latent heat of vapourisation is supplied to a liquid to convert it to its vapour. Latent heat provides sufficient energy to increase the thermal agitation to overcome the intermolecular interaction. Latent heat of vapourisation can be used to measure the intermolecular energy in liquid molecules.

Example 5.1. The latent heat of vapourisation for water is 22.6×10^5 Joule/kg. Calculate intermolecular binding energy for water? (Avogadro's number $N_A=6\times 10^{26}$ K—mole^{-1}
1 Joule$=0.62\times 10^{19}$ ev.)

Solution. Molecular weight of water$=18$.

Therefore, the number of molecules in 1 kg of water,

$$N=\frac{N_A}{18}=\frac{6\times 10^{26}}{18}=\frac{1}{3}\times 10^{26}\ \text{kg}^{-1}$$

Given, the energy required to free molecules in 1 kg of water

$$=22.6\times 10^5\ \text{Joules}=22.6\times 10^5\times 0.62\times 10^{19}\ \text{ev}.$$

$$=1.4\times 10^{25}\ \text{ev}.$$

Inter molecular binding energy

$$=\frac{1.4\times 10^{25}\ \text{ev}}{N}=\frac{1.4\times 10^{25}\times 3\ \text{ev}}{10^{26}}=0.4\ \text{ev}.$$

5.3. Surface Phenomena

If we take a capillary tube and immerse one end of it in water, we see water rises inside the capillary tube. But if the same tube is dipped in mercury, the level of mercury inside the capillary tube is lower than the level outside. A slightly oiled steel needle placed carefully on the surface of water floats though the density of iron is nearly eight times the density of water. We also observe that when we take water in a beaker the free surface of water is horizontal in the interior, but near the surface it rises slightly. If instead of a beaker a glass tube is taken, the free surface will be concave. For a glass tube dipped in mercury the free surface of mercury inside the tube is convex. Tiny drops of liquid placed on a solid surface such as mercury on glass, assume spherical shape instead of spreading over the surface. Raindrops, fog drops etc, assume spherical shape as they fall through the air. If we dip a loop of wire in soap solution and then take it out we will find the loop holds a thin film of a solution. A liquid flowing slowly from a narrow jet such as a medicine dropper emerges as a succession of drops instead of a continuous stream. A thin glass slide cannot be easily lifted from the surface of another similar slide if there is a thin film of water between them. All these phenomena, and many other of similar nature indicates that liquid molecules at a boundary surface between a liquid and some other substance behaves differently from the liquid molecules elsewhere inside the liquid.

The adhesive and cohesive forces discussed earlier can account for this behaviour. Consider two molecules *A* and *B* of a liquid (Fig. 5.2). The dotted lines around the molecules indicates the spheres of influence of the molecules. The sphere of influence of molecule *A* is completely inside the liquid. So the force of attraction (cohesion) due to similar molecules is symmetrical on all sides; the net force of cohesion on the molecule is zero and moves inside the liquid without doing any work. All molecules inside the liquid are more or less under this condition. But the situation is different for the molecule *B* on the surface. A part of its sphere of influence lies inside the liquid and the rest lies outside the liquid. There is negligible attractive force from above the surface by the molecules in liquid vapour. So the net force of attraction on *B* is directed inside the liquid only. All such forces can be resolved into horizontal components and

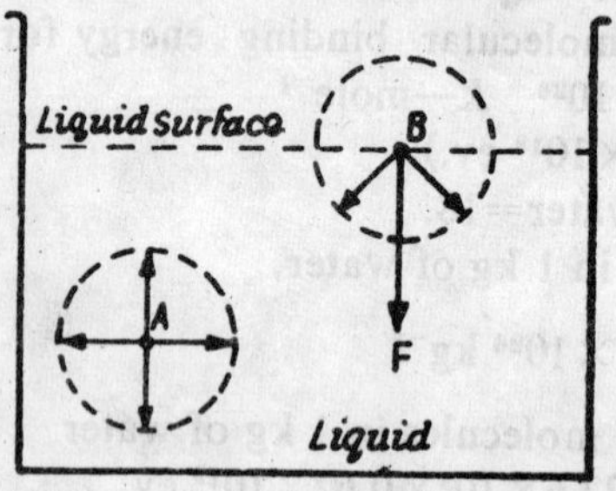

Fig. 5.2. Forces on molecules in a liquid (vertical section). The spheres in influence and molecules have been enlarged for clarity. F—Net downward force on a molecule on the surface.

vertically downward components. For reasons of symmetry net horizontal force is zero,* but there is a net resultant downward force on *B*. Every molecule on the surface is likewise acted upon by a vertically downward pull inside the liquid.

Let us explain the phenomena in more detail. To remove the molecule *A* from inside we have to do work against the attraction of all its neighbours. Since work has to be done on the molecule against all these forces to remove it from liquid, the molecule inside the liquid will have quite a large negative potential energy, say E_A. But the molecule *B* on the surface is surrounded by less number of liquid molecules. Hence less work will have to be done to remove this molecule. Thus the molecule *B* possesses less negative energy say $-E_B$.

where $|E_B| < |E_A|$ or $-E_B > -E_A$

This means the energy of the molecules on the surface is positive as compared to the energy of the molecules inside. It is known that physical system always tends to attain a state of minimum potential energy. Therefore the surface of a liquid tends to contract as much as possible (reducing the number of molecules with highest potential energy). This is also the reason for a liquid drop tending to have a spherical shape as given volume of liquid has least surface area in the form of a sphere.

5.4. Surface Tension

We have seen that the surface of a liquid tends to contract as much as possible. In other words, the surface of a liquid behaves like a stretched membrane and there exists a tension in the surface of liquid. A simple experiment can be performed to demonstrate the existence of this tension. Take a thin metallic wire and bend it into a form, say a circular ring. Tie a thread with a small loop across the wire frame (Fig. 5.3). Then dip the ring into soap solution. On taking it out a thin film of the solution will cover the whole frame and the loop may be of any shape (Fig. 5.3*a*). If, however, the film is ruptured inside the loop, the loop immediately becomes circular in shape (Fig. 5.3*b*).

It is evident, therefore, that the surface film exerts a force or tension perpendicular to any imaginary line drawn on the surface of the film. If the surface is flat, the force across the line in one direction is just equal in magnitude to the force in opposite direction and there is equilibrium as in Fig. 5.4. But as the film is ruptured inside the loop, the force inside the loop disappears, and the unbalanced forces act as indicated in Fig. 5.3*b*. This tension

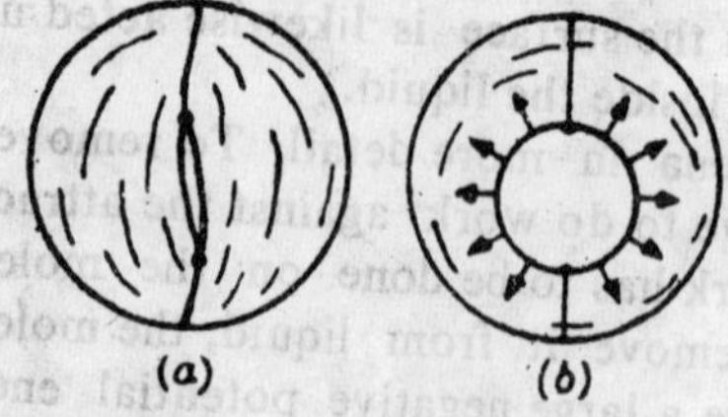

Fig. 5.3. Force on the surface film at right angles to the boundary.

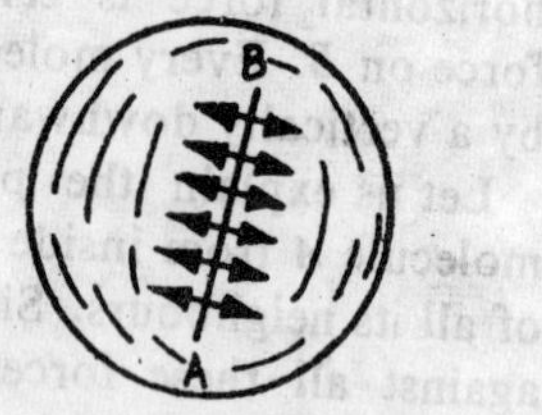

Fig. 5.4. Force of surface tension across an imaginaryline in the surface of a liquid.

measured in a stretched film across unit length of an imaginary straight line on the surface is termed as surface tension. Therefore, the *surface tension* of a liquid is defined as the force of attraction acting perpendicular to unit length of an imaginary line drawn tangential to the surface of the liquid.

$$T=\frac{F}{L} \qquad \ldots(5.1)$$

where F is the force across the line of length L. The unit of the surface tension is dynes per cm in CGS units and Newton per meter in SI units. Since the forces involved are small, usually CGS unit is used for expressing surface tension. It can be shown that the SI unit of surface tension is 10^3 times its CGS unit. The dimension of surface tension are MT^{-2}.

The surface tension of a liquid ordinarily refers to its surface tension when it is exposed to air. The surface tension is a characteristic constant of the two surfaces depending on their physical conditions. Its value decreases with rise of temperature.

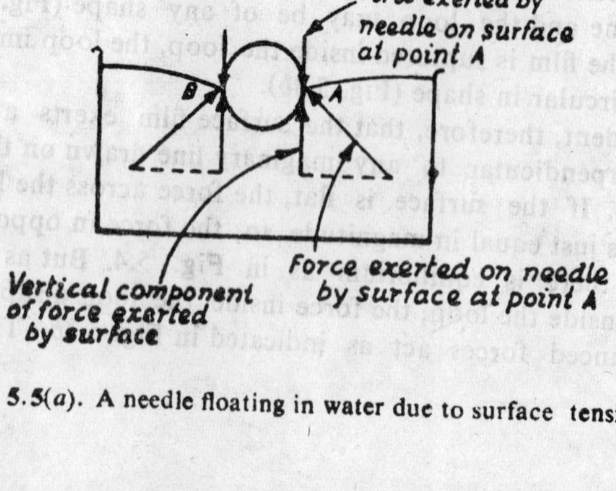

Fig. 5.5(a). A needle floating in water due to surface tension.

Now we can explain how a needle can float in water. The surface under the needle is depressed slightly so that the film is able to exert an upward force (Fig. 5.5*a*). The vertical component

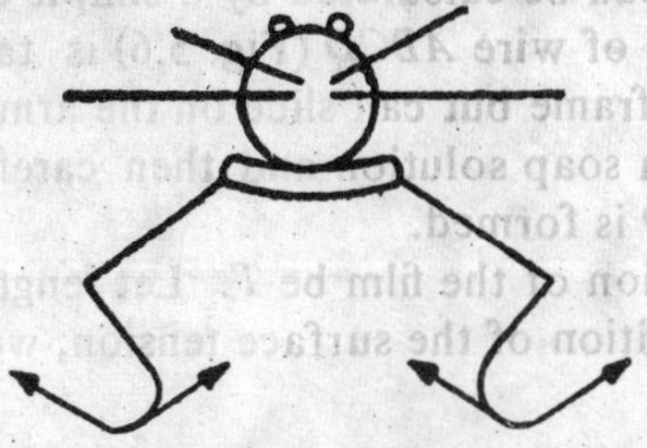

Fig. 5.5. (*b*) **A bug walking on water due to force of surface tension.**

supports the weight of the needle. A bug also floats on water due to surface tension. The bug bends (Fig. 5.5*b*) its legs on the surface of water. There is slight depression on the water surface. The force of surface tension as indicated in Fig. 5.5*b* act on the surface of liquid. The vertical components of these forces support the bug.

TABLE 5.1. Surface Tensions of some Liquid in Contact with Air.

Liquid	*Temperature* *K*	*T-Newton/Meter* $\times 10^{-3}$
Water	293	72.8
Aniline	293	42.9
Benzene	293	27.6
Ethyl alcohol	293	22.3
Methyl alcohol	293	22.6
Mercury	293	475
Glycerine	293	63
Liquid lead	673	445

5.5. Surface Tension and Surface Energy

While dealing with surface phenomena, we have seen that a liquid molecule on the surface of the liquid has more potential energy than a similar molecule inside. Therefore to transfer a liquid molecule from inside to the surface of the liquid, some amount of work must be done on the molecule. But if a surface is to be created, number of molecules from inside must be transferred to the surface. Consequently, the potential energy of the surface shall increase. *The energy required to create a surface is known as surface energy.*

The surface energy can be calculated by a simple experiment.

A rectangular frame of wire *ABCD* (Fig. 5.6) is taken. The arm *BC* is not fixed to the frame but can slide on the arms *AB* and *DC*. The frame is dipped in soap solution and then carefully lifted so that a thin film *ABCD* is formed.

Let the surface tension of the film be T. Let length of BC be l.

Then from the definition of the surface tension, we get, the force

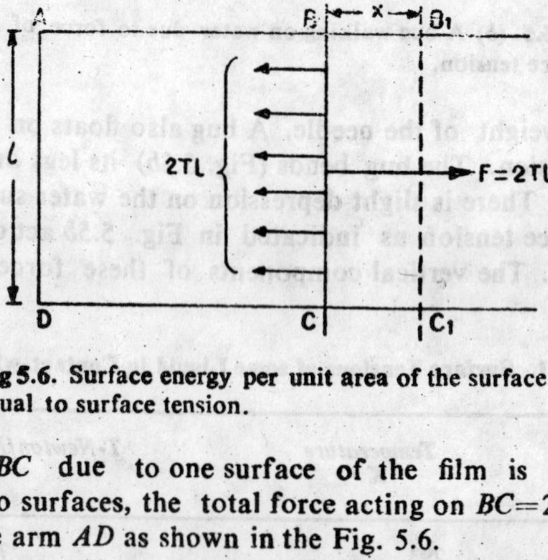

Fig 5.6. Surface energy per unit area of the surface is equal to surface tension.

acting on *BC* due to one surface of the film is Tl. Since the film has two surfaces, the total force acting on $BC=2\,Tl$ directed towards the arm *AD* as shown in the Fig. 5.6.

If *BC* is moved to a new position B_1C_1 through a distance X, then additional work has to be done against this force. The work done in creating the new surfaces is given by:

$$W=F\,.\,X=2TlX \qquad ...(i)$$

This work done in increasing the surface is stored as potential energy of the surface. So W is the surface energy of the additional surface. But the area of the new surfaces

$$=2lX \qquad ...(ii)$$

Thus, $W=T\times$ increase in surface area. If increase in surface area is one unit, then $W=T$ ...(*iii*)

Hence from (*iii*), we find surface energy per unit area is equal to surface tension, T. The surface energy can therefore be expressed in Joules/m² in SI units or erg/cm² in CGS units. These units are equivalent to N/m or dyne/cm.

The force acting on wire *BC* can be measured by a hydrostatic balance (Fig. 5.7). In this arrangement *BC* is fixed to the frame unlike the arrangement in Fig. 5.6 where *BC* could slide. The frame is suspended from one arm of the balance and counterpoised by equal weights on the otherside. Then a container with the liquid is kept on the hydrostatic bench below *BC* without touching it. The bench is slowly raised till liquid just touches *BC*. Then it will be observed that balancing will be disturbed; the side *BC* will be pulled down. To restore the balance, we have to add weights on the other side. In principle, this additional weight is the force,

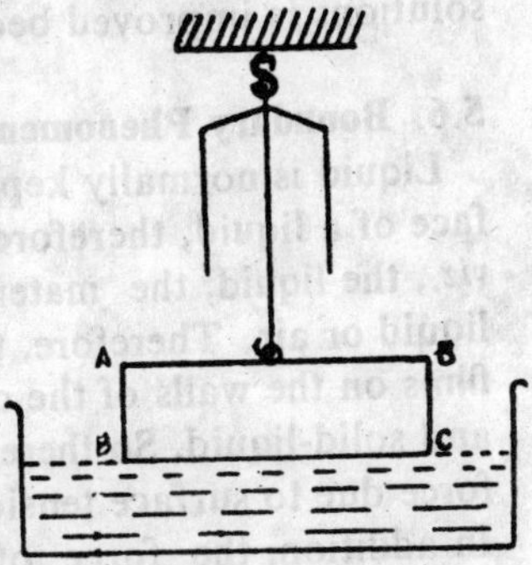

Fig. 5.7. Measuring the force due to surface tension.

$F=2Tl$ due to surface tension.

But in practice, the determination of this weight gives an approximate value only as liquid particles sticking to *BC* will indicate greater weight than the value necessitated by surface tension alone. Hence, after determining the approximate weight, we may proceed for a more accurate determination. To start with, the system may be kept in unbalanced condition, the side *BC* being lighter by the approximate weight determined earlier. The liquid is brought very close to *BC* till it touches liquid surface. The side *BC* will be pulled down but it will remain still lighter. Carefully remove weights from the other side till both sides are balanced.

From these observations, one can calculate the exact value of weight to counteract the force due to surface tension.

When two small globules of mercury are brought into contact, they form a bigger globule. This happens because when both globules unite, the surface area of the bigger globule is less than the combined surface area of individual globules. So the surface energy of the resultant big globule is less then the combined surface energy of the small globules. As every physical system tries to attain a state of minimum energy the globules of mercury coming into contact results in a single globule.

A liquid with lower surface tension spreads easily than a liquid with higher surface tension as the spreading is easier due to low surface energy. Therefore, a liquid with low surface tension can penetrate into narrow spaces or pores. Hence liquids like benzene, CCl_4 etc are better cleaning agents.

Water has relatively high surface tension and soap solution has much less surface tension. Detergent added to water also reduces the surface tension appreciably. The cleaning property of such solutions is improved because of this factor.

5.6. Boundary Phenomena

Liquid is normally kept in a vessel. At the boundary of the surface of a liquid, therefore, there are three substances in contact *viz.*, the liquid, the material of the vessel and the vapour of the liquid or air. Therefore, there are three different types of surface films on the walls of the container *viz.*, liquid-vapour, solid-vapour and solid-liquid. So there is a surface effect at each interface. *viz.*, force due to surface tension which acts parallel to each surface. In addition, the force of adhesion between the molecules of the liquid and those of the solid (vessel) acts perpendicular to the walls of the vessel. The curvature of liquid surface touching the solid depends on the combined effect of these four forces. The shape of a liquid drop on a solid surface is also regulated in the same manner. All the three surface films meet on the walls of the vessel. If we choose an elementary area of the film common to all of them, the portion must be in equilibrium under the action of the four forces mentioned earlier. Therefore, the liquid surface at the junction will have such a curvature that this condition *i.e.* equilibrium under the action of forces is realised.

In Fig. 5.8 (*a*) all these forces are shown when water is kept in a glass vessel. Surface tension at solid-vapour interface, T_{SV} and

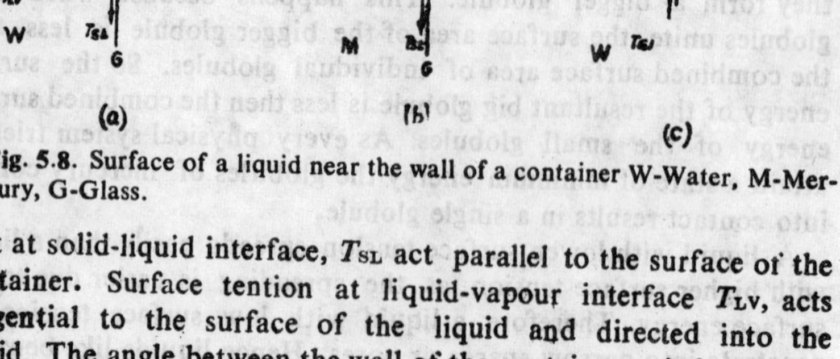

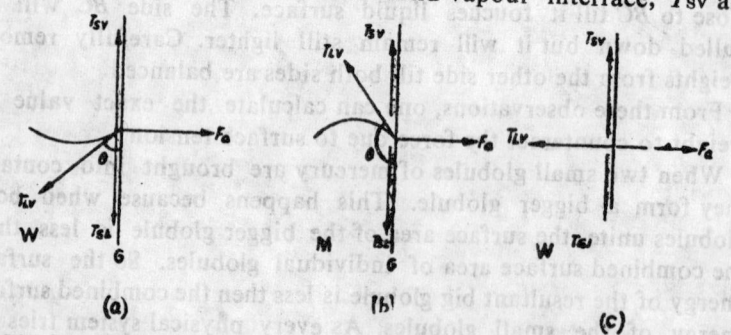

Fig. 5.8. Surface of a liquid near the wall of a container W-Water, M-Mercury, G-Glass.

that at solid-liquid interface, T_{SL} act parallel to the surface of the container. Surface tention at liquid-vapour interface T_{LV}, acts tangential to the surface of the liquid and directed into the liquid. The angle between the wall of the container and the tangent drawn to the surface of the liquid at the point of contact with the wall is called *angle of contact*. In Fig. 5.8 (*a*), θ is the angle of contact between glass and water taken inside the liquid. Since under equilibrium conditions there should not be any net force either in horizontal or vertical direction, we get

$$T_{LV}\sin\theta = F_a \quad \text{...(i)}$$

$$T_{LV}\cos\theta = T_{SV} - T_{SL} \quad \text{...(ii)}$$

From (*ii*) we get:

$$\cos\theta = \frac{T_{SV}-T_{SL}}{T_{LV}} \quad \text{...(iii)}$$

If $T_{SV} > T_{SL}$, cos θ is +ve, so θ lies between 0° and π/2. In this case (Fig. 5.8*a*) the liquid wets the surface of solid. The meniscus

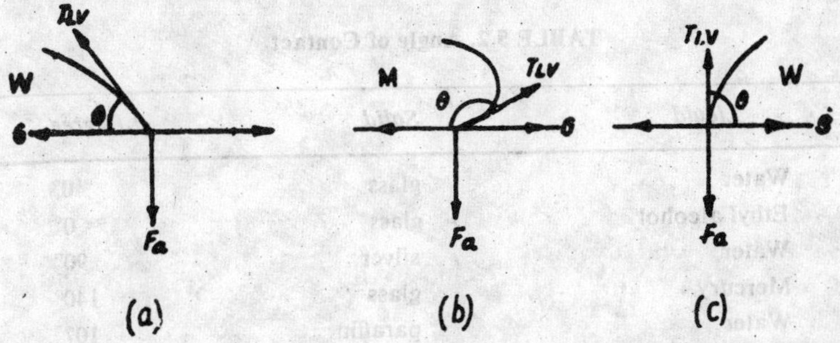

Fig. 5.9. Drop of a liquid on a solid surface. W-Water, G-Glass, M-Mercury, S-Silver.

is concave. If a drop of liquid is put on the surface, the liquid tends to spread (Fig. 5.9 *a*). If $T_{SV} < T_{SL}$, cos θ is —ve and θ lies between π/2 and π (Fig. 5.8 *b*). In this case liquid does not wet the solid. The liquid meniscus is convex. A small drop of liquid put on the solid does not spread in this case (Fig. 5.9 *b*).

However, when $T_{SV} = T_{SL}$ cos θ=0, so θ=π/2 (Fig. 5.8 *c*). The liquid meniscus is horizontal and the drop takes a semispherical shape (Fig. 5.9*c*).

It will be shown later that for a pair of liquid-solid, for which θ is acute, the liquid rises inside a narrow tube made of the corresponding solid when the tube is dipped in the liquid. But if θ is obtuse, the liquid level is lowered inside the tube and for θ=90°, there is neither rise nor fall of liquid level inside the tube.

It is to be noted that if $T_{SV} > (T_{LV} + T_{SL})$ the value of cos θ is greater than unity (from *iii*) as shown below.

Since $T_{SV} > (T_{LV} + T_{SL})$

$$(T_{SV} - T_{SL}) > T_{LV}$$

$$\frac{T_{SV}-T_{SL}}{T_{LV}} > 1$$

$\therefore \quad \cos\theta > 1.$

This is not possible and hence there will be no equilibrium in such a case. The liquid in this case spreads completely over the solid surface.

In systems like glass-water, glass-alcohol etc, the angle of contact is acute and the meniscus is concave. The angle of contact between clean glass and pure water is 0°, but if the surface of the glass is greased or water contains impurities, the angle increases considerably. But for glass-mercury, the angle of contact is obtuse and the liquid meniscus is convex. But for water-silver, angle of contact is 90° and the water miniscus is horizontal.

TABLE 5.2. Angle of Contact.

Liquid	*Solid*	*θ,Degree*
Water	glass	0°
Ethyl alcohol	glass	0°
Water	silver	90°
Mercury	glass	140°
Water	paraffin	107°
Glycerine	glass	0°

5.7. Pressure Difference Across a Surface Film

The surface tension acts tangential to the surface of liquid. If the free surface is plane (Fig. 5.10 *a*) the forces due to surface tension act on the horizontal plane and can cancel each others effect. Hence no net force act on the molecules of the liquid. But the situation is different if the liquid surface is curved. Fig. 5.10 (*b*) shows a surface which is convex upwards. Force of surface tension indicated as T can be resolved into horizontal and vertical components. The horizontal components will balance out each other. But the vertical components add together so that there is a resultant force normal to the surface and directed into the liquid. Therefore, the pressure on the liquid side of the surface must be greater than the pressure on the vapour side in order to keep the surface in equilibrium. *i.e.* $P_L > P_V$ (otherwise, the surface should continue to be depressed till it is horizontal under the action of the resultant force). Similarly, it can be shown from Fig. 5.10 (*c*) that the pressure on the vapour side is greater than the pressure on liquid side *i.e.* **In both cases, we find that the pressure on the concave side of the film is always greater** than the pressure on the convex side. This applies to any curved liquid surface.

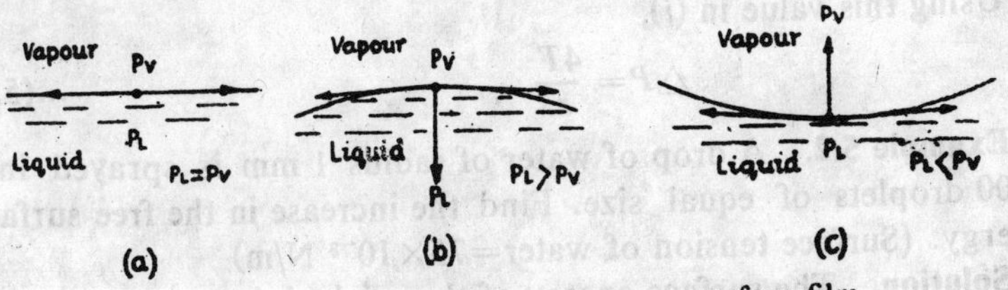

Fig. 5.10. Pressure difference across a surface film.

Let us consider the case of a spherical liquid drop of radius r. The pressure inside the spherical drop will be greater than the outside atmospheric pressure. Let the excess pressure inside the drop be $\triangle P$. Due to this excess pressure, a force acts on each element of the surface of the drop in a direction perpendicular to the element of surface (Fig. 5.11). If we take a cross section of (shaded portion of Fig. 5.11) the drop through the centre, the area of cross section is πr^2. So the resultant perpendicular force on this area due to excess pressure $\triangle P$ is given by,

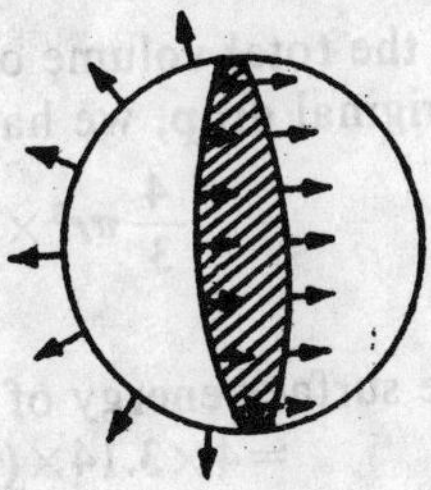

Fig. 5.11. Excess pressure inside a droplet.

$$F = \triangle P \times \pi r^2 \qquad \left(\because \text{Pressure} = \frac{\text{Force}}{\text{Area}}\right) \qquad \ldots(i)$$

This force is provided by the force of surface tension pulling across the perimeter of this cross section *i.e.*, $2\pi r$. If surface tension is T,

$$F = 2\pi r T \qquad \ldots(ii)$$

From (*i*) and (*ii*)

$$\triangle P \times \pi r^2 = 2\pi r T$$

or

$$\triangle P = 2T/r \qquad \ldots(5.2)$$

From Equation 5.2 it follows that the excess pressure is larger for smaller spheres when the surface tension is constant. Due to this reason it is easier to divide a big drop as compared to further division of small drops. A very tiny drops offer considerable resistance to further division.

The above discussion applies to spherical bubbles also. Since there is air inside a soap bubble there are two surfaces; one external and the other internal. So, the perimeter of the cross section is $2\times 2\pi r=4\pi r$ and as such the magnitudes of force given in (*ii*) is,

$$F=4\pi r.T$$

Using this value in (*i*),

$$\triangle P=\frac{4T}{r} \qquad ...(5.3)$$

Example 5.2. A drop of water of radius 1 mm is sprayed into 1000 droplets of equal size. Find the increase in the free surface energy. (Surface tension of water $=75\times 10^{-3}$ N/m)

Solution. The surface energy of the original drop

$=$Surface area $\times$ surface tension $=4\pi R^2T$ Joules.

Here $R=1\times 10^{-3}$ m, $T=75\times 10^{-3}$ N/m

So surface energy of the drop $=4\times 3.14\times 10^{-6}$ m$^2\times(75\times 10^{-3}$ N/m)

$=0.942\times 10^{-6}$ J.

Since the total volume of the 1000 droplets is the same as that of the original drop, we have,

$$\frac{4}{3}\pi r^3\times 10^3=\frac{4}{3}\pi(1\times 10^{-3})^3$$

or $$r=10^{-4}\text{ m}$$

Free surface energy of the 1000 droplets $=4\pi r^2\times 1000\times T$

$=4\times 3.14\times(1\times 10^{-4}\text{m})^2\times 10^3\times 75\times 10^{-3}$ N/m

$=9.42\times 10^{-6}$ J.

Hence increase in surface energy

$=(9.42-0.942)\times 10^{-6}$ J $=8.478\times 10^{-6}$ J.

Example 5.3. Calculate the amount of energy spent in enlarging a soap bubble of diameter 8 mm to one of diameter 1 cm. Surface tension of soap solution is 25×10^{-3} N/m.

Solution. Since it is a bubble, it has two surfaces.

Increase in surface area $=2\times 4\pi$ $(r_2^2-r_1^2)$ where r_1 and r_2 are the initial and final radii.

Here $r_2=1\times 10^{-2}$ m, $r_1=0.8\times 10^{-2}$ m.

So increase in area $=8\times 3.14\times(1\times 10^{-4}$ m$^2-0.64\times 10^{-4}$m^2)

$=9.0432\times 10^{-4}$m^2.

So the energy spent to increase in surface energy

$=$increase in surface area $\times$ surface tension.

$=(9.0432\times 10^{-4}\text{m}^2)\times(25\times 10^{-3}\text{ N/m})=22.61\times 10^{-6}$ J.

5.8. Capillarity

It is a common experience that when one end of a capillary tube is dipped in water, water rises to a certain height in the capillary tube. The narrower the tube, the greater is the height to which water rises. In fact, the origin of the term 'capillarity' lies in the nature of such tube *viz.*, capillary or hair like. In liquids like water that wet the glass the angle of contact is less

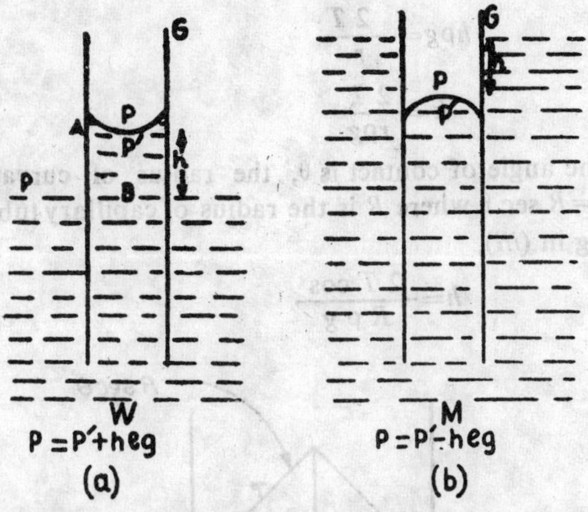

Fig. 5.12. (*a*) Rise of liquid level in a capillary tube. (*b*) Depression of liquid level in a capillary tube G-glass, W-water, M-mercury.

than 90° and liquid rises inside the tube. But in liquids like mercury which does not stick to glass the angle of contact is more than 90° and level of mercury inside the tube is lower than the level outside. Both the effects are due to surface tension.

Consider the capillary tube dipped in water, Fig. 5.12(*a*). Water rises to a height *h* inside the tube before equilibrium is reached. The liquid meniscus is concave, the pressure on the convex side of the meniscus (inside the liquid) is less than the pressure in its concave side (atmospheric pressure.). This condition is satisfied with rise of water in the tube as explained below.

Let the atmospheric pressure be P. Then the pressure inside the liquid at B (at the level of the free surface of liquid) is also P.

Consider the pressure inside the liquid at a point just near the meniscus. Let it be P^1. If the height of this point from free surface of liquid *i.e.*, from B, is h, then

$$P - P^1 = h\,\rho g \qquad ...(i)$$

Where ρ is the density of the liquid and g is the acceleration due to gravity.

But from Equation 5.2, the excess pressure on concave sides of the meniscus is,

$$\Delta P = \frac{2T}{r} \qquad ...(ii)$$

where r is the radius of curvature of curved surface.

From equation (*i*) and (*ii*),

$$h\rho g = \frac{2\,T}{r}$$

or

$$h = \frac{2\,T}{r\rho g} \qquad \ldots(iii)$$

When the angle of contact is θ, the radius of curvature of the surface, $r = R \sec \theta$ where R is the radius of capillary tube (Fig. 5.13), substituting in (*iii*),

$$h = \frac{2\,T \cos}{R\,\rho\,g} \qquad \ldots(5.4.)$$

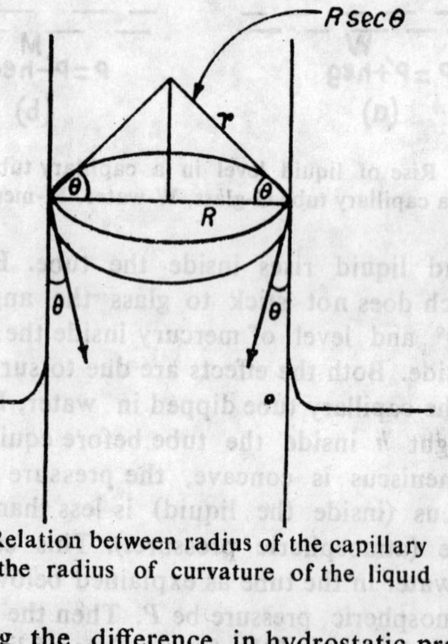

Fig. 5.13. Relation between radius of the capillary tube and the radius of curvature of the liquid surface.

While considering the difference in hydrostatic pressure inside the liquid *i.e.* $h\rho g$, we have not taken into account the liquid in the curved portion of the meniscus (shaded portion in Fig. 15.2*a*). For a narrow tube, it can be shown that the hydrostatic pressure difference is nearly $(h+R/3)\,\rho g$ instead of $h\rho g$. Thus substituting $(h+R/3)$ in place of h in equation 5.4 and rearranging,

$$T \cos \theta = \frac{R\rho g}{2}\,(h+R/3) \qquad \ldots(5.4\,a)$$

However, for narrow tubes the correction term is very small, hence Equation 5.4 is used in most cases.

Therefore, we can determine the surface tersion of a liquid if we know the angle of contact or vice-versa. A method based on this principle is described in the next article.

Equation 5.4 is also applicable for capillary depression such as in case of mercury in glass. Since the meniscus is convex, pressure P^1 inside the liquid is higher than the outside atmospheric pressure, the hydrostatic pressure difference, $P^1 - P = h\rho g$ [Fig. 5.12(*b*) (compare with (*i*) above]. Equating this with $2T/r$, leads to Equation 5.4.

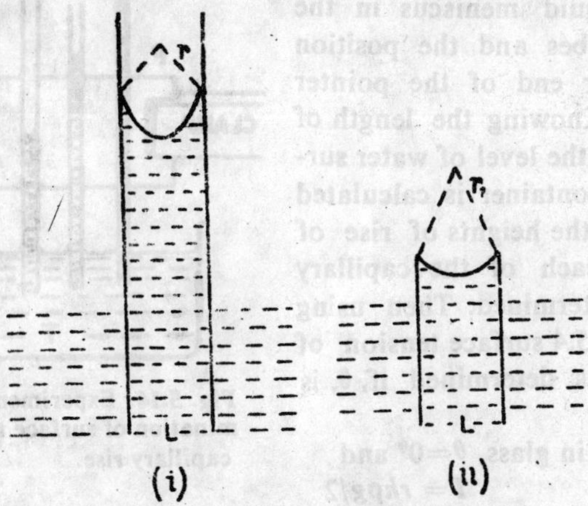

Fig. 5.13 (*a*). (*i*) Tube of sufficient height.
(*ii*) Tube of insufficient height.

The phenomenon of rise of underground water through capillary roots of a tree, soaking of ink by blotting paper and rise of oil in a wick stove etc are familiar examples of capillarity. It may be noted that if the length of the tube is less than h given by Equation 5.4, liquid rises only up to the top of the tube. At the top, liquid spreads so that the radius of curvature 'r' increases to a higher value say, r_1 so that the pressure difference $2T/r$ reduces to $2T/r_1$. This reduced pressure difference is equal to the hydrostatic pressure difference $h_1 \rho g$ where h_1 is the length of the tube ($h_1 < h$).

5.9. Experimental Determination of Surface Tension by Capillary Rise

At least three thickwalled capillary tubes are taken and they are cleaned and dried thoroughly. Their radii are measured by weighing a thread of mercury of known length in each of tube using the relation $m = \pi R^2 \rho l$
where R is the radius of the tube, ρ is density of mercury and l is the length of mercury thread.

The capillary tubes and a sharp needle, N, are fixed on a plane metal plate so as to keep them parallel to each other. The plate is fixed to a clamp (Fig. 5.14) in such a manner that the tubes and the needle N, are held perfectly vertical. The lower ends of capillary tube dip inside the liquid kept in a container while the lowerend of the needle, N, just touches the surface of the liquid. With the help of a travelling microscope the level of liquid meniscus in the capillary tubes and the position of the upper end of the pointer are noted. Knowing the length of the pointer, the level of water surface in the container is calculated from which the heights of rise of liquid in each of the capillary tubes are determined. Then using the relation 5.4 surface tension of the liquid is determined if, θ, is known.

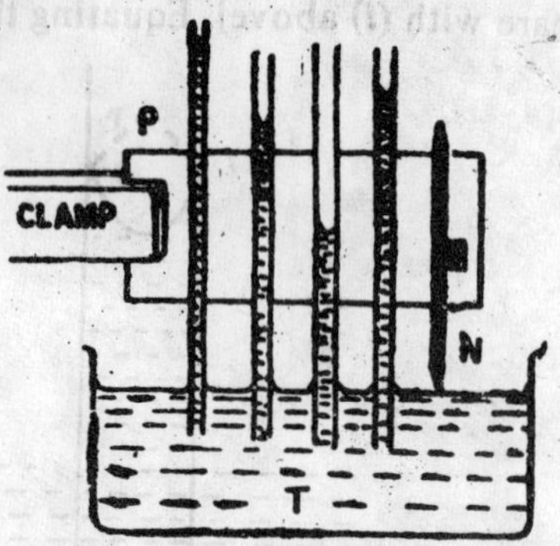

Fig. 5.14. Experimental determination of surface tension by capillary rise.

For water in glass, $\theta = 0^0$ and

$$T = rh\rho g/2$$

Since all quantities are measured, the value of T is calculated.

Example 5.4. The tube of a mercury barometer is 4 mm in diameter, What error is introduced into the reading because of surface tension ? Given, angle of contact for mercury with glass $= 140°$, and surface tension of mercury in glass$= 465 \times 10^{-3}$ N/m.

Solution. For $\theta = 140°$, $\cos\theta = -.766$

The capillary rise of mercury in the tube

$$= \frac{2T\cos\theta}{r\rho g} = \frac{2 \times (465 \times 10^{-3}\ \text{N/m})\ (-.766)}{2 \times 10^{-3}\text{m}\ (13{\cdot}6 \times 10^{3}\ \text{kg/m}^3)\ (9.8\ \text{m/s}^2)}$$

$$= -.267 \text{ cm.}$$

The negative sign indicates that there is depression. Hence the reading of the barometer is lower than the correct reading by this amount.

5.10. Flow of Liquids

Steady and turbulent. The flow of liquid is a complex problem. A steady flowing river turns turbulent with eddies and whirlpools during floods. So the liquid flows differently under different conditions. But careful consideration shows that the flow of liquids can be described in terms of the familiar principles of mechanics.

The flow of liquid can either be *steady* or *unsteady*. Let us first consider steady flow. For such a flow the velocity at any point remains constant both in magnitude and direction. In Fig. 5.15 an example of steady flow is shown. For all particles passing through the point '*A*', the velocity is $\vec{v}_a$. So when a particle leaves the point with velocity $\vec{v}_a$, a second particle comes to that position and leaves with velocity $\vec{v}_a$. In this manner, a steady stream of particles pass through the point '*A*' with velocity $\vec{v}_a$. Similarly all particles cross the point *B* with velocity $\vec{v}_b$, point *c* with velocity $\vec{v}_c$ etc. Thus in a steady flow, the motion of particles is predicatable. If a liquid particle after passing the point *A* has gone to *B* and then to *C*, then the particle following it will also follow the same path. The particle preceeding it also followed the same path.

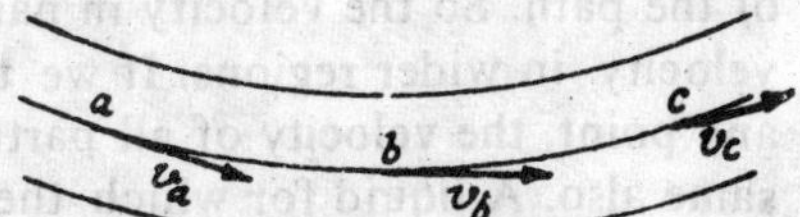

Fig. 5.15. Steady flow of liquid. Path *a b c* is a stream line.

The line *ABC* which represents one such path followed by a particle is called a *stream line*. It represents the fixed path followed by steady flow of particles and tangent to it at any point gives the velocity of liquid, at that point. Thus a stream-line is parallel to the velocity of liquid particles. It is evident therefore, that two stream-lines can not cross each other. The steady or *stream-line motion* is possible only when the liquid velocity is low, and there is no sudden change in the velocity of flow along the path. If the liquid flows with very high velocity or there is frequent or sudden changes in velocity, the flow is unsteady or *turbulent*. In this case, the path of the particle changes contineously; there are eddies and, whirlpools in the liquid.

Sometimes, it is found useful to introduce the idea of tube of flow. The whole region in which flow occurs is imagined to be divided into tubes and the streamlines make the surface of such a tube Fig. 5.16. In other words, a tube of flow of is made up of a bundle of stream-lines of flow. The liquid in a tube of low remains confined to that tube. It is assumed that all particles passing a given cross section in a tube of flow has the same velocity. In reigons where liquid flows through a constriction, the streamlines are crowded together, the tube of flow becomes narrower and the speed of flow is increased.

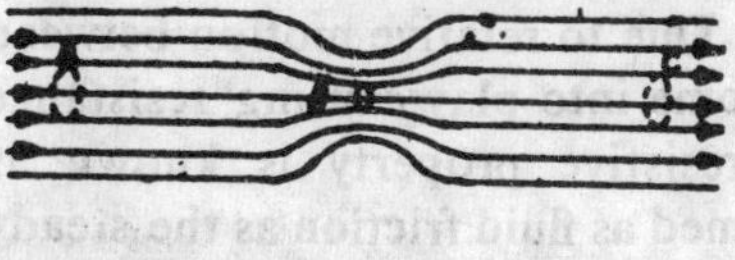

Fig. 5.16. Tube of flow. Volume *ABC* is a stream line.

5.11. Viscosity

In a steady flow there is no accumulation of liquid at any point of the path. So the velocity in narrow regions is greater than the velocity in wider regions. If we take a cross section of the path at any point, the velocity of all particles on this section need not be same also. A liquid for which the speed of all particles in the section are same, is a *non-viscous* liquid. But in all common liquids, it is observed that particles in a cross section move with different velocities. If the flow is through a pipe, velocity of particles is maximum at the centre of the section and decrease towards the wall. Fig. 5.17 shows a velocity profile of flow in a pipe in both cases. Fig. 5.17 (*a*) shows the case where the velocity of all particles on the section are the same and liquid advances as one rigid unit along the pipe. The heads of velocity vectors are in one plane and the flow of liquid may be determined by the advancement of the head of any one velocity vector.

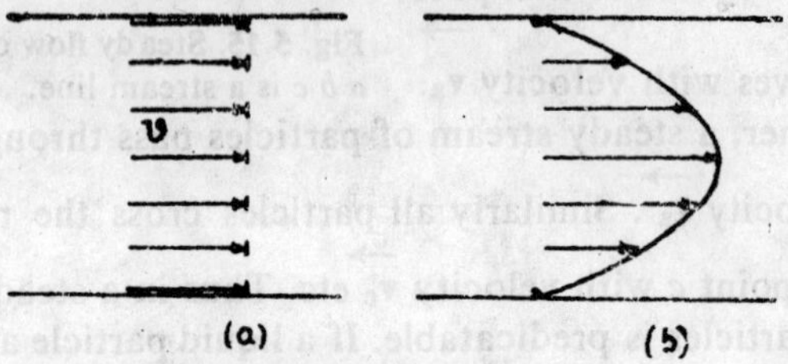

Fig. 5.17. Velocity profile of flow (*a*) Non-viscous liquid (*b*) Viscous liquid.

But in Fig. 5.17 (*b*), situation is different. The velocity profile is no more plane, but Curvilinear as the velocity is highest along the axis and minimum along the walls. Thus liquid can be imagined to be divided into different layers or laminas and each such layer or lamina advances with different velocities. This type of flow is called *laminar flow*. Such a liquid is called a *viscous liquid*.

As the liquid layers move with different velocities, it is evident, they move relative to each other. Due to relative motion between different layers, frictional forces come into play causing resistance to the motion of the liquid. This resistive property is known as viscosity. Thus viscosity can be termed as fluid friction as the steady flow of a liquid and gas follow the same principles. Therefore, an external force must be exerted to overcome this resistance so that a liquid layer slides over another with a steady motion. In a non-viscous liquid, such a force is not necessary.

The reistance offered to relative motion of liquid layers *i.e.*, the viscous properties of all liquids are not same. Consider the motion

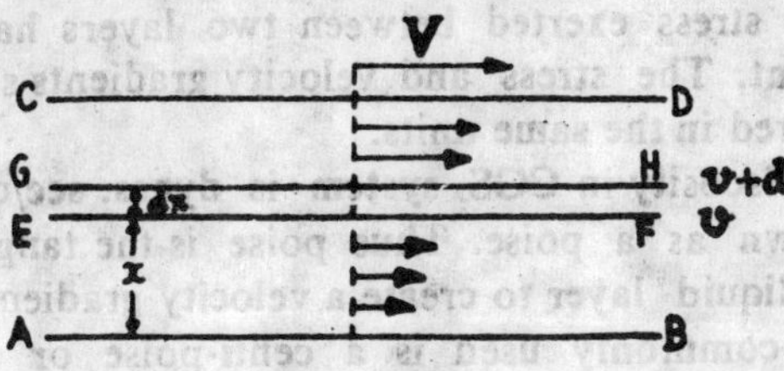

Fig. 5.18. Flow of viscous liquid over a surface.

of a liquid between two parallel plates *AB* and *CD* (Fig. 5.18) where *AB* is fixed and *CD* is moving with a certain velocity $\vec{v}$. If the motion of the liquid is slow and steady, it will be in a direction parallel to *AB*. The layer in contact with *AB* will be at rest and that in contact with *CD* will be moving with the velocity of *CD*.

The intermediate layers will move with velocity proportional to their distances from *AB* as shown in Fig. 5.18. Suppose the particles in the layer *EF* at a distance x from *AB* move with the velocity v and those in the layer *GH* at a distance 'dx' from *EF* move with a velocity $(v+dv)$. Hence the particles in the layer *EF* having comparatively smaller velocity tends to retard the motion of particles of *GH* layer moving with greater velocity. Thus a retarding force acts on *GH* layer, This applies to all such layers to which liquid is divided. This is equivalent to the statement that the flow of liquid is opposed by a tangential force F offered along the surface of the layer.

Consider the layers *EF* and *GH*. The velocity gradient between these two layers is dv/dx. If P is the tangential force acting on an area A of the layers in contact, the tangential stress is F/A.

The tangential stress according to Newtons hypothesis (confirmed by experiment) is proportional to the velocity gradient dv/dx in a direction perpendicular to the direction of motion.

$$\text{Thus} \qquad \frac{F}{A} \propto \frac{dv}{dx} \qquad \text{or} \qquad \frac{F}{A} = \eta \frac{dv}{dx}$$

$$\text{or} \qquad F = \eta A \frac{dv}{dx} \qquad \ldots(5.5)$$

where the constant of proportionality, η (eta) is called the coefficient of viscosity or simply the viscosity of the liquid. From Equation 5.5 we get

$$\eta = \left(\frac{F}{A}\right) \Big/ \left(\frac{dv}{dx}\right)$$

If dv/dx is one unit, $\eta = F/A$. The viscosity, therefore, is equal to the tangential stress exerted between two layers having a unit velocity gradient. The stress and velocity gradients should, however, be measured in the same units.

The unit of viscosity in CGS system is dynes. sec/cm^2 which is generally known as a poise. Thus poise is the tangential stress offered by a liquid layer to create a velocity gradient of 1 cm/s. The unit most commonly used is a centi-poise or cp *i.e*, 0.01 poise. The unit of viscosity in SI units is NS/m^2. Since this unit is quite high, normally it is not used.

It can be shown one unit of viscosity in SI system is equal to 10 poise, The dimensions of η are calculated easily.

$$\text{Dimensions of } \eta = \frac{\text{dimensions of } (F/A)}{\text{dimensions of } dv/dx}$$

$$= \left[\frac{MLT^{-2}}{L^2}\right] \Big/ \left[\frac{L}{T.L}\right] = ML^{-1}T^{-1}.$$

Viscosity is a characteristic constant for the liquid depending on its physical conditions. With the rise of temperature viscosity decreases for liquids.

The viscosity of some liquids are given in Table 5.3.

TABLE 5.3. Viscosity of some Liquids at 20°C.

Liquids	*Viscosity in Cp*
Alcohol (Ethyl)	1.200
Benzene	0.652
Glycerene	1.490
Mercury	1.554
Methyl alcohol	0.597
Castor oil	986
Olive oil	36.3
Water	1.002

5.12. Viscosity and Flow of Liquid

It is evident from earlier discussion that the viscosity of a liquid is related to the velocity gradient and the tangential force F acting along a layer of the liquid to oppose its motion. When the flow is laminar, therefore, a tangential force of equal magnitude should be provided to overcome this force so that laminar flow is maintained. This force is provided by hydrostatic pressure difference between the two ends of a tube.

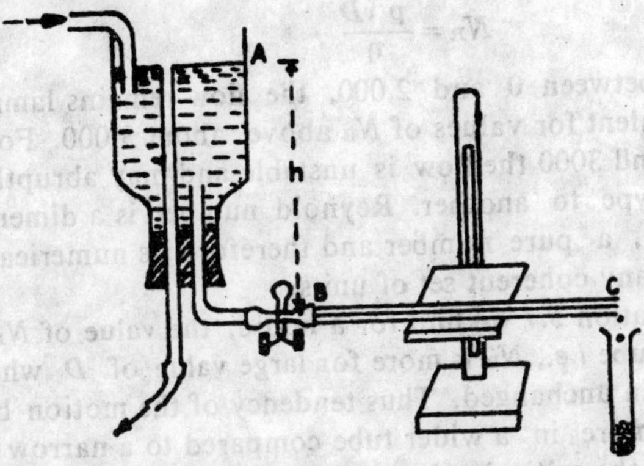

Fig. 5.19. Determination of viscosity of a liquid.

In Fig. 5.19, a liquid is maintained at a level h above the level of a narrow capillary tube BC of length l and radius r. When BC is placed horizontally a constant pressure difference of $P=h\rho g$ is maintained between the inlet end B and the outlet end C of the tube (Atmospheric pressure), when ρ is the density of the liquid and g is acceleration due to gravity. This pressure difference provides the tangential force to maintain laminar flow of liquid of viscosity η through a tube of length l and radius r is given by

$$P=\frac{8\eta lV}{\pi r^4} \qquad ...(5.6)$$

where V is the volume of liquid flowing per second. This is poiseuille's equation for flow of a liquid through a narrow tube.

The Equation 5.6 written in the form,

$$\eta=\frac{\pi Pr^4}{8lV}$$

enables us to determine the viscosity of a liquid if we can measure the values of P, l, V and r.

5.13. Critical Velocity and Reynold's Number

When liquid flows through a long tube at a slow rate, the motion remains streamlined. When, however, the average velocity of liquid exceeds a certain velocity called the *critical velocity*, the flow becomes turbulent. It was observed experimentally by

Osborne Reynold that the combination of four factors, namely viscosity η, density ρ, the average speed v of the liquid and diameter D of the pipe, determines the nature flow of viscous liquid through a pipe. This combination is known as the Reynold number, N_R, and is defined as

$$N_R = \frac{\rho\, vD}{\eta} \qquad \text{...(5.7)}$$

For N_R between 0 and 2,000, the flow remains laminar. The flow is turbulent for values of N_R above about 3,000. For N_R between 2000 and 3000 the flow is unstable and may abruptly change from one type to another. Reynold number is a dimension-less quantity *i.e.*, a pure number and therefore its numerical value is the same in any coherent set of units.

From Equation 5.7 we find for a liquid, the value of N_R is more for a wider tube *i.e.*, N_R is more for large value of D when other factors remain unchanged. Thus tendency of the motion becoming turbulent is more in a wider tube compared to a narrow tube. In other words, the critical velocity in a wider tube is less. Similarly, for a particular tube, N_R decreases with increase of viscosity. Hence flow of a more viscous liquid remains streamlined for comparatively higher velocity, that is, critical velocity for a more viscous liquid is less.

5.14. Motion in a Viscous Medium—Stoke's Law

When a body moves through a viscous liquid at rest, the liquid in contact with it also moves with the velocity of the body. But liquid at a distance remains practically at rest. So layers of liquid at different distances from the body moves with different velocities. Hence there is a relative motion of the layers of liquid near the body and a resisting force acts on the body due to viscosity of the the liquid.

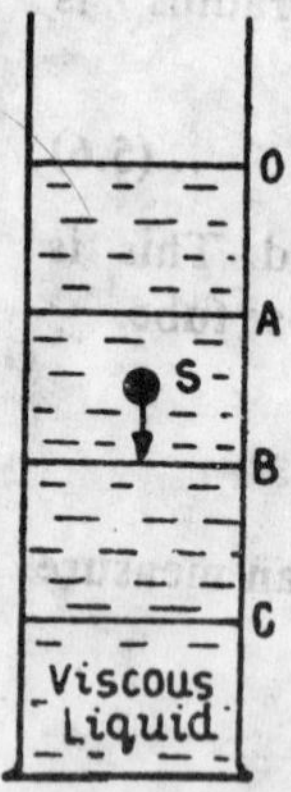

Fig. 5.20. Motion of a body in a viscous medium.

Suppose a small steel ball is dropped into a tall and long vertical glass jar containing a very viscous liquid like mobile, Castor oil or Glycerine (Fig. 5.20). *A*, *B* and *C* are three marks on the tube such that $AB = BC$. As the ball drops through the liquid, time taken to cover the distance *AB* will be found to be equal to the time taken to cover the distance *BC*. Thus we conclude that the ball moves at a constant velocity after a short distance inside the liquid. This velocity is called *terminal velocity*.

How do we explain it? The force due to weight (mg) acts on the ball downwards. Simultaneously the force due to buoyancy and the resisting force due to viscosity both act upwards. When the upward and downward forces become equal to each other, the net force on the body is zero and the body moves with a uniform or terminal velocity.

For a small sphere of radius r moving with small terminal velocity v in an infinite and homogeneous liquid of viscosity η, from experiment stokes established that the resisting or retarding force:

$$F=6\pi\eta rv \quad ...(5.8)$$

This is known as Stoke's law. Stoke's law can also be derived from dimensional analysis in the following manner.

When a spherical body moves through an infinite and homogeneous viscous liquid at rest, it is expected that the resisting force will depend upon (*i*) the viscosity of the liquid, (*ii*) the size of the body, (*iii*) the relative speed of the body with respect to the liquid. Thus the resisting force,

$$F=\text{Constant } \eta^a r^b v^c \quad ...(i)$$

where a, b, and c are respectively the powers of η, radius r and terminal velocity v. The dimensions on left hand side of this relation must be equal to the dimensions on the right hand side.

Substituting the dimensions, therefore, in (*i*)

$$MLT^{-2}=(ML^{-1}T^{-1})^a\,(L)^b\,(LT^{-1})^c$$
$$=(M)^a(L)^{-a+b+c}T^{-a-c}$$

Equating the dimensions of similar quantities on both sides of the equation

$$a=1;\ c+b-a=1:\ a+c=2$$

Solving these relations,

$$a=b=c=1$$

So $\quad F=\text{Constant } \eta rv.$

The value of the constant is experimentally determined to be 6π.

Hence $\quad F=6\pi\,\eta rv.$

5.15. Experimental Determination of η Using Stoke's Law

The equation (5.8) is utilised to determine the viscosity of liquid in the following manner in an experimental set up given in Fig. 5.20.

When a spherical body is moving inside the liquid under gravity the following forces are acting on the body.

(*i*) Weight of the sphere (acting downward) is:

$$mg=4/3\pi r^3\rho g$$

where ρ is the density of the sphere.

(*ii*) Force of buoyancy (acting vertically upwards)

$$B=4/3\pi r^3\rho_o g$$

where ρ_o is the density of the liquid.

(*iii*) Resisting force of liquid due to viscosity acting upwards

$$F=6\pi\eta rv.$$

When the body is first dropped in the liquid, it starts initially moving with an acceleration. But as the body gathers speed, the resistive force due to viscosity comes into play and its magnitude increases with increase in velocity till the magnitude of F equals the magnitude of $(mg-B)$. Thereafter the body moves with constant (terminal) velocity v. So under equilibrium condition,

$$F=mg-B$$

$$\therefore \quad 6\pi\eta rv=4/3\pi\eta^3\rho g-4/3\pi r^3\rho_o g$$

$$=4/3\pi r^3 g(\rho-\rho_o)$$

So the terminal velocity,

$$v=\frac{2}{9}\frac{r^2 g}{\eta}(\rho-\rho_o) \qquad \ldots(5.9)$$

Equation 5.9 holds good provided the terminal velocity is small so that no turbulence sets in. The resistance offered to the motion of a sphere in a viscous medium is more than that given by Stoke's law when turbulence is present. A small raindrop falling through air also attains a terminal velocity as Stoke's law also applied in that case.

Equation 5.9 can be written in the form

$$\eta=\frac{2}{9}\frac{r^2 g}{v}(\rho-\rho_o) \qquad \ldots(5\ 9a)$$

Thus the value of η of a viscous liquid can be determined experimentally if ρ, ρ_o, r and v are known.

Example 5.5. Using the Equation 5.9 determine the terminal velocity of raindrops given η for atmospheric air near the clouds $=1\times10^{-4}$ NS/m² and the average radius, of the drop$=1\times10\text{m}^{-3}$.

Solution. $v=\frac{2}{9}\frac{(1\times10^{-3}\text{m})^2\times9.8\text{m/s}^2\times10^3\text{kg/m}^3}{1\times10^{-4}\text{ NS/m}^2}=21.8$ m/s.

Example 5.6. What should be the maximum average velocity of water in a tube of diameter 0.5 cm so that the flow is laminar? The viscosity of water is 0.001 NS/m².

Solution. The maximum value of Reynold number for laminar flow is 2000. Let v be the maximum average velocity of water for laminar flow.

Thus, $$R_n=\frac{\rho v D}{\eta}=2{,}000$$

or $$v = \frac{2000 \times \eta}{\rho D}$$

Substituting the values,

$$v = \frac{2000 \times 0.001 \text{ N.m}^{-2}. \text{ S}}{10^3 \text{ kg. m}^{-3} \times 0.5 \times 10^{-2} \text{ m}}$$

$$= 0.4 \text{ m/s}$$

5.16. Flow through a Constriction

When a liquid which is practically incomperessible flows steadily through a pipe of varying cross-sectional area (Fig. 5.21) there can be no accummulation of liquid anywhere. Therefore, the mass of liquid passing at A during time interval t will be equal to the mass of liquid flowing at B during the same time interval.

Let, the area of cross-section of the pipe at A be S_1 and that at B be S_2. Let the speed of flow of liquid at A be v_1 and the speed at B be v_2.

So, mass of liquid flowing during time interval Δt at A,

$$m = S_1 v_1 \rho \Delta t$$

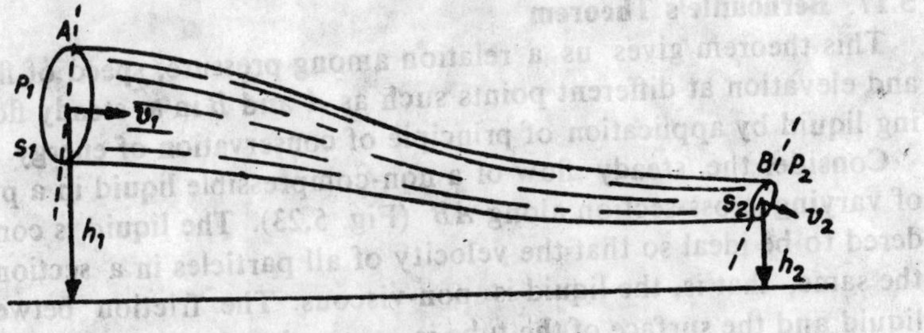

Fig. 5.21. Equation of continuity.

Similarly, mass of liquid flowing during time interval Δt at B

$$m = S_2 v_2 \rho \Delta t$$

where ρ is the density of the liquid that remains constant.

Thus $$S_1 v_1 \rho \Delta t = S_2 v_2 \rho \Delta t \qquad ...(i)$$

This relation can be written in the form,

$$S_1 v_1 = S_2 v_2 \qquad ...(ii)$$

The relation (*ii*) is called equation of continuity. From equation of continuity it follows that the speed of flow in a pipe is greater at a point where the tube is narrower and less where the tube is wider. In a compressible fluid, the density does not remain constant. Since the mass of the fluid flowing past a point is constant, the density at a point will change until the flow becomes steady. The equation of continuity in such case is written in the form

$$S_1 v_1 \rho_1 = S_2 v_2 \rho_2 \qquad ...(iii)$$

where ρ_1 and ρ_2 are the densities of fluid at *A* and *B* respectively.

Referring to Fig. 5.21 we see that the speed v_2 at *B* is greater than the speed v_1 at *A* ($\because S_1 > S_2$). So liquid experiences an acceleration between *A* and *B*. Thus an accelerating force must be acting between *A* and *B*. It is obvious, this accelerating force can be present only when the pressure at *A* is greater than the pressure at *B*. Hence, in a steady flow the pressure is lower where the speed is greater. Thus at a constriction in a pipe, the speed of liquid is maximum and the pressure is minimum. In Fig. 5.22 manometers connected to various locations in the pipe shows that the pressure is lowest at the constriction.

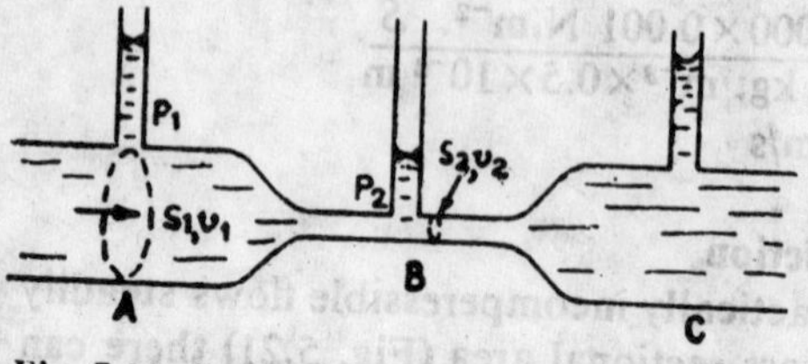

Fig. 5.22. Pressure is minimum where speed is maximum.

5.17. Bernouille's Theorem

This theorem gives us a relation among pressure, speed of flow and elevation at different points such as *A* and *B* in a steady flowing liquid by application of principle of conservation of energy.

Consider the steady flow of a non-compressible liquid in a pipe of varying cross-section along *AB* (Fig. 5.23). The liquid is considered to be ideal so that the velocity of all particles in a section is the same, that is, the liquid is non-viscous. The friction between liquid and the surface of the tube is assumed to be zero.

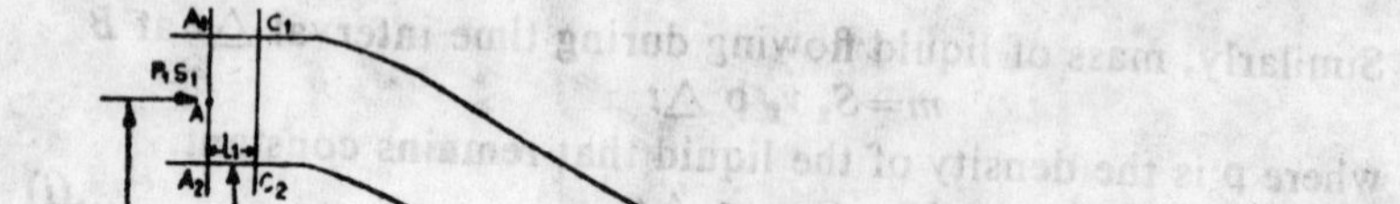

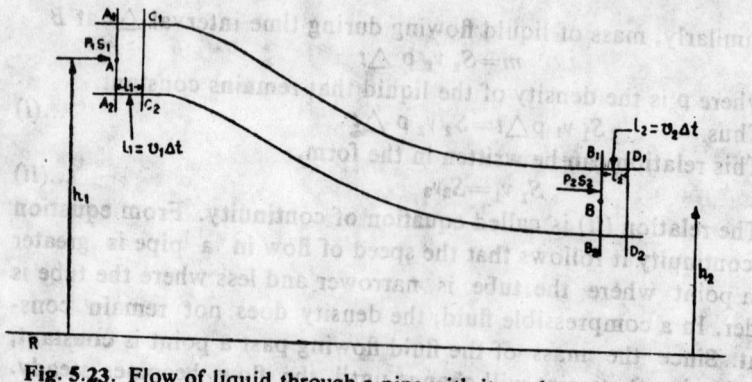

Fig. 5.23. Flow of liquid through a pipe with its ends at different levels.

At *A*, let the speed of liquid be v_1, area of cross section of the pipe be *S* and the height be *h* with reference to some arbitrary reference level RL. Similarly, these quantities at B are v_2, S_2 and h_2. In any small interval of time $\triangle t$ the volume $\triangle v$ that flows through S_1 is the same as that flows through S_2.

Let the pressure acting on S_1 be P_1. So the layer A_1A_2 is pushed to a position C_1C_2 in a short interval $\triangle t$ under the action of the force P_1S_1. The displacement of A_1A_2 is $l_1=v_1\triangle$ So the work done on liquid entering the section during interval $\triangle$,

$$W_1=P_1S_1l_1=P_1S_1v_1 \ \triangle t=P\triangle v=P_1\ m/\rho$$

where m is the mass and $\triangle v$ is the volume of liquid flowing during the interval. Since l_1 is very small, S_1 is regarded as constant over the displacement.

This work done on the system is stored in the mass of liquid as potential energy. This potential energy of liquid mass is known as potential energy associated with pressure or simply 'pressure energy'. Besides pressure energy, liquid mass m has gravitational potential and kinetic energy.

Gravitational potential energy of the mass m at $A=mgh_1$ and the kinetic energy of the mass m at $A=\frac{1}{2}\ mv_1^2$

$\therefore$ Total energy of the mass m entering at A during time

$$\triangle t=m\ P_1/\rho+mgh_1+\tfrac{1}{2}mv_1^2 \qquad ...(i)$$

Similarly at B, the work done on mass m during the same time interval

$$\triangle t=P_2\ m/\rho$$

Gravitational potential energy of the mass

$$m=mgh_2$$

and kinetic energy of mass

$$m=\tfrac{1}{2}mv_2^2$$

Total energy of the mass m flowing at B during time

$$\triangle t=m\ P_2/\rho+mgh_2+\tfrac{1}{2}mv_2^2 \qquad ...(ii)$$

According to principle of conservation of energy, total energy at A and that at B are equal.

Hence from (*i*) and (*ii*),

$$m\ P_1/\rho+mgh_1+\tfrac{1}{2}mv_1^2=m\ P_2/\rho+mgh_2+\tfrac{1}{2}\ mv_2^2$$

or

$$\frac{P_1}{\rho g}+h_1+\frac{v_1^2}{2g}=\frac{P_2}{\rho g}+h_2+\frac{v_2^2}{2g}$$

$$\frac{P}{\rho g}+h+\frac{v^2}{2g}=\text{constant} \qquad ...(5.10)$$

This equation is known as Bernouille's equation. Each term in this equation has the dimension of a length. In relating hydrostatic pressure with depth, h is called the head. In analogy, each term in Equation 5.10 is called a head: $P/\rho g$, the pressure head; h, the elevation head and $v^2/2g$, the velocity head.

Equation 5.10 can also be written in the form

$$\frac{P}{\rho}+hg+\tfrac{1}{2}\,v^2=\text{constant} \qquad \ldots(5.10a)$$

Each term in this equation refers to the three types of energy for unit mass. So sum of pressure energy, gravitational potential energy and kinetic energy is constant.

Equation 5.10 (*a*) can also be written in the form,

$$P+h\rho g+\frac{v^2}{2}\rho=\text{Constant} \qquad \ldots(5.10b)$$

Hence, Bernouille's theorem may alternatively be stated:

At any two points along a streamline in an ideal fluid in steady flow, the sum of the pressure, the potential energy per unit volume and the kinetic energy per unit volume have the same value.

Bernouille's theorem is accurate only for incompressible, non-viscous liquids. But it is also applied to ordinary liquids with sufficient accuracy for many practical applications.

5.18. Some Illustrations and Applications of Bernouille's Principle

(*i*) **Venturimeter.** This is a practical device based on Bernouille's Principle used for measuring the rate of flow of liquid generally water in a pipe. Two horizontal conical tubes T_1 and T_2 joined end to end at O, and three vertical manometer tubes connected at A, B, C together from a venturi-tube (Fig. 5.24). It is placed horizontally and connected with the flow tube. If p_1 and p_2 are the pressures and v_1 and v_2 the velocities of flow of the liquid of density ρ at A and B, then according to Equation 5.10(*b*)

$$p_1\times\tfrac{1}{2}\rho v_1^2=p_2\times\tfrac{1}{2}\rho v_2^2 \qquad \ldots(i)$$

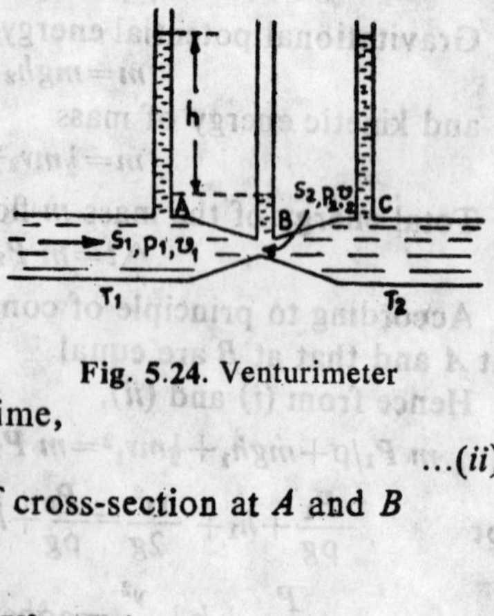

Fig. 5.24. Venturimeter

Volume of liquid flowing per unit time,

$$V=v_1S_1=v_2S_2 \qquad \ldots(ii)$$

Where S_1 and S_2 denote the area of cross-section at A and B

From (*ii*) $v_1=V/S_1$ and $v_2=V/S_2$.

Substituting in (*i*),

$$(p_1-p_2)=\tfrac{1}{2}\rho\left(\frac{V^2}{S_2^2}-\frac{V^2}{S_1^2}\right) \qquad \ldots(iii)$$

Referring to Fig. 5.24 we can get,

$$p_1-p_2=h\rho g \qquad \ldots(iv)$$

where h is the difference between the levels of liquids in the manometer tubes at A and B.

Hence, $h\rho g=\frac{1}{2}\rho V^2\left[\frac{S_1^2-S_2^2}{S_1^2S_2^2}\right]$

or $$V=\frac{S_1S_2\sqrt{2gh}}{(S_1^2-S_2^2)^{1/2}} \quad ...(5.11)$$

Since S_1, S_2 and g are constants, the difference in the level in the manometer tubes is a measure of volume of liquid flowing in unit time in a tube. Suitable mechanical device coupled to the venturi-tube records the total amount of liquid flowing in a tube. This device is called a *venturimeter*.

(*ii*)**Velocity of efflux of a liquid—Toricellis' theorem.** When liquid escapes from a small hole or orifice in a vessel (Fig. 5.25), the velo-

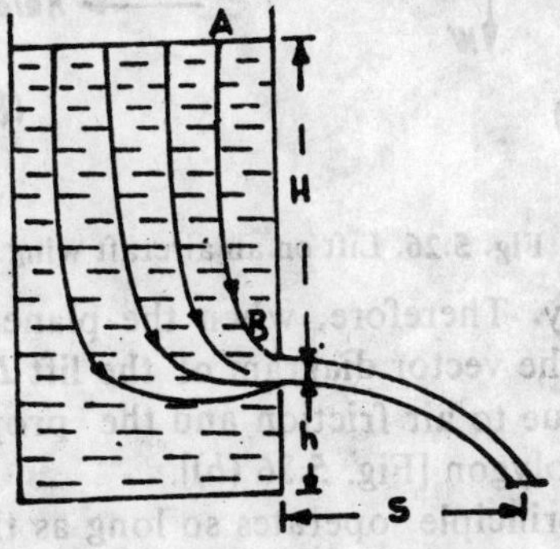

Fig. 5.25. Velocity of Efflux.

city of discharge of the liquid can be calculated from Bernouille's theorem. Suppose in a tall vessel (Fig. 5.25) a small orifice is made at B, at a height h above the bottom. The level of liquid in the vessel is maintained at A, which is vertically H above the orifice. Applying Bernouille's theorem at A and at B,

$p_1+\frac{1}{2}\rho v_1^2+g\rho\,(h+H)=p_2+\frac{1}{2}\rho v_2^2+g\rho h$; taking the bottom of the vessel as reference for calculating hydrostatic pressure.

Since $p_1=p_2=$ atmospheric pressure, and $v_1=0$

$$v_2=\sqrt{2\rho gH} \quad ...(5.12)$$

This gives the velocity with which liquid will emerge out of the orifice.

(*iii*) *Lift on an aircraft wing.* When the wing of an aircraft is inclined upwards a few degree with respect to the wind direction, Fig. 5.26(*a*) air will be deflected from the lower surface. So air speed above the wing will be higher than the speed below it. Therefore, according to Bernouille's theorm, the pressure below will be higher than the pressure above. The pressure below the wing also increases due to the reacting forces set up due to deflecting air. Both effects but mostly the first produce an unbalanced upward lifting force on the wing. The lifting force L can be resolved into vertical and horizontal components; the vertical component supports the weight of the wing and the horizontal component overcome the drag D due to air friction, when the wing is moving with

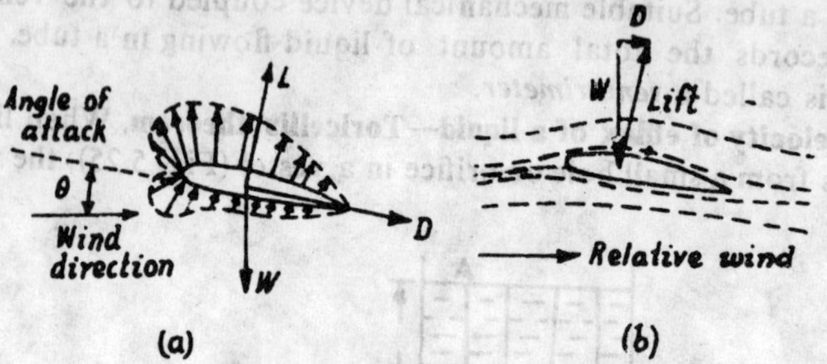

Fig. 5.26. Lift on an aircraft wing.

a constant velocity. Therefore, when the plane is moving with a constant velocity the vector diagram of the lift L, weight W of the wing, the drag D due to air friction and the propeller thrust add to form a closed polygon [Fig. 5.26 (*b*)].

However, this principle operates so long as the angle of attack (Fig. 5.26) is small. If the angle of attack is too great, the streamline flow of wind above and below the wing change into turbulence and eddies that engulfe the wings. The Bernouille's equation no longer holds good and there may not be any lift on the wings.

The kite-fliers also take advantage of this for giving lift to the kites. Rudders and propellers also partly operate on this principle.

(*iv*) **Principle of spray-jet or atomizer.** In an atomizer or sprayer, spray of oil or scent emerges when the bulb is squeezed (Fig. 5.27). Partial vacuum created inside is partly responsible for the rise of liquid in the tube dipped in liquid. But the major part of the rise is explained by Bernouille's principle. When the bulb is squeezed, the air blowing through the central tube with high speed reduces the inside pressure to a value lower than atmospheric pressure on the liquid. Thus the liquid in the tube rises and then blows oil as sprays along with air stream in the nozzle. Instead of the bulb, if one blows by mouth, the same effect will also be observed.

(*v*) **Pitot tube.** This is an arrangement to determine the velocity of flow of a liquid in a horizontal pipe. Two vertical tubes *A*

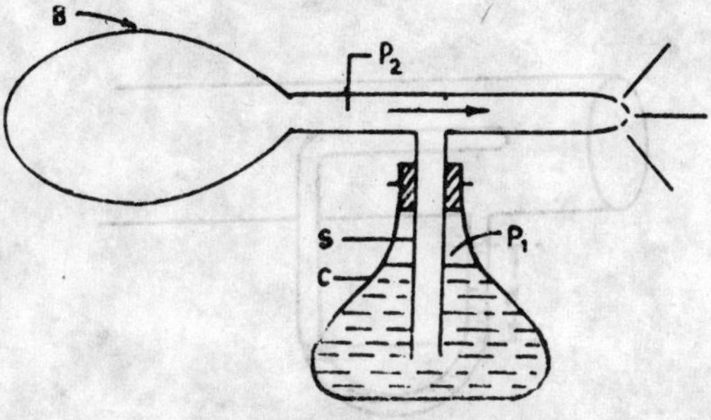

Fig. 5.27. An atomiser *B*-bulb P-atmospheric pressure P_2-pressure inside; *C*-container.

and *B* are inserted in a horizontal pipe through which liquid is flowing (Fig. 5.28). Notice the lower ends of the tubes *A* and *B*. The end of the tube *B* has been bent at right angles such that liquid flows into the and come to rest. So the speed of liquid at *C* is higher than the speed at *D* in the tube *B*. According to Bernouille's principle, the pressure at *D* is higher than the pressure at *C*. So liquid in *B* rises to a higher level than in *A*.

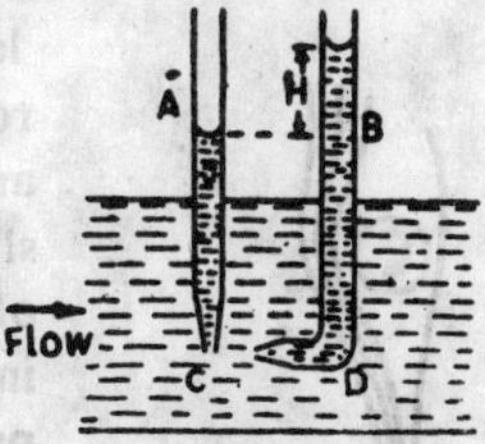

Fig. 5.28. Pitot tube.

Therefore, $p_C+\frac{1}{2}\rho v^2=p_B+0$

or $\quad p_B-p_C=\frac{1}{2}\rho v^2$

or $\quad \rho gh=\rho v^2$

or $\quad v=\sqrt{2gh}$ ...(5.13)

Here of course, it has been assumed that the diameter of the tubes *A* and *B* are too small to affect the stremline flow in the pipe.

Pitot tubes can also be used to measure the flow speed of gases or even speed of a high speed vehicle by mounting it outside its body. The difference in construction of the Pitot tube for such use is shown in Fig. 5.29. Here the pressure difference is measured by liquid manometer.

Many other examples can be cited to show the application of Bernouille's principle. During wind storm, sometimes asbestos or thatched roofs are blown off without damaging the other parts of this house. One of the reason is the low pressure created over the roof due to high speed of the wind.

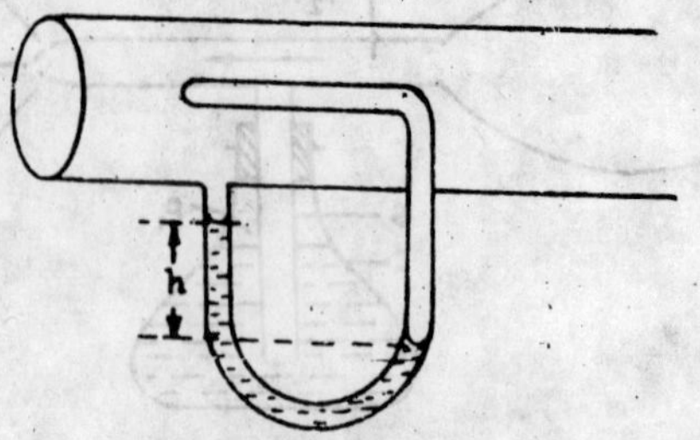

Fig. 5.29. Speed of gas by pitot tube.

If a small pingpong ball is placed in a vertical jet of air or water, it will rise to a certain height above the nozzle, and stay at that level. The ball will be spinning and turning round (Fig. 5.30). The streamlines passing around the ball are altered by it. If the ball slightly moves to one side of the jet, speed of jet on that side decreases and pressure increases. The ball is pushed back to the path of the jet. Like this it continues turning and spinning from one side to the other.

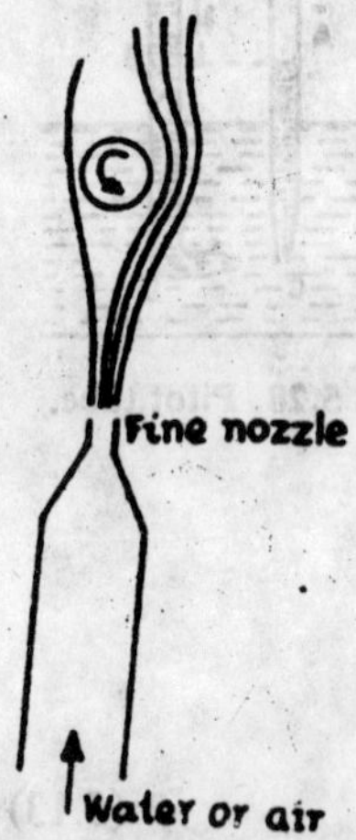

Fig. 5.30. Spining of a pingpong ball on a water jet.

Moderately low pressure can be produced by allowing water to flow through a narrow jet in the enclosed vessel. The steam injection used for accelerating the injection of exhaust steam from the cylinder of a steam engine also works on this princple. In bunsen burner, the gas enters into the burner through a nozzle and hence the pressure inside reduces. The air required for combustion is, therefore, sucked through small holes provided in the burner.

Example 5.7. Water flowing at 1 m/sec in a horizontal pipe passes into a constriction whose area is one tenth the normal pipe cross-section. What is the decrease in water pressure in the constriction?

Solution. Let the cross-sectional area of the pipe be A. So the cross-sectional area at the constriction is $A/10$.

The pressure difference

$$=\frac{1}{2}\rho V^2\left[\frac{1}{(A/10)^2}-\frac{1}{A^2}\right]$$

$$=\frac{1}{2}\rho V^2 \, 99/A^2$$

where V is the volume of water flowing per second.

[Ref. (*iii*) in venturimeter].

But $V = v \times A$ and $\rho = 1000$ kg/m³.

So the pressure difference

$$= \tfrac{1}{2} . 1000 \text{ kg/m}^3 . v^2 A^2 \frac{99}{A^2}$$

$$= 4{,}95{,}00 \text{ kg/m}^3\ v^2.$$

$$= 4{,}95{,}00 \text{ kg/m}^3 \times (1 \text{ m/s})^2$$

$$= 4{,}95{,}00 \text{ N/m}^2$$

$$= 4.95 \times 10^4 \text{ N/m}^2.$$

QUESTIONS

1. The two limbs of an U-tube containing water have unequal diameters. If both limbs are exposed to atmosphere, will the level of water in the limbs be same? Why?

2. If two soap bubbles of radii 0.5 and 1 cm could be joined by a tube without bursting, what would happen? Explain.

3. A small iron needle was placed carefully to float on the surface of water. Then few drops of liquid detergent were added and the needle sank. Why?

4. What part does surface tension play in nature?

5. Pieces of camphor move about on the surface of water but their motion slows down when a finger is immersed in water. Explain.

6. It is difficult to introduce mercury in a fine glass tube whereas water can be introduced into that tube with comparative ease. Why?

7. Can two streamlines intersect each other? If not, why?

8. It is difficult to lift a thin flat metal plate from the surface of another similar plate if there is a layer of oil between them, but the oil layer helps the plate slide easily over the other. Explain.

9. A tall jar is filled with water and kept on ground. If similar narrow holes are made 1/5, 2/5, 3/5, of the way down sides of the container, compute the relative distances from the jar when the jets strike the ground.

10. The wings of slow flying air crafts have very curved upper surfaces whereas a new high-speed aircrafts have wings which are very thin and almost flat. Analyse the reason.

11. A one metre tall glass jar is filled with a mixture of castor oil, water and kerosine. A small lead globule is dropped into it. Describe its motion through the column. What would happen if the globule is made of a material of specific gravity 0.9?

12. Apply dimensional analysis to test the correctness of Poiseuille's equation.

PROBLEMS

1. Calculate the work done in blowing a soap bubble of radius [illegible], surface tension of soap solution being 30×10^{-3} N/m. What additional [illegible] will be performed if the radius is to be doubled ?

[*1[illegible].52 × 10^{-7} J, 565.[illegible] × $10^{-[illegible]}$ J*]

2. If a number of little droplets of water, all of them having same radii, r cm coalesce to a single drop of liquid R cm radius, show that the rise in temperature of water will be given by $3T/J\ (1/r-1/R)$ where T is the surface tension of water and J is mechanical equivalent heat.

3. The two limbs of a U-tube have radii of 0.1 mm and 0.5 mm respectively. If water is put into this tube, what will be the difference in level of water in both limbs ? If water is substituted by mercury, what will be the difference?

Given: surface tension of water $=30\times10^{-3}$ N/m
surface tension of mercury $=500\times10^{-3}$ N/m
Angle of contact of mercury $=135°$
Specific gravity of mercury $=13.6$

[*2.45 cm. —2.122 cm*]

4. At what speed will the velocity head of a stream of water be equal to 10 cm of mercury ?

[*1.40 m/s*]

5. An air bubble of 1 mm radius rises steadily through water of viscosity 72 poise. Neglecting the density of air, calculate the velocity with which the buble rises.

[*3.025×10⁻³ cm/s*]

6. Determine the radius of a drop of water falling through air when the terminal velocity of the drop, co-efficient of viscosity of air and the density of air are 1.2 cm/s, 1.8×10^{-4} poise and 1.2×10^{-3} gm/cc. respectively.

[*0.001 cm*]

6

SOLIDS

We are familiar with three states of matter—solids, liquids and gases. The term solid is used to describe the form of matter which possesses the tendency of maintaining a definite shape and definite volume. But the liquid takes the shape of the container and the volume of the gas is determined by the external conditions like pressure and temperature. Ice, water and its vapour are examples of the three states of matter where the basic molecules involved are same but the physical properties exhibited are different. Therefore, the theories explaining the nature of liquids and gases were found inadequate to explain the properties of the solids. Thus developed a new branch of Physics called 'Solid State Physics'.

6.1. Structure of Solids

Many crystalline materials and gems, particularly quartz, have been known for several centuries. They exhibit regularity in their shape and appearance. Quartz crystals are hexagonal in shape. Later it was found that on sufficiently cooling a liquid, crystals could be grown in the laboratory. When a crystal grows in the laboratory the shape often remains unchanged during growth (the size increases) as if identical building bolcks or units were added continuously to the growing crystal. This led to examine the structure of a crystal.

It is known that common salt (NaCl) crystals can be grown from saline water. After powdering a grain of common salt and examining the tiny particles under a reading glass, one finds them to have the shapes of regular cubes. In otherwords, tiny cubes of common salt similar in shape and size have been arranged in a regular manner to produce common salt as we find it. This symmetry in the external structural pattern suggested internal symmetry in them. Internal symmetry means atoms or molecules in a crystal are not oriented at random but are packed together with some order and regularity in a three dimensional pattern. Most of the solids have this type of structural arrangement called *Crystalline Structure* though to a naked eye it may not be apparent. However, materials like glass, pitch, rubber, celluloise, plastic etc., possess different structural pattern. The structural pattern in this group resembles that of a viscous liquid where no elementary building blocks or units are arranged in a regular fashion. These solids are known as *amorphous solids*. Hence all solids belong to one of the following distinct form—(*a*) Crystalline, (*b*) Amorphous.

(*a*) **Crystalline solids.** Since the arrangement of atoms or molecules are repeated throughout the bulk of a crystalline solid, we

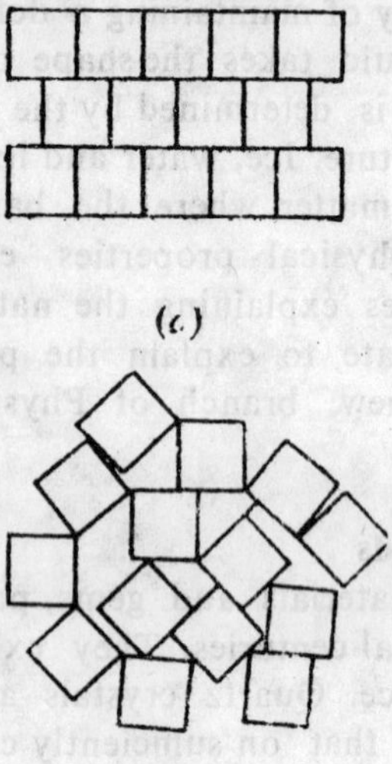

(*a*)

(*b*)

Fig. 6.1 (*a*) Crystalline solids, (*b*) Amorphous solid.

say it has a repetitive pattern. Thus having observed the pattern in a small region of the crystal, it is possible to predict accurately the position of atoms or molecules in any other region of the crystal. Since the structure is maintained even at a longer distance or range, the crystalline solids are said to have *long range order*.

(*b*) **Amorphous solids.** Since there is no regularity in the arrangement of the molecules as in crystalline solids, the structural pattern in amorphous solids is similar to the structural feature of liquids. In a small region, there may appear some pattern of arrangement of molecules *i.e.*, structural unity but when we observe at a distance, the pattern may be different. Thus the amorphous solids possess *short range order*. Liquids also exhibit short range orderliness. but the short range structural units are free to move while they are rigidly held in the case of amorphous solids. Fig. 6.1 illustrates the difference between the two types of solids in two dimensions. The difference between the crystalline and amorphous solid is the way in which the cubes are arranged; they may be carefully packed row upon row, Fig. 6.1(*a*) or may be simply dumped upon one another in a heap, Fig. 6.1(*b*).

Crystalline solids have a definite *melting point* while amorphous solid melts over a small range of temperature. This happens as the structural pattern is same over the entire bulk in a crystalline solid so that bonds between atoms have equal strength and they all break simultaneously melting the solid. On the other hand, in amorphous solid the structural pattern is of short range order. So the bonds between atoms vary in strength and hence break at different temperature so that the melting occurs over a range. The significance of the bonds will be clear later in Art. 6.6 of this chapter.

Amorphous solids differ from crystalline solids and resemble liquids in another important respect. The properties such as electrical conductivity, thermal conductivity, mechanical strength and refractive index are the same in all directions. Amorphous substances are, therefore, said to be *isotropic*. Liquids and gases are also isotropic. Crystalline solids, on the other hand, are *anisotropic i.e.*, their physical properties are different in different directions. The phenomenon of *double refraction* in quartz crystals is one such example.

To sum up, the existence of ordered molecular arrangements in crystalline solids give rise to the following characteristics properties (*i*) long range order or (*ii*) sharp melting point (*iii*) anisotropy.

6.2. Basis of Crystal Structure

(*a*) Basic building block—The unit cell. The symmetrical shape of a crystal is a consequence of internal symmetry or orderliness in arrangement of molecules or atoms inside the crystal. The basic arrangement of molecules or atoms or the structural unit that forms the basis of crystal structure is known as *unit cell* or the lattice unit.

Hence a unit cell is that particular arrangement of the minimum number of molecules or atoms, the regular repetition of which in three dimensional space builds up the whole crystal. Thus the unit cell is the smallest sample that represents the picture of the entire crystal.

A tiny grain of NaCl have cubic shape; so also its basic unit contains Na^+ and Cl^- ions in the form of a cube. The basic unit of quartz (SiO_2) similarly has hexagonal structure.

(*b*) **Crystal lattice.** To understand the crystal structure, a geometrical concept called *space lattice* is introduced at this stage. In

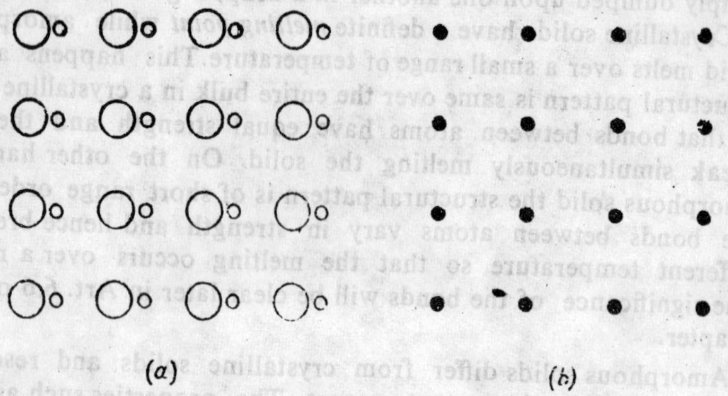

Fig. 6.2. (*a*) Arrangement of a group of atoms. (*b*) Representation in a two dimensional lattice.

Fig. 6.2(*a*) we have got a small group consisting of one big and a small circle which are arranged in a regular pattern (repetitive pattern) on the plane of this paper. If we associate a geometrical point with each group of circles and replace all groups by more points, then we obtain a collection of points in two dimentional space (on the plane of the paper) as shown in Fig. 6.2(*b*). This arrangement of points can be termed as a two dimentional lattice. On repeting the arrangement of the small group in three dimensions (on the plane of the paper and perpendicular to it) and replacing the groups by geometrical points, we get a three dimenstional net work of points called *crystal lattice*. The points in the crystal lattice are called lattice points. It should be remembered that lattice is a mathematical idea only to describe the arrangement of atoms on a crystal.

The groups attached to each lattice point forms the basis of the crystal structure and the lattice shows the relative arrangement of this basis in the actual crystal. So the crystal structure is formed only when a basis of atoms similar or units is attached to each lattice point.

Therefore, to describe a crystal strutcure, we should have knowledge of (*a*) lattice structure and (*b*) basis of atoms or similar units at lattice points.

(*c*) **Lattice parameters—Lattice type.** (Two dimensional). The crystal lattice can be better understood if we examine how a repetitive structure based on a basic unit is built in two dimensions. Fig. 6.3 shows a two dimensional lattice. Each dot or lattice point represents a basic unit. The parallelogram represented by the four lattice points (*OACB*) forms the unit of pattern or *unit cell*. The entire pattern on the plane of the paper can be generated by moving the parallelogram parallel to its sides. The unit of pattern is describeed in terms of two vectors

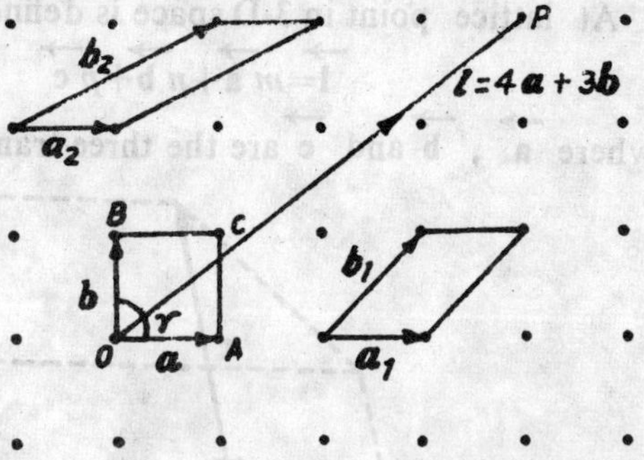

Fig. 6.3. Unit cell in two dimensions.

$\vec{\mathbf{a}}$ and $\vec{\mathbf{b}}$, which form the two sides of the parallelogram. Starting from an arbitrary lattice point *O*, the lattice points along *OA* are situated at equal distance *a* and these along *OB* are lying at equal distances *b*. Suppose the number of lattice points along *OA* are $(m+1)$ and those along *OB* are $(n+1)$. The lattice point located at *P* is represented by the tip of the vector $\vec{\mathbf{l}}$ which is given as—

$$\vec{\mathbf{l}} = m\ \vec{\mathbf{a}} + n\ \vec{\mathbf{b}} \quad \ldots(6.1)$$

In Fig. 6.3, $m=4$ and $n=3$.

The two axes *OA* and *OB* are known as lattice axes and the distances *a* and *b* are known as lattice parameters The two vectors $\vec{\mathbf{a}}$ and $\vec{\mathbf{b}}$ are known as translational vectors and the angle between them is denoted by γ.

There is an unlimited number of possible lattice types because there is no natural restriction on the lengths *a*, *b* of the lattice translation vectors or on the angle between them. The choice is arbitrary. In the Fig. 6.3, three different sets of translational vectors $(\vec{\mathbf{a}}, \vec{\mathbf{b}})$; $(\vec{\mathbf{a}}_1, \vec{\mathbf{b}}_1)$, $(\vec{\mathbf{a}}_2, \vec{\mathbf{b}}_2)$ are choosen. Any of these sets can generate the complete pattern. In fact, in two dimensions there are five possible lattice types or units. These fundamental types of lattices are also called Bravais lattice. But in actual crystal structure, a two dimensional lattice has no relevance.

(*d*) **Lattice in three dimensions—crystal structure**. The repetitive pattern in three dimensions is generated by adding an additional translational vector $\vec{\mathbf{c}}$ out of the plane of the first two (*i.e.* plane of the paper). These three vectors define a parallelopiped which constitutes the unit of pattern or unit cell in 3-dimensions just as a parallelogram is the unit in a 2-dimensional lattice.

At lattice point in 3-D space is defined by a vector $\vec{l}$ such that

$$\vec{l} = m\,\vec{a} + n\,\vec{b} + p\,\vec{c} \qquad ...(6.2)$$

where $\vec{a}$, $\vec{b}$ and $\vec{c}$ are the three translational vectors along the

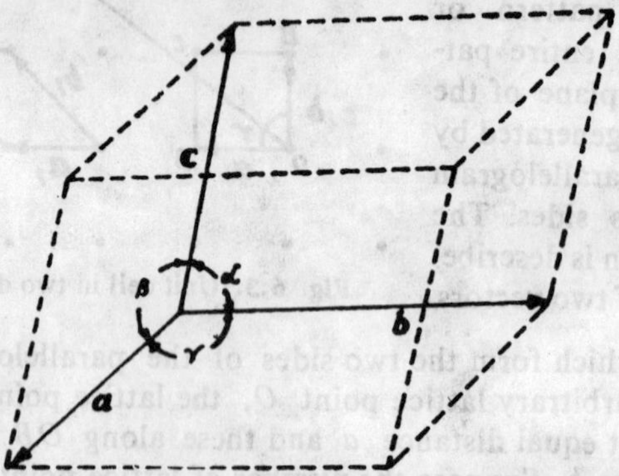

Fig. 6.4. Crystal axes in three dimensions.

three lattice axes *OA*, *OB* and *OC* and *m*, *n* and *p* are integers. The angle between $\vec{b}$ and $\vec{c}$ is α ; the angle between $\vec{c}$ and $\vec{a}$ is β and the angle between $\vec{a}$ and $\vec{b}$ is γ as shown in Fig. 6.4.

Like 2-D lattice, the choice of unit cell is also arbitrary in 3-D lattice. The lattice points can be grouped in fourteen different lattice types. However, the fourteen lattice types are conveniently grouped into seven systems according to seven types of conventional unit cells. The characteristic features of these seven types are given in Table 6.1. and Fig. 6.5 represent these cells in 3-D.

TABLE 6.1. Lattice Systems in 3-dimensions.

System	*Number of lattices in the system*	*Lattice symbols*	*Restrictions on lattice tanslational vector and angles*
Triclinic	1	*P*	$a \neq b \neq c$ $\alpha \neq \beta = \gamma$
Monoclinic	2	*P, C*	$a \neq b \neq c$ $\alpha = \gamma = 90° \neq \beta$
Orthorhombic	4	*P, C, I*	$a \neq b \neq c$ $\alpha = \beta = \gamma = 90°$
Tetragonal	2	*P, I*	$a = b \neq c$ $\alpha = \beta = \gamma = 90°$
Cubic	3	*P* or Sc *I* or bcc *F* or occ	$a = b = c$ $\alpha = \beta = \gamma = 90°$
Trigonal	1	*R*	$a = b = c$ $\alpha = \beta = \gamma < 120° \neq 90°$
Hexagonal	1	*P*	$a = b \neq c$ $\gamma = 120°$, $\alpha = \beta = 90°$

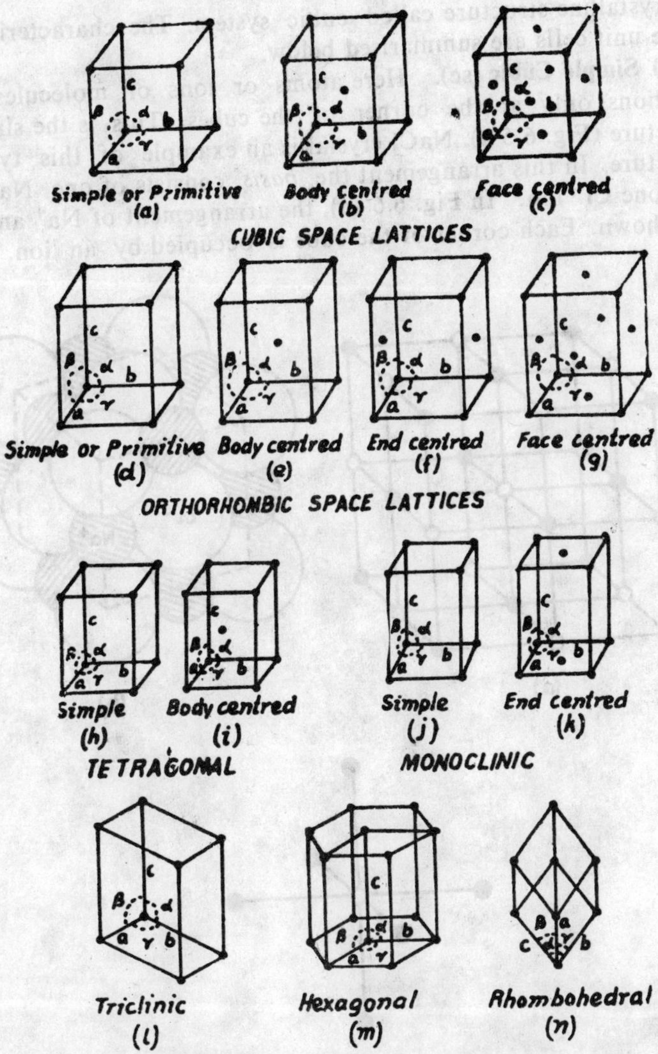

Fig. 6.5. The fourteen space lattices.

6.5. The Cubic System

Study of all types of crystalline structures is beyond the scope of this book. However, we will discuss the most common system—the cubic system, to which large number of crystalline substances belong.

In Fig. 6.6 three types of cubic cells have been shown. All these three units have cubic cells have shape and thus they form the basis of crystalline structure called cubic system. The characteristic of three unit cells are summarised below.

(*a*) **Simple Cubic (sc).** Here atoms or ions or molecules take positions only at the corner of the cubes. This is the simplest structure (Fig. 6.5 *a*). NaCl crystal is an example of this type of structure. In this arrangement the *basis* consists of one Na^+ ion and one Cl^- ion. In Fig. 6.6 (*a*), the arrangement of Na^+ and Cl^- are shown. Each corner of the cube is occupied by an ion. Each

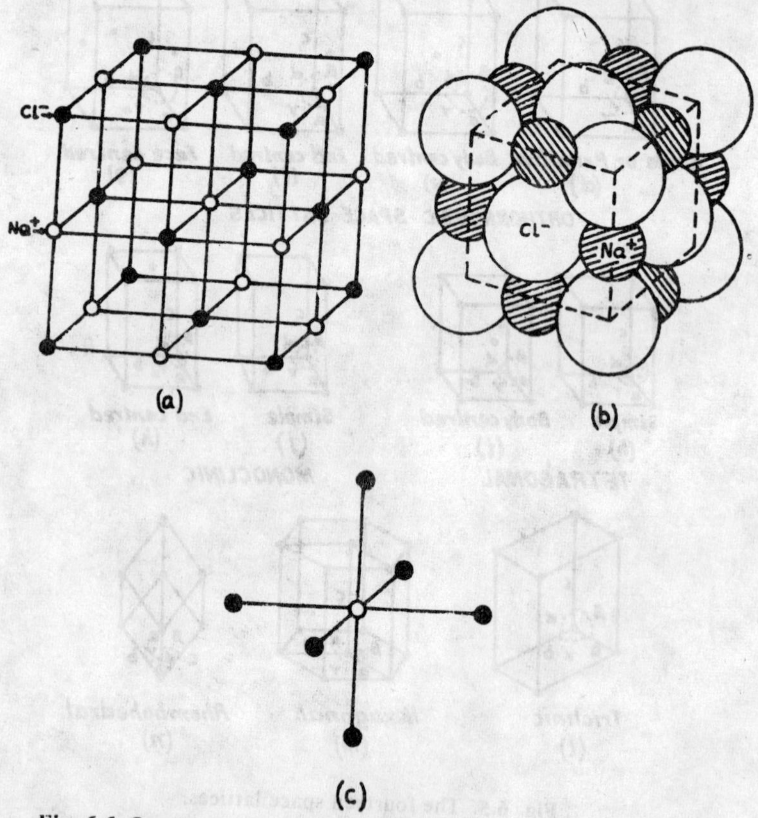

Fig. 6.6. Octahedral structure of NaCl crystal. Hollow circles, sodium ions. Shaded circles, chlorine ions.

Na^+ ion has 6 Cl^- ions as its nearest neighbours and similarly each Cl^- ion has 6 Na^+ ions as its nearest neighbours [Fig. 6.6 (*c*)]. The space-filling model of NaCl crystal represent its actual structure as shown in Fig. 6.6 (*b*). All alkali halides except those of cesium, have this structure. PbS also have this structure.

This type of structure where lattice points are situated at corners is also known as primitive cell. All the seven crystal system have a primitive cell each.

(*b*) **Body centred cubic (bcc).** In addition to the lattice points at the corner of the cube, it has got one more lattice point at the centre of the cell as shown in Fig. 6.5(*b*). Cesium chloride crystal has this structure, Fig. 6.7. The Cl^- ions may be regarded as lying at the corners of the cube and Cs^+ ion to body centre of the cube. Each Cs^+ ion has 8 Cl^- ions as nearest neighbours. Each Cl^- ion is also surrounded by 8 Cs ions.

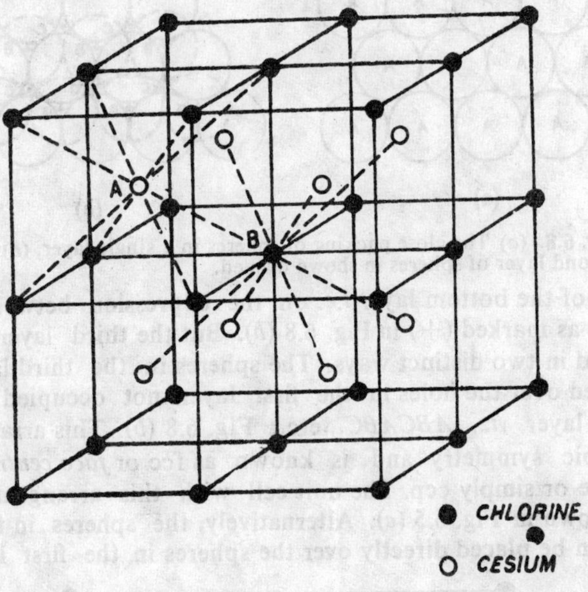

Fig. 6.7. Body centred cubic structure of cesium chloride crystal.

Bromides and iodides of cesium also have this structure. Elements like barium, cesium, chromium, iron, etc also have this structure.

(*c*) **Face centred cubic (fcc).** Besides the lattice points at the corners, this structure has one additional lattice point at the centre of each face as shown in Fig. 6.5 (*c*). Elements like aluminium, copper, silver, calcium, lead, nickel etc., have this type of structure. Each atom in this case has 12 neighbouring atoms.

6.5. Close Packed Structure

In a crystalline structure when atoms are arranged as close to each other as possible so that the volume of unutilised space in the cell is reduced to minimum, the structure is called a close packed structure. The idea can be better understood by imagining each atom as a rigid and impenetrable sphere. Spheres may be arranged in a single close-packed layer by placing each sphere in contact with six others Fig. 6.8 (*a*). A second similar layer is packed on top of first layer by placing each sphere in contact with three

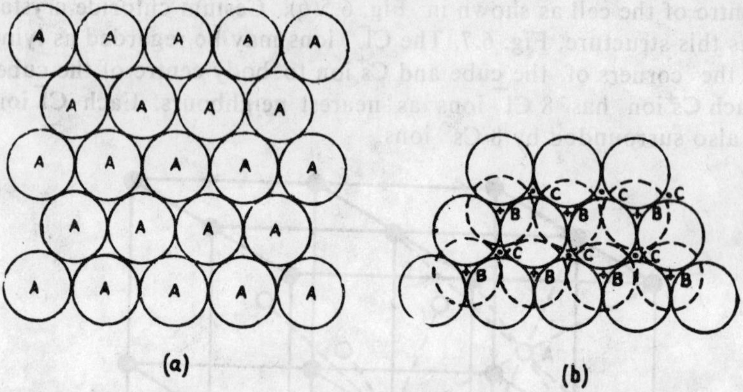

Fig. 6.8. (*a*) The close packing of spheres in a single layer, (*b*) The second layer of spheres in shown dashed.

spheres of the bottom layer *i.e.* on the depression between three spheres as marked (+) in Fig. 6.8 (*b*). But the third layer can be arranged in two distinct ways. The spheres in the third layer can be placed over the holes in the first layer not occupied by the second layer *viz. ABCABC*...etc., Fig. 6.8 (*b*). This arrangement has cubic symmetry and is known as fcc or *face centred cubic* structure or simply ccp. The unit cell with this arrangement has been shown in Fig. 6.5 (*c*). Alternatively, the spheres in the third layer can be placed directly over the spheres in the first layer *viz.*

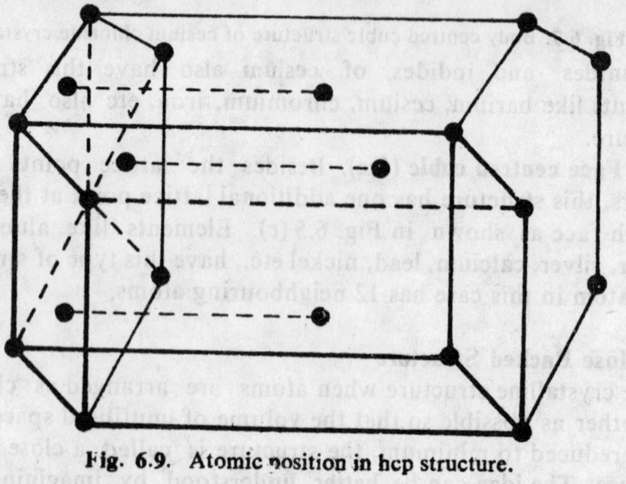

Fig. 6.9. Atomic position in hcp structure.

ABABAB...etc., Fig. 6.8 (*b*). The arrangement is known as hcp or *hexagonal* close packed structure. The unit cell of this structure is a hexagonal primitive cell shown in Fig. 6.9. A typical example of hcp structure is magnesium. In this type of structure, each atom has 12 nearest neighbours, 6 in its plane 3 in the plane above and 3 in the plane below.

Most of the crystals have got closepacked structure. However, crystals of carbon, silicon, germanium etc., having covalent bonds have loose-packed structure.

6.6. Some Characteristics of Cubic Cell

The unit cells of all types are characterised by certain properties. We discuss them briefly.

(*a*) **Volume of the unit cell.** For all the three types of cubic cells, we have the lattice parameters $a=b=c$.

So the volume of the cubic cell, $V=a^3$.

(*b*) **Atoms per unit cell.** (*i*) **scc.** In a simple cubic structure, an atom situated at the corner of each unit cell is common to a total number of eight unit cells. Thus each unit cell can effectively claim 1/8th share of every corner atom. There are eight corners of a cubic cell. Thus the total contribution of all eight corner atoms to each unit cell$=8\times 1/8$th atom$=1$ atom. This means, number of atoms per unit cell is 1.

(*ii*) **bcc.** There is one additional atom at its centre besides having one atom each at its corners in a bcc unit cell. The central atom is not shared by the neighbouring cells and thus the contribution of this atom is limited to one cell. Taking into consideration the contribution of corner atoms as 1 (as in scc), the total number of atoms in every bcc unit cell$=1+1=2$.

(*iii*) **fcc.** There is an additional atom at the centre of each face besides the corner atoms. The atom at the centre of a face of unit cell is shared by only two adjacent unit cells. Hence contribution of one face-atom to the cell is 1/2 atom. There are six faces of a cube and as such there are 6 face-centred atom contributing $6\times 1/2=3$ atoms to a cell. So, the total number of atoms per fcc cell$=1$ (due to corner atoms)$+3=4$ atoms.

(*c*) **Co-ordination number.** The total number of atoms which are nearest neighbour of a particular atom is defined as its co-ordination number. In a scc cell, the nearest neighbour of a corner atom are six atoms as discussed earlier. Hence the co-ordinate number is 6.

In a bcc cell, the co-ordinate number is 8 and that for a fcc, it is 12. The co-ordination number in hcp is also 12.

(*d*) Distance between the nearest-neighbour and atomic radius. In the close-packed arrangement atoms in the crystal touch each other. Thus if the atoms are assumed to be rigid spheres of same size, the distance between the centre of two immediate neighbours is twice the radius of each atom, Fig. 6.10. We deduce below the atomic radius in specific cases.

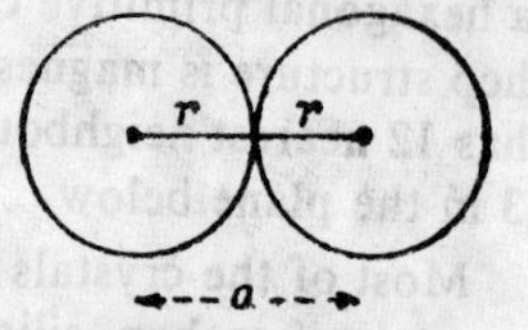

Fig. 6.10. $A=2r$. Atomic radius for scc.

(*i*) **scc.** The distance between two neighbouring atoms (any two corner atoms on one face) is *a i.e.* the side of the unit cell. Hence for scc,

$$2r=a \qquad \ldots(6.3a)$$

and

$$r=a/2 \qquad \ldots(6.3b)$$

where r is the radius of the atom.

(*ii*) **bcc.** Here one atom is located at the centre of the cell. Thus, the two atoms situated at the corners B and E, and the

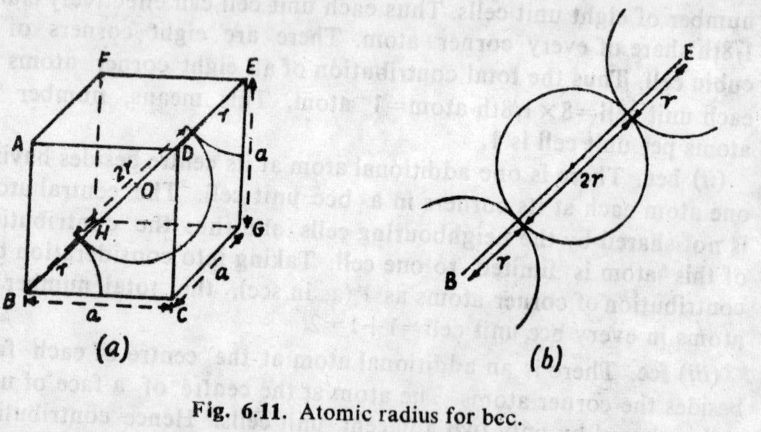

Fig. 6.11. Atomic radius for bcc.

central atom situated at O (at the middle of the diagonal of BE) touch each other in the manner shown in Fig. 6.11. It is evident, $BE=4r$.

In the right angled triangle BGE,

$$BE^2=BG^2+GE^2=(BC^2+CG^2)+GE^2=a^2+a^2+a^2=3a^2$$

Substituting $4r$ in place of BE.

$$16\,r^2=3a^2$$

or

$$r=\frac{\sqrt{3}}{4}\,a \qquad \ldots(6.4a)$$

Further, the distance between the neighbouring atoms

$$2r = \frac{\sqrt{3}}{2} a \qquad ...(6.4b)$$

(*iii*) **fcc.** In this arrangement, one additional atom is situated at the centre of each face. Consider the face *ABCD*, Fig. 6.12.

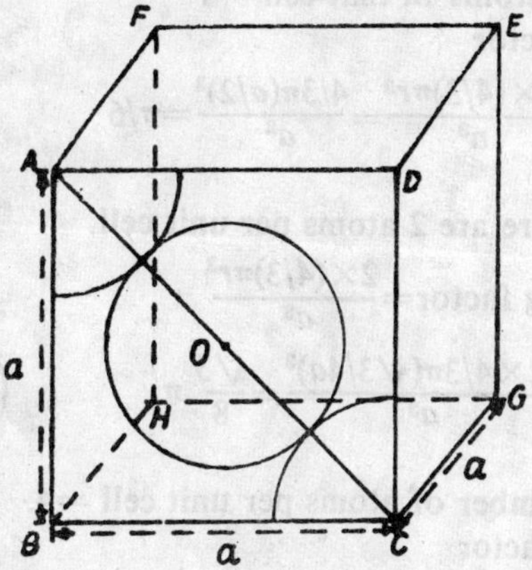

Fig. 6.12 Atomic radius for fcc.

There are four atoms at the corners *A*, *B*, *C*, *D* and the fifth atom at the centre, *O*.

It is evident from the figure,

$$AC = 4r.$$

From the right angled triangle, *ABC* we have,

$$AC^2 = AB^2 + BC^2$$

or $$(4r)^2 = a^2 + a^2 = 2a^2.$$

So $$r = \frac{a}{2/\sqrt{2}} \qquad ...(6.5a)$$

The distance between two nearest neighbours

$$= 2r = \frac{a}{\sqrt{2}} \qquad ...(6.5b)$$

(*e*) **Atomic packing factor.** Atoms considered as rigid spheres touch each other in a close-packed structure, yet they leave appreciable amount of void or empty space. The empty space relative to the total space occupied by a cell is difierent for different structures. This is described by a physical quantity called Atomic packing factor.

Atomic packing factor

$$= \frac{\text{Volume occupied by the atoms in the cell}}{\text{Volume of the unit cell}}$$

We compute below the packing factor for three cubic cells.

(*i*) scc. Let the radius of the atom be *r*.

The volume of one atom

$$=\left(\frac{4}{3}\right)\pi r^3$$

We know, the volume of unit cell $=a^3$
and number of atoms in that cell $=1$

So packing factor

$$=\frac{1\times(4/3)\pi r^3}{a^3}=\frac{4/3\pi(a/2)^3}{a^3}=\pi/6 \qquad \left(\because\ r=\frac{a}{2}\right) \quad ...(6.6a)$$

(*ii*) **bcc.** There are 2 atoms per unit cell.

$$\therefore \quad \text{Packing factor}=\frac{2\times(4/3)\pi r^3}{a^3}$$

$$=\frac{2\times 4/3\pi(\sqrt{3}/4a)^3}{a^3}=\frac{\sqrt{3}}{8}\pi \qquad \left(\because\ r=\frac{\sqrt{3}}{4}a\right) \quad ...(6.6b)$$

(*iii*) **fcc.** Number of atoms per unit cell $=4$.

$\therefore$ Packing factor

$$=\frac{4\times 4/3\ \pi r^3}{a^3}=\frac{4\times 4/3\ \pi\left(\dfrac{a}{2\sqrt{2}}\right)^3}{a^3}$$

$$=\frac{\sqrt{2}}{6}\pi \qquad \left(\because\ r=\frac{a}{2\sqrt{2}}\right) \quad ...(6.6c)$$

(*f*) **Axial ratio (c/a ratio),** Axial ratio is c/a where c is the lattice parameter along z axis and a is the lattice parameter along x and y axis. For cubic structure it is 1. But for hcp structure it is 1.633.

Example 6.1. Aluminium has fcc structure with lattice parameters $a=4.04$ A°. Calculate atomic volume of aluminium.

Solution. For fcc structure, we have $r=a/2\sqrt{2}$

Putting $a=4.04$ A°, $r=1.4283$ A°$=1.4283\times 10^{-10}$ m.

$$\text{Volume}=\frac{4}{3}\pi r^3=\frac{4}{3}\times 3.14\times(1.4283\times 10^{-10}\text{ m})^3$$

$$=0.122\times 10^{-28}\text{ m}^3.$$

Example 6.2. Aluminium has fcc structure and its atomic radius is 1.43×10^{-10} m. Find out its atomic weight when the density of aluminium is 2.70×10^3 kg/m³.

Solution. Since aluminium has fcc structure, atomic radius r is given by.

$$r=\frac{a}{2\sqrt{2}}$$

$$a=2\sqrt{2}\,r=4.04\times 10^{-10}\text{ m}.$$

So volume of the cell $V=a^3=(4.04\times 10^{-10}\text{ m})^3$.

Density of a crystalline solid,

$$\rho = \frac{\text{mass of the unit cell}}{\text{Volume of the unit cell}}.$$

$$= \frac{nA/N}{V} = \frac{nA}{NV}$$

where n is the total number of atoms in a unit cell, A is the atomic weight, N is the Avogadro's number and V is the volume of the unit cell.

In aluminium for fcc structure,

$$n=4,\ V=a^3,\ \rho=2.70\times10^3\ \text{kg/m}^3$$

$$A=\rho\frac{NV}{n}$$

$$=\frac{(2.70\times10^3\ \text{kg/m}^3)(6.02\times10^{26}\ \text{kg}^{-1})(4.04\times10^{-10}\ \text{m})^3}{4}=26.8$$

Example 6.3. Show that the axial ratio c/a for an ideal hcp structure is 1.633.

Solution. In Fig. 6.8(b), it has been shown that the *ABABAB*... type of packing leads to hcp structure.

Joining the centres of the three adjacent atoms of the middle layer *i.e.*, *A*, *B*, *D* to the centres of the atoms *C* and *E* of the top and the bottom layer (Fig. 6.P(b) we get two tetrahedrons *ACBE* and *ADBE* (Fig. 6.P(a) with common base *AB*.

Now $CE=c$.

$$AB=BC=CA=CD=AD=DB=a$$

Let Q be the mid point of BD. Draw a median AQ in the $\triangle ADB$. If P is the centroid of the $\triangle ADB$, we have $AP=(2/3)\ AQ$ (by the properties of triangle).

In the right angled $\triangle AQB$, we have

$$AQ^2=AB^2-BQ^2=a^2-\left(\frac{a}{2}\right)^2=\frac{3a^2}{4}$$

$$AP=\frac{2}{3}AQ=\frac{2}{3}\times\frac{\sqrt{3}\,a}{2}=\frac{a}{\sqrt{3}} \quad ...(i)$$

Again in right angled $\triangle APC$, having right angle at P we have,

$$CP^2=AC^2-AP^2=a^2-\frac{a^2}{3}=\frac{2a^2}{3}$$

$$CP=\sqrt{\frac{2}{3}}\,a \quad ..(ii)$$

But $CP=\dfrac{C}{2}$ as P is the mid point of CE....(iii)

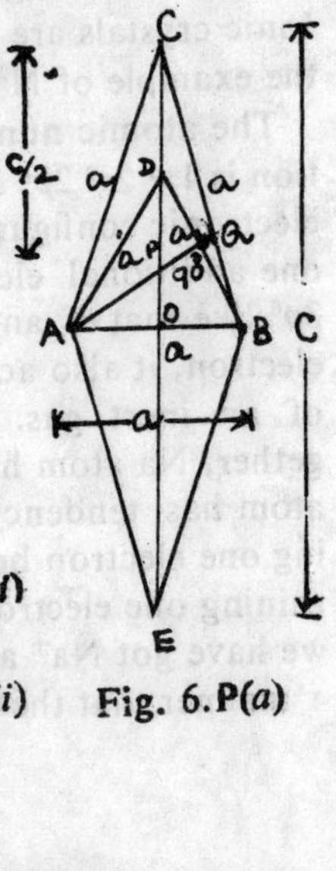

Fig. 6.P(a)

from (*ii*) and (*iii*), we get

$$\frac{C}{2}=\sqrt{\frac{2}{3}}\,a$$

$$\therefore \quad \frac{C}{a}=2\sqrt{\frac{2}{3}}=1.633.$$

Fig. 6.P(*b*)

6.6. Crystal Bonds

The obvious question that arises now is, what holds a crystal together? It has now been established that the electrostatic force of attraction between the negative charge of the electron and the positive charge of the nuclei is entirely responsible for the cohesion of solids. Magnetic forces have only a weak effect on cohesion and gravitational forces are negligible.

The electrostatic link that is formed between the oppositely charged particles to form a stable crystal is called the bond. When free neutral atoms form bond with neighbouring atom, they share a part of their energies to the bond formation. Hence a crystal can be stable only when its total energy is lower than the sum of energy of the constituent atoms of molecules when they are free. The difference *i.e.* (free atom energy)—(crystal energy) is defined as the *cohesive energy*. The stability of the crystal depends on its cohesive energy.

The properties exhibited by the crystal differs due to different types of bonds involved in their formation. There are four types of bonds:—Ionic, Covalent, Metallic and Vanderwall-London.

(*a*) **Ionic Bond**. Crystal of common salt (NaCl), cesium chloride, lithium fluoride, Na_2SO_4, $BaSO_4$, $CuSO_4$ etc are ionic crystals. Ionic crystals are made of positive and negative ions. Let us take the example of NaCl crystal.

The atomic number of sodium is 11 and its electronic configuration is $1s^2\ 2s^2\ 2p^6\ 3s^1$. The atomic number of chlorine is 17 and its electronic configuration is $1s^2\ 2s^2\ 2p^6\ 3s^2\ 3p^5$. Thus if chlorine gets one additional electron it will acquire a stable state $1s^2\ 2s^2\ 2p^6\ 3s^2\ 3p^6$ like that of an inert gas On the other hand if sodium loses one electron, it also acquires a stable configuration $1s^2\ 2s^2\ 2p^6$ like that of an inert gas. Hence when Na and Cl atoms are brought together, Na atom has tendency of losing one electron and chlorine atom has tendency to acquire one electron. Neutral Na atom losing one electron becomes positive Na^+ ion. Similarly chlorine atom gaining one electron becomes $-$ve Cl ions. Thus in NaCl crystal we have got Na^+ and Cl^- ions. The ions arrange themselves in such a manner that the electrostatic attraction between ions of opposite sign is stronger than the repulsive force between similar ions. The obvious arrangement is such that each ion has an oppositely charged ion as its neighbour. The simple lattics of NaCl is composed of Na and Cl placed alternatively at lattice points as shown in Fig. 6.6 (*a*). The ionic bond is the bond resulting from the electrostatic interaction of oppositely charged ions.

The cohesive energy of ionic crystal is very high. Therefore, they are hard and brittle. They also have high melting point and high latent heat of fusion. Since no free electrons are available, ionic crystals are bad conductors of electricity.

(*b*) **Covalent bond.** Covalent bond exists in tetratetravalent atoms like carbon, silicon, germanium etc. The atoms in the crystals having covalent bonds are held together by sharing of electrons with their neighbours.

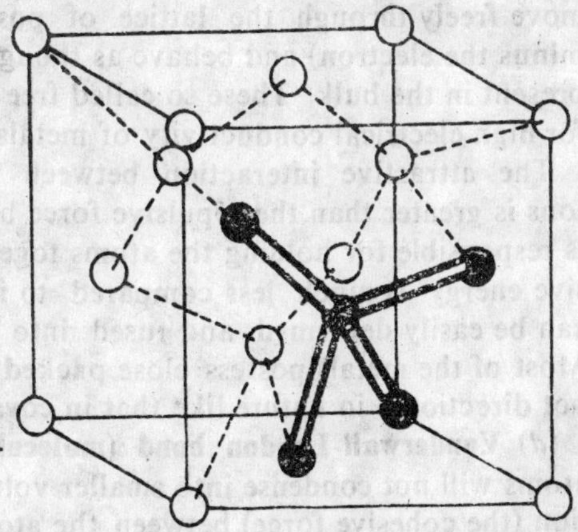

Fig. 6.13. Crystal structure of germanium (diamond structure). Double lines between the neighbouring germanium atoms indicate covalent bond by the two shared electrons.

Fig. 6.13 shows the arrangement of germanium atoms. Germanium has four electrons in its outer most shell. By putting four additional electrons in the outermost shell, the atom will have a stable configuration like that of inert gas. To acquire the stable state, each germanium atom shares one electron each from four neighbouring atoms. Further, the four valence electrons of this Ge atom are also shared by those four neighbouring atoms that contributed electrons to it. In this manner, all germanium atoms acquire stable configuration by co-sharing of valence electrons. These two shared electrons form strong directional bonds called covalent bonds.

Crystals with covalent bond have tetrahedron structure also known as 'diamond structure'. These crystals have loose packed structure and the co-ordination number is 4. They are very hard and they resist deformation. The cohesive energy for covalent bond is quite high but less than that of ionic crystals. These solids have high melting point but not as high as ionic solids have. They act as insulators at absolute zero as no free electrons are available. But due to thermal agitation with increase of temperature some of the covalent bonds break making available free electrons. Hence it acquires some amount of conductivity which increases with temperature. Silicon and germanium belonging to this category are better known as semi-conductors.

(*c*) **Metallic bond.** In metals, the valence electrons *i.e.* the electrons in the outermost shell are very loosely bound with nucleus. In a lump of the metal where large number of such atoms come together, these valence electrons leave the atoms, lose their identity and move freely through the lattice of positive ions (neutral atom minus the electron) and behave as though they belong to all atoms present in the bulk. These so called free electrons are responsible for high electrical conductivity of metals.

The attractive interaction between the electrons and positive ions is greater than the repulsive force between the electrons. This is responsible for holding the atoms together in a metal. The cohesive energy is much less compared to ionic crystals. These solids can be easily deformed and fused into each other to form alloys. Most of the metals possess close packed structure as the bonds are not directional in nature like that in covalent solids.

(*d*) **Vanderwall-London bond (molecular solids).** Any system of atoms will not condense into smaller volume unless there is attraction (the cohesive force) between the atoms. Inert gases like neon, argon, krypton and xenon etc. have their shell completely filled and are supposed not to interact with each other. Yet they also condence at sufficiently low temperature. This is also supported by indirect experimental evidences. The ideal gas law ($PV=RT$) is not obeyed by a real gas. To account for this deviation, Vanderwall thought of intermolecular forces and modified the equation of state taking into account the effect of intermolecular attraction and of finite size of the molecules. The bond resposible for this interaction are known as Vanderwall's bond. Besides inert gas crystals other crystals like paraffin, benzene, chlorine, iodine etc, are also formed due to Vanderwall's bonding. These types of solids are known as *Molecular Solids.*

Since the cohesive energy is very small, this bond is very weak. That is why molecular solids are soft, easily compressible and possess low melting point. They are poor conductors of electricity as no free electrons are available.

Given below are the characteristics of the crystals formed by different bonds.

TABLE 6.1. Characteristics of Various Type of Crystals.

Characteristics	*Ionic crystals*	*Covalent crystals*	*Metallic crystals*	*Molecular crystals*
1. Basis that occupy lattice points	+ve and —ve ions	Atoms	Positive ions in a 'sea' of electrons	Molecules
2. Binding force	Electrostatic attraction	Shared electrons	Electrostatic attraction between +ve ions and electrons.	1. Vanderwalls 2. Dipole-dipole
3. Physical properties	Quite hard and brittle	Very hard	Hard or soft	Very soft
	Fairly high melting point	Very high melting point	Moderate to high melting points	Low melting points
	Semicondutors due to crystal imperfections.	Non-conductors	Good conductors	Good insulators
4. Examples.	NaCl, KNO_3, $NaSO_4$, etc.	Dimond, quartz.	Na, Cu, Fe	NH_3, H_2O, CO_2

6.7. Electrons in Solids

Energy bands. In Chapter 14 we will discuss the structure of an individual atom and the location of the electrons with respect to

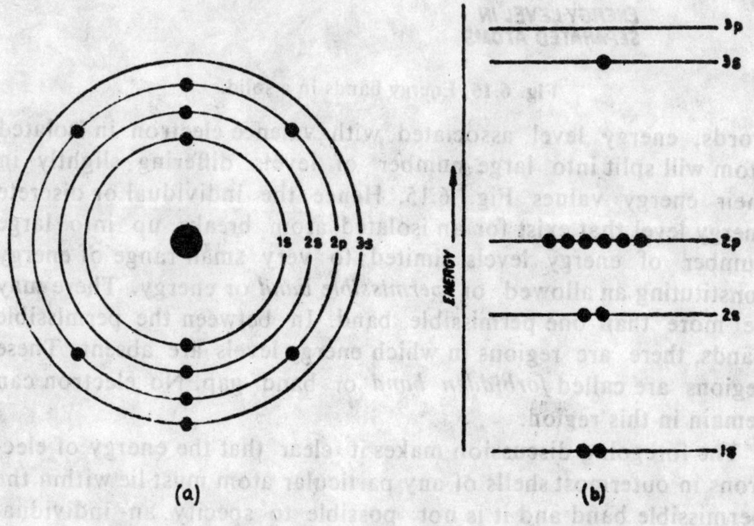

Fig. 6.14. Energy level in an isolated Na atom.

the nucleus. According to Bohr's theory an electron in a particular shell of an atom is associated with some difinite energy. Therefore, the position of the electron is described by the energy associated with it. For convenience, the energy associated with various electrons are described by energy level diagrams. Fig. 6.14.(*b*) shows the energy level diagram in an isolated atom whose shell structure to given in Fig. 6.14 (*a*).

In gases, there is practically no interaction between the electrons of one atom with the electrons of other neighbouring atoms. In solids, however, the forces that bind atoms together greatly modify the behaviour of the associated electrons. The electrons in the outermost shells of one atom (valence electron) are affected by the electrons in the outermost shells of neighbouring atoms. Therefore, the energy associated with individual electrons in the solid are different from that in the valence electron of isolated atom. In other

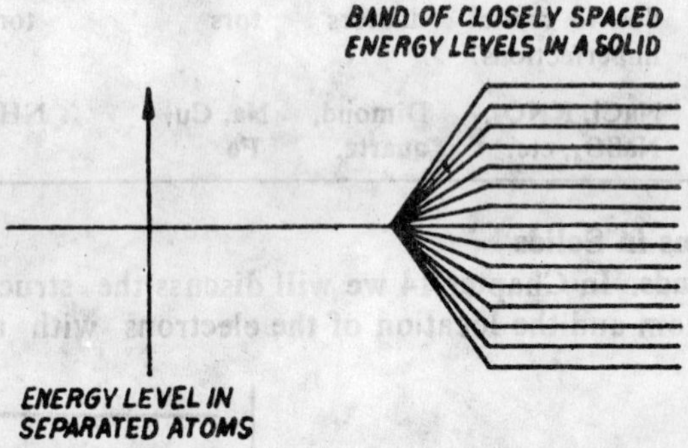

Fig. 6.15. Energy bands in a solid.

words, energy level associated with valence electron in isolated atom will split into large number of levels differing slightly in their energy values Fig. 6.15. Hence the individual or discrete energy level that exist for an isolated atom breaks up into large number of energy levels limited to very small range of energy constituting an allowed or *permissible band* or energy. There may be more than one permissible band. In between the permissible bands, there are regions in which energy levels are absent. These regions are called *forbidden band* or band gap. No electron can remain in this region.

The foregoing discussion makes it clear that the energy of electrons in outermost shells of any particular atom must lie within the permissible band and it is not possible to specify an individual level any more (as in the case of isolated atom).

Let us take example of carbon. The electronic configuration of carbon atom is $1s^2\ 2s^2\ 2p^2$. Since each electron is associated with

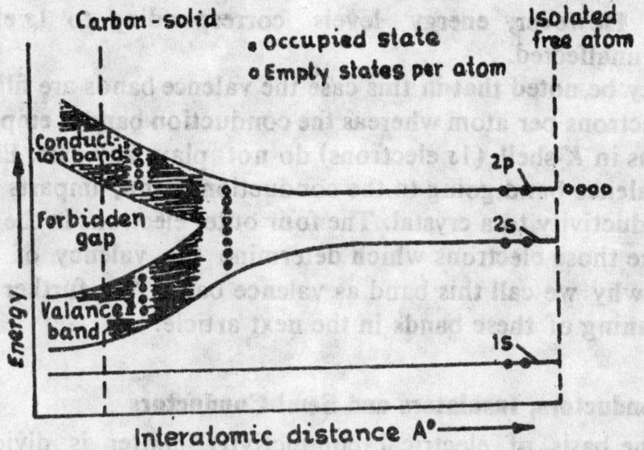

Fig. 6.16. The energy levels in carbon crystal as a function of interatomic distance. The dashed line indicate the interatomic separation in carbon.

one energy level, there are six energy levels in an isolated carbon atom *i.e.* 2 for *K* shell and 4 for *L* shell. But in *L* shell, the maximum number of electrons to accommodate is 8. So in *L* shell four permissible energy levels are still empty. As the inter atomic distance is decreased, the interaction between the atoms increase. The discrete energy level of the isolated atoms splits into bands and the bands corresponding to 2*s* and 2 *p* shell overlap. Further reduction in interatomic distance results in separation between

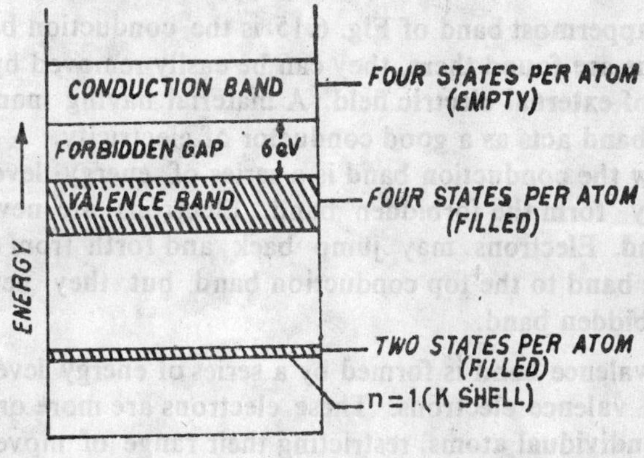

Fig. 6.17. Energy bands in carbon crystal.

the upper four energy levels and lower four energy levels (of *L* shell), Fig. 6.16. The upper levels form the *conduction band* and the lower levels from the *valence band*. Both the bands are permissible. The energy band diagram for carbon crystal is shown in Fig. 6.17, which is a section of the Fig. 6.16, after separation into energy bands. However, energy levels corresponding to $1s$ electrons remain unaffected.

It may be noted that in this case the valence bands are filled with four electrons per atom whereas the conduction band is empty. Two electrons in K-shell ($1s$ electrons) do not play any part. Electrons from valence band going to the conduction band, imparts electrical conductivity to a crystal. The four outer electrons in the valence band are those electrons which determine the valency of carbon. This is why we call this band as valence band. We further explain the meaning of these bands in the next article.

6.8 Conductors, Insulators and Semi-Conductors

On the basis of electrical conductivity matter is divided into three categories—Insulators, semi-conductors and conductors. The conductivity of insulator is very poor ($\approx 10^{-17}$ mho/metre), for conductors it is of the order of 10^7 mho/metre and for semi-conductors it lies in between these extreme values. All metals are good conductors. Germanium, silicon are examples of semi-conductors. Insulators are, quartz, mica, ebonite, etc. Conductivity of silver is 6×10^7 mho/meter ; that of quartz is 2×10^{-17} mho/meter and for germanium it is nearly 2 mho/meter. These are known facts. Here, however, we will explain this property on the basis of energy band.

The uppermost band of Fig. 6.15 is the conduction band. When electrons are found there, they can be easily removed by the application of external electric field. A material having many electrons in this band acts as a good conductor of electricity.

Below the conduction band is a series of energy levels that collectively form the forbidden band. Electrons are never found in this band. Electrons may jump back and forth from the bottom valence band to the top conduction band but they never stay in the forbidden band.

The valence band is formed by a series of energy levels containing the valence electrons. These electrons are more or less bound to the individual atoms, restricting their range of movement compared to the movement of electrons in the conduction band. Electrons can be moved from the valence band to the conduction band by application of energy. The amount of energy required to transport an electron from the valence band to the conduction band indicates the width of the valance band. In otherwords, the wider this band, more energy is required to raise an electron from the valence band to the conduction band where it can become a carrier of electricity. Thus the width of the forbidden band determines whether a substance is an insulator or a semi-conductor or a good conductor of electricity.

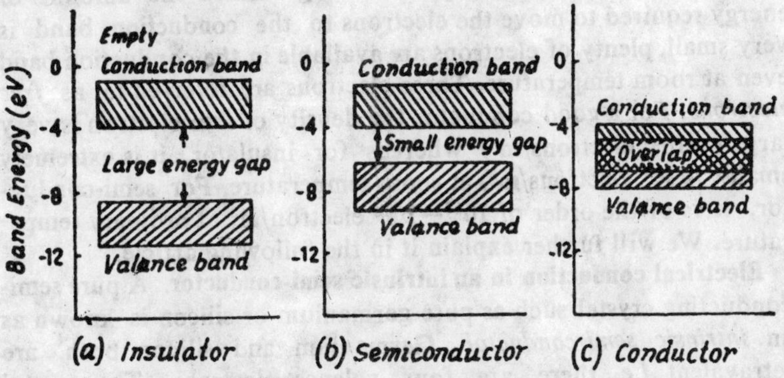

Fig. 6.18. Energy bands in (*a*) insulator (*b*) semi-conductor (*c*) conductor.

Fig. 6.18 shows the difference between insulators, semi-conductors and conductors in terms of energy bands. The forbidden band is widest in insulators. In terms of energy it is more than 3 ev. For diamond at room temperature, it is about 6 ev. In insulators, the valence band is completely occupied and the conduction band is normally empty. However, at higher temperature some of the electrons in the valence band may acquire energy to jump to the conduction band. So the conductivity of an insulator increases marginally with increase of temperature. Since the forbidden band is very wide, large applied voltage is necessary to transport an electron from valence to conduction band. That is why the insulators have very negligible conductivity.

In a semi-conductor, the forbidden band is narrower. This means, comparatively less energy is needed for transporting electron from valence to the conduction band. Hence, in semi-conductors, comparatively more current will flow for a certain applied voltage, but this current will not be as large as we would obtain in a conductor. The width of forbidden band in silicon is 1.1 ev and that in germanium is 0.7 ev. Like insulators, the valence band is full and the conduction band is empty. But due to narrow width of the forbidden band, some electrons jump over to the conduction band even at room temperature. Thus its conductivity is more compared to that of an insulator. With increase in temperature, number of such electrons increase. So the conductivity of the semi-conductor increases appreciably with rise of temperature.

In a conductor, the valence band and the conduction band overlap *i.e.* the forbidden band is absent. So we need only a small amount of energy, of the order of 0.01 ev, to move electrons into the conduction band. Consequently, electricity flows through the conductors with very little applied voltage. Since the amount of energy required to move the electrons to the conduction band is very small, plenty of electrons are available in the conduction band even at room temperature. These electrons are referred to as *free electrons*. For a good conductor the density of free electron is very large ($\approx 10^{28}$ electrons/m^3) whereas for insulator, it is extremely small ($\approx 10^7$ electrons/m^3) at room temperature. For semi-conductor, it is of the order of $10^{13}-10^{15}$ electron/m^3 at ordinary temperature. We will further explain it in the following article.

Electrical conduction in an intrinsic semi-conductor. A pure semi-conducting crystal such as pure germanium or silicon is known as an *intrinsic semi-conductor*. Germanium and silicon both are tetravalent *i.e.* there are four valence electrons. The crystal structure of Ge or Si is known as diamond structure. In this structure the co-ordination number is four *i.e.* each atom has four neighbouring atoms. All these atoms are equidistant from each other. Fig. 6.19 represents a two dimensional lattice for silicon crystal. Here each Si atom is surrounded by four Si atoms and they are all equidistant from each other. we have discussed in Art. 6.6 that for tetravalent atoms, each atom attains stable configuration by co-sharing its valence electrons with four neighbouring atoms. Thus the binding force or the bond between the neighbouring atoms arise due to the sharing of valence electrons. These bonds are shown as dashed lines in Fig. 6.19. Since all the valence electrons of silicon atoms in a silicon crystal form covalent bonds, free electrons are not available and pure silicon crystal acts as an insulator so long as this crystal structure remains unaffected. But the crystal structure as depicted in Fig. 6.19 remains true only at very low temperature (near 0°K). So silicon will be a perfect insulator near 0°K. At room temperature however, some of the electrons may acquire sufficient thermal energy to break the covalent bond and jump from valence band to the conduction band making available free electrons for electric conduction. The minimum energy required to break a covalent bond and therefore to remove an electron from the valence band to the conduction band is equal to forbidden energy gap which is 1.1 ev for silicon.

Only 1 or 2 electrons out of 10^8 electrons may acquire this amount of energy at room temperature to break the bond. But considering the large number of atoms present in 1 cc of silicon crystal, it is expected that comparatively a large number of free electrons will be available in silicon crystal at room temperature *i.e.*

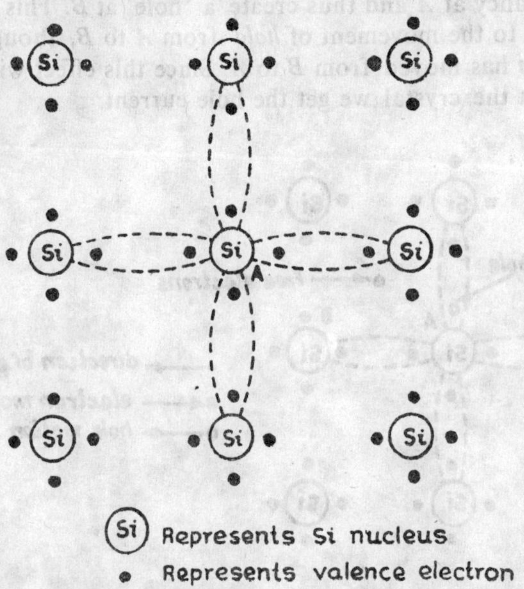

Fig. 6.19. Two dimensional lattice for silicon crystal.

of the order of 10^{13} to 10^{15} electrons/m^3. So silicon crystal shows considerable amount of conductivity at room temperature.

When a covalent bond is broken, an electron jumps from the valence band to the conduction band. This creates a vacancy of one electron in the valence band. This electron vacancy in the valence band is called a *Hole*. This is shown in Fig. 6.20 where one covalent bond of Si atom *A* is shown as broken. The 'Hole' and the free electron both are indicated there.

The importance of the holes lies in fact that they take part in the conduction of electricity through the semi-conductor like the free-electrons. A 'Hole', created due to the vacancy of an electron is imagined to have a positive charge. So when an electric potential difference is applied across a silicon crystal the free electrons carrying negative charge ($-q$) will move in the direction from lower to higher potential. But the 'holes' with positive charge ($+q$) will move from higher to lower potential. So, together they will constitute the current.

Total current=Electronic current in the conduction band+hole current in the valence band.

The motion of holes is actually the motion of electrons in the valence band. Suppose initially a 'hole' is present at *A* and then a field is applied as shown in Fig. 6.20. It is possible that the electron forming the covalent bond at the next atom *B* may jump in the vacancy at *A* and thus create a 'hole' at *B*. This in effect an equivalent to the movement of *hole* from *A* to *B*, though actually an *electron* has moved from *B* to *A*. Since this effect will continue thruoghout the crystal we get the hole current.

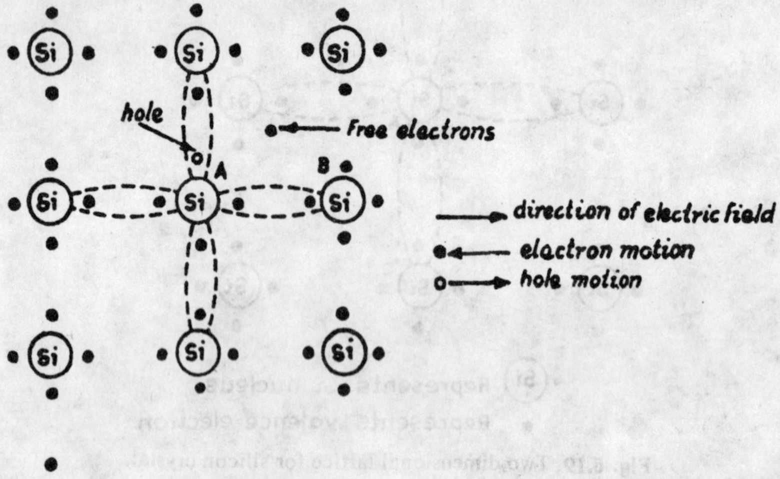

Fig. 6.20. Breaking of covalent bond produces electron-hole pair.

The number of free electrons and holes are equal to each other at any time. But the free electrons move faster in the conduction band and the holes (actually electrons) move comparatively at a slow velocity in the valence band. Therefore, the current in a semi-conductor is largely due to free electrons and a very small fraction is due to holes.

As the temperature is increased more number of covalent bonds break and more free electrons and holes are available. Hence the conductivity of a pure semi-conducting material increases with increase of temperature *i.e.* it has got negative temperature co-efficient of resistance. Since the conductivity of pure semi-conducting material depends on 'free electrons' and 'holes' created within itself, they are called intrinsic semi-conductors.

Charcteristic properties of semi-conductors. From the earlier discussions, we can summarise the following properties of a semiconductor.

1. A pure semi-conducting crystal at absolute zero temperature

behaves as a perfect insulator as it contains no free electrons.

2. It becomes conducting as the temperature is increased due to free hole or free electron creation.
3. Its conductivity is mostly due to electrons.
4. Its conductivity increases with temperature *i.e.* it has got negative temperature co-efficient of resistance.
5. Since the number of free electrons available in semi-conductor is much less than the number of free electrons available in a conductor, the conductivity of semi-conductor is less than that of a conductor. But it is much higher than that of an insulator.
6. The semi-conductor crystal is electrically neutral inspite of the fact that it contains free charge carriers and ionised atoms.

Extrinsic semi-conductors. *N* and *P* type semi-conductors. The condutivity of a pure semi-conductor can be increased considerably by adding few atoms of a trivalent substance like indium, boron, gallium or aluminium or few atoms of a pentavalent substance like arsenic, phosphorous or antimony to a pure semi-conductor. The foreign material called impurity is added in a very small quantity say 1 to 2 atoms for 10^6 Si atoms. This process is called doping.

The degree of doping controls the conductivity of the semi-conductor. We will see how the doping makes more free electrons and holes available for conduction of electricity. Hence these types of semi-conductors are known as **extrinsic semi-conductors.**

Suppose a pentavalent impurity such as arsenic is added to a silicon crystal. Consider Fig. 6.21 where one arsenic impurity atom has taken the place of one Si atom. So the arsenic atom is surrounded by four silicon atoms. Four out of five valence electrons of arsenic form covalent bonds with the four neighbouring silicon atoms and the fifth valence electron does not take part in binding process. So this fifth electron moves very easily to conduction band. Thus a free electron is available.

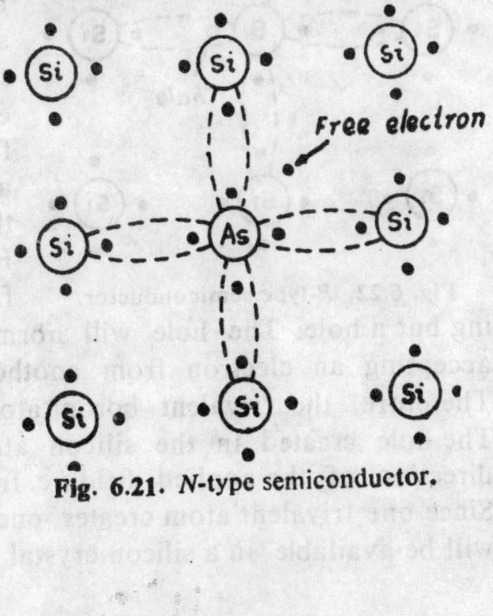

Fig. 6.21. *N*-type semiconductor.

Hence addition of one impurity atom donates one free electron to the silicon crystal and therefore large number of free electrons are available for conduction of electricity in a crystal. Since the arsenic atom donates free electron to the silicon crystal, it is called a *donor atom*.

It is to be remembered that free electron and hole pairs are also formed due to thermal agitation as in case of intrinsic semi-conductors. But due to availability of large number of electrons donated by the impurity arsenic atom, most of these holes recombine with electrons and hence very few holes are present in silicon crystal.

Therefore, in silicon crystal doped with arsenic the conductivity is mostly due to electrons. As electron has negative charge, this type of semi-conductors are known as *N*-type semi-conductors. Since in a *N* -type semi-conductors electrons are the principal charge carriers, they are known as *majority carriers*. There will be a very small current due to small number of holes still present. The holes are, therefore, known as '*minority carriers*' in a *N* type semi-conductors.

Let us now consider the effect of doping the silicon crystal by a trivalent impurity say boron. In Fig. 6.22 one boron atom has taken the place of a silicon atom. So one boron atom is surrounded by four silicon atoms. The three valence electrons of boron from covalent bonds with three neighbouring silicon atoms, but for the fourth silicon atom, no electron can be contributed by the boron atom for bond formation. Therefore, a vacancy of electron exists in the fourth band as shown in Fig. 6.22. The missing electron from the covalent bond is nothing but a hole. The hole will normally be filled by snatching or accepting an electron from another neighbouring silicon atom. Therefore, the trivalent boron atom is called an *acceptor atom*. The hole created in the silicon atom continues to travel in the direction of the applied field *i.e.* from higher to lower potential. Since one trivalent atom creates one hole, large number of holes will be available in a silicon crystal doped by boron. In addition

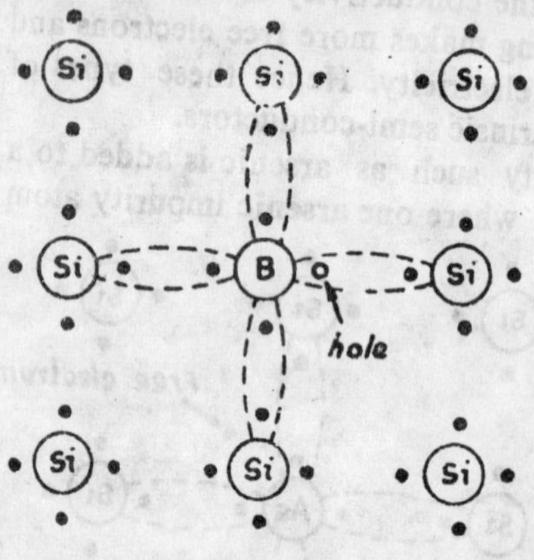

Fig. 6.22. *P*-type semiconductor.

to the holes created by the impurity atoms, free electrons and hole pairs created by thermal agitation are also present as in case of intrinsic semi-conductor. But most of the free electrons are neutralised by recombination with some of the holes created by the acceptor atoms. So conductivity of this type of semi-conductor is due to mainly positively charged holes. Hence they are called *P*-type semi-conductors. The holes are the majority carriers in a *P*-type semi-conductor and few free-electrons present due to thermal agitation are *minority carriers*.

An extrinsic semi-conductor behaves like a resistor. The resistance of the extrinsic semi-conductor is known as bulk-resistance. If doping is more, the bulk resistance is less since more majority carriers are created. But with excessive doping, the conductivity may be quite high. In such case, it will not act as a semi-conductor.

From the foregoing discussions we can sum up these relevant features of extrinsic semi-conductors.

1. A tetravalent substance like silicon or germanium doped with few atoms of a pentavalent element like phosphorous, arsenic etc. results in a *N*-type extrinsic semi-couductor.
2. But the tetravalent substance doped with few atoms of a trivalent element like boron, indium etc, creates a *P*-type extrinsic semi-conductor.
3. Electrons are majority carriers of electricity in a *N*-type semi-conductor whereas holes are majority carriers of electricity in *P*-type semi-conductor. Holes are minority carriers in *N*-type and electrons are minority carriers in *P*-type semi-conductor.
4. An extrinsic semi-conductor behaves like a resistor *i.e.* that it obeys Ohm's law when the applied voltage is low. Its conductivity increases if doping is increased.

QUESTIONS

1. Amorphous solid is isotropic whereas a crystalline solid is anisotropic. Explain.
2. Show that the void in a hcp structure is 26% of the total volume.
3. Find the distance between the nearest neighours in hcp structure.
4. Copper has fcc structure. If the atomic radius of copper is 1.27 A°, estimate the lattice parameter. (3.6 A°)
5. How many lattice points are there in a bcc unit cell?
6. What do you mean by the coordination number in a crystal structure? How do you find the co-ordination number in hcp structure?
7. Which of the following elements do not form a close packed structure? Al, Zn, Cu, Si, B, Ge, Pb, Mg, Cd.

8. What is an ionic crystal? Give five examples of ionic caystals not mentioned in this book.

9. On the basis of the structure of metals how would you account for the following :

1. High electrical conductivity, 2. Thermal conductivity, 3. Ductility, 4. Softness, 5. Tensile strength, 6. Malleability.

10. Covalent crystals have very high melting points, but the molecular crystals have low melting points. Why ?

11. Molecular crystals are good insulators. Explain the reason.

12. What will happen if silicon is doped with a divalent impurity?

13. What are the 'basis' of crystal srructure in different type of crystalline solids ?

14. The temperature of a bulk of pure germanium is increased. What happens to its resistivity?

15. While doping silicon with boron, the quantity of boron exceeded the prescribed value. How will the semi-conducting property be affected?

16. If degree of doping is equal, will two similar pieces of *P* and *N* type semi-conductor have same conductivity?

17. Increase in temperature increases resistivity in metallic conduction but decreases in semi-conductors. Explain.

7

PROPERTIES OF MATERIALS

The advancement in technology is only possible through our detailed understanding of the properties of materials, in connection with their mechanical, thermal, electrical and magnetic behaviour. By such understanding methods can be devised to develop materials of specific properties to meet our various needs. The Physicist finds it a challenge to account for the observed properties of materials in terms of atomic arrangement and inter atomic forces. We will discuss different properties of materials and explain them in terms of atomic structure and interatomic forces in this chapter.

7.1 Mechenical Properties of Solids

Mechanical properties of solids in general deals with the deformations due to forces applied on the materials. An ideal rigid body does not change its shape or size under the influenc of forces of any magnitude. But such a body is never realised in practice. On the other hand various material bodies, we see around us are always deformed under the action of external forces. They undergo a change in shape or size when subjected to external forces. These external forces are called '*Deforming forces*'.

If the forces are not too great, a body returns to its original condition, when they are removed. This property of the body by which it experiences a change in shape or size, when deforming forces act upon it and by which it returns to its original size or shape, when the deforming forces are removed, is called '*Elasticity*'.

7.2. Stress and Strain

As a result of deforming forces applied to a body forces of reaction come into play internally in it, due to the relative displacement of its atoms. These forces tend to restore the body to its original conditions. The restoring or recovering force per unit area over which the force is distributed, is called 'Stress'. But the restoring force can not be measured directly. Therefore the stress is measured by the deforming force applied per unit area of the body since the magnitude of restoring force is equal to the magnitude of deforming force. The stress may be normal or tangential depending whether deforming force acts normally or tangentially. Stress being force per unit area, the unit and dimensions of it are same as those of pressure *viz.* N/M^2 and $ML^{-1}T^{-2}$ respectively.

A body under deforming forces undergoes a change in length, volume or shape. Then the body is said to be under 'strain'. The strain produced in the body is measured in terms of the change in some dimensions of the body such as its length or volume and so on, divided by the original dimension (measure). According to the different kinds of deformations, the strains produced are differently named. As strain is a ratio of two like quantities, it is a pure number and has no unit for it.

7.3. Elastic and Plastic Bodies

All bodies do not behave in the same way, when they are deformed and then deforming forces are withdrawn. Some of them get back to their original condition soon after the deforming forces are withdrawn but some others, once deformed, do not show any tendency to return back. On the other hand they retain their altered state completely. The substances belonging to the former category are known as 'Perfectly Elastic'. And the substances belonging to latter class are called 'Perfectly Plastic'. But in practice no substance is perfectly elastic, or perfectly plastic. The nearest approach to perfectly elastic body is 'Quartz-fibre' and to perfectly plastic body is 'Putty'.

7.4. Elastic Limit

A body behaves perfectly elastic as long as stress on it does not exceed a certain maximum or limiting value. This maximum value of the stress is known as 'Elastic Limit'. It is found that elastic bodies have very large elastic limit where as plastic bodies have very small elastic limit.

7.5. Hooke' Law

It is a fundamental law of elasticity and was established by Robert Hooke of England. It is stated as, *the deformation produced in an elastic body is directly proportional to the deforming force provided that the elastic limit is not exceeded.*

A simple experiment can be performed to test the validity of this law. This can be done as follows.

A helical spring of length l_0 is taken and suspended from a rigid support, Fig. 7.1 (*a*). Weights are successively added to the lower end of the spring. As a result of this, the spring elongates. Then

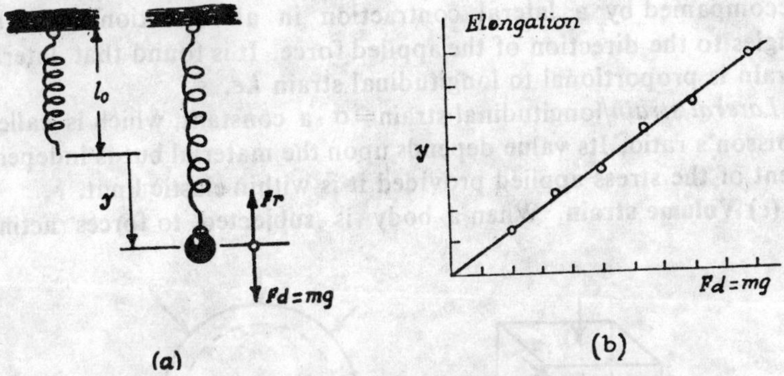

Fig. 7.1. (*a*) The weight of the ball elongates the spring. The elastic force, F_r is the reaction of the spring on the ball, (*b*) The deformation y is proportional to the applied force F_d.

the corresponding elongations y are noted. It is found that the deforming force F_d is proportional to the lengthening of the spring.

Hence $\quad F_d \propto y$ or $\quad F_d = ky$ ...(7.1)

The restoring force F_r set up in the spring being equal and opposite to the deforming force is given by

$$F_r = -ky \qquad ...(7.2)$$

The constant k is called the force constant of the spring. In the Equation 7.1 if y=unit lenthg, then $F_d = k$ *i.e.* k is the restoring force per unit length which means k is the magnitude of force necessary to stretch the spring for one unit length.

The act of stretching, stores potential energy in the spring. It can do work through the action of the restoring force, when the stress is taken off. The potential energy stored for a displacement y is given by

$$P = \frac{1}{2} ky^2 \qquad ...(7.3)$$

A graph can be drawn between F_d and y which will come out a straight line as shown Fig. 7.1 (*b*). This graph indicates that restoring force is directly proportional to increase in length.

7.6. Different Kinds of Strain

(*a*) **Tensile (longitudinal) strain.** When a body is acted upon by a deforming force along its length a change in length is produced. The ratio of the change in length to the original length is called longitudinal or Tensile strain. The corresponding stress is known as Tensile stress. It is measured by the deforming force per unit area of cross section.

(*b*) **Poisson's ratio.** When the body is acted upon by stretching force, the extention in the direction of the applied force is always accompanied by a lateral contraction in all direction at right angles to the direction of the applied force. It is found that lateral strain is proportional to longitudinal strain *i.e.*

Lateral strain/longitudinal strain=σ, a constant, which is called poisson's ratio. Its value depends upon the material but is independent of the stress applied provided it is within elastic limit.

(*c*) **Volume strain.** When a body is subjected to forces acting

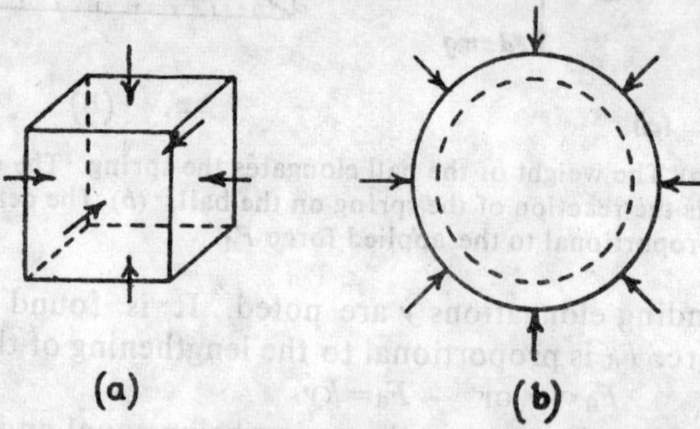

Fig. 7.2. Volume stress. Normal forces on (*a*) The faces of cube, (*b*) The surface of the sphere.

uniformly and normally over entire surface of the body as shown in Fig. 7.2 (*a*) and 7.2 (*b*), the body undergoes a change in volume but not in the shape. The ratio of the change in volume to the original volume is known as 'Volume Strain' and corresponding stress is known as 'Volume Stress' or 'normal stress'. Such a stress is sometimes called hydrostatic pressure for the fact that a body feels such stress when dipped in liquid.

(*d*) **Shearing strain or shear.** Tangential forces on acting on the body change the shape of the body but not its volume and thickness. Such a change is known as shear and the strain produced is

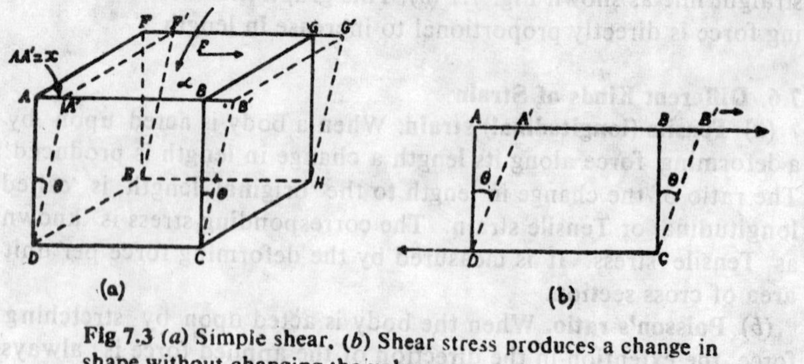

Fig 7.3 (*a*) Simple shear, (*b*) Shear stress produces a change in shape without changing the volume.

known as shearing strain or simply shear. It is a special property of the solids only because the solids have definite shape of their own.

To have an idea of shear deformation, consider a cube *ABCD EFGH* having its bottom face *CDEH* fixed to horizontal platform. Let the surface area be a and a tangential force F acts tangentially on the upper face as shown in the diagram, Fig. 7.3 The force F causes the consecutive horizontal layers of the cube to be slightly displaced or sheared relative to one another The cube will take the rhombic form as shown in the Fig. 7.3(*b*). The material of the block suffers a change in shape only without any change in volume. The strain produced in this case is a case of shear and is measured by the angle ADA' or BCB' which is called the angle of shear. Let AA' be equal to x and AD be equal to b then as θ (in radian) is small.

Shearing strain $\theta = \tan\theta = \dfrac{x}{b}$

$$= \frac{\text{relative dlsplacement of two planes of the body}}{\text{distance of separation of the two planes}}$$

The corresponding stress which is tangential to the surface called the shearing stress and is given by F/a.

7.7. Elastic Modulii

Young's definition of Hooke's law. It states that *within elastic limit stress is proportional to strain i.e., within elastic limit,*

$$\frac{\text{stress}}{\text{strain}} = \text{constant} = E.$$

The constant E is called modulus of elasticity or the co-efficient of elastitcity of the material of the body. This law holds good for all cases of elasticity such as: tensile, volume, twisting, bending etc.... For different types of strain, the modulus has different value and is also differently named. But we will consider three types of modulus of elasticity here, because they are simple.

These are

(*a*) Young's modulus $Y = \dfrac{\text{tensile stress}}{\text{tensile strain}}$

(*b*) Bulk modulus $B = \dfrac{\text{volume stress}}{\text{volume strain}}$

(*c*) Rigidity of modulus

$$\eta = \frac{\text{shearing stress}}{\text{shearing strain}}$$

(*a*) **Young's modulus.** The ratio of the tensile stress to the tensile strain within elastic limit is called Young's modulus. It is denoted by Y.

Suppose a wire of length L and area of cross section A is stretched by a force F, which acts in the direction of length. Let the increase in length be $\triangle L$. Then the tensile stress is given by F/A and the corresponding tensile strain is given by

$$\frac{\triangle L}{L}$$

Hence Young's modulus $Y=\dfrac{FA}{\triangle L/L}=\dfrac{F\times L}{A.\triangle L}$ ...(7.4)

(*b*) **Bulk modulus.** The ratio of the volume stress to the volume strain is known as Bulk modulus.

Consider a body of volume V, subjected to a uniform pressure P which acts on the body from all sides. Let the decrease in the volume of the body due to application of the pressure be $\triangle V$.

Then the volume strain is given by $-\dfrac{\triangle V}{V}$ and the volume stress is given by P. According to definition Bulk,modulus

$$B=\frac{P}{-\dfrac{\triangle V}{V}}=\frac{-PV}{\triangle V} \quad ...(7.5)$$

The values of Bulk modulus are always positive. The negative sign in the equation compensates the fact that an increase in pressure leads to decrease in volume. Therefore $\triangle V$ is negative and B is positive.

(*c*) **Elasticity of shear or modulus of rigidity.** It is the co-efficient of shear elasticity and is equal to $\dfrac{\text{shering stress}}{\text{shearing strain}}=\eta$

Thus in the case of shear as described in art (7.6*d*),

Strain$=\theta$ (in radians) $=\dfrac{x}{b}$

and stress $=\dfrac{F}{a}$

Therefore

$$\eta=\frac{F/\alpha}{\theta}=\frac{F/a}{x/b} \quad ...(7.6)$$

7.8. General Elastic Behaviour of Solids: Stress-Strain Diagram

The elastic behaviour of a solid particularly that of a metal can conveniently be illustrated by a diagram called stress-strain diagram. In order to obtain such a diagram; a metal in the form of a wire is taken and is subjected to an increasing tensile stress.

The process is continued till the wire breaks. Noting each time the stress applied and the corresponding strain produced a graph is plotted. The curve so obtained is known as stress-strain diagram (Fig. 7.4*a*).

To study the stress-strain diagram to obtain necessary informations regarding the elastic behaviour of solids, the curve can conveniently be divided into four parts such as *OA*, *AB*, *BC*, and *CD*.

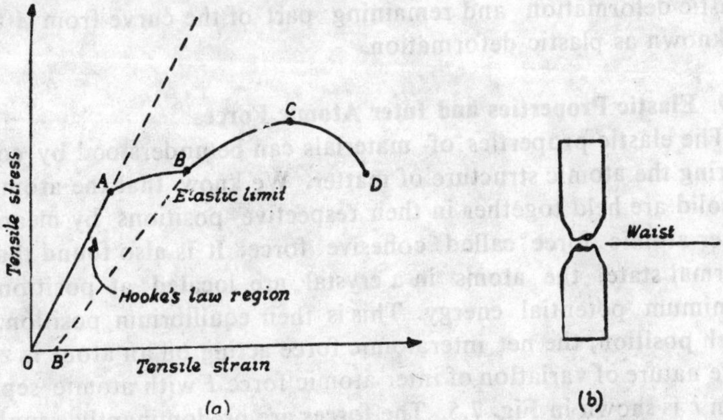

Fig. 7.4(*a*) Stress-strain graph. (*b*) Nocking or waist.

Part OA. The part *OA* of the curve is a straight line. That is, in this region there is a linear relationship between stress and strain and the material obeys Hook's law. If the stress is not carried beyond *A*, the specimen returns to its original length, when the stress is removed. In other words the portion of the curve from *O* to *A* is the region of perfect elasticity. The stress corresponding to the point '*A*' is called elastic limit.

Part AB. After *A*, the curve takes a slight bend and continues upto *B*. As seen in the curve the part *AB* is almost parallel to strain axis. This indicates that in this region a slight increase in stress produces a larger strain in the material. The point *B* is known as yield point. At any point between *A* and *B*, if the stress is taken off the specimen will never returns back to original length, but acquires a permanent set or strain as shown by the dotted line in the diagram.

Part BC. After passing point *B*, the specimen seems to regain strength some what. Here more stress than in the region *AB* is necessary to produce same strain. This phenomenon will continue upto *C*. The stress corresponding to point *C* is called maximum or ultimate stress, because beyond this point the specimen elongates rapidly. A waist or local contraction develops at some part of the material where finally fracture occurs. The position of the material where it will occur is however unpredictable.

Part CD. *CD* is the portion where local contraction or necking occurs and breaking may occur at any point between *C* and *D*. The stress for which the specimen breaks is called breaking stress.

The part of the extension curve upto the elastic limit '*A*' is called elastic deformation and remaining part of the curve from *A* to *C* is known as plastic deformation.

7.9. Elastic Properties and Inter Atomic Forces

The elastic properties of materials can be understood by considering the atomic structure of matter. We know that the atoms in a solid are held together in their respective positions by means of inter-atomic force called cohesive force. It is also found that in normal state, the atoms in a crystal are located at positions of minimum potential energy. This is their equilibrium position. At such position, the net interatomic force acting on an atom is zero. The nature of variation of inter atomic force *F* with atomic separation *r* is shown in Fig. 7.5. The forces are predominantly repulsive for small separation but becomes predominantly attractive for large separation. In the normal state, these two forces balance and the net force is zero. The separation r_0 in this state is called equilibrium distance. But when a tensile stress is applied, to a body the tensile stress increases the separation of the atoms and the atoms will experience a force of attraction. If the external force is removed, the attractive force between the atoms bring them back to their equilibrium positions. On the other hand if a rod is compressed, the separation *r* of the atoms will reduce and as a result a force of repulsion will develop between them On removal of the external force, this force of repulsion again takes the atoms back to their equilibrium positions.

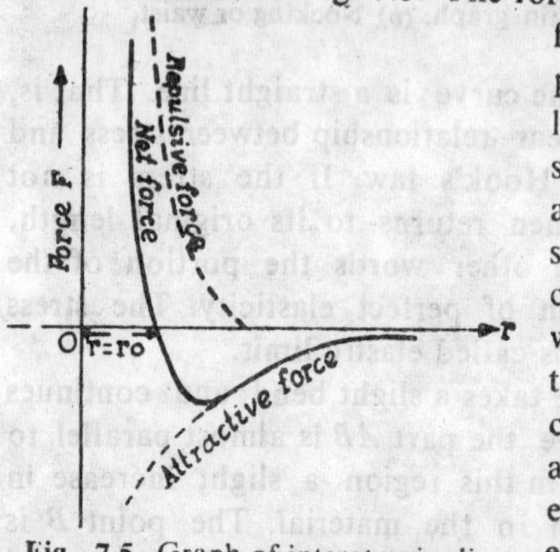

Fig. 7.5. Graph of interatomic distance vs interatomic force.

The displacement suffered by individual atoms for particular tensile force depends upon the strength of the interatomic forces.

Stronger the interatomic forces, smaller will be the displacement. Hence for particular stress the corresponding strain will be less and the value of Young's modulus will be larger.

In the case of a shear stress the distance between atoms in a plane remains the same, but neighbouring planes of the atoms in a

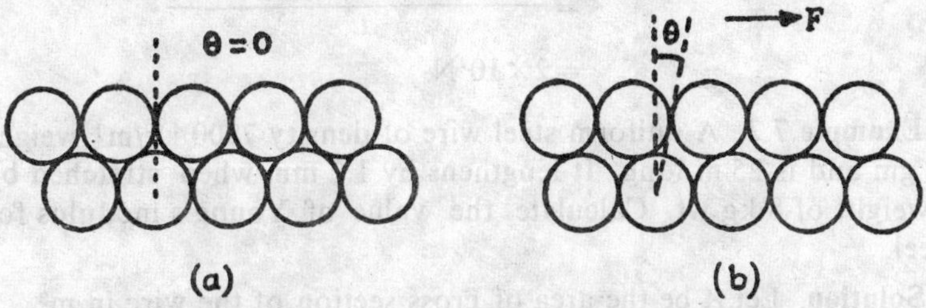

Fig. 7.6. (*a*) Location of atoms in neighbouring planes in the absence of any stress. (*b*) Shear deformation makes one plane slide past each other.

solid slide past each other which results in shear strain. The situations have been shown in the Figures 7.6 (*a*) and 7.6 (*b*).

TABLE 7.1. Elastic Modulus (Approximate Volume).

Meterial	*Young's modulus* 10^{10} N/m^2	*Shear modulus* 10^{10} N/m^2	*Bulk modulus* 10^{10} N/m^2
Aluminium	7.0	2.6	7.7
Brass	10	3.5	11
Copper	13	4.8	14
Glass	6.0	3.1	3.7
Lead	1.6	0.5	4.6
Iron	20	8.0	17
Steel	20	8.4	17

Example 7.1. What force is required to stretch a steel wire of 1 sq cm cross section to double its length?

(Young's modulus of steel$=2\times10^{11}$ N/m^2)

Solution. Let L be the length of the wire. Since its length is doubled by stretching force, the change in length is L.

$$\text{Young's mdulus } Y=\frac{F\times L}{A\times\Delta L}$$

Here $\triangle L = L \quad Y = 2 \times 10^{11}\ \text{N/m}^2$

$A = 1\ \text{sq cm} = 10^{-4}\text{m}^2$

$$\therefore \quad F = \frac{Y \times A \times \triangle L}{L}$$

$$= \frac{2 \times 10^{11}\ \text{N/m}^2 \times 10^{-4}\text{m}^2 \times L}{L}$$

$$= 2 \times 10^7 \text{N}$$

Example 7.2. A uniform steel wire of density 7800 kg/m³ weighs 16 gm and is 25 m long. It lengthens by 1.2 mm when stretched by a weight of 8 kg wt. Calculate the value of Young's modulus for steel.

Solution. Let A be the area of cross section of the wire in m².

Length of the wire $L = 2.5$ m.

Mass of the wire $m = 16\ \text{gm} = 16 \times 10^{-3}\ \text{kg}$

Density of wire $= 7800\ \text{kg/m}^3$.

The volume of the wire $= (A \times L)\ \text{m}^3$

$= A \times 2.5\ \text{m}^3$ ...(*i*)

But the volume $= \dfrac{\text{mass}}{\text{density}} = \dfrac{16 \times 10^{-3}}{7800}\ \text{m}^3$ ...(*ii*)

From equation (*i*) and (*ii*) we get,

$$2.5\,A = \frac{16}{78} \times 10^{-5}\ \text{m}^3$$

$$A = \frac{16}{78} \times \frac{10^{-5}}{2.5}\ \text{m}^2$$

Deforming force $= 8\ \text{kg wt} = 8 \times 9.8$ N.

Increase in length $\triangle L = 1.2\ \text{mm} = 12 \times 10^{-4}$ m.

Young's modulus $Y = \dfrac{\text{stress}}{\text{strain}} = \dfrac{F \times L}{A \times \triangle L}$

$$Y = \frac{8 \times 9.8\text{N} \times 2.5\ \text{m}}{\frac{16}{78} \times 10^{-5} \times \frac{1}{2.5}\ \text{m}^2 \times 12 \times 10^{-1}\ \text{m}}$$

$$= \frac{8 \times 9.8\ \text{N} \times 2.5 \times 78 \times 2.5\ \text{m}}{16 \times 10^{-5} \times 12 \times 10^{-4}\ \text{m}^3}$$

$$= 1.99 \times 10^{11}\ \text{N/m}^2$$

Example 7.3 A cube of aluminium 10 cm side is subjected to a shearing force of 100 Newton. The top face of the cube is displaced by 0.01 cm with respect to the bottom. Calculate the shearing stress, the shearing strain and shear modulus.

Solution. Shearing stress

$$=\frac{\text{Tangential force}}{\text{Area of the face}}$$

$$=\frac{F}{A}=\frac{100\ \text{N}}{(10\times10^{-2})\ \text{m}^2}=10^4\ \text{N/m}^2$$

Shearing strain $\theta=\dfrac{\text{Displacement}}{\text{Side of the cube}}$

$$=\frac{0.01\ \text{cm}}{10\ \text{cm}}=10^{-3}$$

Shearing modulus $\eta=\dfrac{\text{Shearing stress}}{\text{Shearing strain}}=\dfrac{10^4\ \text{N/m}^2}{10^{-3}}$

$\therefore$ $\eta=10^7\ \text{N/m}^2$

Example 7.4. What will be the density of lead under a pressure of 20,000 N/cm² ?

(Density of lead is 11.4 gm/cm³ and bulk modulus of lead is

$0.80\times10^{10}\ \text{N/m}^2$

Solution. Neglecting the sign, Bulk modulus

$$B=\frac{V\cdot P}{\Delta V}$$

Here $B=0.8\times10^{10}\ \text{N/m}^2$

$P=20.000\ \text{N/cm}^2=20{,}000\times10^4\ \text{N/m}^2$

Let V be the initial volume of the lead block and ΔV be the decrease in volume of the block when subjected to pressure P

$$\frac{V}{\Delta V}=\frac{B}{P}=\frac{0.80\times10^{10}\ \text{N/m}^2}{20{,}000\times10^4\ \text{N/m}^2}=\frac{80\times10^8\ \text{N/m}^2}{2\times10^8\ \text{N/m}^2}=40$$

$\therefore$ $\Delta V=\dfrac{V}{40}$

Let new volume be V' and its new density be ρ'

New volume $V'=V-\Delta V=V-\dfrac{V}{40}=\dfrac{39V}{40}$

Density of the lead $\rho=11.4\ \text{gm/cm}^3$.

Since mass will remain same and mass is equal to volume $\times$ density,

$\therefore$ $V\cdot\rho=V'\rho'$

$$V\times11.4\ \text{gm/cm}^3=\frac{39V}{40}\rho'$$

$$\rho' = \frac{V \times 11.4 \text{ gm/cm}^3}{\frac{39V}{40}},$$

$$\rho' = \frac{V \times 11.4 \text{ gm/cm}^3 \times 40}{39V}$$

$$\rho' = \frac{11.4 \times 40 \text{ gm/cm}^3}{39} = 11.622 \text{ gm/cm}^3$$

Example 7.5. A cube of aluminum 10 cm on a side is subjected to a shearing force of 100 N. The top face of the cube is displaced by 0.01 cm with respect to the bottom. Calculate the shearing stress, the shearing strain and the shear modulus ?

Solution. (*i*) Shearing stress

$$= \frac{\text{tangenial force}}{\text{area of the face}}$$

$$\frac{F}{A} = \frac{100 \text{ N}}{(10 \times 10^{-2})^2 \text{ m}^2} = 10 \text{ N/m}^2$$

Shearing strain $\theta = \frac{\text{displacement}}{\text{side of the cube}} = \frac{0.01 \text{ cm}}{10 \text{ cm}} = 10^{-3}$

Shear modulus $\eta = \frac{\text{shearing stress}}{\text{shearing strain}} = \frac{10^4 \text{ N/m}^2}{10^{-3}} = 10^7 \text{ N/m}^2$

7.10. Thermal Properties of Solids

Heat energy. The X-ray studies have revealed that the atoms in solids are arranged in arrays. They occupy lattice sites in their equilibrium state by interatomic forces exerted on each other. These atoms are not at rest. On account of thermal agitation, they vibrate perpetually about the equilibrium position. However, the atomic vibrations are not independent. On the otherhand the displacement of any atom in one direction affects the vibrations of neighbouring atoms. Hence the motions of atoms are interdependent. These facts suggest a mechanical model for the vibrations of atoms. In such a mechanical model solids may be regarded as an assembly of mass points linked by springs and are oscillating about a periodical array of equilibrium positions as shown in Fig. 7.7. The motion of such an assembly is highly complicated and the vibrations are not simple harmonic. When a solid is heated the atoms vibrate with greater amplitude and since the energy of vibrations is directly proportional to the square of amplitude, this results in the increase of kinetic energy.

Thus the applied heat energy, causes an increase in the kinetic energy of the atoms. But we know that the increase in the kinetic

Fig. 7.7. Mechanical model for a vibrating crystal lattice.

energy causes the rise of temperature. Therefore solid shows rise of temperature when heated. We have outlined above, the broad mechanism of interaction of heat with matter in solid state. With help of it, we shall explain some thermal properties such as, specific heat, latent heat, thermal expansion and thermal conductivity of solids in this part of the unit.

7.11. Specific Heat

If a certain quantities of heat is applied to equal mass of different substances rise of temperature will be different in each case. To discuss this property of matter the term specific heat is introduced.

"The specific heat of a substance is the amount of heat needed to raise the temperature of the unit mass of that substance through one degree. Specific heat is the characteristic property of a substance and varies from substance to substance."

If Q is the quantity of heat added to a mass m to raise its temperature by $\triangle t$ then the specifie heat, C is given by

$$C = \frac{Q}{m.\triangle t} \qquad ...(7.7)$$

In the SI system it is expressed in joule per kilogram per kelvin degree [j/(kg) (k)] and in the metric system in calories per gram per centigrade degree [Cal/(gm) (C)]. It can also be expressed in kilo calories per kilogram per centigrade degree [Kcal/(kg)(C)] in S.I. system.

The specific heat is not strictly a constant but varies some what with temperature. Its value depends upon the range of temperature of observation. However at ordinary temperature and over temperature interval which is not too great, specific heat is usually considered a constant. The specific heat of some substances are given in Table 7.2.

Knowing the specific heat of a material, one can calculate the quantity of heat Q necessary to change the temperature of a mass from an initial value t_1 to final value t_2 from the relation,

$$Q=mC\triangle t=mC\,(t_2-t_1) \qquad \ldots(7.8)$$

7.12. Molar Specific Heat

The amount of heat required to raise the temperature of one mole of a substance through one degree is called molar specific heat of the substance. It is equal to the product of the specific heat and the molecular weight. But in case of solid elements the molar specific heat is given by the product of the specific heat of the element and its atomic weight. The unit of molar specific heat is J/(mole) (k) in SI system and cal/(mole) (°C) in metric system.

7.13. The Specific Heat of Solid Elements

Dulong and Petit conducted a series of experiments on the determination of specific heats of various elements in the solid state. They came to the conclusion that the product of the specific heat and the atomic weight, called molar specific heat of elements is constant for nearly all elements in the solid state. This rule is known as Dulong and Petit's law. Experimental measurements show that near room temperature, molar specific heat (specific heat × atomic weight) of all solids is nearly 25 J/(mole) (degree) or 3 R where R is universal gas constant. This fact can be explained by classical theory considering the degrees of freedom and equipartition of energy of vibrating atoms in solids.

But the empirical law of Dulong and Petit is only a first approximation and does not hold good for all temperatures because it does not take into account the variation of specific heat with temperature. The variation of specific heat of a typical solid at constant volume *i.e.* C_V can be studied by plotting C_V as a function of temperature. In general the specific heat increases with rise of temperature. At lower temperature, the specific heat decreases rapidly as shown in the Fig. 7.8.

It is not possible to furnish any explanation for the drop of specific heat with lowering of temperature on the basis of classical theory. However a satisfactory explanation can be furnished by quantum theory which is beyond the scope of this book.

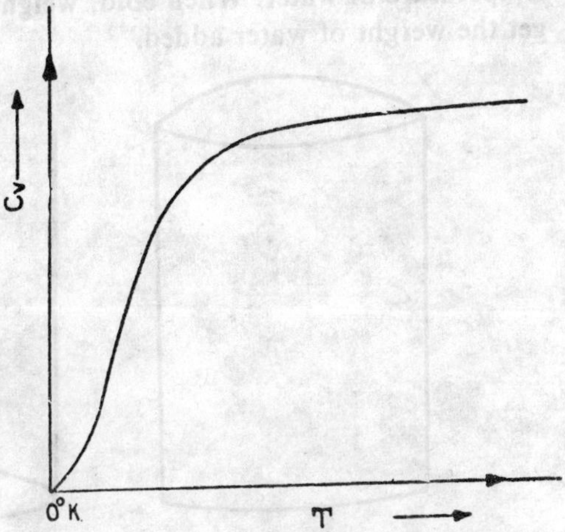

Fig. 7.8. The specific heat of a typical solid at constant volume C_V as a function of temperature T.

7.14. Water Equivalent of a Body

Water equivalent of a body is the mass of water which will be heated through 1° by the amount of heat required to raise the temperature of the body through 1°.

If m kg be the mass of a body c its specific heat, the amount of heat required to raise the temperature of the body through 1°C $=mc$ kcal. This amount of heat will raise the temperature of m kg of water through 1°C.

So water equivalent body $= m \times c$ kg

$= $ mass of the body in kg $\times$ its specific heat ...(7.9)

Determination of water equivalent of calorimeter. The vessels in which the measurement of heat is carried out are called calorimeters. These vessels are usually made of copper. Every calorimeter is provided with a stirrer made of same material. The stirrer is nsually taken in the form of wire endning in a loop. The calorimeter and stirrer are shown in the Fig. 7.9(a) and 7.9(b).

To determine the water equivalent of calorimeter dry the calorimeter and weigh it along with the stirrer. Fill the calorimeter to about one third with cold water, note its temperature and weigh it again. From this note the weight of the water taken. To this add quickly about an equal quantity of hot water. Temperature of the hot water should not be very high, otherwise loss due to radiation has to be accounted for. Now stir the mixture and note the final temperature of water. When cold, weigh the calorimeter again to get the weight of water added.

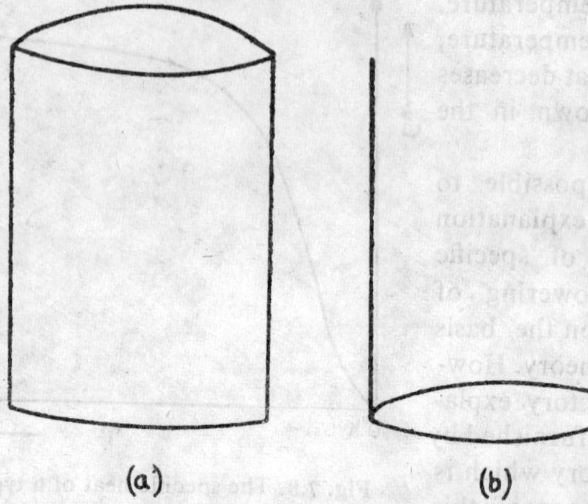

Fig. 7.9. (a) Calorimeter, (b) Stirrer.

Let the mass of cold water $= m$ kg. Mass of hot water $= m'$ kg
Temperature of the cold water $= t_1$°C
Temperature of the hot water $= t_2$°C
Common tempature of the mixture $= t$°C
Water equivalent of the calorimeter and stirrer $=$ w kg
Heat lost by the m' kg of hot water in cooling through $(t_2 - t_1)$°C

$$= m'(t_2 - t) \text{ kcal}$$

Heat gained by m kg of water in rising $(t - t_1)$°C

$$= m(t - t_1) \text{ kcal.}$$

Heat gained by calorimeter and stirrer in rising through $(t - t_1)$°C $= w(t - t_1)$ kcal. Now we have
total heat lost $=$ total heat gained.

So $\quad m'(t_2 - t) = w(t - t_1) + m(t - t_1)$

$$w = \frac{m'(t_2 - t)}{(t - t_1)} - m \qquad \text{...(7.10)}$$

7.15. Determination of Specific Heat of a Solid (Metal)

It is determined by the method of mixture. This essentially consists in adding a known mass of metal at known high temperature to a known mass of water at a known low temperature and then determining the equilibrium temperature that results. The heat absorbed by water and the calorimeter is equal to the heat given by the hot metal. From this equation, the unknown specific heat can be computed.

Let the mass and the specific heat of the solid be m_1 and c_1 respectively and its temperature be t_1. Suppose the mass of the water in the calorimeter is m_2 and c_2 its specific heat and t_2 is initial temperature of water.

If W=Water equavalent of calorimeter and stirrer and t is the final temperature of the liquid after the solid is added into it, we have applying the principle of heat lost is equal to the heat gained,

$$m_1c_1(t_1-t)=(m_2c_2+w)(t-t_2)$$

or

$$c_1=\frac{(m_2c_2+w)(t-t_2)}{m(t_1-t)} \quad ...(7.11)$$

All quantities on the right side of the equation are known

So c_1 can be calculated from it.

Example 7.6. A 0.450 kg cylinder of lead is heated to 100°C and then dropped into 50 gm of copper calorimeter containing 0.100 kg of water at 10°C. Water is stirred until equilibrium is established at which time, the temperature of the whole system is 21.1°C. Find dhe specific heat of lead.

Solution. The water equivalent of calorimeter=.050×.093 kg, as the specific heat of copper=0.093 kcal/(kg)(°C). Let C is the specific heat of lead.

Heat lost=Heat gained

$$\therefore \quad 0.450\bar{C}(100-21.1)=[0.100\times 1+0.050\times .093][21.1-10]$$

$$35.5\bar{C}=1.16$$

$$\therefore \quad \bar{C}=0.033 \text{ kcal/(kg)(°C)}$$

Example 7.7. Eighty grams of iron shot at 100°C is dropped into 200 grams of water at 20°C contained in an iron vessel of mass 50 grams. Find the resulting temperature, specific heat of iron is 0.12 cal/(gm)(°C).

Solution. In the mixture heat is lost by the shot and heat is gained by water and calorimeter. Then equilibrium is established and they attain a common temperature. Let this temperature be t.

Heat lost=Heat gained.

Heat lost by the shot=Heat gained by water+Heat gained by vessel.

(80 gm)[0.12 cal/(gm)(°C)](100°C—t)=200 gms×
[1.00 cal/(gm) (°C)](t—20°C]+(50 gm)[0.12 cal/(gm)(°C)](t—20°C)

Solving we get, t=24°C

TABLE 7.2. Specific Heats of the some of the Elements.

Element	Atomic weight	Specific heat kcal/(kg) (°C)	Specific heat J/(kg) (°C)	Heat Capacity kcal/(kmole) (°C)	Heat Capacity J/(kmole) (°C)
Aluminium	27	0.22	920	5.9	25,000
Titanium	47.9	0.14	590	6.7	28,000
Iron	55.8	0.115	460	6.1	26,000
Copper	63.5	0.093	390	5.9	25,000
Tin	118.7	0.054	230	6.4	27,000
Lead	207.2	0.031	130	6.4	27,000
Platinum	195.09	0.032	135	6.3	27,000

7.16. Latent Heat

It is general experience that whenever heat is given or taken out, a substance shows rise or fall in temperature. However this fact is not true at the time when the substance undergoes a change of phase *i.e.* when a substance changes from solid to liquid, or liquid to vapour or vice versa. During this time until the whole substance undergoes a change of phase, the temperature remains constant, although heat is continuously being given. The heat involved in this case, which does not cause temperature rise is called latent heat. It is measured by the heat absorbed or given out per unit mass of the substance when the substance changes its phase. It is characteristic of a substance and varies from substance to substance.

The word Latent means hidden, that is the heat which has no external manifestation such as rise of temperatute is called Latent Heat, but when it raises temperature of a substance it is called sensitive heat.

Latent heat of fusion. The amount of heat absorbed or given out per unit mass at constant temperature, when a material, changes from its solid to liquid state and vice versa is called latent heat of fusion. It is denoted by L_f and given by Q/m where Q is the amount of heat necessary for a substance of mass of m to change from solid phase to liquid phase and vice versa at constant temperature. Its units is given by cal/gm or Kcal/kg.

As $$L_f = \frac{Q}{m},$$

$$Q = mL_f \qquad ...(7.12)$$

7.17. Latent Heat and Interatomic Forces

We will now look at latent heats in the light of what we know about the interatomic forces and explain where the heat energy goes at the change of phase. We know that atoms or molecules of a substance are held together by inter atomic forces. When a substance undergoes a change of phase, energy has to be supplied to overcome the attractive interatomic force. So the entire energy supplied in that stage *i.e.*, at the change of phase is spent in separating the molecules to a larger distance. The energy we supply is stored as potential energy instead of being converted in to kinetic energy. So there is no change in the kinetic energy and the total kinetic energy remains constant. Therefore, as the change in kinetic energy is directly proportional to change in temperature, the temperature remains constant at the change of phase.

7.18. Determination of the Latent Heat of Fusion of Ice

Weigh a calorimeter and a stirrer (W gm) and fill half of it with warm water at about 5°C above the room temperature. Weigh the calorimeter with its contents again when the weight of water added is found (m gms). Note with a sensitive thermometer, the initial temperature (t_1°C) of water in the colorimeter. A block of ice is broken into small fragments which are washed with clean water and dried by means of blottiug paper. Get some of them and drop them into the calorimeter holding them with blotting paper. Stir well until all the ice is melted. Note the lowest temperature attained by the mixture (t_2°C) which should be about 5°C below the room temperature, weigh the calorimeter and its contents again from the weight of ice added is found (M gm).

The gain of heat takes place in two parts, (a) an amount of heat is necessary to melt the ice at 0°C to water at 0°C. (b) a further amount of heat is required to raise the ice cold water to t_2°C.

Heat lost by calorimeter and stirrer and water contained in it

$$=(WC+m)(t_1-t_2) \text{ cals.}$$

where C=specific heat of the material of the calorimeter. Heat gained by ice in melting and by ice cold water in rising to

$$t_2°\text{C}=(ML_f+M\times 1\times t_2) \text{ cal}$$

where L_f is latent heat of fusion of ice.

Since heat gained=Heat lost.

$$ML_f+Mt_2=(WC+m)(t_1-t_2)$$

whence
$$L_f=\frac{(WC+m)(t_1-t_2)}{M}-t_2 \tag{7.12}$$

Errors and precautions. (1) Fingers should not be used at the time of dropping the ice pieces for by so doing some ice will melt and the melted ice *i.e.*, water if added along with pieces of ice, will appreciably affect the accuracy of the result. For example if only 0.1 gm of water (and not ice) is added, there will be an error of 0.1×80 or 0.8 calories of heat in the calculation.

(2) The initial temperature of water is taken 5°C above the room temperature and final temperature 5°C below it in order that any gain of heat from the surroundings by the calorimeter after addition of ice may be exactly compensated for by the loss of heat due to radiation by the calorimeter, before addition of ice (Rumford's method of compensation).

(3) The ice, during the process of melting, should be kept below the surface of water and not allowed to float, otherwise the portion above the water surface will absorb heat from the outside air instead of from the water in the calorimeter and the calculations adopted above will not apply. For this, use of a wire-gauge stirrer is adviced. Care should be taken so that no water particle accompanies the thermometer while removing it.

Example 7.8. In an experiment designed to measure the heat of fusion, 25 gm of ice at 0°C was dropped into 195 gm of water at 30°C contained in a copper calorimeter of mass 100 gm. The final temperature was 18°C. Find the heat of fusion of ice. Specific heat of copper=0.093 kcal/(kg)(°C).

Solution. Here ice gains heat and the calorimeter and the stirrer lose it. So applying the principle Heat gained=Heat lost, we get

$$0{\cdot}25\ L + 0.025 \times 1 \times (18-0)$$
$$=(0.195)(1)(30-18)+0.100(0.093)(30-18)$$

where L is the latent heat of fusion of ice and $0.025\,L$ is the heat needed to melt the ice.

$$0.025\ L + 0.450 = 2.45$$
$$L = 80 \text{ kcal/kg}$$

Example 7.9. A well-insulated copper calorimeter cup of mass 100 gm contains 400 gm of water at 5°C. When 40 gm of icc at −8°C is added, it is found that 23.6 gm of ice melts. What is latent of fusion of ice. (The specific heat of ice is 0.50 cal/(gm)(°C) and the specific heat of copper is 0.093 cal/(gm)(°C).

Solution. Since not all the ice melts, we know that the final temperature is 0 C with 16.4 gm of ice floating.

Heat gained=Heat lost.

∴ (Heat absorbed to raise 40 gm of ice to 0°C+Latent heat absorbed to meet 23.6 gm of ice.)

=(Heat given off bywater cooling to 0°C+Heat given off by copper cup cooling to 0°C.)

$$(40 \text{ gm})(0.50 \text{ cal/gm C°}) [(0°C)-(-8°C)]+(23.6 \text{ gms})(L)$$

$$=(400 \text{ gm})(1 \text{ cal/gm C°})(5-0)°C+(100 \text{ gm})(0.093 \text{ cal/(gm)(C°)}) \times (5-0)°C$$

or $$160 \text{ cal}+23.6 \text{ gm } L = 2000 \text{ cal}+46 \text{ cal}$$

$$L = 80 \text{ cal/gm or } 80 \text{ kcal/kg}$$

7.19. Thermal Expansion

Most solids expand when heated. According to this property, such solids can expand in length,in area or in volume. The expansion in length, width or thickness called linear expansion, the expansion in area is called area or superficial expansion and expansion in volume is called volume or cubical expansion.

7.20. Linear Expansion

Co-efficient of Linear Expansion or Linear Expansivity :

As stated before solids increase in length, when heated, it has been found experimentally that the increase in length is directly proportional to the (*i*) original length at 0°C and (*ii*) rise in temperature. If L_0 is the length of a solid at 0°C and L_t is the length when heated to t°C, then the increase in length *i.e.*, $L_t - L_0 = \triangle L$.

Thus $\triangle L \propto L_0$, $\triangle L \propto (t°C - 0°C)$

$\therefore$ $\triangle L \propto L_0 \, t°C$

or $$\triangle L = \alpha L_0 t°C. \quad ...(7.13a)$$

α is the constant of proportionality called co-efficient of linear expansion or expansivity.

From Equation 7.13(*a*):

$$\alpha = \frac{\triangle L}{L_0 \, t°C} = \frac{L_t - L_0}{L_0 \times t°C°}$$

$$= \frac{\text{Increase in length}}{\text{Original length at °C} \times \text{rise in temp.}} \quad ...(7.13b)$$

Thus the co-efficient of linear expansion α may be defined as the ratio of increase in length to the original length at 0°C for 1 °C rise of temperature. The unit of α is/°C or°C^{-1}

Putting the value of $\triangle L = L_t - L_0$ in the Equation 7.13*b*,

we get $$L_t - L_0 = \alpha \, L_0 t°C$$

or $$L_t = L_0 + \alpha L_0 t°C$$

$$= L_0 (1 + \alpha \, t°C) \quad ...(7.14)$$

This is the relation between initial length L_0 at 0°C and length L_t and t°C.

7.21. Superficial Expasion

The co-efficient of superficial expansion (area expansion) is the ratio of change in area to the original area of a surface at 0°C for 1°C change in temperature.

If S_0 be the initial area at 0°C and S_t be the final area at t°C of a surface when t°C is the rise of temperature, then the co-efficient of superficial expansion

$$\beta = \frac{S_t - S_0}{S_0\ t} = \frac{\triangle S}{S_0\ t\ ^\circ\mathrm{C}} \qquad \text{...(7.15)}$$

Its unit is/°C or °C⁻¹

From 7.15 we get increase in area

$$\triangle S = S_t = S_0 = \beta\ S_0\ t^\circ\mathrm{C}.$$

$$S_t = S_0\ (1 + \beta t) \qquad \text{...(7.16)}$$

7.22. Cubical Expansion

The co-efficient of cubical expansion for a material is the change in volume per unit volume at 0°C per 1°C rise in temperature.

If V_0 and V_t are the volumes of body at 0°C and t°C respectively, then the co-efficient of cubical expansion

$$\gamma = \frac{V_t - V_0}{V_0 \times t^\circ\mathrm{C}}$$

$$= \frac{\text{Change in volume}}{\text{Original volume at 0°C} \times \text{rise in tepmperature}} \qquad \text{...(7.17a)}$$

γ has same unit *i.e.*/°C as in other two cases of expansion.From this relation also we get change in volume,

$$\triangle V = V_t - V_0 = \gamma\ V_0 t$$

and $$V_t - V_0(1 + \gamma\ t) \qquad \text{...(7.17b)}$$

7.23. Co-efficient of Expansion at Different Temperature

Measurements show that the body expands more of high temperature and less at low temperature for same rise of temperature. So the value of co-efficient of expansion will be different unless the original dimensions involved are measured at a standard temperature. A fixed or standard temperature therefore chosen is 0°C and all initial dimensions are taken at 0°C. But it has been found out that such variations are small in case of solids if the temperature range is moderate. Therefore the original dimension may be taken to be their value at room temperature for all practical purposes. In the modified form

$$\alpha = \frac{L_{t2} - L_{t1}}{L_{t1} \times (t_2 - t_1)°C} \quad ...(7.18a)$$

$$\beta = \frac{S_{t2} - S_{t1}}{S_{t1}(t_2 - t_1)°C} \quad ...(7.18b)$$

$$\gamma = \frac{V_{t2} - V_{t1}}{V_{t1}(t_2 - t_1)°C} \quad ...(7.18c)$$

where L_{t1}, S_{t1} and V_{t1} are the initial length, area and volume at t_1°C and L_{t2}, S_{t2} and V_{t2} are the corresponding values at temperature t_2°C. It may be noted here that as the values of the co-efficients of expansion varies with temperature, the values of α, β, γ given in the Equation 7.18*a*, 7.18*b*, and 7.18 *c* may be taken to be the average value over the temperature range of observation.

7.24. Relation among the Three Co-Efficient of Expansion

(*a*) Relation between α and β. Consider a square sheet of solid material Fig. 7.10 whose side measures L_0 at 0°C therefore surface area at 0°C *i.e.* $S_0 = L_0^2$. If the square is heated to t°C, let the length

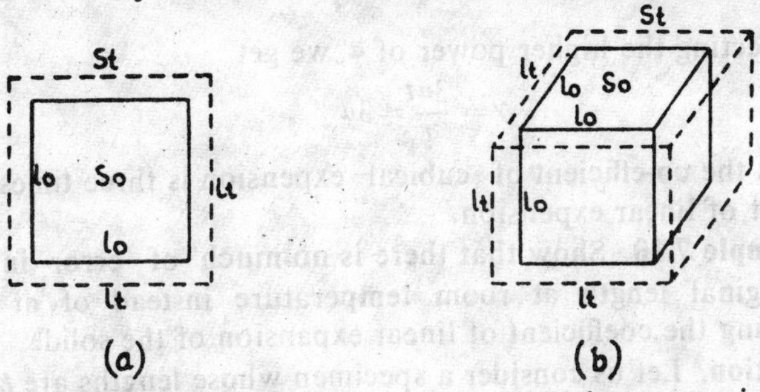

Fig.. 7.10. (*a*) Area expension. (*b*) Volume expension.

of each become L_t and therefore its surface area is represented by $S_t = L_t^2$

$$S_t = L_0^2 (1 + \alpha t)^2 \qquad [\because L_t = L_0 (1 + \alpha t)]$$

So

$$\beta = \frac{S_t - S_o}{S_o t°C} = \frac{L_0^2(1+\alpha t)^2 - L_0^2}{L_0^2 t}$$

$$= \frac{L_0^2[(1+\alpha t)^2 - 1]}{L_0^2 t}$$

$$= \frac{L_0^2[1 + \alpha^2 t^2 + 2\alpha t - 1]}{L_0^2 t°C}$$

$$= \frac{\alpha^2 t^2 + 2\alpha t}{t}$$

As α is very small, α^2 is much smaller. The terms containing α^2 can be neglected.

$$\therefore \qquad \beta = \frac{2\alpha t}{t} = 2\alpha \qquad ...(7.19)$$

Thus the co-efficient of area expansion is two times the co-efficient at linear expansion.

(*b*) **Relation between α and γ.** Let us consider a cube of solid material each side of which is L_0 at 0°C. Each side increases in length to L_t when the temperature is increased to t°C. Hence volume of the cube at 0°C *i.e.*

$$V_0 = L_0^3.$$

Volume of the cube at

$$t°\text{C} = V_t = L_t^3 = L_0^3\ (1+\alpha t)^3$$

Hence

$$\gamma = \frac{V_t^3 - V_0^3}{V_0^3 t} = \frac{L_0^3(1+\alpha t)^3 - L_0^3}{L_0^3 t}$$

$$= \frac{L_0^3[(1+\alpha t)^3 - 1]}{L_0^3 t}$$

$$= \frac{1+3\alpha^2 t^2 + 3\alpha t^2 + \alpha^3 t - 1}{t}$$

$$= \frac{3\alpha^2 t^2 + 3\alpha t^2 + \alpha^3 t^3}{t}$$

Neglecting the higher power of α, we get

$$\gamma = \frac{3\alpha t}{t} = 3\alpha \qquad ...(7.20)$$

Thus the co-efficient of cubical expension is three times the co-efficient of linear expension.

Example 7.10. Show that there is no much of error in taking the original length at room temperature instead of at 0°C in determing the coefficient of linear expansion of the solid.

Solution. Let us consider a specimen whose lengths are l_0, l_1 and l_2 at, 0°C, t_1°C and t_2°C respectively.

According to Equation 7.13.

i.e. $\qquad L_t = L_0(1+\alpha t)$

We have then $\qquad l_1 = l_0(1+\alpha t_1) \qquad ...(i)$

$$l_2 = l_0(1+\alpha t_2) \qquad ...(ii)$$

$$\therefore \qquad \frac{l_2}{l_1} = \frac{1+\alpha t_2}{1+\alpha t_1} = (1+\alpha t_2)(1+\alpha t_1)^{-1}$$

$$= (1+\alpha t_2)(1-\alpha t_1) = 1+\alpha(t_2 - t_1)$$

neglecting the higher power of α as α is very small.

$$\therefore \qquad l_2 = l_1[1+\alpha(t_2 - t_1)]$$

or

$$\alpha = \frac{l_2 - l_1}{l_1(t_2 - t_1)},$$

which is the modified form of Equation 7.13(*b*).

The modified form of equations (7.15 and 7.17*a*) can be obtained following the same procedure as that of the Equation (7.13 *b*).

7.25. Origin of Thermal Expansion

The origin of thermal expansion can be well understood if we take the atomic structure into picture. In this picture the thermal expansion is considered to be the increase in the average distance between the atoms as the temperature increases. When a solid is heated, the amplitude of vibrating atoms at lattice, points increases. Consequently the average distance of inter atomic separation also increases. This in fact leads to the expansion of solids. However, increase, in amplitude does not always result in increase of length. This increase in length depends upon whether the oscillatory motion is harmonic or inharmonic. To understand this fact we discuss below the possible potential energy curves of the atomic oscillators in solids.

Let us suppose that the oscillatory motion of an atom in solid is simple harmonic. In that case, the potential energy curve would take the shape of a parabola as illustrated in Fig 7.11*a*. The curve is symmetric about the position of minimum potential. When the solid is heated the amplitude of vibration increases, but the mean

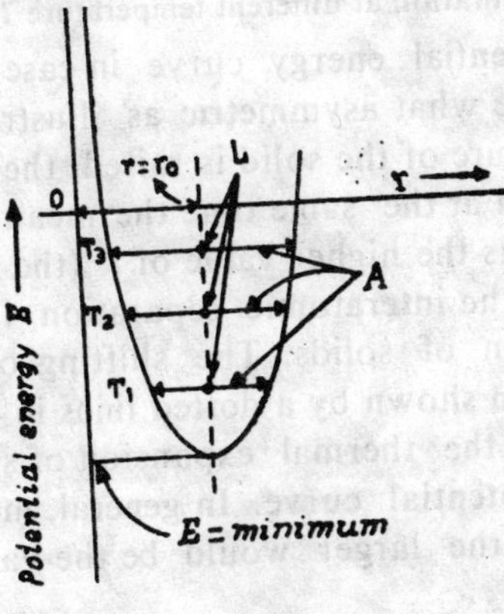

(a)

Fig. 7.11. (*a*) Vibrational motions in solid for a symmetrical interatomic potential energy (*E*), *r*—interatomic sepatration, *L*—Average interatomic separation, *A*—Amplitude of vibrated on at different temperature $T_1 T_2 T_3$.

position of equilibrium remains unaltered. So the average distance of separation between two atoms remain same what ever might be the temperature of the solid. Thus it follows that if the potential energy curve is symmetric about the point of minimum potential, there is no increase in the interatomic separation and hence no thermal expansion.

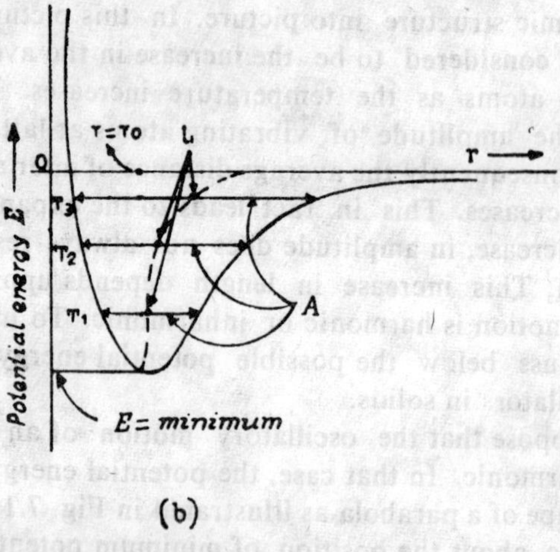

Fig. 7.11. (*b*) The interatomic potential (*E*) for a real crystal symmetrical, *r*—Interatomic separation, *L*—Average interatomic separation, *A*—Amplitude of vibration at different temperature T_1 T_2 T_3.

In fact the potential energy curve in case of real solid is not symmetric but some what asymmetric as illustrated in Fig. 7.11*b*. When the temperature of the solid is raised, the amplitude of vibration increases and at the same time the mean position of equilibrium shifts towards the higher value of r (the distance of separation). Due to this, the interatomic separation increases and leads to over all expansion of solids. The shifting of mean position of equilibrium has been shown by a dotted lines in the Fig. 7.11*b*.

Thus we see that the thermal expansion of solid depends upon the shape of the potential curve. In general, more asymmetric is the potential curve the larger would be the value of the thermal expansion.

Example 7.11. The length of a steel span of a bridge is 0.5 km and it has to withstand temperature from 30°C to 70°C. What allowance should be kept for its expansion if the co-efficient linear expansion of steel 10^{-5} per C°?

Solution. The length of the bridge is

$= L_{t1} = 0.5$ km $= 500$ m

Initial temperature $= 30°C$

Final temperature $= 70°C$

The allowance=the increase in length

$$=l_{t2}-l_{t1}$$

Since $\alpha=\dfrac{l_{t2}-l_{t1}}{l_{t1}(t_2-t_1)}$

$$10^{-5}/°\text{C}=\frac{l_{t2}-l_{t1}}{(500\text{ m})(70-30)°\text{C}}$$

or $l_{t2}-l_{t1}=10^{-5}/°\text{C}\times5\times10^2\text{ m}\times4\times10^1°\text{C}$

$$=10^{-2}\times20\text{ m}=\frac{2}{10}\text{ m}=0.2\text{ m}.$$

Example 7.12. A brass rod at 30°C is observed to be 1 m, long when measured by a steel scale which is correct at 0°C. Find the correct length of the brass rod at 0°C. (Co-efficient of linear expansion of steel=0.000012/°C and co-efficient of linear expansion of brass=0.000019/°C).

Solution. The length measured by the steel scale is 100 cm. This reading of a scale is true only at 0°C but measurement has been done at 30°C. The actual length of the steel scale

$$l_t=l_o+(l_t-l_o)=l_o+l_o\alpha(t_1-t_o)$$

$$=l_o[(1+\alpha(t_1-t_o)]$$

Here $\alpha=0.000012/°\text{C}$

and $t_1-t_o=(30-0)°\text{C}=30°\text{C}$

$$l_o=100\text{ cm}$$

$$l_t=100\text{ cm }(1+.000012/°\text{C}\times30°\text{C})$$

$$=100.036\text{ cm}.$$

Thus the length of the brass rod at 30°C is 100.036 cm.

Its length at °C would be given by :

$$l_t=l_o\ [(1+\alpha\ (t_1-t_o)]$$

$$100.036\text{ cm}=l_o\ (1+0.000019/°\text{C}\times30°\text{C})$$

$$\therefore\quad l_o=\frac{100.36\text{ cm}}{1\times0.0057}=99.98\text{ cm}$$

Length of the brass rod at 0°C

$$=99.98\text{ cm}.$$

Example 7.13. What should be the length of a steel and copper rod at 0°C so that the length of the steel rod is 5 cm, longer than a copper rod at any temperature. (Co-efficient of linear expansion of copper=1.7×10^{-5}/°C and the co-efficient of linear expansion of iron=1.1×10^{-5}/°C).

Solution. Let L_{so} and L_{co} be the length of the rods of steel and copper respectively at 0°C and l_{st}, l_{ct} be their values at t°C respectively and Δt be the temperature difference.

Now $l_{so}-l_{co}=5\text{ cm}$ ...(*i*)

$l_{st}-l_{ct}=5\text{ cm}$ ...(*ii*)

$\because \quad l_t = l_o\,(1+\alpha\triangle t)$

$l_{st} = l_{so}\,(1+\alpha_s\triangle t)$

and $\quad l_{ct} = l_{co}\,(1+\alpha_c\triangle t)$

$l_{st} - l_{ct} = l_{so}\,(1+\alpha_s\triangle t) - l_{co}\,(1+\alpha_c\triangle t)$

or $\quad l_{st} - l_{ct} = (l_{so} - l_{co}) + \triangle t\,(\alpha_s l_{so} - \alpha_c l_{co})$

or $\quad 5\text{ cm} = 5\text{ cm} + \triangle t\,(\alpha_s l_{so} - \alpha_c l_{co})$

or $\quad \triangle t\,(\alpha_s l_{so} - \alpha_c l_{co}) = 0$

$\because \quad \alpha_s l_{so} = \alpha_c l_{co}$

$$\frac{l_{so}}{l_{co}} = \frac{\alpha_c}{\alpha_s} = \frac{1.7\times10^{-5}}{1.1\times10^{-5}} = \frac{17}{11} \qquad \ldots(i)$$

Subtracting 1 from both sides of the equation (*i*), we get

$$\frac{l_{so}}{l_{co}} - 1 = \frac{l_{so} - l_{co}}{l_{co}} = \frac{17}{11} - 1 = \frac{6}{11}.$$

But we have $\quad l_{so} - l_{co} = 5\text{ cm}$

$$\therefore \quad \frac{5}{l_{co}} = \frac{6}{11}$$

and $\quad l_{co} = \dfrac{11\times5}{6} = 9.16\text{ cm.}$

$\therefore \quad l_{so} = 9.16 + 5 = 14.16\text{ cm.}$

Example 7.14. A steel girder is 50 cm long and has a cross section of 250 cm². What is the force exerted by the girder when it is heated from 5°C to 25°C (co-efficient of linear expension of steel $= 11\times{}^{-6}/°C$ and Young's modulus of steel $= 20\times10^{11}$ dynes/cm²).

Solution. Here $\quad \alpha = 11\times10^{-6}/°C$

Young's modulus $\quad Y = 20\times10^{11}$ dynes/cm²

Area of cross section $\quad A = 250$ cm²

Rise of temperature $\quad \triangle t = 25°C - 5°C = 20°C$

Change in length of the steel girder

$\triangle L = L\alpha\triangle t$ where L is the length at 5°C.

$\therefore \quad \triangle L = 50\times10^2\text{cm}\times11\times10^{-6}/°C\times20°C = 1.1\text{ cm}$

The force F required to prevent any change in length *i.e.*, $\triangle L$ equals the force required to produce this elongation at constant temperature.

$$\text{Young's modulus} = \frac{\text{longitudinal stress}}{\text{longitudinal strain}}.$$

$$Y = \frac{F/A}{\triangle L/L} \quad \text{or} \quad F = YA\frac{\triangle L}{L} \qquad \ldots(i)$$

Substituting the value of

$Y, A, \triangle L, L$ in (*i*)

we get $\quad F = 20\times10^{11}\times250\times(1.1/50)$ dyne.

$=11\times10^{12}$ dyne.
$=11\times10^{7}$ N.

Example 7.15. The brass scale of barometer gives correct readings at 0°C. Co-efficient of thermal expansion of brass is 0.00002/°C. The barometer reads 75 cm at 27 °C. What is the atmospheric pressure at 0°C?

Solution. If H is the barometer reading at 0°C, then H is in fact the correct value of the division upto which the mercury stands in the tube only at 0°C at which the scale was calibrated. If α is the co-efficient of the linear expansion of brass, the length H cm brass scale at 0°C would become H_t given by

$$H_t=H(1+\alpha t) \text{ cm at } t^\circ\text{C} \quad \ldots(i)$$

$$H_t=75 \text{ cm}, \; t=27^\circ\text{C}$$

and $\alpha=0.00002/\text{C}^\circ \quad H=?$

In equation (i)

$$75=H_0(1+0.00002/^\circ\text{C}\times27^\circ\text{C})$$

$$\therefore \quad H=\frac{75}{(1+0.000054)}$$

$$=75(1+0.000054)^{-1}=75(1-0.000054)$$

$$=74.96 \text{ cm of Hg.}$$

7.26. Thermal Conductivity

When a metal rod is heated at one end, the heat gradually spreads along the rod and the other end also becomes hot after some time. During this process of heat-flow no material particle of the medium is bodily transferred from one part of the body to the other. The process in which heat is transmitted from one point to the other through the substance without any transference of material particle is known as conduction.

Let a rod of uniform cross-section be heated at one end as in Fig. 7.12. Heat starts flowing from the hot end to the cold along the rod. Consider a thin element A_1A_2 of the rod at some distance from the hot end, with faces perpendicular to the length of the rod. The face nearer the hot end will be obviously at a higher temperature than the other face farther away. This means more heat enters the element than leaves it. The amount of heat thus left with the element is used partly to warm up the element and partly lost in radiation from the surface of the element. As the element gets hotter, the portion radiated to the surrounding increases, while the portion spent in heating the element assumes practically a constant value. Soon a stage is reached, when the amount of heat radiated becomes exactly equal to the excess left with the element, over and above the heat absorbed to raise its temperature which has now reached a constant value depending on the thermal capacity of the material. Such a stage is known as the steady state, when the temperature of each element in the rod

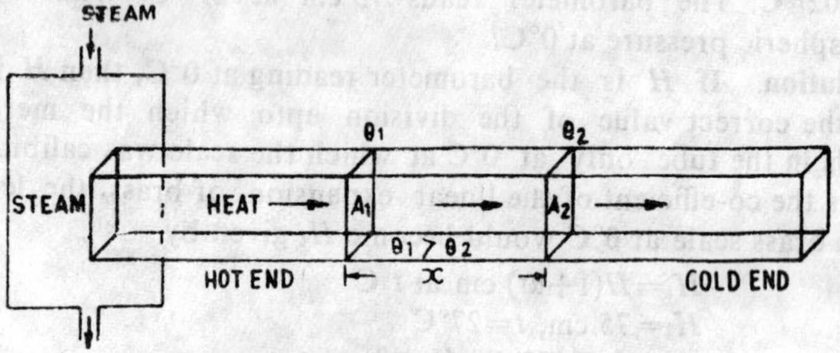

Fig. 7.12. Thermal conduction of a solid.

attains a constant value, which decreases gradually from the hot to the cold end. In other words, when the steady state is reached the temperature of different points are different but they remain constant with the passage of time. It may be noted also that in the steady state, the heat flow and the temperature distribution along the rod become independent of the specific heat of the material of the rod, since no further change in temperature takes place at any point.

Let θ_1 and θ_2 be the constant temperatures of the two faces of an element of the rod of thickness x and of cross-sectional area A, in the steady state. It is found experimentally that the amount of heat Q flowing through the element in a time t is proportional directly to A, $(\theta_1-\theta_2)$ and t, and inversely to x

$$Q \propto A,\ Q \propto (\theta_1-\theta_2),\ Q \propto t \text{ and } Q \propto \frac{1}{x}$$

$$\therefore \qquad Q = \frac{KA(\theta_1-\theta_2)t}{x} \qquad \ldots(7.30a)$$

where K is a constant whose value depends upon the nature of the material of the rod and is called the co-efficient of thermal conductivity or simply thermal conductivity.

If $\quad A=1,\ \theta_1-\theta_2=1,\ t=1$

and $\quad x=1$, then $Q=K$,

this enables the co-efficient of thermal conductivity to be defined as:

The quantity of heat flowing per second through unit area of cross-section of an element of the material, of unit thickness when the difference of temperature between its faces is unity. The SI unit of K is $JS^{-1}\ m^{-1}\ {}^\circ C^{-1}$.

The quantity $\frac{\theta_1-\theta_2}{x}$, which represents the rate of fall of temperature with distance in the direction of the heat flow, is known as the temperature gradient. As there are many cases in which the temperature gradient is not uniform it becomes necessary to reduce the quantity to the limiting value $\frac{d\theta}{dx}$, and rewrite relation 7.30(*a*).

as $$Q=-KA\frac{d\theta}{dx}t \qquad ...(7.30b)$$

The relation is of fundamental importanee in the theory of heat conduction. Usually a negative sign is attached to the RHS of the equation, to indicate that $\frac{d\theta}{dx}$ is negative, since it represents the rate of fall of temperature with distance in the direction of the heat flow.

A material which has high value for its co-efficient of thermal conductivity (k) is regarded as good conductor of heat. All metals are good conductor of heat. On the other hand if material has a low value for its k, it is classified as bad conductors of heat or insulator. Glass wood, mica, porcelain etc., are examples of bad conductors of heat.

7.27. Searle's Method of Determination of Thermal Conductivity of Good Conductors

(G.F.G. Searle of Cambridge University devised a method to measure thermal conductivity of good conductor. The method is known Searl's method after his name).

Searle's method. Co-efficient of thermal conductivity of good conductors can readily be determined by means of Searle's appa- ratus shown in Fig. 7.13. *AB* is a bar of material whose co-efficient of conductivity is required. It is well covered with some bad con- ductors like wool or felt to minimise heat losses from the surface. The end *A* is heated to steam temperature by placing it in a cham- ber through which steam is passed. Over the end *B* a coil *C* having two openings is wound. Through this coil a steady flow of cold water is maintained. The temperature θ_1 and θ_2 are noted by the two thermometers T_1 and T_2 placed in position as shown. Two other thermometers are placed at a known distance apart preferably 8 to 10 cm in two cavities *E* and *H* in the bar. To ensure good con- tact a little mercury is placed in these cavities. The flow of water is regulated in the coil *C* so that a steady state of temperaturse reached. At this stage the quantity of heat flowing through any section of the bar is the same.

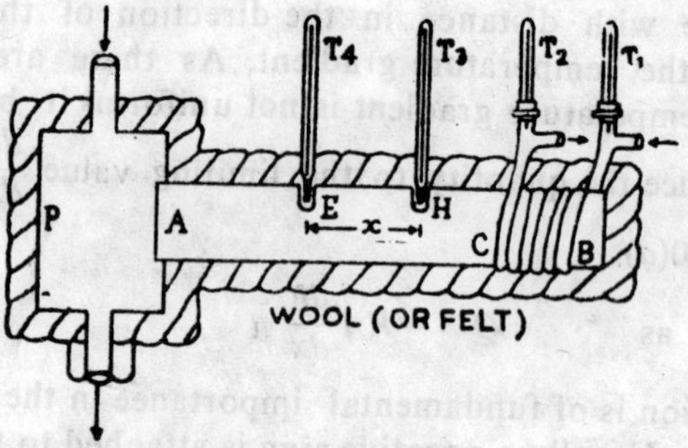

Fig. 7.13. (*a*) Searle's apparatus for measurement of thermal conductivity of metals.

The quantity of water flowing through the coil C (when steady state is reached) in a given time is collected and weighed. The temperature of thermometers T_1, T_2, T_3 and T_4 are also noted. Let m be the mass of this water in kilogram which flows in the given time t and θ_1 and θ_2 be the temperatures of water at the entrance and the axit respectively. Then the amount of heat Q which reaches end B and is absorbed by water in t seconds is given by

$$Q = m(\theta_2 - \theta_1) \quad \text{...(7.23.)}$$

Let x be the distance between the thermometers T_3 and T_4. measured in meters and let θ_3 and θ_4 be the temperature recorded by them. If A is the area of the cross section of the bar in square meter and K be the co-efficient of thermal conductivity, then amouut of heat Q passing through these points in the given time t is given by

$$Q = \frac{KA\,(\theta_4 - \theta_3)t}{x}$$

$$\therefore \quad m(\theta_2 - \theta_1) = \frac{KA\,(\theta_4 - \theta_3)}{x}\,t \quad \text{...(7.24)}$$

From this equation K is found out as all other quantities involved are known.

7.28. Thermal Conduction and Atomic Structure

The process of heat conduction through a solid both insulator and conductor can be explained on the basis of atomic theory of matter. All solids consist of atoms which are linked up by inter-atomic forces. These atoms are not at rest, but vibrate with some amplitude about lattice point, depending upon the temperature of the solid. When one part of a solid is heated, the atoms in this region vibrate vigorously, with larger amplitude. As all atoms are linked, an atom vibrating with larger amplitude will try to increase the amplitude of vibration of its neibhour which is vibrating with a smaller amplitude and this gives energy to it. This neighbour in turn gives energy in the same way to its neighbour and so on. In this way atoms at the colder end will receive energy from atoms at hotter end of the material. This is the common mechanism of heat transfer from hotter part to the colder of the solid of all kinds.

But in case of metals, heat energy can also be transferred by electrons in addition to the above said mechanism. Metals have sufficient free electrons where as insulators do not have. Since these electrons are quite mobile, they are much more effective in transporting energy from hotter to colder, parts of a solid compared to the mechanism of lattice vibration. Actually speeking in metals the thermal couductivity is governed by free electrons. In insulators mobile electrons are few. So electrons have no effect on thermal conductivity. So thermal conductivity is less in case of insulators than conductors.

7.29. Thermal Conductivity and Electrical Conductivity

All good conductors of heat are also good conductors of electricity. The examples are all metals. The reason is that the free electrons in the metal which are carriers of electricity are also carrier of thermal energy. So there is an intimate relationship between thermal and electrical conductivity. It has been found that the ratio of electrical conductivity to thermal conductivity is a constant for metals. So if σ is the electrical conductivity of metal then $K/\sigma=$constant.

This law is known as Weidman-Franz Law.

TABLE 7.3. Thermal Conductivity at 20°C.
1 J–m/(s) (m²) (C°)=1 W/(m) (C°)

Substance	*Thermal conductivity W/(m) (C°)*	*Thermal conductivity Cal/ (cm)²(sec)/(C°/cm)*
Aluminium	230.0	0.504
Air	0.023	0.000053
Brick	0.6	0.0017
Brass	100.0	0.26
Cork	0.5	0.0001
Copper	400.0	0.92
Glass (crown)	1.0	0.002
Ice	1.7	0.004
Iron (cast)	72.0	0.12
Soil (dry)	0.14	0.0004
Water	0.60	0.0014
Wood	0.13	0.0003

Example 7.16. Assume that thermal conductivity of copper is 4 times that of brass. Two rods of copper and brass having the same length and cross section are joined end to end. The free end of the copper rod is kept at 0°C and the free end of the brass rod is kept at 100°C. Calculate the temperature of the junction of the two rods at equilibrium. Neglect radiation losses. (IIT Entrance Examination, 1976).

Solution. Let θ°C is the temperature of the junction, Thermal conductivity of brass$=K$. Thermal conductivity of copper$=4\,K$.

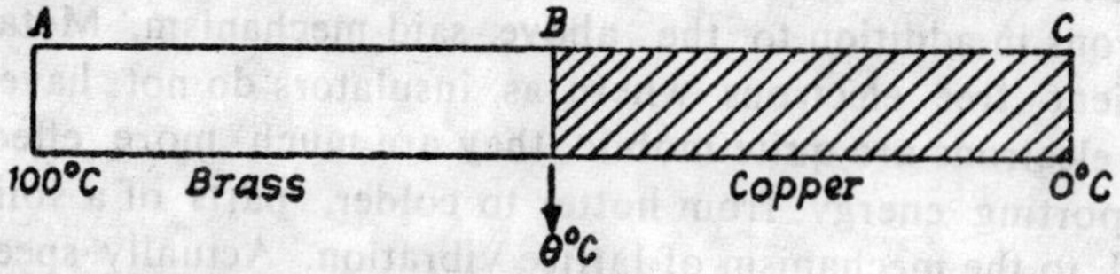

Fig. 7.13. (*b*) Copper and brass rods are joined end to end.

Area of cross section$=A$ (for both)

Length of the each rod$=x$.

At equilibrium for a junction of two materials

$$\frac{K_1A_1\,(\theta_1-\theta)t}{d_1}=\frac{K_2A_2\,(\theta-\theta_2)t}{d_2}$$

Here $A_1=A_2=A$, $K_1=K$ and $K_2=4K$, $\theta_1=100°\text{C}$, $\theta_2=0°\text{C}$ $d_1=d_2=x$, $\theta=?$

Putting the above values, we have

$$\frac{KA\,(10-\theta)t}{x}=\frac{4\,KA\,(\theta-0)t}{x}$$

or $$100-\theta=4\,\theta.$$

$\therefore$ $$\theta=20°\text{C}.$$

Example 7.17. A layer of ice, 20 cm thickness is formed on a pond. The temperature of air is -10°C. Find how long it will take for the thickness of ice for increase by 1 mm. (The thermal conductivity of ice is=0.005 CGS units, density of ice is 0.9 gm/cm^3.)

Solution. Let A be the surface area of the layer of thickness 20 cm. The temperature of ice is 0° C and the temperature of air is-10°C. We have :

$$Q=\frac{KA\,(\theta_1-\theta_2)t}{x}$$

Here $$\theta_1-\theta_2=0-(-10)=10°\text{C},\ x=20\text{ cm}.$$

$$Q=\frac{.005\text{ cal cm}^{-1}\,(\alpha)^{-1}(\text{sec})^{-1}\times A\text{ cm}^2\times 10°\text{C}\times t\text{ sec}.}{20\text{ cm}}$$

$$Q=\frac{.005\times A\times 10\times t}{20}\text{ cal} \qquad ...(i)$$

Since the thickness of the ice is to be increased by 1 mm the mass of ice formed in t sec is given by:

$$m = \text{Volume} \times \text{density}$$
$$= A \times 0.1 \times 0.9 = 9A/100 \text{ gm}$$

$$\therefore \quad m = \frac{9A}{100} \text{ gm} \qquad ...(ii)$$

The amount heat conducted into air by ice formation

$$Q = m \times L_f$$

where L_f is the latent heat of fusion.

$$Q = m \text{ gm} \times 80 \text{ cal/gm} = 80 \text{ m. cal}$$
$$= 80 \times \frac{9A}{100} \text{cal} \qquad ...(iii)$$

Now putting this value of Q in equation (i) we get,

$$\frac{0.005 \times A \times 10 \times t}{20} = \frac{9A \times 80}{100}$$

$$\because \; t = \frac{9 \times 80 \times 20}{100 \times 0.005 \times 10} \text{ sec}$$

$$t = 2880 \text{ sec} = 48 \text{ minutes.}$$

Example 7.18. If a glass pane of thickness 7 mm is used in the windows of a room instead of a common window glass of thickness 3 mm and if temperature of the inside of a glass is 10°C higher than that of the outside how much saving of heat per hour will result from this arrangement? Total area of window glass is 5 square meters and the thermal conductivity of glass is 0.0025.

Solution. Here $K = 0.0025 \text{ cal cm}^{-1} (°\text{C})^{-1} (\text{sec})^{-1}$

$$A = 5 \text{ m}^2 = 5 \times 10^4 \text{ cm}^2$$
$$\theta_1 - \theta_2 = 10°\text{C}, \; t = 1 \text{ hour} = 3600 \text{ sec}$$

(a) For a common glass pane $x = 3 \text{ mm} = 0.3 \text{ cm}$

The heat Q_1 which is conducted through common glass pane is given by

$$Q_1 = \frac{KA(\theta_1 - \theta_2)t}{x}$$

$$= \frac{0.0025 \text{ cal cm}^{-1} (°\text{C})^{-1} (\text{sec})^{-1} \times 5 \text{ m}_2 \times 10°\text{C} \times 3600 \text{ sec}}{0.3 \text{ cm}}$$

$$= \frac{0.0025 \text{ cal cm}^{-1} (°\text{C})^{-1} (\text{sec})^{-1} \times 5 \times 10^4 \text{ cm}^2 \times 10 \text{ °C} \times 3600 \text{ sec}}{0.3 \text{ cm}}$$

$$= \frac{.0025 \times 5 \times 10^4 \times 10 \times 3600}{0.3} \text{cal}$$

$$= 15 \times 10^6 \text{ cal.} \qquad ...(i)$$

(b) For the new glass pane

$$x = 7 \text{ mm} = 0.7 \text{ cm}$$

The amount of heat Q_2 conducted through this pane is

$$Q_2=\frac{0.0025\ \text{cm}^{-1}\ (^\circ\text{C})^{-1}\ \text{sec}^{-1}\times 5\times 10^4\ \text{cm}^2\times 10^\circ\text{C}\times 3600\ \text{sec}}{0.7\ \text{cm}}$$

$$Q_2=\frac{0.0025\times 5\times 10^4\times 10\times 3600}{0.7}\ \text{cal}$$

$$=6.43\times 10^6\ \text{cal} \qquad \ldots(ii)$$

Saving of heat per hour Q_1-Q_2

$$=(15-6.43)\times 10^6=857\times 10^6\ \text{cal}$$

7.30. Electrical Properties

Dielectrics. Conductors possess a huge amount of free electrons. So when a conductor is subjected to electric field the electrons move opposite to the direction of electric field, constituting what is known as electric current. But on the other hand the dielectric (insulaters) do not have appreciable amount of free electrons. So no current is possible, when they are subjected to any electric field. But it does not mean that electric field has no effect on dielectric when dielectric is placed in an electric field. The electric field in fact affects the dielectric and the dielectric tends to reduce the strength of the original field. This reduction in electric field is useful in the designing capacitors. To understand the mechanism of reduction in the original field strength by dielectric, we have to consider few facts which are discussed below.

7.31. Polarisation by Induction

We know, atoms are electrically neutral, which means they have got equal number of positive and negative charges. Moreover as

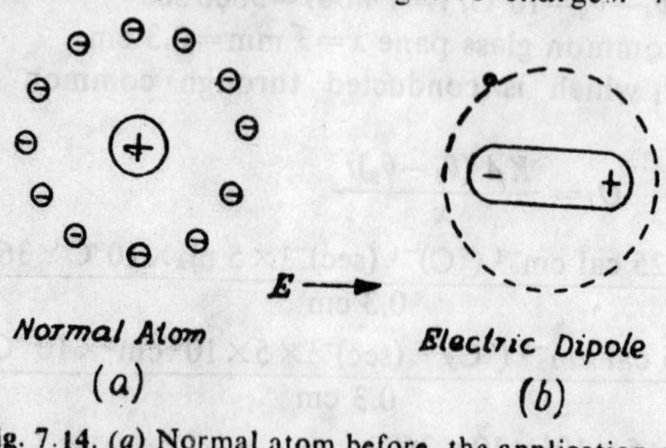

Fig. 7.14. (*a*) Normal atom before the application of electric field centre of the charge and centre of negative charge fall or each other. (*b*) Atom after application of the electric field. The centre of positive charge and the centre of nagative charge are separated to form *electric dipole*.

the centre of positive charges and centre of negative charges fall on each other the atom does not have any external electric effect. But when such an atom is subjected to an electric field, the positive charges would be displaced in the direction of the field and negative charges would be displaced in the opposite direction as shown in the Fig. 7.14. Consequently the atom is no longer electrically neutral, and has excess of positive charges on one side and excess of negative charges on the other side. This state of the atom is called polarised state and the atom is now called an electric dipole. This phenomena of formation of dipoles due to the electric field is known as polarisation by Induction and dipoles are called Induced dipoles. When a bulk of dielectric is subjected to an electric field, each atom of the dielectric behaves in the manner described above to form dipoles. These dipoles tend to align themselves along the applied etectric field.

7.32. Polar and Non-polar Molecule

There are some molecules in which centre of positive charge and centre of negative charge coincide, under normal condition (when no field is applied). Such molecules are called non-polar molecules. But there are some other kind of molecules in which the centre of positive charges and centre of negative charges do not coincide due to non-symmetrical distribution of charges. Such kind of molecules are known as polar molecule.

In polar molecules the centre of positive charges and the centre of negative charges remain separated by a distance and form an electrical dipole. The product of the charge and the distance of of separation in a dipole is known as dipole moment which is a vector quantity polar molecules possess permanent dipole moments. In the normal state of a dielectric, the polar molecules have nearly random orientation and are kept in that state by thermal agitation. Due to this arrangement the aggregate electric effect is zero.

7.33. Effect of Electric Field on Dielectrics

When a dielectric is placed in an electric field, the electric field exerts an influence on the molecules of the dielectric. As a result, if the molecules are already polar, the centre of positive charge and centre of negative charge get further separated. Also the dipoles tend to align themselves with the field direction. The extent of further displacement of charge and the degree of alignment of dipole depend upon the intensity of the electric field and nature of the dielectric. But even when the constituent molecules are not polar and as such do not possess permanent dipole moment, the field also induces dipole moments in them by causing separation between the centres of positive and negative charges. The induced moment will also be lined up along the direction of the field because the induced dipole moment is always along the direction of the field. Therefore an electric field always polarises the dielectrics.

7.34. Dielectric Slab between the Plates of a Capacitor (The Reduction of Original Electric Field

When a slab of dielectric is put between the plates of a parallel plates capacitor, the electric field between the plates polarises the dielectric. This means negative charges of the molecules shift slightly towards the positively charged plate and the positive charges shifts slightly towards the negatively charged plate as shown in Fig. 7.15. These displacements are very small and are of the order of the molecular diameter *i.e.* 10^{-10} m. It can be seen that in the bulk of the slab there will be no net charge but at the top surface, Fig. 7.16 there would be an excess of negative charge and at the bottom there would be an excess of positive charge due to polarisation of the material of the dielectric. Because of the charges on the plates of the capacitors, an electric field is set up in the region between the plates. We call this field as external field and denote it by $\vec{E}_0$. The induced charges on the surface of the dielectric also set up an electric field $\vec{E'}$ in a direction of opposite to $\vec{E}_0$, Fig. 7.16. Thus it follows that when a dielectric is placed in an electric field, the surface charges are induced

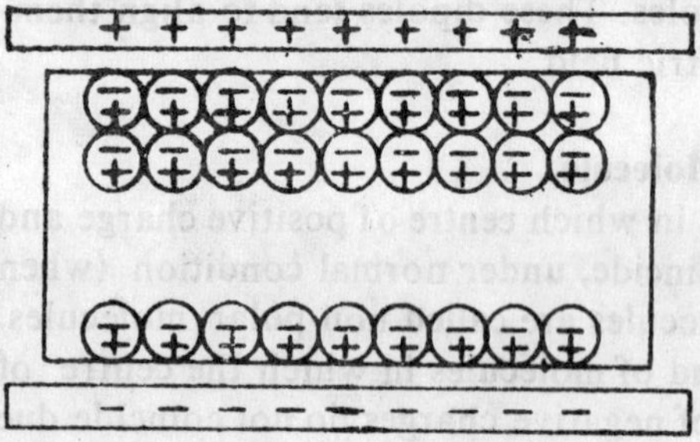

Fig. 7.15. Capacitor with a slab of dielectric place between the plates.

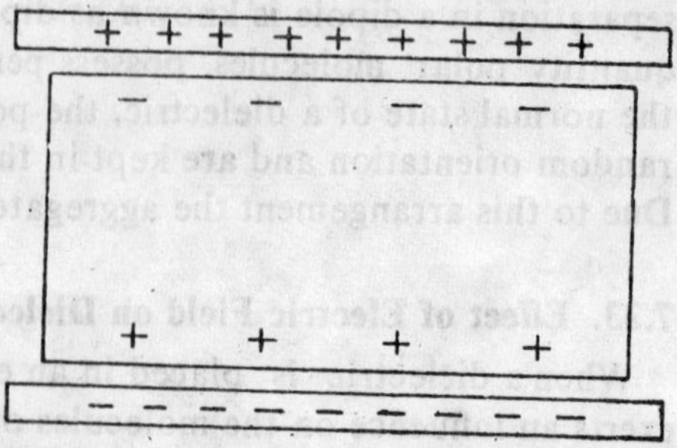

Fig. 7.16. Dielectric slab between two charged plates.

in the dielectric which tendes to reduce the original field as shown in Fig. 7.16.

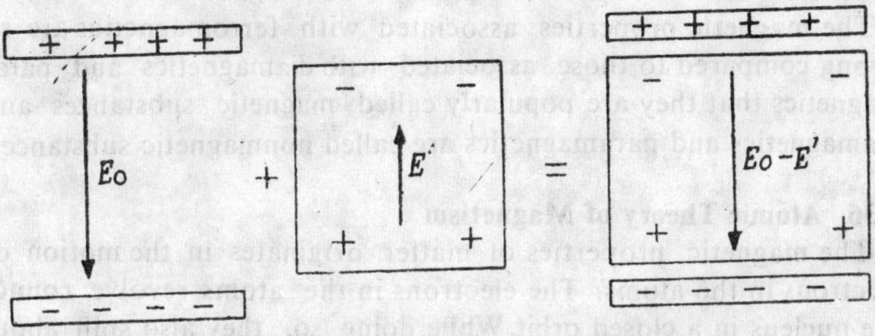

Fig. 7.17. Reduced electric field due to dielectic.

The ratio

$$\frac{\overrightarrow{E_o}}{\overrightarrow{E_o}-\overrightarrow{E'}}$$

is called dielectric constant of the dielectric. If the dielectric consists of polar molecules, dipoles orienting in the applied would give a larger value of $\overrightarrow{E'}$, therefore $\overrightarrow{E_o - E'}$ is small. The dielectric constant of such a dielectric is therefore relatively large.

7.35. Magnetic Behaviour of Solids

All materials are magnetic in nature. They exhibit magnetic behaviour, when put in a magnetic field. From the point of view of their magnetic behaviour, we can classify all materials into three groups, which are as follows.

(*a*) **Diamagnetic**. These are the materials which are repelled by magnets. In a magnetising field, they acquire such magnetism that they move away from the stronger to the weaker part of the field. If such a material is freely suspended, it will set itself at right angles to the applied magnetic field. The materials which come under this class are Bi, Sb, Hg, Ag, Zn, Cu, Pb, etc.

(*b*) **Paramagnetics**. These substances are fully attracted by a magnet. In a magnetising field they acquire such a magnetism that they move away from weaker to the stronger part of the field. If such a material is freely suspended it will set itself parallel to the magnetic field. The substances which show paramagnetic behaviour are aluminium, chromium, salts of transition metals or rare earth like chrome alum etc.

(*c*) **Ferromagnetics**. These substances are strongly attracted even by a weak magnetic field. When they are freely suspended they set themselves parallel to the magnetic field. Iron, nickle and cobalt are the examples of ferromagnetics.

The magnetic properties associated with ferromagnetics are so strong compared to those associated with diamagnetics and paramagnetics that they are popularly called magnetic substances and diamagnetics and paramagnetics are called nonmagnetic substances.

7.36. Atomic Theory of Magnetism

The magnetic properties of matter originates in the motion of electrons in the atoms. The electrons in the atoms revolve round, the nucleus in a closed orbit. While doing so, they also spin about an axis passing through itself. Infact the rotational and spinning motions of electrons are responsible for the magnetic behaviour of materials. But it is not possible to present an accurate and detailed picture of magnetism without using quantum mechanics, because spin of an electron as well as its motion in a solid can be described accurately only with help of quantum mechanics.

Therefore we shall give here only a qualitative picture of diamagnetism, paramagnetism and ferromagnetism on the basis of classical theory.

(*a*) **Diamagnetism**. In an atom the electrons move in circular orbits around the nucleus. The rotating electrons behave like a current loop. This current loop is considered to be equivalent to a

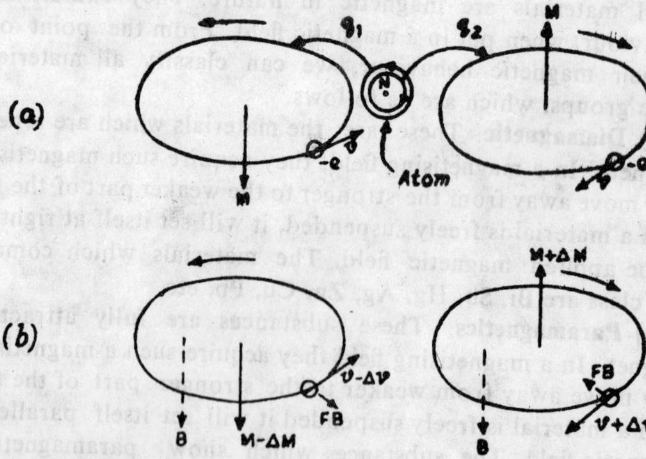

Fig. 7.18. (*a*) Motion of two electrons q_1 and q_2 in a normal atom. The direction of motion and the magnetic moments are shown in (*a*). (*b*) Motion of two electrons inside an atom, when subjected to external field *B*. The direction of motion and resulted magnetic moment are shown in Fig. (*b*).

magnet, whose magnetic moment is equal to the product of the current and the area of the loop (here the electron orbit). The direction of the moment is normal to the plane of the orbit and follow the right hand corkscrew rule, as shown in the Fig. 7.18.

In a material of diamagnetics, there are very large number of such electrons and the orbits of electrons are oriented at random. So the magnetic moments due to orbital motion of electrons cancel each other and there is no net magnetism in the bulk of the material.

Let us examine what happens when such a material is subjected to magnetic field. Consider first, a model of an atom with single electron of charge e and mass m, moving with speed v in a circular orbit of radius r, once in time t. The orbital motion of the electron produces a magnetic moment M, equal to the product of the current and the area of the orbit.

$$\therefore \quad M=\frac{ve}{2\pi r}(\pi r^2)=\frac{ver}{2}$$

Since $\frac{2\pi r}{v}=t$ and $\frac{e}{t}=\text{current}=\frac{ve}{2\pi r}$

when a magnetic field is applied, the electron experiences a force

$$F=-e(\vec{v} \times \vec{B})$$

along the radius at right angles to the direction of its motion. As a result the net centripetal force increases or decreases, depending upon its direction of circulation. But in this interaction radius of the orbit remains constant. Hence the speed of the electron gets increased or decreased which results in an increase or decrease in the circulation current. We know that magnetic moment depends upon the strength of the current. Therefore the magnetic moment of an orbiting electron increases or decreases due to application of external magnetic field.

Let us consider an atom with two electrons circulating in opposite directions but have their speed and orbits equal. For convenience let us take their orbits to be in the plane of the paper and designate the electrons as q_1 and q_2. Let q_1 move in anti-clockwise direction and q_2 in clockwise direction.

Due to above assumptions it can be seen that, the magnetic moments arising out of circulation are equal in magnitude but are opposite in direction (since electrons are circulating in opposite direction). Hence the net magnetic moment sums up to zero Fig. 7.18(*a*).

Let us apply a magnetic field of strength $\vec{B}$ at right angles to the plane of the orbit and in downward direction as in Fig. 7.18 (*b*). It is obvious that due to the application of this field, the speed of the electron q_1 which is circulating in an anti-clockwise direction would decrease by $\triangle v$ and that of the electron q_2 which is circulating in clockwise direction would increase by $\triangle v$. As a result of this change in speed, the magnetic moment of elctron q_1 would decrease and would become $M-\triangle M$ and that of q_2 would increase and become $M+\triangle M$. Therefore the net change in magnetic moment of the atom would be,

$$(M+\triangle M)-(M-\triangle M)=2\triangle M$$

and has the direction opposite to the applied field $\vec{B}$.

Consider an atom with number of electrons having orbits so oriented that there is no net magnetic moment. When a magnetic field is applied to such an atom, the orbital magetic moments of the electrons circulating in opposite direction would no longer cancel each other. On the other hand a resultant magnetic moment is induced whose direction is opposite to the applied field $\vec{B}$ (as discussed above in case of two electrons). Such an atom would show diamagnetic behaviour and sets itself at right angles to applied magnetic field. A material made up of such atoms is called diamagnetic material, and behaves in the same way as its constituent atoms in pointing at right angles to applied magnetic field.

The above fact accounts for the diamagnetism of the materials. This also shows that diamagnetism must be present in all materials because the orbital motion of electrons in atoms is a universal phenomena. This theory was given by P. Longvin in 1905. So it is called Longevin Theory of diamagnetism.

The example of diamagnetic materials are Antimony, Bismuth, Carbon, Copper, Silver, Mercury and Water.

(*b*) **Paramagnetism**. Paramagnetism occurs in those materials where the individual atoms, ions, or molecules in it possess a permanent magnetic dipole moment. The electrons of an atom in fact possess two kinds of magnetic moments. In addition to the orbital magnetic moment described earlier the electron has also an additional magnetic moment which is due to its spin motion about an axis through its centre. In some atoms containing many electrons these moments are so oriented that they can not cancel completely but give the atom a net magnetic moment. This happens in atoms whose inner electron shells are not completely filled, such as the atoms of transition elements like iron, nickel, cobalt, palladium and rare earth atoms like galadium and they exhibit paramagnetic behaviour.

In the absence of any magnetic field, these atomic magnetic

moment vectors are oriented in a random fashion due to thermal agitation. This results in the cancellation of the effect of each other. Thus a piece of paramagnetic material as a whole does not show any magnetic behaviour. But when this piece is subjected to external magnetic field $\vec{\mathbf{B}}$, the magnetic moment vector of atoms of such material experience torque tending to align along the direction of external field. But the tendency of alignment along the field direction is hampered by the thermal motions of the atoms. When the applied field is quite strong and the temperature of the material is low, the alignment of atomic dipole along the field direction does take place in some measure and the material starts showing the magnetic behaviour known as paramagnetism. The alignment of atomic dipoles along the direction of the external field increases with the increase in the field strength and decreases due to increase in thermal motion with the rise of temperature. So paramagnetism is temperature dependent.

The paramagnetic property is better understood by introducing a new term called intensity of magnetisation (I). The intensity of magnetisation I is defined as the magnetic moment per unit volume of the substance. If M is the resultant magnetic moment at any stage, then

$$I=\frac{M}{V} \qquad \ldots(7.25)$$

where V is the volume of the specimen. Suppose the specimen contains N atoms, each with magnetic moment m. Then maximum possible magnetic moment is equal to Nm. Hence if all the magnetic moments align in the direction of field, the maximum possible intensity of magnetisation known as saturation intensity is given by Nm/V. But for small magnetic field the intensity of magnetisation I of a material is found to be directly proportional to the strength of the applied field B and inversely proportional to the absolute temperature T. Symbolically we can write

$$I \propto B$$

$$\text{and } I \propto \frac{1}{T}$$

$$I \propto \frac{B}{T}$$

or

$$I=K\frac{B}{T} \qquad (7.26)$$

where K is a constant of proportionality. This relation is known as Curie's law.

At a given temperature the intensity of magnetisation I of a material increases directly with the increase of external applied field strength if the field strength is low.

If the field strength is gradually increased then after a certain stage it is found that I attains a maximum value and does not increase any more. At this stage all atomic moments inside the

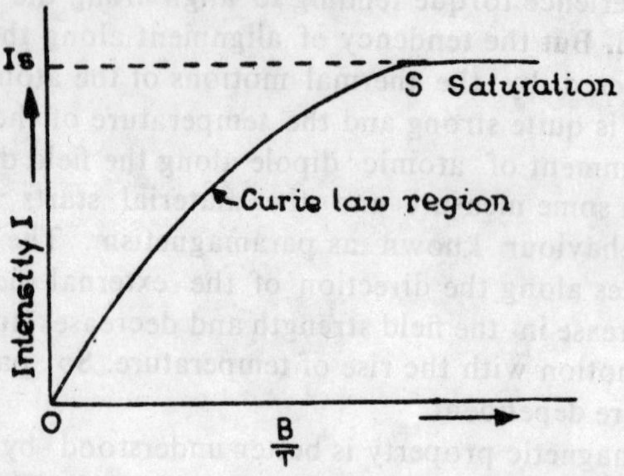

Fig. 7.19. Intensity of magnetisation of paramagnetic solid as a function of B/T. OS represnts region where Curie law holds good. Beyond S, the region belongs to saturation state.

material have been aligned along the applied field direction and state of saturation is reached. This has been illustrated graphically by plotting I versus B/T in Fig. 7.19. Here OS represents the region where Curie's law holds good, beyond S the region belongs to saturation stage.

Thus from above discussion it is evident that for a solid to be paramagnetic, the atoms or ions should have permanent magnetic moments and paramagnetism is temperature dependent *Examples*: copper, aluminium, oxide, manganese, oxygen, and platinum are paramagnetic substances.

(*c*) **Ferromagnetism materials**. These materials are always solids and exhibit marked magnetisation even in a weak magnetic field. They are attracted more than the paramagnetic substances by a magnet. The materials which come under this group are iron, cobalt, nickel, gadolinium, dysprorium and number of alloys.

The ferromagnetic material contains atoms with permanent magnetic moment, as in a paramagnetic material. But due to a special form of interaction called exchange interaction, the magnetic moment of neighbouring atoms are coupled strongly to form a groups, called domain.

In each domain the atomic dipoles are aligned in the same direction and have a resultant magnetic moment. That is to say, the domain is spontaneously magnetised even in the absence of the magnetic field. In the unmagnetised material, the direction of magnetisation in different domains are different. So, on the average the resultant magnetisation is zero. But when such a material

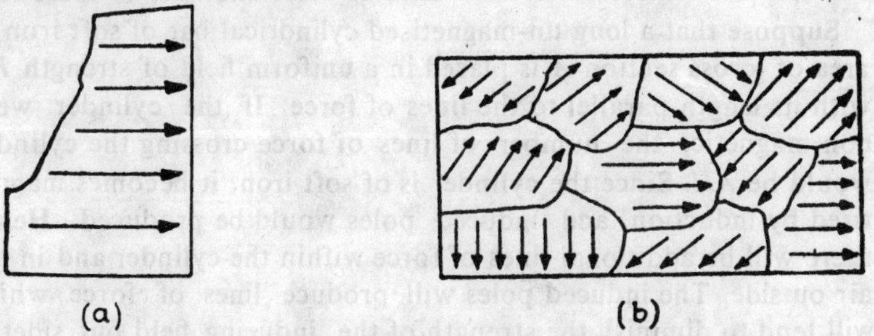

Fig. 7.20. (*a*) A single domain, (*b*) Random orientations of the domains a ferromagnetic substance.

is placed in an external magnetic field, it becomes magnetised due to the alignment of domains along direction of the magnetic field. Stronger the field, greater will be the magnetisation due to the alignment of more number of domains along the field.

7.37. Magnetic Permeability

When a magnetisable material is placed in a magnetic field the magnetic dipoles of magnetisable material get aligned parallel to the field and the material becomes magnetised. Then within the magnetisable material we will have got two sets of lines of force. Those which are due to magnetising field and which would exist if the magnetic material were replaced by a non-magnetic material,

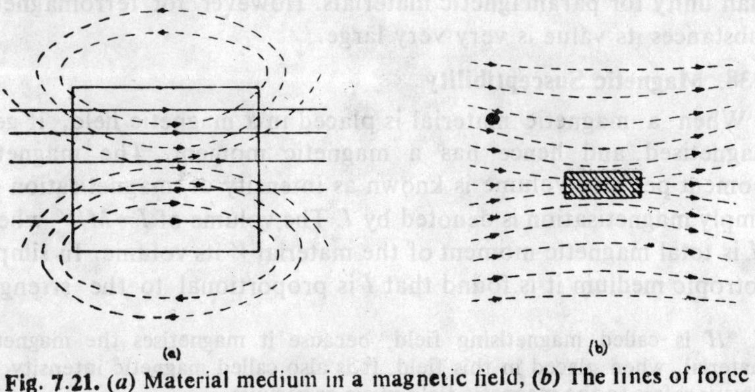

Fig. 7.21. (*a*) Material medium in a magnetic field, (*b*) The lines of force gets concentrated inside the magnetic material such as iron.

and those due to the magnetisation of the material itself when placed in the magnetic field, Fig. 7.21. The latter are called lines of magnetisation. The lines of force and the lines of magnetisation together are known as the lines of induction. The strength of the magnetic field within magnetic material is due to the lines of induction *i.e* the combined effect of lines of force and lines of induction.

Suppose that a long un-magnetised cylindrical bar of soft iron of area of cross section A is placed in a uniform field of strength H* with its length parallel to the lines of force. If the cylinder were non-magnetic, the number of lines of force crossing the cylinder would be AH. Since the cylinder is of soft iron, it becomes magnetised by induction and induced poles would be produced. Hence there will be additional lines of force within the cylinder and in the air outside. The induced poles will produce lines of force which will tend to diminish the strength of the inducing field out sidethe cylinder as shown in Fig. 7.21 (*a*). Inside the cylinder however the field is concentrated and in this condition the lines of force crossing the cylinder is given by AB instead of AH, where B is the field due to lines of induction inside the material.

As lines of force per unit normal area of cross-section measures the magnetic field, magnetic field B in this case would be greater than H. If the cylinder were diamagnetic then B would have been less than H as induced magnetisation would have been opposite to the applied field H. However, to compare the magnetic property of the material we introduce a quantity called permeability of the material, which is given by B/H and is denoted by μ. Therefore

$$\mu = B/H \qquad \ldots(7.27)$$

The permeability of a medium is different for different materials. Its value is less than unity for diamagnetic and slightly greater than unity for paramagnetic materials. However for ferromagnetic substances its value is very very large.

7.38. Magnetic Susceptibility

When a magnetic material is placed in a magnetic field, it gets magnetised and hence has a magnetic moment. The magnetic moment per unit volume is known as intensity of magnetisation or simply magnetisation is denoted by I. The volume of $I = M/V$ where M is total magnetic moment of the material V its volume. In simple isotropic medium it is found that I is proportional to the strength

*H is called magnetising field, because it magnetises the magnetic material, when placed in this field. It is also called magnetic intensity. H always refers to the magnetic field in air or vacuum.

H of the magnetising field. So we may write :

$$I \propto H$$

$$I = \chi_m H \qquad \ldots(7.28)$$

Here χ_m is a dimensionless quantity called magnetic susceptibility of the material, and is a characteristic property of the material. In many cases χ_m is negative and such substances are called diamagnetic and many other substances for which χ_m is positive are called paramagnetic and ferromagnetic substances.

QUESTIONS

1. A uniform spring with force constant K is cut in half what is spring constant for one of the halves?
2. Explain why iron is called more elastic than rubber?
3. Discuss how the shearing strain is explained in terms of atomic structure of the solids?
4. Graphite consists of planes of carbon atoms. Between the atoms in the planes there are strong inter atomic forces. Between the atoms in different planes, there are only very weak forces. What kind of elastic properties you expect for graphite?
5. A cable is replaced by another one of same length and same material but of twice its diameter. How will it affect elongation under a given load?
6. Why is there a gap between two spans of the concrete bridge? Why is there a gap between two rails at the joint?
7. Why do you take room temperature as original temperature instead of 0°C in determining the various co-efficients expansions of solids? Explain your answer by considering the area expansion and volume expansion.
8. What do you mean by bi-mettallic element? Explain your answer by mentioning some of its uses?
9. A hollow sphere and a solid sphere of same material and same diameter have been raised the temperature by equal amounts. How are their changed volumes related with each other?
10. What kind of thermal conductivity and specific heat requirements would you suggest for a cooking utensil?
11. Explain why good conductors of heat are also good conductor of electricity?
12. Water is a dielectric and is available in plenty in nature. Why is it not used as dielectric medium in capacitors?
13. Distinguish between diamagnetism and paramagnetism. Why do you say that diamagnetism is a universal phenomena?
14. Distinguish between permeability and susceptibility of a medium.

PROBLEMS

1. A copper wire of 3 m long and 2 mm^2 in cross sectional area hangs from a ceiling. What will be its elongation if a 2 kg mass is suspended from the lower end. *[Ans. 0.235 mm]*

2. Find the elongation of a tendon 4 cm, long and 6 mm in diameter if it is put under tension of 1600 N. Assume Young's modulus to be 1.6×10^{10} N/m^2.

[*Ans. 0.000141m*]

3. An aluminium rod of 300 cm long and 0.508 cm in diameter should normally contract 1.32 cm in cooling from 225°C to 25°C. What force would be required to prevent this contraction. [*Ans. 6.15×10^3 Newtons*].

4. How much force is required to punch a hole 8 mm in diameter in a metal sheet 1.5 mm thick if the minimum shearing stress needed to rupture the metal is 25×10^7 N/m^2.

[*Ans. 12.6×10 N*]

5. A sphere of certain material is immersed in a fluid and the normal force per unit area on its surface is increased by 5.0×10 N/m^2. The volume is thus reduced by 5.0 per cent. What is the bulk modulus of the material.

[*Ans. 1.0×10^3 N/m^2*]

6. In to a 100 gm copper calorimeter containing 180 gm of water at 10 °C there are dropped simultaneously 70 gm of silver at 100°C, 60 gm of iron at 85°C and 20 gm of platinum at 90°C. What is the resulting equilibrium temperature?

[*Ans. 14.5°C*]

7. A 30 gm piece of ice at—20°C is dropped into a 25.0 gm calorimeter of specific heat 840 joules/(kg) (°C) containing 100 gm of water at 35.0°C. The final equilibrium temperature corrected for thermal leakage is found to be 7.2°C. What is the specific heat of the ice? [*Ans. 2.1×10 Joules/(kg)(°C)*]

8. 400 gm lead shot is heated to 100°C and dropped into 100 gm of glass beaker containing 200 gm of water at 20°C. The specific heat of glass is 0.20/gm-°C. The mixture is stirred until temperature equilibrium is reached, the final temperature is 24.2°C. What is the specific heat of lead.

[*Ans. 0.0305 cal/gm-°C*].

9. A well insulated copper calorimeter cup of mass 100 gm contains 400 gm of water at 5°C. When 40 gm of ice at—8°C is added, it is found that, 23.6 gm of the ice melts. What is the latent heat of fusion of ice? Given specific heat of ice is 0.50 cal/gm-°C. [*Ans. 79.9 cal/gm*]

10. How much heat is required to change 5 gm of ice at—5°C into water at 50°C? Given specific heat of ice is 2.11 J gm^{-1} deg^{-1}, specific heat of water is 4.186J gm^{-1} deg^{-1} and latent heat of ice is 334 J gm^{-1}

[*Ans. 2773 J*].

11. Railway lines are laid with gaps to allow for expansion. If the gap between steel rails 66 m long is 3.63 cm at 10°C at what temperature will the lines just touch? Co-efficient of linear expansion for steel is 11×10^{-6} per °C

12. A steel wire 2 mm in diameter is just stretched between two fixed points at temperature of 20°C. Determine its tension when temperature falls to 10°C. Given co-efficient of linear expansion of steel=11×10^{6}/°C. *Y* for steel=2×10^{12} dynes/cm^2.

[*Ans. 69.14 N*]

13. A solid occupies 1000 ml at 20°C. Its volume becomes 1016.2 ml at 320°C. What is the value of the co-efficient of linear expansion?

[*Ans. 18×10^{-6}/°C*].

14. A brass scale measures true centrimeter at 10°C. The length of a copper rod measured by the same scale is found to be 100 cm at 20°C. Find the real length of the rod at 0°C (the co-efficient of linear expansion of copper is 0.000017/°C and that of brass 0.000019/°C) [*Ans. 99.985 cm*]

15. The volume of a lead bullet at 0°C is 25 cc. The volume increases at

98°C by 0.021 cc. Find co-efficient of linear expansion of lead.

[*Ans.* $28.6 \times 10^{-6}/°C$].

16. An iron tyre 0.6 m in radius at 20°C is to be put on a wooden wheel 3.825 m in circumference. Up to what temperature should it be heated? (Co-efficient of linear expansion of iron is $= 0.000012/C°$).

17. Taking the Young's modulus of brass to be 10^{11} N/m². Calculate the increase in energy of a bar 0.1 m long and of cross section area 1.0 cm, when compressed with a load of 10 kg wt along to its length ($g = 10$ ms^{-2})

[*Ans.* 0.5×10^{4} *J*]

18. A copper rod whose diameter is 2.0 cm and length 50 cm has one end in boiling water, the other end in a jacket cooled by flowing water, which enters at 10°C. The thermal conductivity of copper is 0.102 kcal/(m²) (sec) (°C/m),. If 0.20 kg of water flows through the jacket in 6.00 min by how much does the temperature of this water increases? [*Ans. 10°C*].

19. The thermal insulation of a woolen glove may be regarded as being essentially a layer of quiscent air 3.0 cm thick of conductivity 5.7×10^{-6} kcal/(m³) (sec) (C°/m). How much heat does a person lose per minute from his hand of area 200 cm² and skin temperature 34°C on a winter day at —5°C?

[*Ans. 91 cal/min*]

20. A compound wall consists of parallel layers of two different materials 5 cm brick and 1 cm wood. If the difference of temperature across the compound wall is 20°C, calculate the heat current per square metre of the wall.

Given thermal conductivity of brick $= 0.6$ w m^{-1}deg^{-1}and thermal conductivity of wood $= 0.15$ w m^{-1} deg^{-1}. [*Ans. 133.32 w* m^{-2}]

21. A hollow glass cube with walls 1.5 mm thick and edge 10 cm is filled with ice and placed in a container maintained at 100°C. At what rate will the ice melt. When the flow of heat across the walls has become steady? Thermal conductivity of glass $= 1$ wm^{-1} deg^{-1}and latent heat of ice $= 334.8$ J g m^{-1}. [*Ans. 11.95 gm/s*]

22. How much water would be evaporated per hour per square meter by the heat, which flows through a steel boiler plate 2.5 mm thick when there is a difference of temperature of 2.5°C between the faces of the plate and water enters at 20°C? Assume the pressure remains 1 atm and that it takes 540 kcal to evaporate 1 kg of water at 100°C. [*Ans. 89.3 kg*]

23. A glass window is 4.5 mm thick, 1.2 m wide and 1 m high. The temperature of the inner surface is 14.1°C and that of the outer surface is 12.9°C. How much of heat is lost per hour through the window? [*Ans. 726 K J*].

24. Find the heat conducted in 10 minute through an area of 2 m² of layer consisting 20 mm of cork and 75 mm of wood. The outer surface of the wood is at 25°C and the opposite cork surface is at 0°C. What is the temperature of the cork wood interface? [*Ans. 26.8 kg, 12.1°C*]

8

ELECTRONIC DEVICES

The phenomenal growth of technology in the present century owes much to the development of electronics. Manipulation of flow of electrons in a vacuum tube, known as 'electronic value', enabled us to perform many functions such as rectification of A.C. to D.C., amplification of small signals, generation of high frequency signals in an oscillator; radio communication and transmission of television signal etc. With the invention of transistors and subsequent developments of integrated circuits and microprocessors the electronic valves are being rapidly replaced by these solid state devices. We will deal with the basic principles involved in transistors and valves and some of their common uses in this chapter.

8.1. P-N Junction

We have discussed earlier the formation of *N*-type semiconductor and *P*-type semiconductor by doping a silicon or germanium crystal with suitable impurities. It is, however, possible to dope a silicon or germanium crystal with a trivalent impurity at one end and with a pentavalent impurity at the other end. So one side of the crystal will be a *P*-type semiconductor and the other side will be a *N*-type semiconductor. Such an arrangement is called a *P-N* junction. The actual process of manufacturing *P-N* junction is quite complex. But for understanding the basic principles involved, *P-N* junction is digramatically represented as a *P*-type semiconductor placed by the side of an *N*-type semiconductor as shown in Fig. 8.1.

The *P*-type crystal contains more holes and the *N*-type contains excess of electrons. Therefore, when both *P*-type and *N*-type material remain on the same crystal, some of the holes from the *P*-type will move towards the *N*-type. Simultaneously, some of the free electrons from the *N*-type will also move towards the *P*-type. In the junction region, the recombination of electrons with holes takes place. But this flow of charge carriers (electron and hole) does not continue indefinitely, otherwise both the *P* and *N* type material would lose all free electrons and holes due to recombination. Actually what happens is explained below.

Either a *P*-type or an *N*-type material is electrically neutral. But in the *P-N* junction, when electrons diffuse to the *P*-side of the

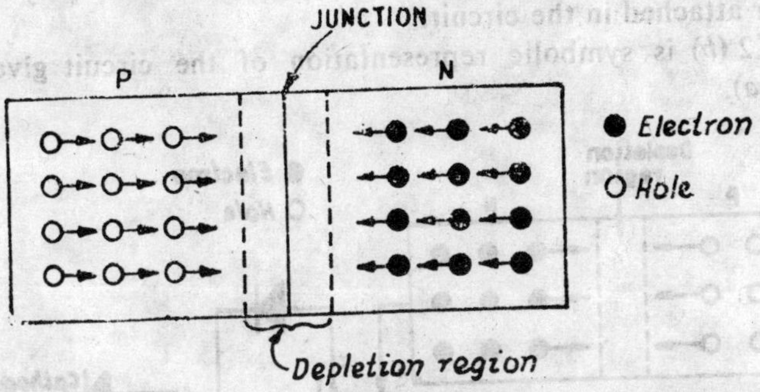

Fig. 8.1. *P-N* junction.

junction, the *P* region near the junction become electrically negative due to excess electrons. As the diffusion continues, the side becomes more and more negative. So it prevents further flow of electrons from *N*-side to *P*-side. It may be remembered that these electrons are no more available as free electrons near the junction as they have already recombined with holes.

Similarly, due to diffusion of holes to the *N*-side of the junction, the region near the junction on the *N*-side becomes electrically positive and prevents further flow of holes from *P*-side to the *N*-side. In this region, free holes are also not available as they have recombined with electrons.

Hence there is small area on either side of the junction where the carriers of electricity are absent. This region is known as *depletion region.* The thickness of this region is of the order of 10^{-4} cm. The thickness of the depletion region changes when potential or bias is applied across the region. We will consider it here.

8.2. Forward Biased P-N Junction

The manner in which the electric potential is applied in Fig. 8.2*a* is called forward biasing. Here the positive terminal of the battery is connected to the *P*-side and negative terminal is connected the *N*-side of the *P-N* junction. The holes in the *P* region will be pushed towards the depletion region in the direction of the applied field. Similarly the electrons in the *N* region will be pushed towards the depletion region in the opposite direction of the field. As these carriers will move into depletion region, the depletion region will be practically absent. So the holes and electrons will cross the junction in opposite directions. The flow of these charge carriers wiil constitute a relatively large current indicated by milli-ammeter attached in the circuit.

Fig. 8.2 (*b*) is symbolic representation of the circuit given in Fig. 8.2(*a*).

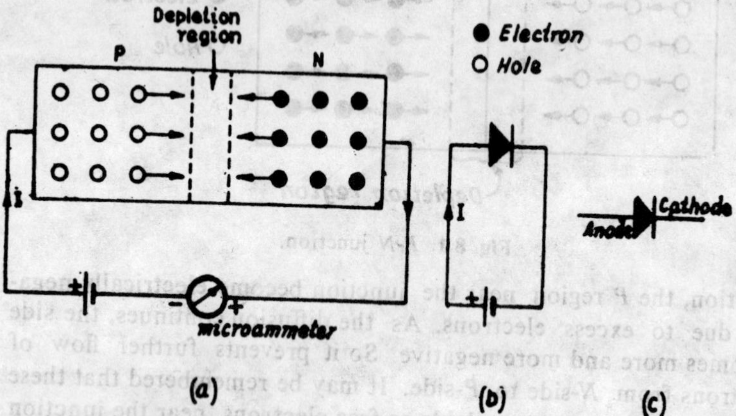

Fig. 8.2. (*a*) Forward biased *P-N* junction, (*b*) Symbolic repersentation of (*a*), (*c*) Symbolic representation of a diode.

8.3. Reverse Biased P-N Junction

The circuit arrangement of Fig. 8.3 (*a*) shows a reverse biased *P-N* junction. In the reverse biasing, the negative terminal of the battery is connected to the *P*-side and the positive terminal is

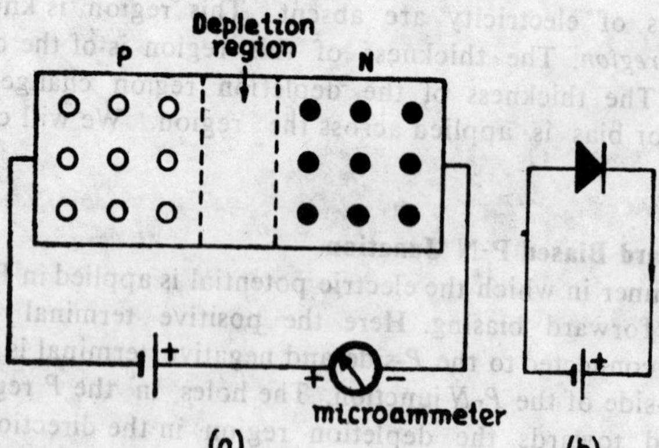

Fig. 8.3. (*a*) Reverse biased *P-N* junction, (*b*) Symbolic representation of (*a*).

connected to the *N*-side of the *P-N* junction. Due to the applied field, holes in the *P*-side and electrons in the *N*-side will be pulled away from near the depletion region. So the depletion region widens. Hence it will be relatively difficult for the majority carriers to cross over the junction and there will be practically very little current in the external circuit. The small current, in this arrangement is due to minority carriers (electrons in the *P* region and holes in the N region for an intrinsic semiconductor). Generally it is of the order of a few microamperes. The Fig. 8.3 (*b*) is symbolic representation of circuit given in Fig 8.3 (*a*).

8.4. Characteristic Curve of P-N Junction

The nature of variation of current in the cricuit with the applied bias voltage is called characteristic of a *P-N* junction. Fig. 8.4 shows a typical characteristic curve for a *P-N* juction.

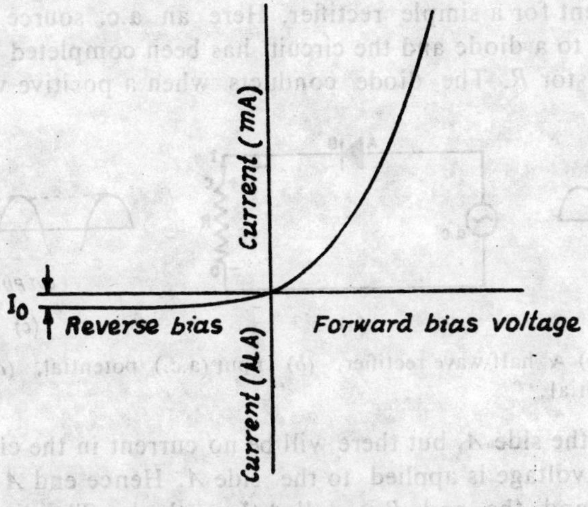

Fig. 8.4. Characteristic curve of *a P-N* junction. I_0=Reverse saturation current.

It can be seen from this curve that the current increases nearly linearly for small applied voltage and then very rapidly with the increase of applied forward bias voltage. But the value of current for the reverse bias voltage is very small and increase is small enough to regard the current very nearly constant. This constant reverse bias current is known as *reverse saturation current.*

8.5. P-N Junction Used as a Rectifier

We have seen that in reverse biased conditions practically there is no current in the circuit though in a forward biased arrangement there is appreciable current even for a small applied voltage. In other words, in one direction electrical resistance is negligible whereas in the reverse direction, the electrical resistance is infinite. Hence a *P-N* junction allows current in one direction only or this is an arrangement for unidirectional flow as in case of a valve. We will discuss later how the same objective is attained in a two-electrode electronic valve called a diode. In view of similarity in operation, a *P-N* junction is also called a diode. The symbolic diagram of a *P-N* junction in Fig. 8.2 (*b*) and Fig. 8.3 (*b*) is used to represent a diode valve or semiconductor in electrical circuits.

A diode can be used to convert alternating current (A.C.) to direct current (D.C.). This is called *rectification* and the circuit arrangement using a diode to convert A.C. to D.C. is called rectifier.

We will explain here how rectifier works. Fig. 8.5 (*a*) is the arrangement for a simple rectifier. Here an a.c. source has been connected to a diode and the circuit has been completed by including a resistor *R*. The diode conducts when a positive voltage is

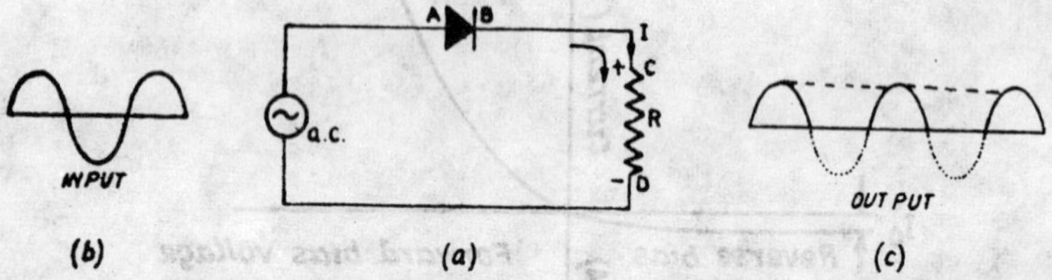

Fig. 8.5. (*a*) A half-wave rectifier, (*b*) Input (a.c.) potential, (*c*) Output (d.c.) potential.

applied to the side *A*, but there will be no current in the circuit if a negative voltage is applied to the side *A*. Hence end *A* is called the anode and the end *B* is called the cathode. The direction of current is from *A* to *B* inside the diode.

Look at the wave-shape of the a.c. voltage applied to the diode, Fig. 8.5 (*b*). Here we have shown one complete cycle of the applied voltage. In the first half cycle (upper half) the anode *A* remains at a higher potential compared to the cathode *B*. So the diode remains in forward biased condition and there is easy flow of electricity. The direction of current in the load resistor *R* is from *C* to *D*. Hence the end *C* of the resistor will remain at a higher potential compared to the end *D*. The value of the current in the resistor vary according to the change in the voltage applied between the end *A* and *B* of the diode. From the characteristic curve of a *P-N* junction, we have seen that for a small applied voltage, current in the diode is proportional to the applied voltage. Hence during the first half cycle, current will continue to be unidirectional *i.e.* from *C* to *D*, but its value will vary in a manner similar to variation of applied voltage. So the potential difference across *R* will also vary similar to variation of applied voltage as shown in Fig. 8.5 (*c*).

In the other half cycle (lower half), the anode end *A* remains negative with respect to the cathode end *B*. During this time, there will be practically no current in the resistor. Hence the potential difference across *R* will be very nearly zero. This has been indicated by dots in the output side.

This will be repeated for each cycle. So at the output end, for every cycle we will get output pulse for a half cycle separated by a half cycle, Fig. 8.5 (*c*). If the frequency of the alternating voltage is high, the gap between two consecutive pulses will be small and the fluctuation will be rapid. Hence the output voltage or current will be continuous far all practical purposes (Indicated by the dashed lines from peak to peak. The frequency of a.c. supply in our country is 50 cps. So we can get d.c. from the line supply using a diode as a rectifier.

8.6. A Full-Wave Rectifier

A rectifier using a single diode is called a half-wave rectifier as it operates only for one half of a cycle. But by using two diodes it is possible to rectify the full or complete cycle. Such a rectifier is called a full wave rectifier as shown in Fig. 8.6 (*a*). The circuit operates in the following manner.

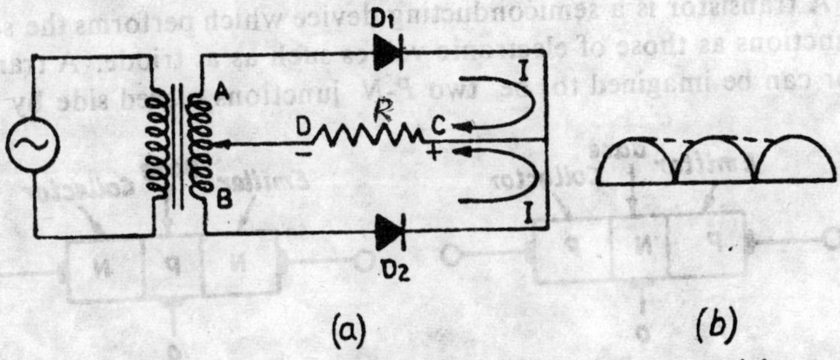

Fig. 8.6. (*a*) A full-wave rectifier, (*b*) Output (d.c.) potential.

The load resistor *R* is connected between the common cathode end of the diodes D_1 and D_2 and the mid-point of the secondary coil of a transformer. The alternating voltage source is connected to the primary of the transformer.

Suppose, during the first half cycle, the end *A* of the transformer is at +ve potential with respect to end *B*. Then the diode D_1 will be forward biased and hence conduct electricity. But the diode D_2 will be reverse biased and wlll be non-conducting. So during this half cycle, there will be current from *C* to *D* in the load resistor *R* due to current in D_1 alone.

In the next half cycle, the end *B* remains at the higher potential compared to the end *A*. So D_2 becomes forward biased and conducts electricity in the direction *C* to *D* in the resistor *R*. D_1 is reverse biased and is non-conducting.

So for one half cycle D_1 functions and for the other half cycle the diode D_2 works. But the direction of current is always from *C* to *D* across the resistor *R*. So the potential difference or voltage developed across *R* is unidirectional as indicated by Fig. 8.6 (*b*).

The voltage across the load is pulsating in nature. In this respect it differs from a steady d.c. obtainable from a battery. The output voltage from a rectifier is smoothened by adding filter circuits containing capacitors and inductors. A device containing a rectifier and a filter is called a power supply. These are beyond the scope of this book.

8.7. Transistors

John Bardeen, William Shockley and W.H. Brattain, three scientists working in Bell telephone laboratory developed transistors in 1948. The basic transistors developed by them has come a long way through a series of developments leading to integrated circuits and microprocessors. We will discuss here the basic principle involved in a transistor.

A transistor is a semiconducting device which performs the same functions as those of electronic valves such as a triode. A transistor can be imagined to be two *P-N* junctions placed side by side

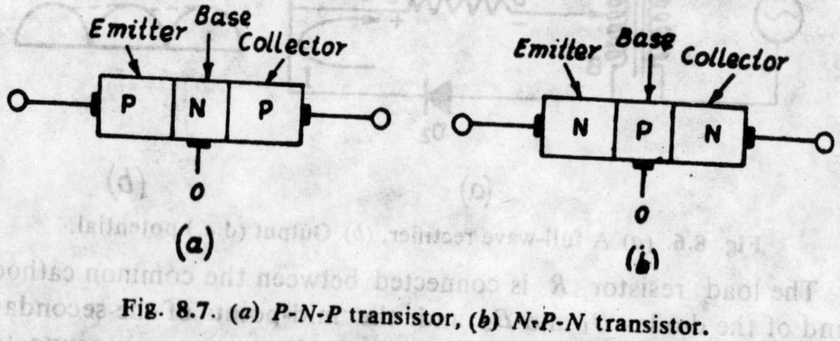

Fig. 8.7. (*a*) *P-N-P* **transistor,** (*b*) *N-P-N* **transistor.**

in oppose directions in the manner shown in Fig. 8.7. If they are placed in the manner *P-N—N-P* (Fig. 8.7 *a*) it is called a *P-N-P* transistor and if placed like *N-P—P-N*, it makes an *N-P-N* transistor Fig. 8.7 (*b*).

In actual transistor manufacture, however, these *N* and *P* semiconductors are created in the same silicon or germanium crystal. The middle semiconductor in a transistor is called **Base** and the two end regions are called as **emitter** and the **collector** respectively as shown in Fig. 8.7. Generally the emitter and collector are heavily doped semiconductors whereas the base is lightly doped and its thickness is very small compared to the thickness of emitter and collector. Base, emitter and collector are connected with suitable conductors called 'legs' for connecting the transistor into electrical circuits. The complete transistor is enclosed in a metal case and looks like one in Fig. 8.8.

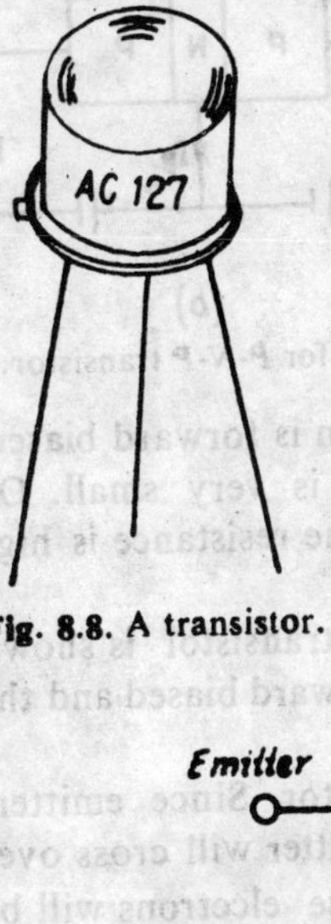

Fig. 8.8. A transistor.

We will show later that the emitter-base diode of a transistor is always forward biased and the collector-base diode is always reverse biased.

The transistors are represented by symbols in circuit diagrams as shown in Fig. 8.9. The arrow mark distinguishes the emitter. The flate line is the base and the other line represents collector. In *P-N-P* transistor, the arrow on the emitter side points towards the base whereas in *N-P-N* transistor arrow is pointed away from the base.

There are various processes for transistor manufacture. Hence different transistors are

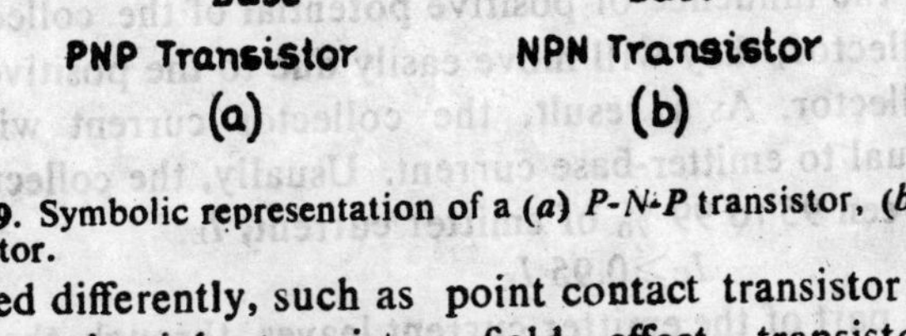

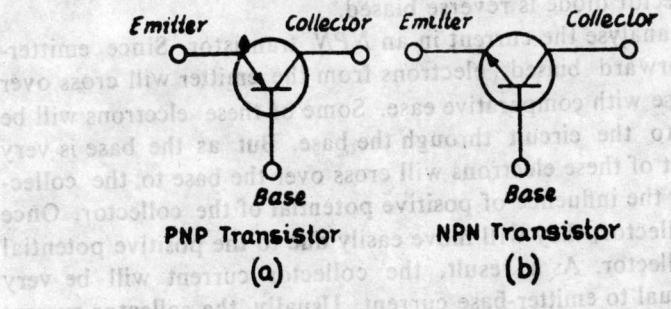

Fig. 8.9. Symbolic representation of a (*a*) *P-N-P* transistor, (*b*) *N-P-N* transistor.

also named differently, such as point contact transistor, junction transistor, epitaxial transistor, field effect transistor (*FET*), metal oxide field effect transistors (*MOSFET*) etc. But the basic ingredients in transistor manufacture in all types of transistors is *N* and *P* semiconductors. We will restrict ourselves to the working of a junction transistor only.

8.8. Biasing a Transistor

The main function of a transistor is to act as an amplifying device. To increase a small input signal to comparatively larger value is called amplification. In order to understand the functioning of a transistor, let us analyse the biasing of the transistor.

Look at the Fig. 8.10 (*a*). In the base-emitter section, the negativ e pole of the battery is connected to the emitter (*N*) and

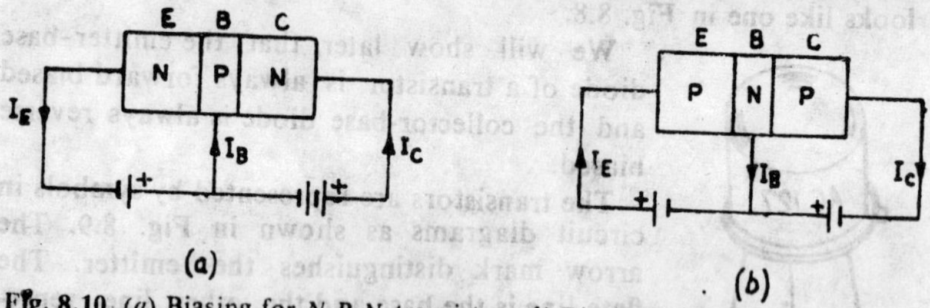

Fig. 8.10. (*a*) Biasing for *N-P-N* transistor, (*b*) Biasing for *P-N-P* transistor.

the positive pole to the base (*P*). So this section is forward biased. Hence the electrical resistance of this section is very small. On the other hand, in the base-collector section, the resistance is high due to reverse biasing.

The battery connection or biasing in a *PNP* transistor is shown in Fig. 8.10(*b*). Here also the base-emitter is forward biased and the base collector diode is reverse biased.

Let us analyse the current in an *NPN* transistor. Since emitter-base is forward biased, electrons from the emitter will cross over to the base with comparative ease. Some of these elcctrons will be diverted to the circuit through the base. But as the base is very thin, most of these electrons will cross over the base to the collector under the influence of positive potential of the collector. Once in the collector, they will move easily due to the positive potential of the collector. As a result, the collector current will be very nearly equal to emitter-base current. Usually, the collector current I_C is between 95 to 99 % of emitter current, I_E.

$$I_C \geqslant 0.95\, I_E \qquad ...(8.1)$$

Since a part of the emitter current leaves through the base, we can see in the external circuit, Fig. 8.10 (*a*).

$$I_C = I_E - I_B \qquad ..(8.2)$$

where I_B is the base current.

Thus I_C will be more than 20 I_B in most cases. Condsidering the equation 8.1, therefore,

$$I_B \leqslant 0.5\ I_E.$$

It may be noted here that we have considered the movement of majority carriers (electrons) only. In the reverse biased collector section there will be some contribution from the minority carriers (holes). The contribution of minority carrier is very small and for the present, we neglect it. Moreover, in the base (*P*) some of the holes recombine with electrons coming from the emitter side (*N*). But the doping of base being very light, this effect we will also neglect.

The charge carriers (electrons) starting from the emitter, travelling through the base and ultimately reaching the collector is the characteristic property of a transistor. The same thing also happens in *P-N-P* transistor, but the charge carriers are holes in that case.

From Ohm's law we know that if there is a current I in a conductor of resistance R then the potential difference between the ends of the conductor is given by $V=IR$. In a transistor, the resistance of the base-collector diode is much higher than the resistance of the base-emitter diode which is forward biased. A given base or emitter current can be caused in the base-emitter side by a small input voltage. We have seen that the collector current is nearly equal to the emitter current and is quite large compared to the base current. Therefore, the collector current will produce a greater output voltage in the collector circuit. Compared to the small input voltage applied in the base-emitter side. This is the basic principle involved in amplification by a transistor. We will further discuss it later.

8.9. Current Amplification Factors

There are two current amplification or gain factors used to describe the working of a transistor.

(*a*) **The emitter-collector current gain factor** α is defined as the ratio of the change in collector current to the change in the emitter current for a constant collector-base potential difference.

Thus,

$$\alpha=\left[\frac{\triangle I_C}{\triangle I_E}\right] V_{CB}\text{constant.} \qquad \ldots(8.3)$$

Since most of the emitter current flows to the collector α lies between 0.95 to 0 99.

(*b*) **The base collector current gain Factor** β is defined as the ratio of the change in collector current to the change in base current with collector to emitter potential difference V_{CE} kept constant. Thus

$$\beta = \left[\frac{\triangle I_C}{\triangle I_B}\right]_{V_{CE}\ \text{constant}} \qquad \ldots(8.4)$$

It should be remembered that α and β are interrelated. From equation 8.2, we get

$$I_B = I_E - I_C$$

$$\therefore \quad \triangle I_B = \triangle I_E - \triangle I_C = \triangle I_E \left(1 - \frac{\triangle I_C}{\triangle I_E}\right)$$

Substituting from equation 8.3,

$$\triangle I_B = \triangle I_E\,(1-\alpha) \qquad \ldots(i)$$

From equation 8.4,

$$\triangle I_B = \frac{\triangle I_C}{\beta}$$

Hence,

$$\frac{\triangle I_C}{\beta} = \triangle I_E\,(1-\alpha)$$

or

$$\beta = \frac{\triangle I_C}{\triangle I_E}\,\frac{1}{(1-\alpha)}$$

$$\therefore \quad \beta = \frac{\alpha}{1-\alpha} \qquad \ldots(8.5)$$

As the value of α tends to unity, β becomes very large. For most of the transistors, the typical value of β lies in the range of 20 to 1000. Since α is very close to 1, its accurate determination is difficult. But value of β can be accurately determined experimentally and α can be calculated using equation 8.5. The terms α and β usually refer to a.c. But the same definitions are also applicable for d.c. To differentiate, usually α_{dc} and β_{dc} are specified. The values of α_{dc} and β_{dc} are almost equal to the values of α and β for the same transistor.

8.10. Characteristic Curves of a Transistor

The working of a transistor depends on the manipulation of current in its various sections. Therefore, a transistor is known as a current regulated device. Hence it is necessary to study the effect of variation of emitter or collector current or emitter or collector voltage by plotting the characteristic curves for a transistor.

There are three types of circuit arrangement or configuration to study the characteristics: 1. common base configuration (Fig. 8.11). 2. common emitter configuration (Fig. 8.14) and 3. common collector configuration.

In the common base configuration, both the emitter and collector voltages are measured relative to the base. In the common emitter configuration, the emitter is common to both base and collector voltage measurement and in the common collector arrangement collector is common to both sides.

Two basic sets of characteristic curves for transistors are studied in these arrangements. They are, (*a*) Output characteristics and (*b*) Input characteristics. The output characteristic curves show the variation of collector current with collector voltage for a constant value of emitter current (for common base configuration) or for constant value of base current (for common emitter configuration). Similarly, the input characteristic curves show the variation of emitter current with emitter voltage (for common base) or variation of base current with base voltage (for common emitter) for a constant value of collector current. We will see later that a common collector circuit has no amplifying property. Hence we will not discuss the characteristic curves for common collector circuits here.

Common base characteristics. The circuit arrangement is shown in Fig. 8.11. The emitter voltage and the collector voltage is vari-

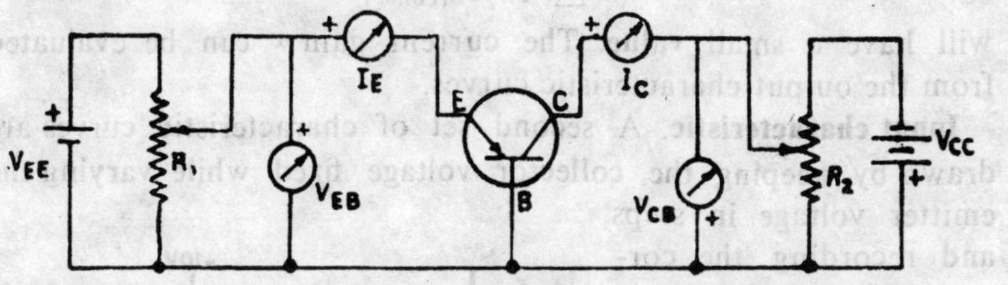

Fig. 8.11. Circuit for study of common base characteristic (*P-N-P* transistor used).

ed by the rheostats R_1 and R_2. The emitter-base voltage V_{EB} and the collector base voltage V_{CB} are measured by suitable millivoltmeter and voltmeter respectively. Similarly the emitter current I_E and collector currents I_C are measured by a suitable milliammeters.

Output characteristics. With the collector voltage at the minimum, the emitter current I_E is kept a constant value say 50 μA by adjusting the rheostat R_1. The collector voltage is gradually increased and the corresponding collector current is noted at each step and the curve between I_C and V_{CB} is drawn as in Fig. 8.12. This is the output

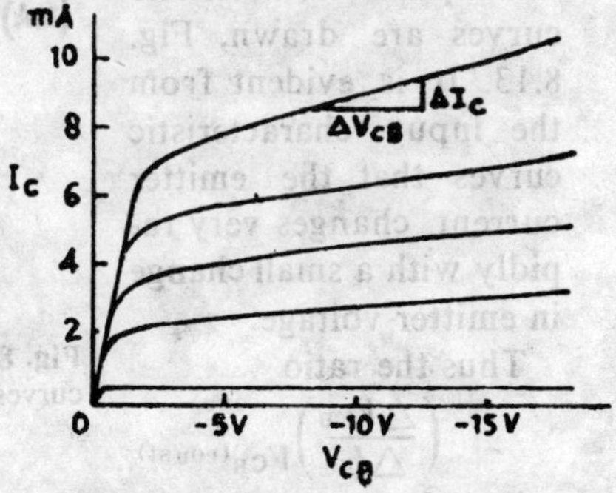

Fig. 8.12. Common base output characteristic curves.

characteristic curve for the given emitter current. Emitter current is then changed to another value and again collector current is noted by varying the collector voltage. Thus a second curve is obtained. Similarly a number of output characteristic curves are obtained for different values of I_E. For a constant emitter current the collector current increases more or less linearly with the collector voltage increase for small value of collector voltage. But for higher value of V_{CB} the curve becomes horizontal *i.e.* the increases in I_C is very small with increase of V_{CB}. Thus the ratio

$$\left(\frac{\Delta V_{CB}}{\Delta I_C}\right)_{I_E(const)}$$

called *output resistance* has a very high value and is nearly constant for the transistor in a particular circuit arrangement. The reciprocal of this term is called *output conductance.*. Thus output conductance

$$\left(\frac{\Delta I_C}{\Delta V_{CE}}\right)_{I_E(const)}$$

will have a small value. The current gain α can be evaluated from the output characteristic curves.

Input characteristic. A second set of characteristic curves are drawn by keeping the collector voltage fixed while varying the emitter voltage in steps and recording the corresponding emitter current. By plotting the emitter current I_E against the emitter voltage V_{EB} for a fixed value of V_{CB} the input characteristic curves are drawn, Fig. 8.13. It is evident from the input characteristic curves that the emitter current changes very rapidly with a small change in emitter voltage.

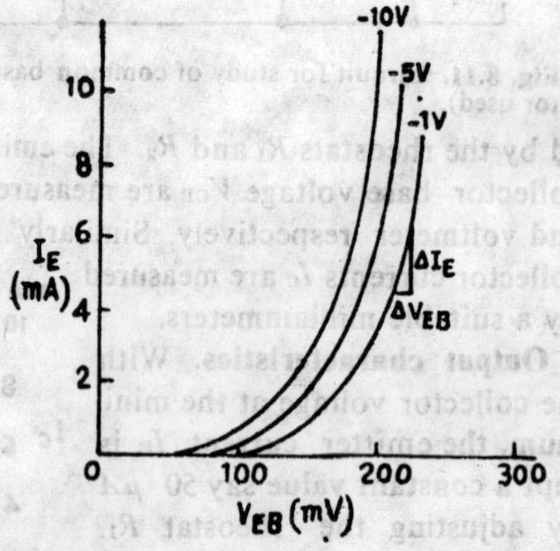

Fig. 8.13. Common base input characteristic curves.

Thus the ratio

$$\left(\frac{\Delta V_{EB}}{\Delta I_E}\right)_{V_{CB}(const)}$$

called *input resistance* is very small.

The fact that the output resistance is much higher than the input resistance is responsible for amplifying action in a transistor.

The relation between the various transistor currents in this arrangement are given by the following equations.

From equation 8.3,

$$I_C = \alpha I_E.$$

But in the collector circuit which is reverse biased there will be small amount of current due to minority carriers. Denoting this

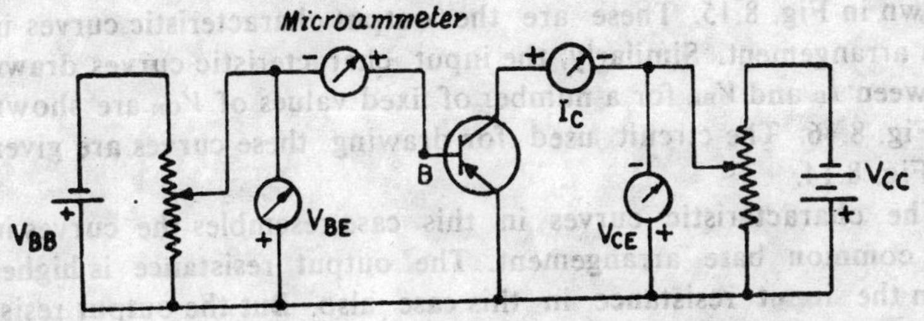

Fig. 8.14. Circuit for study of common emitter characteristics (*P-N-P* transistor used).

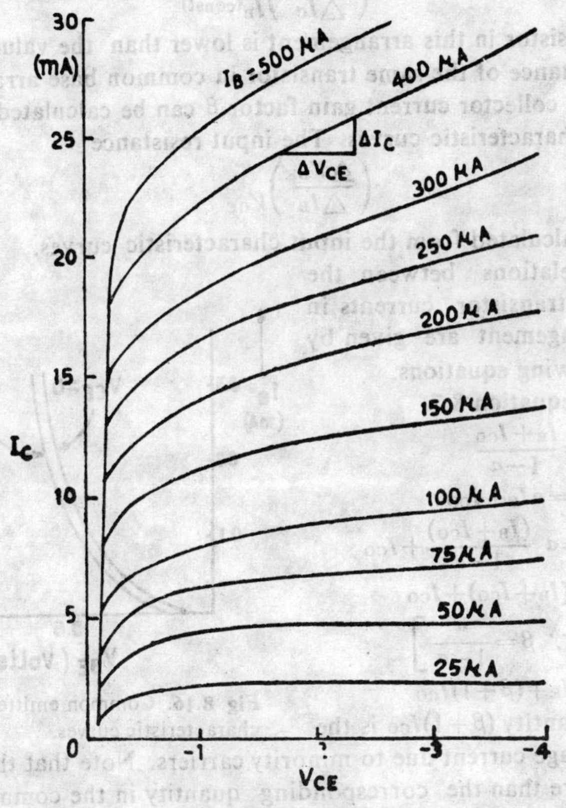

Fig. 8.15. Common emitter output characteristic curves.

current as I_{CO}, we can write,

$$I_C = \alpha I_E + I_{CO} \qquad ...(8.6)$$

But $\qquad I_B = I_E - I_C = I_E - \alpha I_E - I_{CO}$

or $\qquad I_E = (1-\alpha)I_E - I_{CO} \qquad ...(8.7)$

Common emitter characteristics. The set of curves drawn for I_C versus V_{CE} for a number of fixed values of base currents I_B are shown in Fig. 8.15. These are the output characteristic curves in this arrangement. Similarly, the input characteristic curves drawn between I_B and V_{BE} for a number of fixed values of V_{CE} are shown in Fig. 8.16. The circuit used for drawing these curves are given in Fig. 8.14.

The characteristic curves in this case resembles the curves in the common base arrangement. The output resistance is higher than the input resistance in this case also. But the output resistance

$$\left(\frac{\triangle V_{CE}}{\triangle I_C}\right)_{I_B \text{(const)}}$$

of a transistor in this arrangement is lower than the value of output resistance of the same transistor in common base arrangement. The base collector current gain factor β can be calculated from the output characteristic curves. The input resistance

$$\left(\frac{\triangle V_{BE}}{\triangle I_B}\right)_{V_{CE}}$$

can be calculated from the input characteristic curves.

The relations between the various transistor currents in this arrangement are given by the following equations.

From equation 8.7,

$$I_E = \frac{I_E + I_{CO}}{1-\alpha}$$

But $I_C = \alpha I_E + I_{CO}$

So $I_C = \alpha \dfrac{(I_B + I_{CO})}{1-\alpha} + I_{CO}$

$\qquad = \beta(I_B + I_{CO}) + I_{CO}$

$\qquad \left[\because \beta = \dfrac{\alpha}{1-\alpha}\right]$

$\qquad = \beta I_E + (\beta+1)I_{CO}$

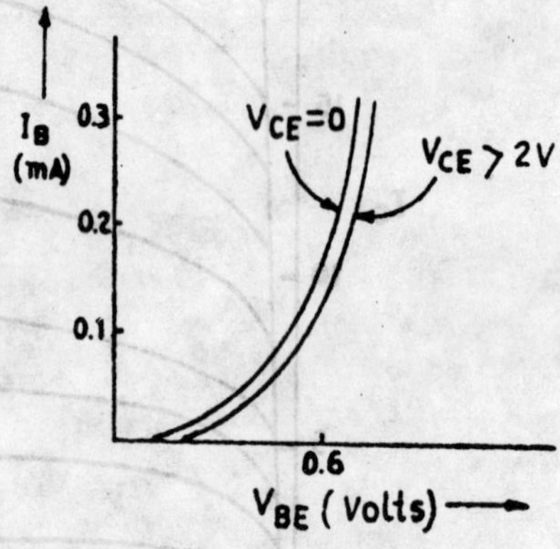

Fig. 8.16. Common emitter input characteristic curves.

The quantity $(\beta+1)I_{CO}$ is the total leakage current due to minority carriers. Note that this quantity is more than the corresponding quantity in the common base arrangement where it is simply I_{CO}. As the semiconductors are sensitive to the fluctuation of temperature, the transistors are very temperature sensitive. So in practical circuits, heat-sinks are attached to the transistors to absorb heat produced. The circuits drawn in this section are for *P-N-P* transistors. The same circuits are also used for *N-P-N* transistors, with the polarity of the battery connection reversed. The nature of characteristic curves for *N-P-N* transistor is similar to those for *P-N-P* transistor.

Example 8.1. A given transistor has a current gain $\beta=40$. If the circuit is connected in a grounded base configuration, how much change in the collector current will be produced for a change of 100 μA in the emitter current ? Assume the collector potential to be constant and neglect I_{CO}.

Solution. We have

$$\beta=\frac{\alpha}{1-\alpha}$$

$$\therefore \quad 40=\frac{\alpha}{1-\alpha}$$

$$\therefore \quad \alpha=\frac{40}{41}$$

Now $$\alpha=\frac{\Delta I_C}{\Delta I_E}$$

$$\therefore \quad \Delta I_C=\alpha\times\Delta I_E=\frac{40}{41}\times 100\ \mu A=97.5\ \mu A.$$

Example 8.2. If the above transistor is connected in a common emitter configuration and change in the base current is 100 μA, what will be the change in the collector current ?

Solution. We have

$$\frac{\Delta I_C}{\Delta I_B}=\beta$$

$$\therefore \quad \Delta I_C=\beta\times\Delta I_B=40\times 100\ \mu A=4\times 10^3\ \mu A$$

Example 8.3. The output characteristic curve in a common emitter configuration shows that at a collector potential of 5.0 V, the collector current changes from 1.0 mA to 2.5 mA when the base current changes from 50 μA to 100 μA. Calculate α and β.

Solution. Here, $\Delta I_B=(100-50)\ \mu A=50\times 10^{-6}$ amp

$$\Delta I_C=(2.5-1.0)mA=1.5\times 10^{-3} \text{ amp}$$

So $$\beta=\frac{\Delta I_C}{\Delta I_B}=\frac{1.5\times 10^{-3}\text{ amp}}{60\times 10^{-6}\text{ amp}}=30$$

Since $\alpha=\frac{\beta}{1\times\beta}$ and $\alpha=\frac{30}{31}=0.97.$

8.11. Transistor as an Amplifier

In the foregoing discussions we have seen that the collector output resistance of a transistor is much greater compared to the input resistance (base or emitter) of the same transistor in either the common emitter or the common base configuration. So if a given value of current flows in both the input and output circuits, the output potential difference ($I_C\ R_{out}$) will be higher than the input potential difference ($I_B\ R_{in}$ or $I_E\ R_{in}$). But we know in a common base circuit, the collector current I_C is about 0.95 I_E and in the common mitter circuit, I_C is nearly 20 I_B. Therefore, a small P.D. applied in the input side (the base emitter section) to cause an input current I_E or I_B, will cause a greater P.D. in the collector circuit. This means a small input voltage change will appear as a large output voltage change. This is the basic principle involved in amplification. There are three types of amplifier circuit, 1. common base, 2. common emitter, 3. common collector. These names are derived from the arrangements described in Art. 8.10. We will discuss here the circuit of a common base amplifier to illustrate the amplification produced in a transistor.

The basic circuit of a common-base amplifier. The basic circuit for a common base amplifier with a *P-N-P* transistor is shown in Fig. 8.17 (*a*). The emitter-base diode is forward biased by D.C. voltage V_{EE} and the collector base diode is reverse biased by a battery of voltage V_{CC}. v_i is a small a.c. voltage or signal to be amplified. ($v_i = v \sin \omega t$, Fig. 8.17 *b*). R_S is the resistance of the source that supplies the a.c. voltages. R_L is the resistance used in the output circuit across which the output voltage v_O will be developed. R_L is known as the load resistance. The load resistance is usually of the order of the output resistance of the particular transistor.

When the signal voltage $v_i = 0$ the base emitter voltage V_{EE} determines the emitter current I_E and hence the collector current I_C and the base current I_B.

We have, $I_E = I_C + I_B$

But when the signal is applied *i.e.* $v_i \neq 0$, the voltage between the emitter and the base varies between ($V_{EE} + v$) and ($V_{EE} - v$), Fig. 8.17 (*c*). At any instant it will be V_{EE} v_i. So emitter current at any instant will be the sum of I_E and the current i_E due to the instantaneous a.c. voltage. The value of i_E goes on changing during the cycle of the a.c. signal. As the emitter current changes the base current and the collector current will also change during the cycle. Thus the emitters current, base current and collector current will have an a.c. component (i_E, i_B, i_C), in addition to the d.c. com-

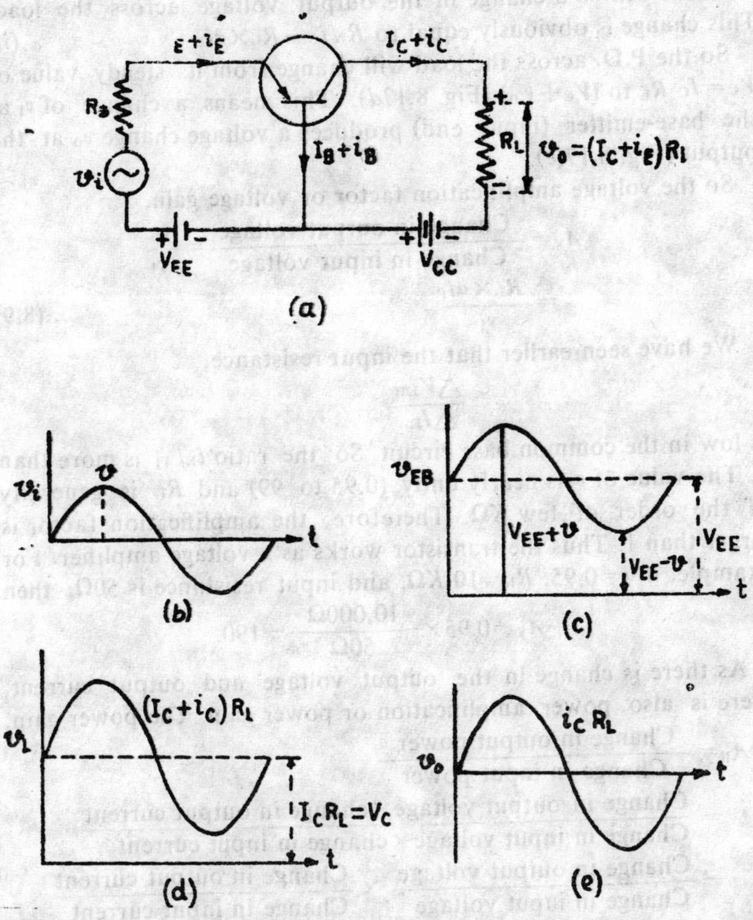

Fig. 8.17. (*a*) Common base amplifier circuit using a *PNP* transistor. (*b*) Input a.c. voltage, (*c*) Variation of applied voltage between emitter and base, (*d*) Variation of voltage across the load, (*e*) a.c. voltage across the load corresponding to input voltage (shown in *b*).

ponent (I_E, I_B and I_C). Since in an amplifier, we are concerned with amplification of a.c. signal, it is sufficient to consider the a.c. component of the different currents.

In the circut given in Fig. 8.17 (*a*), therefore at any instant,

$$\left.\begin{aligned} i_E &= i_C + i_B \\ \text{and} \qquad i_C &= \alpha i_E \end{aligned}\right\} \qquad \ldots(i)$$

Thus the value of the current in the load resistance will change from its d.c. value I_C by the amount i_C at any instant. When the a.c.

signal is not applied, the P.D. across R_L is $I_C R_L$. But with varying i_C, there will be a change in the output voltage across the load. This change is obviously equal to $R_L i_C = R_L \times \alpha i_E$ ...(*ii*)

So the P.D. across the load will change from its steady value of $V_C = I_C R_L$ to $(V_C + v_o)$ (Fig. 8.17*d*). This means a change of v_i at the base-emitter (input end) produces a voltage change v_o at the output (Fig. 8.17(*e*).

So the voltage amplification factor or voltage gain,

$$A_V = \frac{\text{Change in output voltage}}{\text{Change in input voltage}} = \frac{v_o}{v_i}$$

$$= \frac{R_L \times \alpha i_E}{v_i} \qquad ...(8.9)$$

We have seen earlier that the input resistance,

$$\frac{\triangle V_{EB}}{\triangle I_E}$$

is low in the common base circuit. So the ratio i_E / v_i is more than 1. The value of α is nearly unity (0.95 to .99) and R_L is generally of the order of few $K\Omega$. Therefore, the amplification factor is larger than 1. Thus the transistor works as a voltage amplifier. For example, if $\alpha = 0.95$, $R_L = 10\ K\Omega$, and input resistance is 50Ω, then

$$A_V = 0.95 \times \frac{10{,}000\Omega}{50\Omega} = 190.$$

As there is change in the output voltage and output current, there is also power amplification or power gain. The power gain,

$$A_P = \frac{\text{Change in output power}}{\text{Change in input power}}$$

$$= \frac{\text{Change in output voltage} \times \text{change in output current}}{\text{Change in input voltage} \times \text{change in input current}}$$

$$= \frac{\text{Change in output voltage}}{\text{Change in input voltage}} \times \frac{\text{Change in output current}}{\text{Change in input current}}$$

$$= A_V \times \alpha. \qquad ...(8.10)$$

A_P also is much greater than 1.

So we find that for a common base amplifier, though the current amplification α is less than unity, yet there is appreciable voltage and power amplification.

The common emitter amplifier. The common emitter amplifier circuit comfiguration is shown in Fig. 8.18. In this arrangement, the input current is base current. Hence the current amplification, factor β is to be considered here. As β is greater than 1, voltage and power gain are higher than the corresponding values in the common base amplifiers.

The common collector amplifier. Fig. 8.19 is a common collector

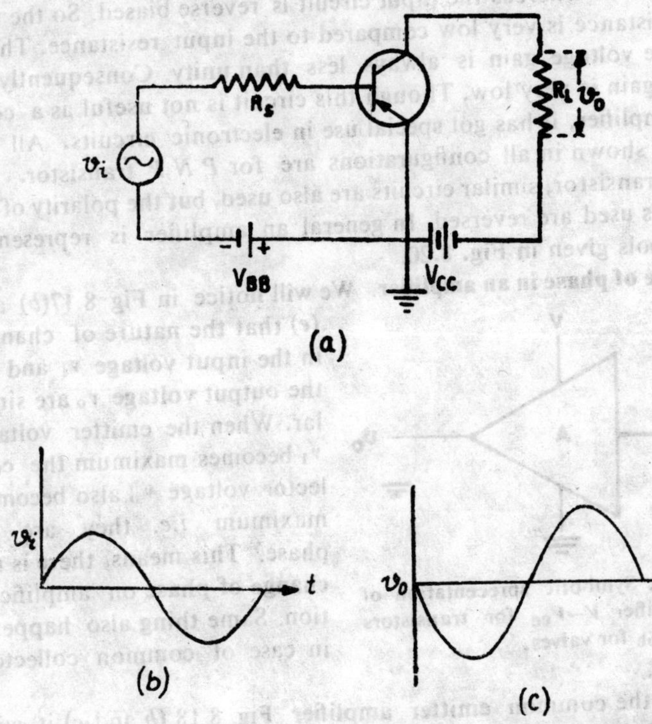

Fig. 8.18. (*a*) Common emitter emplifieir circuit with *PNP* transistors, (*b*) Input a.c. voltage, (*c*) Output a.c. voltage across the load.

amplifier circuit. Here the input current is the base current and the output current is emitter current. The current gain (I_E/I_B) is more than 1 in this arrangement. Note that the output circuit is for-

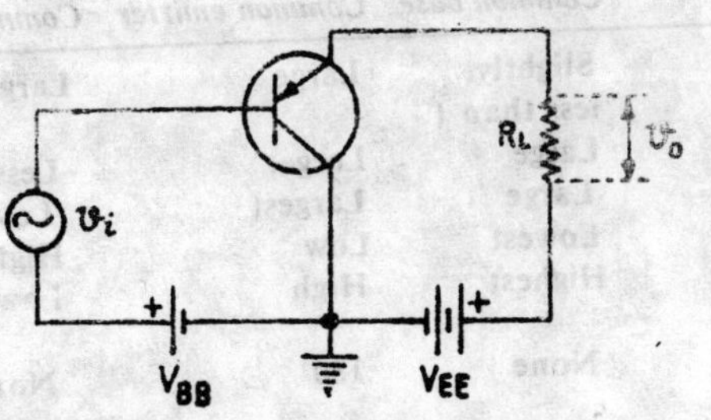

Fig. 8.19. Circuit for common collector amplifier using *PNP* transistor.

ward biased whereas the input circuit is reverse biased. So the output resistance is very low compared to the input resistance. Therefore the voltage gain is always less than unity. Consequently the power gain is very low. Though this circuit is not useful as a common amplifier, it has got special use in electronic circuits. All the circuits shown in all configurations are for *P N P* transistor. For *N P N* transistor, similar circuits are also used, but the polarity of the batteries used are reversed. In general an amplifier is represented in symbols given in Fig. 8.20.

Change of phase in an amplifier. We will notice in Fig 8.17(*b*) and (*e*) that the nature of changes in the input voltage v_i and in the output voltage v_o are similar. When the emitter voltage v_1 becomes maximum the collector voltage v_o also becomes maximum *i.e.* they are in phase. This means, there is no change of phase on amplification. Same thing also happens in case of common collector amplifier.

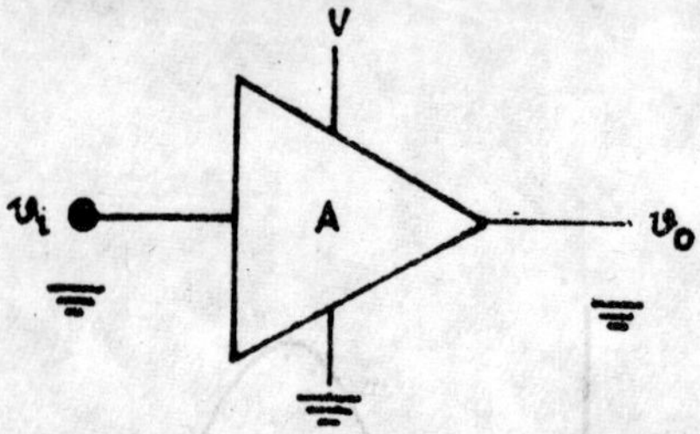

Fig. 8.20. Symbolic sprecentation of an amplifier V—V_{cc} for transistors and V—V_{bb} for valves.

But in the common emitter amplifier Fig. 8.18 (*b* and *c*) it will be seen that when the change in base voltage v_1 reached the positive peak the change in collector voltage v_o reaches the negative peak and vice versa *i.e.* they are in opposite phase.

Hence, there is a phase change of 180° on amplification in a common emitter circuit.

We summarise below the salient features of the three types of amplifiers.

Characteristics	*Common base*	*Common emitter*	*Commmon collector*
Current gain	Slightly less than 1	Large	Large
Voltage gain	Large	Larger	Less than 1
Power gain	Large	Largest	Lowest
Input resistance	Lowest	Low	Highest
Output resistance	Highest	High	Lowest
Change of phase between output and input.	None	180°	None

Considering the relative performance of different configurations as summarised above, the common emitter configuration is most widely used

8.12. Different Types of Amplifiers

The circuits described earlier give only the basic arrangements. In actual practice, number of modifications are made in the circuits. We will not go into those details in this book.

The amplifiers are also designed for specific uses and they are named accordingly. An amplifier used to achieve voltage amplification over a range of frequencies is called an *untuned potential amplifier*. If it is designed to achieve voltage amplification at a specified frequency only it is known as *tuned potential amplifier*. If the tuned frequency lies in the Radio frequency range, it is called *R.F. amplifier*. If the frequencies of interest lies in the audio frequency range, it is an *A. F. amplifier*. Note that the amplifiers used in the Public address systems are in the A.F. Range. There are some amplifiers where power gain is more important. Such amplifiers are known as *Power amplifiers*. Power amplifiers also belong to different categories depending on the frequency and purpose for which they are used. In a category of amplifier, a part of the output is fed back to the input by a circuit arranged. They are known as *Feed-Back amplifiers*. More than one amplifiers can be grouped or *cascaded* to give higher gain. There are different types of cascading or coupling between the different stages such as Resistance-Capacitance coupling (R.C.), transformer coupling etc. These amplifiers are named accordingly. There are many other varieties of amplifiers, we will, however, not discuss them.

8.13. Oscillators

Transistors can also be used to generate a.c. signals of different frequencies. The process is known as oscillation and the device used for this purpose is called an 'Oscillator'.

We have seen that a weak signal applied to the input end of an amplifier gets magnified at the output end. An oscillator is basically a selective amplifier. But no external signal is applied at the input of the oscillator. In every transistor, noise of wide range of frequencies are present. This is inherent with the manufacture of the device. The designers try to keep the noise amplitude to minimum. That is why in an ordinary amplifier it does not create much difficulty though it gets amplified along with the external signal. But if the external circuit connected with the amplifier is tuned to a particular frequency, the noise signal of that frequency only gets smplified. Amplification of noise of other frequencies is negligible. It should be noted here that the noise signal selected by the tuned circuit is internal to the transistor unlike the external signal applied for an amplifier and the frequency is choosen by the external tuned circuit. Hence the noise signal is the input signal in this case.

Inspite of the amplification, the amplified noise signal is still very small to be of any practical use. A part of the amplified signal is fed back from the output end to the input end, Fig. 8.21. The feed back of the signal is so arranged that the input signal and the signal feed-back are in the same phase. So the two signals will add to each other thereby increasing the effective input signal amplitude. So the amplfied signal amplitude increases compared to its previous value. Consequently the feed-back signal amplitude also increases. This in turn enhances the input and then the output. This process continues till a stable condition is reached when the output signal amplitude remains constant. This condition is attained when the energy lost in the circuit is exactly compensated by the feed-back of energy. This device is called an oscillator. Thus we get a fixed-frequency a.c. in an oscillator though we have not applied any external input to it. This is the basic principle involved in an oscillator. It should be noted that the energy of the a.c. generated comes from the d.c. source (battery) in this case. As in case of amplifier, different types of tuned circuits are used for generating signals of different frequencies. Depending on the range of frequency of the signals produced, the oscillators are also variously named: R.F. Oscillator, A.F. Oscillator etc. High frequency signals connected to aerial or antenna spreads in the atomosphere. This has made possible radio communication or transmission of microwave and television signals.

Fig. 8.21. Principle of oscillator feed back takes place in the same phase.

As an example of the principle involved in oscillators, we will describe here a tuned collector oscillator circuit, Fig. 8.22. The tuned circuit contains the secondary of a transformer and a capacitor. The tuned circuit is usually called the tank circuit. If the inductance of this coil is L and the value of capacitance connected is C, then the tank circuit is tuned to the frequency,

$$f=\frac{1}{2\pi}\frac{1}{\sqrt{LC}}\ .$$

The oscillator as given in Fig. 8.22 will generate signals of frequency nearly equal to this frequency. The functioning of the circuit as an oscillator is explained below.

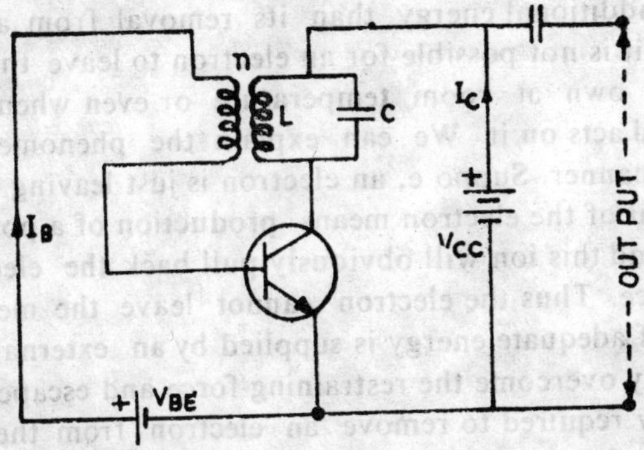

Fig. 8.22. A tuned collector oscillator circuits using *NPN* trasistor.

The primary of the transformer is connected in the base-emitter section *i.e.* in the input side. The tank circuit containing the secondary is connected between the collector and the emitter *i.e.* the output side. The circuit arrangement here is that of a common-emitter amplifier. So the output is 180° out of phase with the input (Art. 8.11). A part of the output is fed-back to the input through the transformer by inductive coupling (mutual induction between the primary and secondary of the transformer). The induced e.m.f. is always in opposite phase relative to the inducing e.m.f. (Lenz's Law). Hence the fed-back signal is 180° out off phase with the output signal which itself is 180° out of phase with the input signal. Thus the phase difference between the input signal and the feed back signal is 360° *i.e.* they are in phase. So the circuit will act as an oscillator. Of course, there are other factors to be taken into account for oscillations to be sustained. But we are concerned only with the basic principles involved as discussed above.

8.14. Emission of Electron from Metals

Electrons moving in different orbits of an atom are associated with energy level that increases as we go to outer orbits. When additional energy is supplied to an electron in a lower orbit, it may jump to a higher orbit (Bohr's theory). If the energy supplied is sufficient the electron may leave the atom altogether. In matter however, the atoms do not have separate existence. The electrons of different atoms form bonds making difficult removal of the electrons.

In metals, sufficient free electrons are, however available. Hence it is relatively easier to remove an electron from a metal by supplying additional energy than its removal from an insulator. Of course, it is not possible for an electron to leave the metal surface on its own at room temperature or even when an average electric field acts on it. We can explain the phenomenon in the following manner. Suppo e, an electron is just leaving the surface. The removal of the electron means production of a positive ion in the metal and this ion will obviously pull back the electron to the metal surface. Thus the electron cannot leave the metal surface. However, if adequate energy is supplied by an external agent, the electron may overcome the restraining force and escape. The minimum energy required to remove an electron from the surface by overcoming the restraining force is called work function. It is measured in electron volts (ev). Therefore, metals with lower work function should be more suitable for emission of electrons.

There are four methods of supplying this energy *i.e.* four methods of electron emission. They are: (*i*) Field emission, (*ii*) Secondary emission, (*iii*) Photo-electric emission, (*iv*) Thermionic emission.

(*i*) *Field emission.* When a high *electrostatic field*, of the order of million volts per centimeter is applied to a metal, electrons are pulled away from the metal surface. Since the use of high voltage is technically difficult and very often dangerous, the field emission is not treated as a suitable method for emission of electrons in common appliances.

(*ii*) *Secondary emission.* In secondary emission, a metal is bombarded with high-speed particles which cause the ejection of electrons from the surface. Depending on the type and energy of the incident particle, two or more number of electrons are emitted for each bombarding particle. Even electrons accelerated by a very high positive potential may cause secondary emission of electrons whose number may exceed the number of the bombarding electrons.

(*iii*) *Photo-electric emission.* While studying the physics of the atoms, we have seen that light or other radiation of suitable frequency falling on a surface causes emission of electrons. Since the energy supplied by light is usually of small order, surfaces with very small work function are suitable for this purpose. Therefore, the current available from photo-electrons is usually small.

(*iv*) *Thermionic emission.* Thermionic emission is the most important method of getting electrons ejected in large quantities. This effect was discovered by Thomas A. Edison during development of light bulbs and later utilised by J.A. Fleming, Lee De Forest and others in the development of electronic valves.

In this method a metal filament is heated to a high temperature by passing a large current through it. Increase in the temperature of the filament increases the energy of the electrons. If the temperature is sufficiently high, number of electrons acquire energy larger than the work function and then they leave the metal. This type of emission is known as thermionic emission and it is possible to get comparatively large current from this type of emission.

Since the metal should be heated to very high temperature in order to get sufficient supply of electrons, the metals used should have high meltingpoints.Tungsten and Molybdenum are usually used for this purpose. But since their work functions are relatively large, they emit electrons only at very high temperature *i.e.*, about 2400-2500°K. So large filament current is necessary for appreciable emission current. In other words, its efficiency of emission is low. Thoriated-tungsten, an alloy of Thorium and Tungsten, has lower work function and it emits electrons at relatively lower temperature of 1800--1900°K. But the most effective emitters are tungsten coated with barium oxide. They are known as oxide-coated filaments. Their work function is much lower and they emit sufficient electrons at about 1150°K. Therefore their efficiency as heaters is very high. In most of the vacuum tubes, oxide coated filaments are used.

8.15. Thermionic Diode

Emission of electrons from metal by heating is utilised in vacuum tubes or valves. Before the invention of semi-conductors, valves were being used in all electronic devices. A valve is essentially a vaccum tube containing a source to emit electrons and some electrodes to which voltage may be applied to regulate the flow of electrons.

The simplest valve is a *diode*. It has an electron emitting filament (tungsten, thoriated tungsten or oxide coated) called a *cathode* and a surrounding metal electrode to collect electrons called *anode* or *plate* both enclosed in an evacuated envelope. Since there are two electrodes in this valve, this is called a diode. Some times the cathode is not heated directly by passing electric current in it. It is heated indirectly by heating a filament enclosed by the cathode. The cathode in this case is known as indirectly heated cathode whereas the former one is known as directly heated cathode. Fig. 8.23 shows the features of a diode. It should be remembered that the symbol used for a *P—N* junction diode is also used for a thermionic diode. When the cathode is heated by the current from battery, electrons are emitted. If the plate *P* is at a positive potential with respect to the cathode, electrons will be attracted to the plate and there will be a current in the tube. Note that electrons travel from the cathode to the plate inside the diode. So outside the diode the direction of conventional current is from cathode to the plate. On the other hand, if the plate is made negative with respect to the cathode, the electrons will be

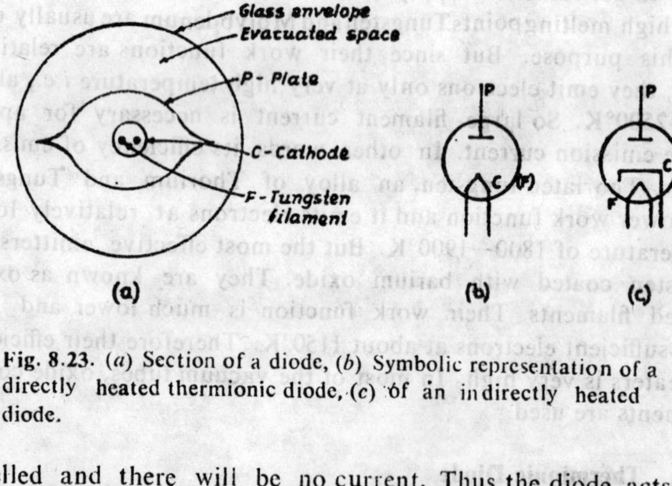

Fig. 8.23. (*a*) Section of a diode, (*b*) Symbolic representation of a directly heated thermionic diode, (*c*) of an indirectly heated diode.

repelled and there will be no current. Thus the diode acts a *valve* permitting the current in one direction only and not in the other direction. Therefore, a valve diode is used as a rectifier to change a.c. to d.c. The principle of rectification has already been explaned in Articles 8.5 and 8.6. The *P-N* junction diodes in Figs. 8.5 and 8.6 may be replaced by valve diodes such that the end *A* is the plate and and *B* is the cathode end of the diode valve.

8.16. Characteristic Curves of a Diode

When electrons are emitted from the cathode, they remain in the space close to the cathode in the absence of an electric field

These electrons form a cloud known as *space charge*. Since like charges repel, this cloud of electrons forces the electrons emitted later to stay still close to the cathode. When a positive voltage is applied to the plate, the electrons from the 'space-charge cloud' are attracted to the plate; thus the current starts. In Fig. 8.24 the change in plate current with plate voltage is shown. All electrons that come out of the cathode will not reach the plate when the plate voltage is low. Therefore as the positive voltage is increased gradually from zero, the current in the diode (known as plate current) also increases gradually. Initially the rise in current value is very slow because of the presence of space-charge cloud. But as the plate voltage is increased, more and more of electrons are pulled to the plate. Thus the plate current increases more rapidly. But when the plate voltage is raised sufficiently, a stage is reached when the accumulated space charge disappears and then the emitted electrons travel directly towards the plate so that the plate collects all electrons emitted during a certain time interval. At a temperature T_1, the portion of the graph from *O* to *A* (Fig. 8.24) shows this condition known as 'Space-charge limited condition'. Increasing the plate voltage beyond this value will not cause further increase in the plate current as there is no scope of increase in emission of electrons. So the current remains nearly constant or saturated. The value of this current is known as *saturation current*. If the cathode temperature is raised to T_2 more electrons are emitted and hence the saturation current is higher. If however, the plate potential is kept constant while the filament temperature is increased, the plate current increases at first but reaches saturation because of increase in space charge near the cathode.

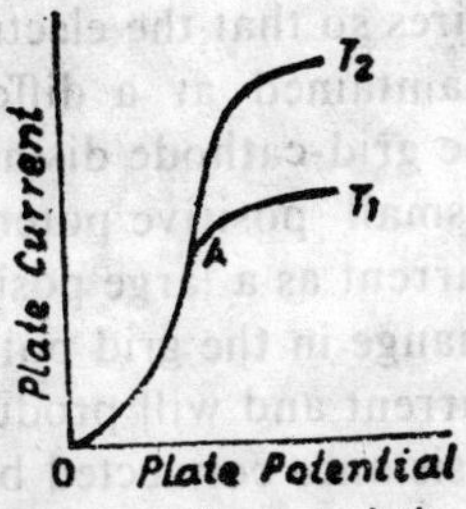

Fig. 8.24. Characteristic curves of a valve diode. $T_2 > T_1$.

A look at the Fig. 8.24 will show that the variation of current with voltage in a diode does not obey Ohm's law. *i.e.* current is not proportional to potential difference. Hence a vacuum diode like a junction diode is a non-linear circuit device.

8.17. Triode

It was found that better control of flow of electrons can be achieved by introducing a third electrode in a valve. This third electrode, called the *control grid*, is inserted into the tube nearer to the cathode. Such a tube is called a *triode*, or three-electrode tube (Fig. 8.25). The grid usually consists of a helix or spiral of fine wires so that the electrons may freely pass through it. The grid is maintained at a different potential negative to the cathode. Since the grid-cathode distance is less than the cathode-plate distance, a small positive potential on the grid will produce the same plate current as a large positive potential of the plate. Therefore a small change in the grid potential will cause a large change in the plate current and will produce large voltage change across an external resistance connected between the plate and the cathode. Therefore a small a.c. signal applied between the grid and cathode of a triode appears amplified across the load. Thus a triode acts as an amplifier. The principle involved in amplification has been explained earlier in connection with transistors. The difference to be stressed here is that the operation of the triode is regulated by variation of voltage whereas in the transistor the operation is

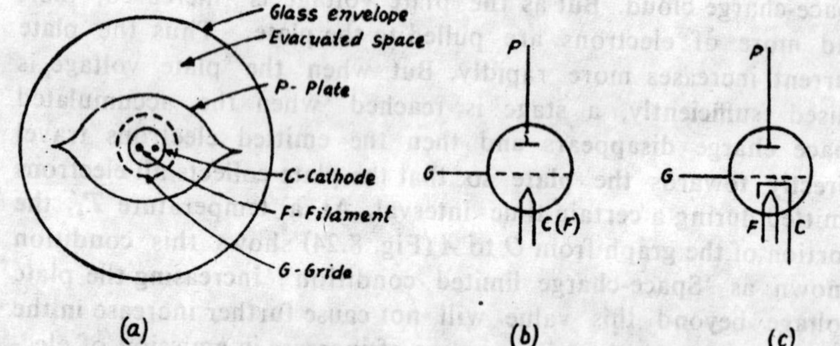

Fig. 8.25. (*a*) Section of a triode, (*b*) Symbolic representation of a directly heated triode, (*c*) Symbolic representation of an indirectly heated triode.

current controlled. The grid of a triode is usually kept negative with respect to the cathode so that it does not attract any electrons towards itself. Hence the grid current is zero and all the electrons move to the plate.

8.17. Characteristic Curves of a Triode

The curves drawn to represent the interdependence of the plate voltage, grid voltage and plate current are known as the characteristic curves of a triode. Usually the filament current is kept fixed as per the manufacturers specification and hence we will consider the filament temperature as constant.

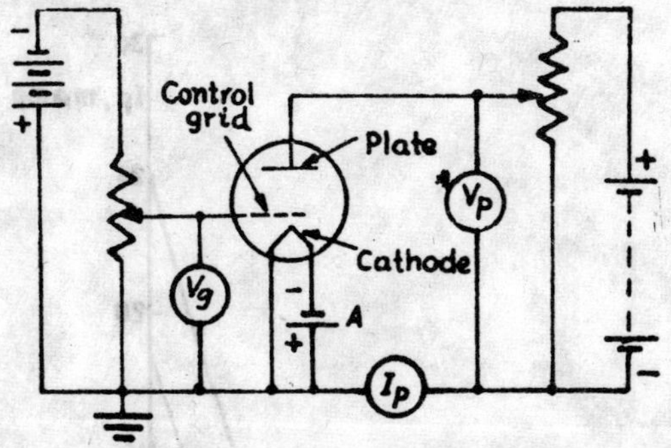

Fig. 8.26. Circuit for drawing the characteristic curves of triode.

The circuit given in Fig. 8.26 is used to draw the characteristic curves. Two sets of curves can be drawn. First, keeping the grid

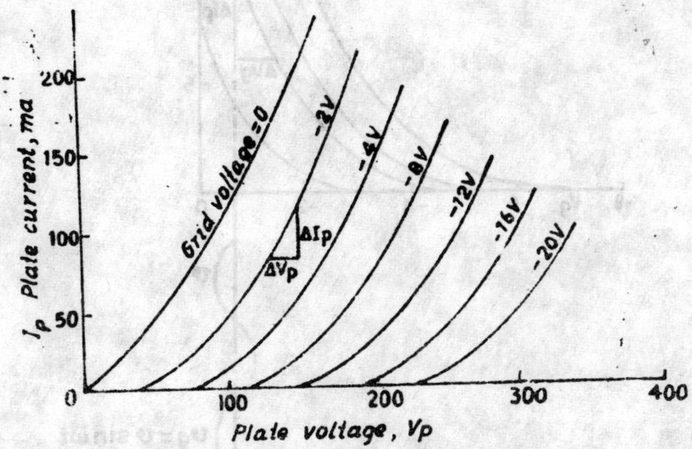

Fig. 8.27. Plate characteristic current of a triode.

voltage V_g constant, if we increase the plate voltage the change in plate current will be as shown in Fig. 8.27. It will be seen that when $V_g=0$, the plate current gradually increases as the plate voltage V_p is increased from zero. But as the grid voltage is made negative, the plate current is initiated only after a certain minimum plate potential. This happens because large positive plate potential should be applied to overcome the effect of negative grid potential on the electrons. This set of curves are called plate characteristic curves.

A second set of curves are drawn by keeping the plate potential fixed and varying the grid potential, Fig. 8.28. The plate current

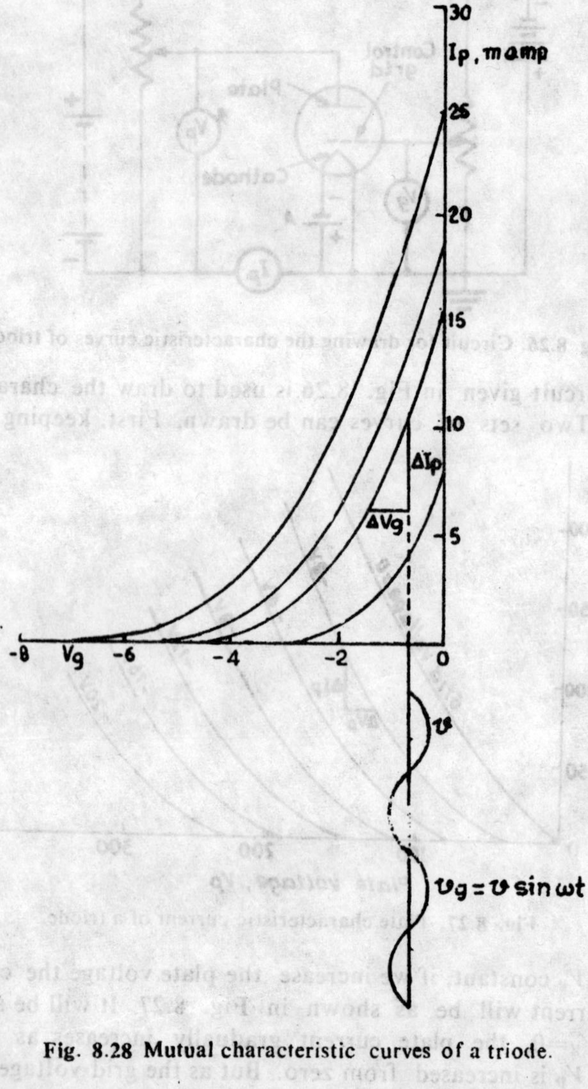

Fig. 8.28 Mutual characteristic curves of a triode.

I_p is plotted against grid potential V_g for a number of plate potentials. This set of curves are called grid-characteristic or mutual-characteristic curves.

Plate-resistance r_p, trans-conductance g_m and amplification factor μ of a triode (*a*) plate. resistance, r_p is defined to be the ratio of a small change in the plate voltage $\triangle V_p$, to the corresponding change $\triangle I_p$, in plate current.

Thus,

$$r_p = \left(\frac{\Delta V_p}{\Delta I_p}\right)_{V_g \text{ const}} \text{ ohms} \quad \ldots(8.11)$$

Though the plate characteristic curves are not linear as a whole yet we can choose a small portion which is linear and calculate r_p as shown in Fig. 8.27.

(*b*) **Mutual conductance or trans-conductance**. At a fixed plate voltage, if a small change ΔV_g, in the grid voltage causes a small change ΔI_p in the plate current, then the mutual conductance g_m is given by,

$$g_m = \left(\frac{\Delta I_p}{\Delta V_g}\right)_{V_p \text{ const}} \text{ mhos} \quad \ldots(8.12)$$

g_m is computed from Fig. 8.28.

(*c*) **Amplification factor** μ. Either a change in the plate voltage or a change in the grid voltage can cause a change in the plate current. But if the grid voltage is made more negative while the plate voltage is made correspondingly more positive, it is possible, to keep the plate current constant. If ΔV_g is a small, change in grid voltage and ΔV_p is the corresponding change in plate voltage so that the plate current remains constant, then the amplification factor,

$$\mu = \left(\frac{\Delta V_p}{\Delta V_g}\right)_{I_1 \text{ const}} \quad \ldots(8.13)$$

The sign of the changes has not been taken into consideration here. Since a small change in the grid potential is equivalent to a large change in the plate potential.

$$\Delta V_p > \Delta V_g \qquad \text{so } \mu > 1$$

Since μ is ratio of two potentials, it is dimensionless. The manufacturers specify the value of μ for the tube. Larger the value of μ, greater is its capacity to amplify.

These three quatities μ, r_p and g_m are inter related.

$$\mu = \left(\frac{\Delta V_p}{\Delta V_g}\right) = \left(\frac{\Delta V_p}{\Delta V_g} \times \frac{\Delta I_p}{\Delta I_p}\right)$$

$$= \left(\frac{\Delta V_p}{\Delta I_p}\right)\left(\frac{\Delta I_p}{\Delta V_g}\right)$$

$$= r_p \times g_m \quad \ldots(8.14)$$

8.18. Triode as an Amplifier

The variation of plate current with grid voltage is parctically linear in certain portions. In Fig. 8.29 we have shown a small a.c. voltage $v_g = v_m \sin \omega t$ is applied to the grid in addition to the d.c. grid voltage V_g. So the grid voltage fluctuate in the manner shown in Fig. 8.28. The grid voltage at any instant therefore, is $V_g + v_g$. In this figure it is shown to be changing between $(-1+v)$ and $(-1-v)$. So the plate current fluctuates accordingly being

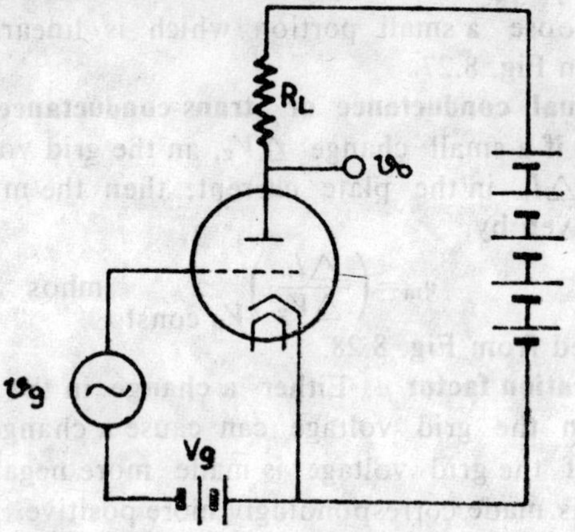

Fig. 8.29. Triode amplifier.

maximum when the grid voltage is $(-1+v)$ and minimum when it is $(-1-v)$. Thus fluctuating plate current flows through the load resistance R_L. So at any instant of time, if the plate current variation is i_p, the P.D. across R_L changes by $i_p R_L$. The magnitude of R_L is usually high. So the voltage change v_0 across it is much higher than v_g. In the circuit given in given in Fig. 8.29, the output voltage change

$$v_0 = \frac{\mu v_g \times R_L}{R_L + r_p} \qquad \text{..(8.15)}$$

So, the voltage amplification,

$$A_v = \frac{v_0}{v_g} = \frac{R_L}{R_L + r_p} \mu \qquad \text{...(8.16)}$$

For $\qquad R_L = \infty \qquad A_v = \mu.$

Thus the amplifier can have a gain equal to the amplification factor if the load resistance is infinity. But such a device is not practicable as the output must develop across a load to be of any practical use. So all practical amplifiers must have gain less than μ. The practical value of μ ranges from 20 to 100. R_L is usually taken equal to or higher than r_p. Thus a triode will function as amplifier.

Triode is also used as oscillator. The principle invovled in oscillation has already been explained in Art. 8.14. Additional grids are also added in vacuum tubes. With addition of one more grid between the central grid and the plate, called the screen grid, we get a *tetrode.* Similarly, addition of a fifth grid between the screen and the plate, makes a five-electrode or *Pentode* tube. There are many other type of tubes.

It is of interest to note that the operation of valves are controlled by variation in grid voltage whereas transistors working is governed by variation of the base or emitter current. That is why valves are known as voltage regulated devices whereas transistors are known as current regulated devices.

Example 8.4. Find the voltage gain of an amplifier using a triode valve in the circuit given in Fig. 8.29.

When $R_L=80\ k\Omega$, $r_p=20\ k\Omega$, and $\mu=20$.

Solution. Amplifier gain

$$A_V=\frac{\mu R_L}{r_p+R_L}$$

$$\therefore \quad A_V=\frac{20\times 80\times 10^3\ \Omega}{(20\times 10^3\Omega)+(80\times 10^3\Omega)}$$

$$=16$$

8.19. Cathode-ray Tubes

In the devices considered so far the principle involved is regulation of the number of electrons or charge carriers to produce effects like rectification, amplification, oscillation etc. Another way of controlling the electron beam is to deflect it from its path by application of a magnetic field or an electric field. This principle is used in cathode-ray oscilloscope tubes, in television and in radar etc. In all these devices, the positional change of the electron beam produces visible picture on a screen.

The cathode-ray tube was first developed by Braun in 1897. The

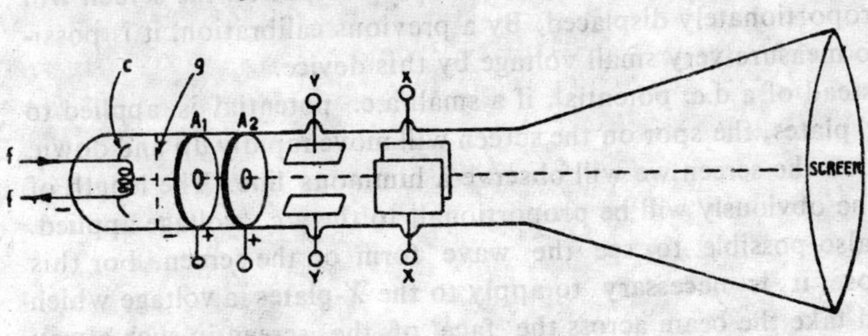

Fig. 8.30. Cathode ray tube.

ff--filament	A_2—focussing anode
c—cathode	*yy*—vertical deflecting plats
g—grid	*xx*—horizontal deflecting plate
A_1—Accelerating anode	*s*—screen

basic features of a cathode-ray tube is shown in Fig. 8.30. As in case of a valve, it is a vacuum tube and it has got a thermionic cathode C as a source of electrons. A grid 'g' infornt of it controls the number of electrons going towards the screen by maintaining the grid at a negative potential with respect to the cathode. A_1 is the accelerating anode with a small aperature in it. The electrons are accelerated by a high positive potential applied to this anode. The electrons passing through the anode have a tendency to diverge from each other. So there is a second focussing anode A_2, placed after A_1. A suitable voltage applied to it keeps the beam from spreading *i.e.* the beam moves in a sharp line.

The electron beam then pass through two pairs of parallel plates. One pair of plates YY is horizontal and the other XX is vertical. These plates are known as deflecting plates. We know that when an electron moves in a magnetic field such that the field is perpendicular to the direction of motion of the electron, the electron is deflected in a direction which is normal to both the field and the motion. The sense of movement of the electron is best given by Right Hand Rule. So if some potential is applied to the Y plates, the electron beam will be deflected in the vertical direction *i.e.* either up or down. Similarly, the suitable potential applied to X plates deflects the electron beam in horizontal direction. In the absence of any field applied on the plates the beam proceeds undeviated and strikes at the centre of a phosphorescent screan S at the end of the tube. The spot where the electron beam strikes the screen becomes luminous point. But when a d.c. voltage is applied to either X or Y plate, the deflection of the beam will be proportional to the voltage and hence the bright spot on the screen will be proportionately displaced. By a previous calibration, it is possible to measure very small voltage by this device.

Instead of a d.c. potential, if a small a.c. potential is applied to the Y plates, the spot on the screen will move rapidly up and down. Thus on the screen we will observe a luminous line. The length of the line obviously will be proportional to the a.c. voltage applied. It is also possible to see the wave form on the screen. For this purpose, it is necessary to apply to the X-plates a voltage which would take the beam across the face of the screen in such a way that the spot on the screen moves in the horizontal direction with a constant veolcity. As a result, the spot will have simultaneous movement in the vertical and horizontal direction and the wave will be traced on the screen. It is, however, necessary to bring the beam back to its original position so that the waveform can be repeatedly treaced on the screen for observation. The voltage applied to the X-plates has a waveform similar to a sawtooth and is shown in Fig. 8.31. The voltage increases steadily from A to B, thus the spot also deflects steadily along the X-axis reaching the

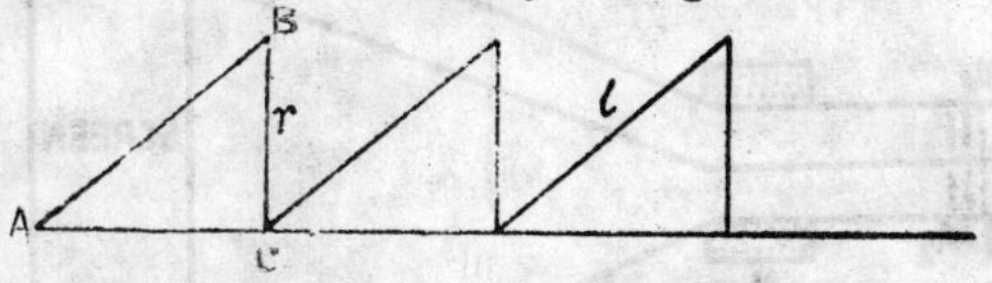

Fig. 8.31. Saw tooth wave form l—linear sweep, r—return of the beam.

maximum for B. After reaching the value at B, the voltage suddenly drop to zero at C. Thus the spot comes back to the original undeviated position. Because of the small mass of electron and its extraordinary speed, the spot can follow the change quickly. The voltage to be studied is applied to the Y-plates.

We have discussed only the basic principles involved. The various controls and other devices used in a cathode ray tube to increase its sensitivity are beyond the scope of this book.

Cathode-ray tubes are used as radar screens. In radar, pulse modulated radiowaves are sent at very short intervals. When these signals meet an obstacle they are reflected and the reflected signals are viewed on the radar screen which is actually the screen of a cathode-ray tube. The type of picture produced on the screen will depend upon the nature of the obstacle. Thus it will be possible to identify the nature of the obstacle say an aeroplane. or an approaching storm etc.

The tube used in a television is also a cathode-ray tube. In the T.V. tube, we have exactly similar arrangements consisting of a cathode, a grid, accelerating and focussing anodes. But here the deflection of the beam is achieved magnetically by a pair of deflection coils kept outside on the neck of the tube as shown in Fig. 8.32. When the T.V. receiver is not receiving any transmission, number of horizontal lines are seen on the screen. This is caused as the electron beam continuously sweep the whole face of the screen. Here both the deflection coils are used to sweep the beam across the T.V. screen, in such a way that the spot starts from the left top corner and moves horizontally across to the other edge. Thus it traces a line. On reaching the other side, it again comes back to left edge to a position slightly lower than the first line and then moves to the right and a second line is traced. Like this, third line and so on till the whole face is covered. After reaching the bottom, the beam starts sweeping again from the top

But the intensity of the electron beam is controlled by the type of signals received. Thus, when signals are received, the screen will

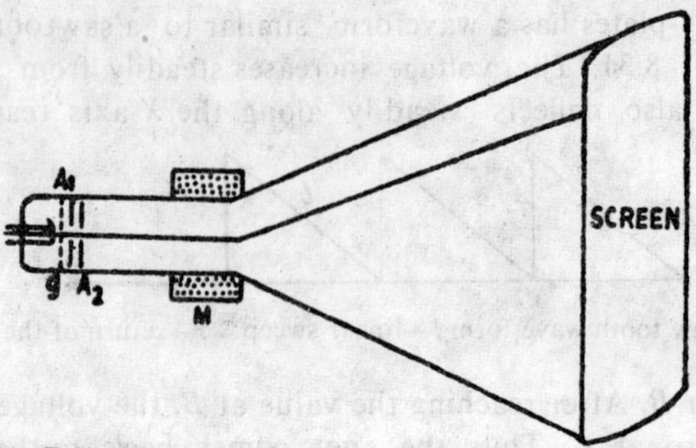

Fig. 8.32. T.V. tube; M—Magnet, G—Gride, A_1—Accelerating anode, A_2—Focussing anode.

have various shades of brightness or darkness.

While televising a picture, the picture is transmitted by scanning it in horizontal lines from top to bottom. Therefore, reproduction is also done in a similar manner.

Television and radar are examples where both intensity and position of the electrons are manipulated.

OUESTIONS

1. The reverse saturation current for a germanium diode is higher thar that for a silicon diode. Offer a reasonable explanation to this statement.
2. Will the working of a silicon diode be affected by change in working temperature? Explain.
3. Using the curves of Fig. 8.12., calculate the output resistance for three different emitter currents.
4. In a common emitter configuration, the reverse saturation current is a nuisance. Explain the statement.
5. Using the curves of Fig. 8.15., calculate β at three different collector voltages.
6. Is Ohm's law obeyed in valves and in transistors?
7. Generally common emitter configuration is choosen for an amplifier circuit? Why?
8. The performance of a transistor amplifier is affected by the change in temperature whereas a triode amplifier remains unaffected. Explain.
9. Suggest some methods for positive feed back to be used in an oscillator circuit.
10. If in a cathode-ray tube, the sawtooth wave is fed to the *Y*-plates and on alternating voltage is applied to *X*-plates, can you observe the wave shape? If so, what is the change?

9

KINETIC THEORY OF GASES

The modern view of matter is that everything in its state of solid, liquid or gas is composed of a very large number of tiny objects called molecules. These molecules can exist in free state and retain the characteristic properties of the parent substance. It is supposed that the molecules of any substance are held together by the inter molecular attractive forces. These forces resist the thermal agitation, which always tends to separate the molecules apart.

If the forces due to thermal agitation are weaker than inter-molecular attractive forces, then the substance will remain a solid and retain both its shape and size even at high temperatures. If the forces due to thermal agitations in a substance are comparable to inter molecular forces; the molecules of substance experience less force of attraction and can slide past one another. Such substances are called liquids. They have definite volume but no definite shape. They take the shape of container in to which they are poured. If on the other hand forces due to thermal agitations are far stronger than the intermolecular forces, the molecules of the substance can fly apart and move independently of each other. Such substances are called gases. A gas can neither retain its volume nor its shape. It fills the entire space, available to it.

The molecules of all substances supposed to be moving about rapidly, colliding frequently and presenting rather a chaotic picture when viewed on a small enough scale. In the case of a solid, only motion permissible to a molecule is oscillation or vibration, about a mean position which is fixed. But in the case of liquid and gas, the molecules can move from one place to another with all possible velocities ranging from zero to infinity inside the container. While in motion they freqnently collide with each other and with the walls of the container. In fact these collisions give rise to pressure in a gas or liquid.

On the basis of molecular theory not only three states of matter but also a great many physical and chemical phenomena like evaporation, surface tension, elasticity and diffusion etc can satisfactorily be explained.

9.1. Evidence of Molecular Motion

(*a*) **Diffusion**. The phenomena of inter-mixing of two liquids or gases even in opposition to gravity is called diffusion. Diffusion provides an evidence in support of the molecular motions in fluids. If a jar containing a light gas like hydrogen is inverted over another containing say Bromine, a heavier gas, a uniform mixture is formed after a while. This happens inspite of gravity under which the heavier gas should remain in the lower jar and lighter one in the upper jar. A similar case happens when a strong copper sulphate solution is kept at the bottom of a cylinder and water is slowly, added from above without causing any agitation. The coloured copper sulphate solution, although heavier than water gradually works up and finally after some time spreads out uniformly, throughout the whole mass.

The above process of intermixing is only possible if the molecules of the substances are in constant and random motion.

(*b*) **Brownian motion**. The Brownian motion is one of the most direct and obvious evidence of molecular concepts. Robert Brown, an English Botanist, in 1827 while observing the suspension of microscopic pollen grain under a highly powerful microscope found the particles moving about in the wildest fashion. Each particle observed, found to perform motion like spinning, sinking, rising and sinking again. The motions were spontaneous and incessant. The motions were also found to be more vigorous in less viscous liquid or when the temperature of the liquid is increased. Such chaotic motions of particles are called Brownian motion. The cause of this phenomena could not be understood for a long time until it was proved that this motion of particles was due to the impacts of the surrounding liquid molecules. Although the molecules strike the particles on all sides, their blows do not balance one another completely. Sometimes the impacts on one side of a particle are greater than on the other sides and this makes the particle move in a certain direction. Next impacts from a different side may predominate and the particle, may begin to move in a new direction. The result is the chaotic motion of the particles. This chaotic motion of the Brownian particles points clearly to a similar irregularity in the motion of the molecules of liquids. This phenomena can also be observed with smoke particles, or fog particles suspended in the air or gas. Thus Brownian motion strongly supports the concept of molecular nature of matter.

(*c*) **Evaporation and vaporisation**. The phenomena of evaporation and vaporisation are intimately connected with molecular motion which can be explained as follows.

(*i*) *Vaporisation.* When a liquid is heated, molecules move rapidly. As a result, their kinetic energy increases. Some of the energetic molecules overcome the inter molecular attractive force and escape out. We call it vaporisation. So long as heating is continued, more and more molecules escape out, ultimately leaving nothing in the container.

(*ii*) *Evaporation.* The phenomena of evaporation is also connected with molecular motion. In ordinary temperature the molecules also possess kinetic energy. Some of the molecules possessing energy much higher than the mean kinetic energy of the molecules, escape from the surface. Since the molecules of higher energy go out of the liquid, the mean kinetic energy of the molecules decreases. This results in the fall of temperature of the liquid. This explains why cooling is produced by evaporation.

The above facts led the scientists to believe that matter consists of molecules and heat energy is stored as kinetic energy of molecules, whenever heat is given to matter.

Based on above two facts namely matter is made up of molecules and heat can be identified with molecular motion, a theory has been developed to explain satisfactorily. Some physical concepts like temperature, pressure, energy etc. This theory is called kinetic theory of matter. Here only the kinetic theory of gases is considered because of its simplicity. The theory thus developed explains familiar laws like (*i*) gas laws, (*ii*) Avogadro's law, (*iii*) Grahm's laws of diffusion, which describe the behaviour of gases. It is to be noted that in the kinetic theory of gases we attempt to relate the microscopic behaviour of individual gas molecules to the macroscopic properties of the gas as a whole such as pressure, temperature and energy etc.

9.2. Fundamental Assumptions for the Development of Kinetic Theory Gases

In building a theory one has to start with a set of assumptions which are considered valid if the results predicted on them, are found correct by experiments. For development of kinetic theory we make the following fundamental assumptions.

(*i*) A gas is composed of a large number of minute discrete particles called molecules. The molecules of a gas are thought to be rigid, perfectly elastic, solid spheres, identical in all respects such as mass and size, but they differ in these, from gas to gas.

(*ii*) The molecules within a container are in a state of ceaseless chaotic motion during which they move in all directions with all possible velocities.

(*iii*) The molecules in their motion collide with each other and with the walls of the container and thus their speeds and directions change continuously. It however, does not affect the molecular density of the gas as the molecules do not accumulate at any place.

(*iv*) The bombardment of the container walls by the molecules gives rise to the phenomena called pressure which is the rate of change of mementum per unit surface area of the walls.

(*v*) All collisions between molecules and with walls are perfe tly elastic so that there is no loss of kinetic energy in the collision.

(*vi*) The molecules exert no forces on each other except when they collide. It means that energy of the gas is wholly kinetic.

(*vii*) The molecules travel in straight paths between collisions. The distance between any two consecutive collisions is called free path and the average distance travelled by a molecules in successive collisions is called *Mean Free Path.*

(*viii*) The time of collisions is negligible as compared with the time taken to traverse the free path.

(*ix*) Since the molecules are small compared with the distances between them, their volumes may be considered negligible compared with the total volume of gas.

(*x*) The inter molecular distances in gas is much larger than that of a solid or liquid and the molecules of a gas are free to move in the entire space available to them.

The gases having above properties are known as i leal gases.

9.3. Kinetic Calculation of the Pressure of a Gas

Pressure in a gas. The frequent collisions of the molecules with the walls of the containing vessel and their reflection from the walls results in the change of momentum of the molecules. According to the Newton's second law of motion, the rate of change of momentum per unit area of the wall surface corresponds to the forces exerted by the gas per unit area. The force per unit area measures the pressure of the gas. In short the continuous impact of the molecules on the walls of the containing vessel accounts for the pressure of the gas.

Let us now calculate the pressure of an ideal gas from kinetic theory. To simplify the matters we consider a gas in a cubical vessel *ABCDEFGH*, whose walls are perfectly elastic. Let each edge be, of length l metre. The volume of the vessel and hence that of the gas is l^3. Let N and M represent respectively the number of molecules and mass of each molecule present in the vessel. The molecules in vessel are moving in all directions with different speeds.

Consider a particular molecule Q which is moving with velocity C_1 in random direction. This velocity can be resolved into three perpendicular components u_1, v_1 and w_1 along x, y and z axis parallel to three adjacent edges of the cube Fig. 9.1 (*a*).

Therefore $C_1^2 = u_1^2 + v_1^2 + w_1^2$

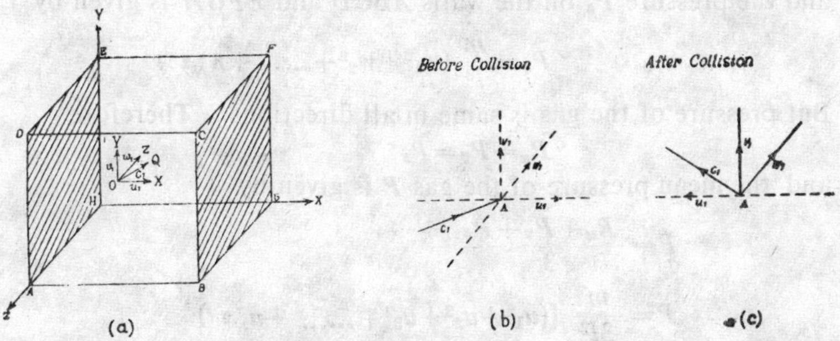

Fig. 9.1. (*a*) Velocity components of a molecule inside vessel, (*b*) Before collision, (*c*) After collision.

The components of the velocity with which the molecule Q will strike the opposite face $BCFG$ is u_1 and the momentum of the molecule is mu_1. This molecule is reflected back with the same momentum mu_1, in opposite direction and after traversing a distance l will strike the opposite face $ADEH$.

The change of momentum produced due to the collision is $mu_1 - (-mu_1) = 2mu_1$. As the velocity of the molecule is u_1 the interval between two successive collisions on the wall $BCFG$ is $2l/u_1$ sec.

Number of collisions per sec is therefore,

$$\frac{1}{\frac{2l}{u_1}} = \frac{u_1}{2l}$$

Change in momentum produced in one second i.e. the force exerted due to the collision of this molecule is

$$2 \times mu_1 \times \frac{u_1}{2l} = \frac{mu_1^2}{l} \quad \ldots (9.1)$$

The force F_x due to the collision of all N molecules having velocity components u_1, u_2..... u_N etc. in one second

$$=\frac{m}{l}\,(u_1^2+u_2^2+\ldots\ldots+u_N^2)$$

Now the force per unit area on the wall *BCFG* or *ADEH* is equal to the pressure P_x.

$$\therefore \quad P_x=\frac{F_x}{l^2}=\frac{m}{l\times l^2}\,(u_1^2+u_2^2+\ldots\ldots+u_N^2) \qquad \ldots (9.2)$$

as; l^2 is the area of the face

Similarly pressure P_y on the walls *CDEF* or *ABGH* is given by

$$P_y=\frac{m}{l\times l^2}\,(v_1^2+v_2^2+\ldots\ldots+v_N^2)$$

and the pressure P_z on the walls *ABCD* and *EFGH* is given by

$$P_z=\frac{m}{l^3}\,(w_1^2+w_2^2+\ldots\ldots+w_N^2)$$

But pressure of the gas is same in all directions. Therefore

$$P_x=P_y=P_z$$

and the mean pressure of the gas P is given by

$$P=\frac{P_x+P_y+P_z}{3}$$

$$\therefore \quad P=\frac{m}{3l^3}\,[(u_1^2+u_2^2+u_3^2+\ldots\ldots+u_N^2)$$

$$+(v_1^2+v_2^2+\ldots+v_N^2)+(w^2+w_2^2+\ldots\ldots+w_N^2)+\ldots]$$

$$=\frac{m}{3l^3}\,[(u_1^2+v_1^2+w_1^2)+(u_2^2+v_2^2+w_2^2)+\ldots\ldots$$

$$+(u_N^2+v_N^2+w_N^2)$$

$$=\frac{1}{3}\,\frac{m}{l^3}\,[C_1^2+C_2^2+C_3^2+\ldots+C_N^2]$$

$$=\frac{1}{3}\,\frac{m}{V}\,\Sigma\, C_N^2 \qquad \ldots (9.3)$$

as $l^3=V$ the volume of the gas.

Let $\bar{C}^2$ be the mean square velocity of the molecule then

$$\bar{C}^2=\frac{C_1^2+C_2^2+C_3^2+\ldots\ldots+C_N^2}{N}$$

or $$N\bar{C}^2=C_1^2+C_2^2+C_3^2+\ldots\ldots+C_N^2=\Sigma_N C_N^2$$

Substituting this value in equation (9.3) we get

$$P=\frac{mN\bar{C}^2}{3V} \qquad \ldots (9.4)$$

But $mN=M$, where M is the total mass of the gas of volume v, m is the mass of each molecule and N is the number of molecules in the volume.

$$\therefore \quad P = \frac{M\overline{C^2}}{3V} = \frac{1}{3}\rho\overline{C^2} \text{ as } M/V = \rho \text{, the density of the gas.} \quad \ldots (9.5)$$

$$\text{From (9.5) } \overline{C^2} = \frac{3P}{\rho} \quad \text{and} \quad \sqrt{\overline{C^2}} = Crms = \sqrt{\frac{3P}{\rho}} \quad \ldots (9.6)$$

where *Crms* is root mean square (*rms*) velocity.

In equation (9.6) we relate a macroscopic quantity, the pressure, to an average value of microscopic quantity i.e. *r.m.s.* velocity *Crms*. The root mean square velocity, *Crms* is the square root of the mean of the squares of the individual velocities. It is different from the average or mean velocity of the molecule. The average of different velocities of molecules will be zero as the molecules are moving with all possible velocities in all possible directions, while the root mean square velocity has a non zero value.

Example 9.1. The density of air is 1.293 kg/m² at STP. Calculate the root mean square velocity of its molecules.

Solution. The standard value of the pressure of air at *STP* is $= 0.76$ m, of mercury $= 0.76 \times 13.6 \times 10^3 \times 9.81 = 1.01 \times 10^5$ N/m²

Putting the value in equation [9.6] i.e.

$$Crms = \sqrt{\frac{3P}{\rho}}, \text{ we have}$$

$$Crms = \sqrt{\frac{3 \times 1.01 \times 10^5}{1.293}} = 485 \text{ m/sec.}$$

This shows that speed of the air molecules is comparable to the speed of the sound in air which is about 350 metre/sec.

9.4. Path of the Molecules and Mean Free Path

From the above example, it is seen that speed of the molecules is comparable to the speed of the sound in air which is about 330 m/sec. Then, the question arises, why the smell of the scent appears to spread in a room at much slower speed than this ? If the speed of the molecules of the scent were of the order 500 m/sec the molecules of the scent should have taken almost no time to reach any corner of the room. The explanation lies in the fact that when the pressure was derived on the basis of the kinetic theory, it was assumed that the molecules are of negligible size, and hence intermolecular collisions will not be possible. But the molecules has a finite size (though small) and they collide with each other.

A molecule in its path under goes a number of collisions so the path traversed by it is not a straight line but some what zig zag as shown in the Fig. 9.2. As a result of this zigzag motion of the molecule, its net displacement per unit time is comparatively small. So the scent takes more time to reach a place than the time predicted by kinetic theory. Between two successive collisions a molecule however travels in a straight line with uniform velocity. But the velocity and the distance between any two successive collisions are not always same. So we introduce the term mean free path as the average distance between two successive collisions. This mean free path reduces if the pressure of the gas is increased and vice versa.

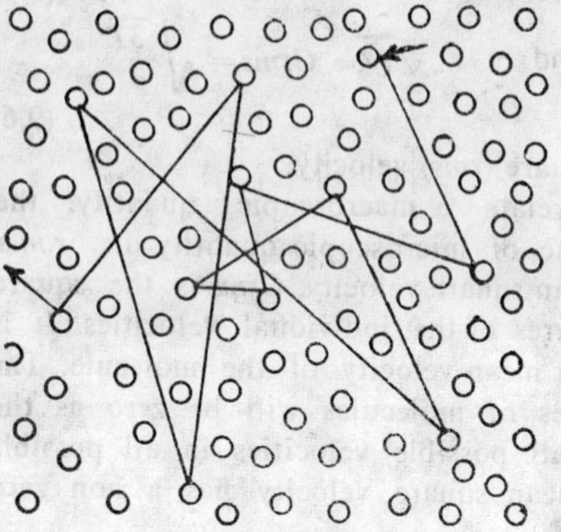

Fig. 9.2. Mean free path.

Example 9.2. Calculate root mean square velocity for (*a*) hydrogen, (*b*) oxygen. (*c*) air at STP when the density under this condition for,

(*a*) Hydrogen 8.99×10^{-2} kg/m^3

(*b*) Oxygen 1.43 kg/m^3

(*c*) Air 1.29 kg/m^3 and pressure 1 atm. $= 1.01 \times 10^5$ N/m^2

Solution. (*a*) We know $C_{rms} = \sqrt{\frac{3P}{\rho}}$,where P is the pressure and ρ is the density.

$$C_{rms} \text{ for hydrogen} = \sqrt{\frac{3 \times 1.01 \times 10^5}{8.99 \times 10^{-2}}} = 1.84 \times 10^3 \text{ m/sec}$$

$$= 1.84 \text{ km/sec.}$$

(*b*)
$$C_{rms} \text{ for oxygen} = \sqrt{\frac{3 \times 1.01 \times 10^5}{1.43}} = 0.459 \times 10^3 \text{ m/sec}$$

$$= 0.459 \text{ km/sec.}$$

(*c*)
$$C_{rms} \text{ for air} = \sqrt{\frac{3 \times 1.01 \times 10^5}{1.29}} = 0.486 \times 10^3 \text{ m/sec}$$

$$= 0.486 \text{ km/sec.}$$

9.5. Relation between Pressure and Kinetic Energy

The pressure of a gas given by the equation 9.5 can be put in the form

$$P=\frac{2}{3}\cdot\frac{1}{2}\rho\bar{C}^2$$

The quantity $\frac{1}{2}\rho\bar{C}^2=\frac{1}{2}\frac{M\bar{C}^2}{V}$ gives mean translational kinetic energy per unit volume, ϵ, of the gas. Then we have

$$P=\tfrac{2}{3}\epsilon \qquad \ldots (9.7)$$

Thus we find that the pressure of a gas is équal to the two thirds of the mean translational kinetic energy of the molecules per unit volume.

9.6. Kinetic Interpretation of Temperature

The pressure of a gas consisting of N molecules having molecular weight m and occupying a volume V is given by the equation:

$$P=\frac{1}{3}\frac{mN\bar{C}^2}{V}$$

or

$$PV=\frac{1}{3}mN\bar{C}^2 \qquad \ldots (9.8)$$

But from the perfect gas equation we know that,

$$PV=n\,RT \qquad \ldots (9.9)$$

where n is the number of kilogram-mole. R is the universal gas constant and T is the absolute temperature.

From equations (9.8) and (9.9) we have:

$$\frac{1}{3}mN\bar{C}^2=nRT$$

or

$$m\bar{C}^2=\frac{3nRT}{N} \qquad \ldots (9.10)$$

Multiplying $\frac{1}{2}$ on both sides we get $\frac{1}{2}m\bar{C}^2=\frac{3}{2}\frac{n}{N}RT$ ... (9.11)

Since 1 kilogram-mole of a gas contains N_A number of molecules where N_A is the Avogadro's number, the number of kilogram-mole in a gas is equal to the total number of molecule N divided by N_A.

Hence $n=\frac{N}{N_A}$ or $\frac{n}{N}=\frac{1}{N_A}$

Thus from equation (9.11) we have

$$\frac{1}{2}m\bar{C}^2=\frac{3RT}{2N_A}=\frac{3}{2}kT \qquad \ldots (9.12)$$

where $k=R/N_A$ is called Boltzman constant and its value is equal to 1.38×10^{-23} J-molecule^{-1} k^{-1}.

Example 9.3. Calculate the effective velocity of nitrogen molecule at 27°C.

Solution. By the effective velocity $\bar{C}_{eff}$ we mean r.m.s. velocity C_{rms}. We know $C^2_{rms}=\bar{C}^2$ where $\bar{C}^2$ is mean square velocity. So,

$$C^2_{eff}=\bar{C}^2$$

One kilo-mole of nitrogen has a mass of 28.0 kg and one molecule of mass of $(28.0/N_A)$ kg. $=28.0\ \text{kg}/6.022\times10^{26}=4.65\times10^{-26}$ kg.

$$T=273+27=300\ K$$

Hence by equation (9.12):

$$\frac{1}{2}m\bar{C}^2=\frac{1}{2}mC^2_{eff}=\frac{3}{2}kT$$

$$\frac{1}{2}(4.65\times10^{-26}\ \text{kg})\ C^2_{eff}=\frac{3}{2}(1.38\times10^{-23}\ \text{J/k})(300°\ \text{k})$$

$$C^2_{eff}=26\ 7\times10^4\ \text{m}^2/\text{s}^2$$

$$C_{eff}=517\ \text{m/s}.$$

Thus from equation (9.12) it is seen that mean kinetic energy of a molecule is directly proportional to the absolute temperature of a gas. When the temperature is increased mean kinetic energy of the molecule increases where as when temperature is reduced, the mean kinetic energy of the gas decreases. It is also obvious from equation (9.12) that $\bar{C}^2 \propto T$. Thus at absolute zero temperature, the kinetic energy should be zero. It means at absolute zero temperature, the molecules are in a perfect state of rest, and have no kinetic energy. Hence according to the kinetic theory absolute zero of temperature is that temperature at which all kinds of motions of a gas reduces to zero. But much before the absolute zero temperature is reached, all gases change their state to liquids and solids. However according to modern quantum theory there is also an appreciable motion even at $T=0$ and molecule possess some energy. The energy possessed by a molecule at absolute zero is called zero point energy.

Example. 9.4. At what temperature, will the mean square velocity of a gas be half of its value at STP, the pressure remaining constant?

Solution. Let C_{rms} of a gas at STP i.e. at $T_0=273°$K be C_1. Let $T°K$ be the required temperature at which the value of C_{rms} be C_2.

We have

$$C_1^2 \propto T_0$$

$$C_2^2 \propto T$$

$\therefore$

$$C_1^2/C_2^2=\frac{T_0}{T} \quad \ldots (i)$$

But given that $\quad C_2=\frac{1}{2}C_1$

Putting this value in equation (*i*) we have:

$$\frac{C_2^2}{(\frac{1}{2}C_1)^2}=\frac{T_0}{T}$$

or
$$\frac{4C_1^2}{C_1^2}=\frac{T_0}{T}=\frac{273}{T}$$

or
$$4=\frac{273}{T}$$

$$\therefore \quad T=\frac{273}{4}\text{ K}=68.25\text{ K}=(68.25-273)°\text{C}=-204.75°\text{C}.$$

9.7. Derivation of Gas Law

(*a*) **Boyle's law.** According to kinetic theory, the pressure P of a gas from equation (9.5) is given by

$$P=\frac{1}{3}\rho\bar{C}^2$$

or
$$P=\frac{1}{3}\frac{M}{V}\bar{C}^2$$

$$\therefore \quad PV=\frac{1}{2}M\bar{C}^2 \qquad .(9.13)$$

where $M=mN$, the mass of the whole gas. At a constant temperature T, $\bar{C}^2$ is constant according to kinetic interpretation of temperature. Therefore at a constant temperature $\frac{1}{2}MC^2$ is constant.

Hence we can write PV=constant ...(9.14)

or
$$V\propto\frac{1}{P} \qquad ...(9.15)$$

Therefore it follows that volume of a given mass of gas is inversely proportional to its pressure provided the temperature remains constant. This is Boyle's Law.

(*b*) **Charle's law.** We know from equation (9.8)

$$PV=\frac{1}{3}Nm\bar{C}^2=\frac{2}{3}N\left(\frac{1}{2}m\bar{C}^2\right)$$

But mean kinetic energy of a molecule is

$$\frac{1}{2}m\bar{C}^2=\frac{3}{2}kT$$

Therefore
$$PV=NkT \qquad ...(9.16)$$

or
$$V=\frac{NkT}{P} \qquad ...(9.17)$$

If N is constant and P is also constant then $V\propto T$ or V/T is constant. This means that for a given mass of gas the volume is directly proportional to its absolute temperature provided the pressure remains constant. This is Charle's Law.

(*c*) **Regnault's law.** From equation 9.16 we have:

$$PV=NkT$$

If N is constant and V is also constant,

$$P \propto T$$

or

$$\frac{P}{T} = \text{constant} \quad \ldots(9.18)$$

This means that for a given mass of gas, the pressure is directly proportional to its absolute temperature provided the volume remains constant. This is Regnault's law.

(*d*) **Avogadro's hypothesis**. Consider two gases A and B at equal pressure P and each having a volume V. Let mass of each molecule of the first gas $= m_1$, number of molecules of the first gas $= n_1$ and mean square velocity of the molecules of the first gas $= \bar{C}_1^2$

For the first gas $P = \frac{1}{3} \rho \bar{C}_1^2 = \frac{1}{3} \frac{m_1 N_1 \bar{C}_1^2}{V} \quad \ldots(9.19)$

Similarly for the second gas

$$P = \frac{1}{3} \rho_2 \bar{C}_2^2 = \frac{1}{3} \frac{m_2 N_2 \bar{C}_2^2}{V} \quad \ldots(9.20)$$

Where m_2 represents the mass of each molecule, N_2 the number of molecules and $\bar{C}_2^2$ the mean square velocity of the second gas.

From equation (9 19) and (9.20) we have:

$$\frac{1}{3} \frac{m_1 N_1 \bar{C}_1^2}{V} = \frac{1}{3} \frac{m_2 N_2 \bar{C}_2^2}{V} \quad \ldots(9.21)$$

$$m_1 N_1 \bar{C}_1^2 = m_2 N_2 \bar{C}_2^2$$

If two gases are at the same temperature T, the mean kinetic energy of the molecules of both the gases is the same

$$\therefore \quad \frac{1}{2} m_1 \bar{C}_1^2 = \frac{1}{2} m_2 \bar{C}_2^2$$

or

$$m_1 \bar{C}_1^2 = m_2 \bar{C}_2^2 \quad \ldots(9.22)$$

From equation (9.21) and (9.22) $N_1 = N_2$

Hence we see equal volume of gases under similar conditions of temperature and pressure have the equal number of molecules. This represents Avogadro's law.

(*e*) **Graham's Law of diffusion** Consider a gas A which is diffusing into the gas B. When the pressure exerted by two gases become equal to one another, that is, when $P_A = P_B$ it follows from the equation (9.5) that

$$\frac{1}{3} \rho_A \bar{C}_A 2 = \frac{1}{3} \tfrac{1}{3} \rho_B \bar{C}_B 2 \quad \text{or} \quad \frac{c_A \text{ rms}}{c_B \text{ rms}} = \sqrt{\frac{\rho_B}{\rho_A}} \quad \ldots(9.23)$$

$$\text{or} \quad \frac{\gamma_A}{\gamma_B} = \sqrt{\frac{\rho_B}{\rho_A}}$$

where γ_a and γ_b represent the rate of diffusion of two cases respectively.

This implies that gases diffuse at different rates, and higher the density of the gas, the lower is the rate of diffusion and vice-versa. This is Graham's law of diffusion.

(*f*) **Dalton's law of partial pressure.** Consider a gas confined in a volume V and containing N_1 molecules. Let m be the mass of each molecule and $\bar{C_1}^2$ be mean square velocity. Then from kinetic theory, the gas pressure P_1 is given by

$$P_1=\frac{1}{3}\ \frac{m_1N_1\bar{C_1}^2}{V}=\frac{2}{3}\ \frac{E_1}{V}$$

where $E_1=\frac{1}{2}\ m_1N_1\bar{C_1}^2$ is the total kinetic energy of translation of all the molecule. If in the same volume V, there is another gas containing N_2 molecules of mass m_2 each and having $\bar{C_2}^2$ as the mean square velocity in place of first gas, then pressure P_2 of the second gas is given by

$$P_2=\frac{1}{3}\ \frac{m_2N_2\bar{C_2}^2}{V}=\frac{2}{3}\ \frac{E_2}{V}$$

In this way pressure P_3, P_4, P_5 of other gases if individually present in the same volume are given by:

$$P_3=\frac{2}{3}\ \frac{E_3}{V}$$

$$P_4=\frac{2}{3}\ \frac{E_4}{V}$$

and so on.

If all the above mentioned gases are simultaneously present and they have no chemical action on one another, then the resultant pressure P is given as:

$$P=\frac{2}{3}\ \frac{E}{V}$$

Where E is the sum of kinetic energies of all the different gases therefore:

$$P=\frac{2}{3}\ \frac{1}{V}\left[E_1+E_2+E_3+\ \ldots\right]$$

$$=\frac{2}{3}\ \frac{1}{V}\left[\frac{1}{2}\ m_1N_1\bar{C_1}^2+\frac{1}{2}\ m_2N_2\bar{C_2}^2\right.$$

$$\left.+\frac{1}{2}\ m_3N_3\bar{C_3}^2+\ldots\ldots\right]$$

$$P=\frac{1}{3}\ \frac{m_1N_1\bar{C_1}^1}{V}+\frac{1}{3}\ \frac{m_2N_2\bar{C_2}^2}{V}+\frac{1}{3}\ \frac{m_3N_3\bar{C_3}^2}{V}$$

$$P=P_1+P_2+P_3+ \qquad \ldots(9.24)$$

which is Dalton's law of partial pressure. It states that in a mixture of gases which has no chemical action with one another total pressure is the sum of individual pressures which the gases would exert, when each of them alone occupies the whole volume.

Thus we see that our concepts of molecular nature of matter has satisfactorily explained the laws describing the behaviour of gases, which has been experimentally established. Hence our picture of the molecular nature of matter is correct.

9.8. (a) Specific Heat of Gases

The specific heat of a substance is defined as the amount of heat required to raise the temperature of unit mass of the substance by one degree. The SI units of specific heat is Joule per kilogram per degree K, when heat is supplied to a body it is used in two ways. A part of it, is used to increase the temperature i.e. the kinetic energy of the molecules by thermal agitation and the remaining part of heat is used in doing work against external pressure as the body expands on heating. However in case of solids and liquids, the expansions are negligible. So the heat is only used to raise the temperature. Thus we speak of only one specific heat for solids and liquids. However the situation is different in case of gas. When we heat a gas the expansion in case of a gas is quite appreciable. So if we have to keep the volume of the gas constant we have to increase the external pressures and entire heat energy is to be used to raise the temperature of the gas On the other hand if pressure is kept constant, there will be expansion in volume. A part of heat supplied is spent in expanding the gas and other part is spent in raising the temperature. So the amount of heat necessary to raise the unit mass of a gas is different under different conditions. Therefore the specific heat of a gas is not a constant but depends upon the conditions under which the gas is heated.

Consider a gas of mass M at a pressure P and volume V. If the gas is compressed, there is a rise in temperature, although no heat is added from outside to raise the temperature.

From the definition of specific heat i.e. $C=\dfrac{Q}{M\triangle T}$,

where Q is the heat supplied, $\triangle T$ is the rise of temperature and C is the specific heat,

we get $C=0$ in this case

as heat supplied, $Q=0$, here

$\therefore$ $C=0$, even when the temperature rises.

On the other hand if heat is applied to the gas and at the same time the gas is allowed to expand such that, there is no rise in temperature, then

$$C=\frac{Q}{M\triangle T}=\frac{Q}{M\times O}=\infty$$

as there is no rise of temperature i.e. $\triangle T=0$.

Thus we see the specific heat of a gas varies from zero to infinity. In order to find the value of the specific heat either the pressure or the volume has to be kept constant. Consequently it leads to two specific heats of gas. i.e. (1) Specific heat at constant volume, Cv, (2) Specific heat at constant pressure Cp.

Cv is defined as the quantity of heat required to raise the temperature of the unit mass of a gas through 1° at constant volume.

Cp is defined as the quantity of heat required to raise the temperature of unit mass of the subsance through 1° at constant pressure.

9.8. (b) C_P is Greater than C_V

When a gas is heated at constant volume the heat supplied to the gas is wholly used up to raise its temperature On the other hand when a gas is heated at a constant pressure, a part of the heat is used to raise its temperature and a part is used to do external work to keep the pressure constant. So heat necessary to raise the temperature of unit mass of gas by 1° at constant pressure is greater than heat necessary to raise the temperature of unit mass of the gas through 1° at constant volume.

Hence $\qquad Cp>Cv$

9.9. Molar Specific Heats C_V and C_P of a Gas and their Relationship

The amount of heat required to raise the temperature of one mole of a gas by one degree is called its molar specific heat. Molar specific heat is two kinds. They are molar specific heat at constant volume C_V and molar specific heat at a constant pressure C_P.

If m is the molecular weight

$$C_V=mCv \text{ and } C_P=mCp$$

where Cv and Cp are specific heats of a gas at constant volume and constant pressure.

Consider one mole of an ideal gas enclosed in a cylinder fitted with a frictionless piston. Since the gas is ideal there is no intermolecular force between its molecules and when such a gas expands no work is done in separating the molecules apart. The whole work done is external i.e. work done against the external pressure.

Let P be the external pressure and A be the cross sectional area of the piston. Then force F which acts on the piston $= P \times A$.

Suppose the gas is heated at constant pressure through 1° and as a result the piston moves through a distance x as shown in Fig. 9.3(c).

Let V_1 be the initial volume and V_2 be the final volume of the gas at temperature T_1 and T_2 of the gas. The work done by the gas

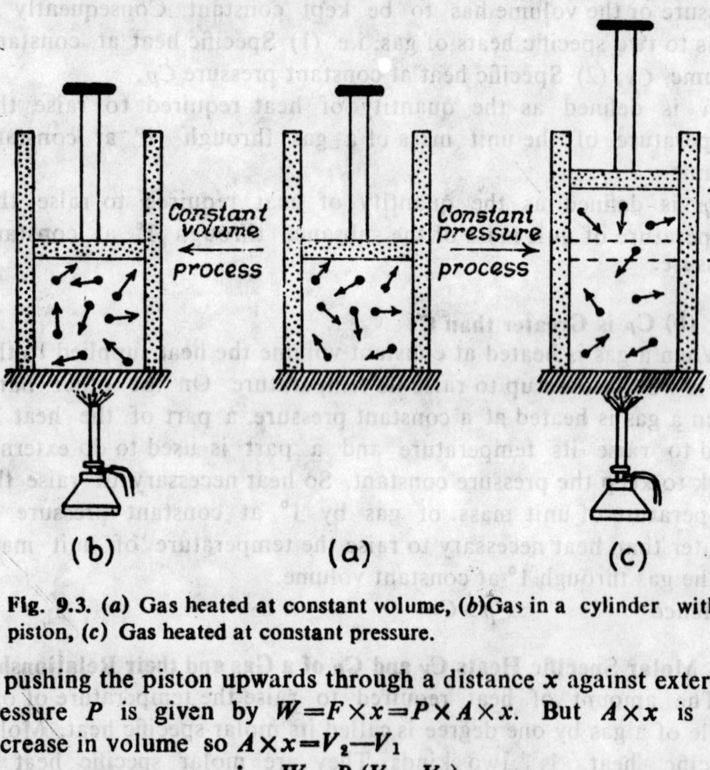

Fig. 9.3. (*a*) Gas heated at constant volume, (*b*)Gas in a cylinder with piston, (*c*) Gas heated at constant pressure.

in pushing the piston upwards through a distance x against external pressure P is given by $W = F \times x = P \times A \times x$. But $A \times x$ is the increase in volume so $A \times x = V_2 - V_1$

$$\therefore \quad W = P\,(V_2 - V_1).$$

But we know that the amount of heat added at constant pressure to raise the temperature of one mole of a gas by one degree is C_P, the specific heat at constant pressure. Now this C_P amount of heat is spent in two ways. A part equal to $P\,(V_2 - V_1)$ is spent in increasing the volume and other part is spent in raising the internal energy. Let us calculate the amount of heat energy which raises the internal energy of the gas in the above process. We know that when a gas is given some heat at constant volume no external work is being done and the given amount heat energy is entirely spent in increasing the internal energy. So if one mole of gas is heated at constant volume by one degree then the heat given$=C_V \times 1 \times 1° = C_V$. This C_V amount of heat raises internal energy of one mole of a gas when heated by 1°.

Therefore
$$C_P = C_V + P(V_2 - V_1)$$
or
$$C_P - C_V = P(V_2 - V_1) \quad \ldots (9.25)$$

But from our gas equation we have,
$$PV_1 = RT \quad \ldots (i)$$
and
$$PV_2 = R(T+1) \quad \ldots (ii)$$

From (*i*) and (*ii*) $P(V_2 - V_1) = R$

Putting this value in equation 9.25 we get
$$C_P - C_V = R \text{ joule/(mole)(k)} \quad \ldots (9.26)$$

If C_P and C_V are measured in calories, then we shall have
$$C_P - C_V = \frac{R}{J},$$
where J is expressed in joule/calorie.

9.10. Degrees of Freedom

The degrees of freedom of a particle indicate the number of independent motions or the number of independent modes of energy exchange which the particle can undergo. The molecule of a mono-atomic gas (such as He, A etc.) consists of a single atom. Its translational motion can take place in any direction in space. Thus it can be resolved along three co-ordinate axes Fig. 9.4 (*a*),

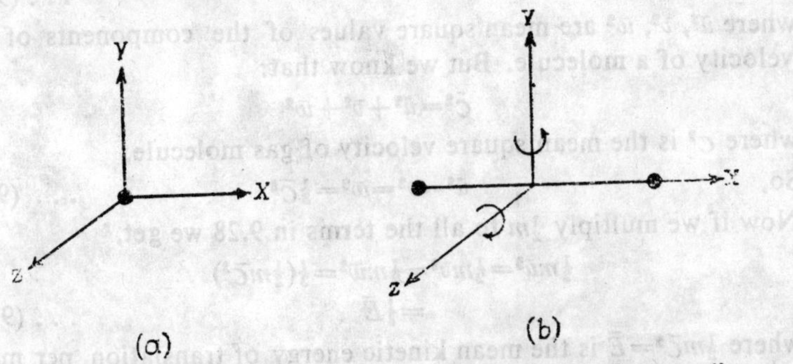

Fig. 9.4. (*a*) A monoatomic molecule can move in any of three directions and has three degrees of freedom. (*b*) A diatomic molecule can have translatory motion along three mutually perpendicular axes X, Y, Z, and the two rotational degrees of freedom as it can rotate about Y and Z axes. Its rotation about X axis does not contribute anything to the rotational energy as moment of inartiq about it is negligibly small.

and can have three independent motions. Hence it has three degrees of freedom, all translational.

The molecules of a diatomic gas such as H_2, O_2 etc, is made up two atoms joined rigidly to one another through a bond as Fig. 9.4 (*b*). This molecule can have three translational motions as usual but in addition also, it rotates about any one of three co-ordinate axes. However its moment of inertia about the axis joining the two atoms is negligible compared to that about the other two axes. Hence it can have practically two rotational motions. Thus a diatomic molecule has 5 degrees of freedom, 3 translational and 2 rotational, Fig. 9.4 (*b*).

A polyatomic molecule (such as CO_2) can rotate about any of three co-ordinate axes. Hence it has 6 degrees of freedom, three translational and three rotational.

If the atoms of the molecule vibrate with respect to each other it would possess a number of degrees of freedom in respect of vibration also.

9.11. Law of Equipartition of Energy

From kinetic theory, we know that the molecules of gas are in random motion with no preference for any particular direction of motion, i.e. all direction of motions are equally probable. If *u*, *v* and *w* are components of velocity of a molecule along *x*, *y* and *z* direction, then in the steady state we can write.

$$\bar{u}^2=\bar{v}^2=\bar{w}^2 \qquad \ldots (9.27)$$

where $\bar{u}^2$, $\bar{v}^2$, $\bar{w}^2$ are mean square values of the components of the velocity of a molecule. But we know that:

$$\bar{C}^2=\bar{u}^2+\bar{v}^2+\bar{w}^2$$

where C^2 is the mean square velocity of gas molecule.

So, $$\bar{u}^2=\bar{v}^2=\bar{w}^2=\tfrac{1}{3}\bar{C}^2 \qquad \ldots (9.28)$$

Now if we multiply $\frac{1}{2}m$ to all the terms in 9.28 we get,

$$\tfrac{1}{2}m\bar{u}^2=\tfrac{1}{2}m\bar{v}^2=\tfrac{1}{2}m\bar{w}^2=\tfrac{1}{3}(\tfrac{1}{2}m\bar{C}^2)$$
$$=\tfrac{1}{3}\bar{E} \qquad \ldots (9.29)$$

where $\frac{1}{2}m\bar{C}^2=\bar{E}$ is the mean kinetic energy of translation per molecule of the gas.

We know from 9.12 the average kinetic energy per molecule

$$\bar{E}=\tfrac{3}{2}\,kT$$

where *k* and *T* have the usual meaning

From equation 9.29 we find that:

$$\tfrac{1}{2}m\bar{u}^2=\tfrac{1}{2}m\bar{v}^2=\tfrac{1}{2}m\bar{w}^2=\tfrac{1}{3}\bar{E}=\tfrac{1}{2}kT$$

This shows that with each independent motion along *x*, *y* and *z* axes, the energy associated is $\frac{1}{2}kT$. In other words the average energy carried by each molecule is $\frac{1}{2}kT$ per ***degree of freedom.***

This is the law of equipartition of energy and states that for any dynamical system the total energy is equally divided among the different degrees of freedom. The average energy carried by each molecule is $\frac{1}{2}\ kT$ per degree of freedom.

For diatomic molecules, which has five degrees of freedom, the total average kinetic energy of a molecule is $5(\frac{1}{2}kT)=\frac{5}{2}kT$ out of this total energy $\frac{3}{2}kT$ would be the kinetic energy of translation of the molecule and rest kT would be kinetic energy of rotation of the molecule.

9.12. Proof of Equipartition of Energy

Mono atomic gas. A monoatomic gas molecule has one atom. Each molecule has three degrees of freedom due to translatory motion only.

But energy associated with each degree of freedom $=\frac{1}{2}kT$.

Therefore energy associated with three degrees of

$$\text{freedom}=\frac{3}{2}kT$$

Consider one kilo-mole of gas having N_A number of molecule where N_A is the Avogadro's number.

The energy associated with N_A number of molecules

$$N_A\times\frac{3}{2}kT=\frac{3}{2}(N_Ak)T=\frac{3}{2}RT$$

(where $N_Ak=R$=universal gas constant).

So $\quad E \quad$ the total energy$=\frac{3}{2}RT$

Now $$C_V=\frac{dE}{dT}=\frac{3}{2}R$$

(dE/dT is the increase in internal energy per unit degree rise of temperature).

But $$C_P-C_V=R$$
$$C_P=C_V+R=\frac{5}{2}R$$

For monoatomic gas

$$\frac{C_P}{C_V}=\gamma=\frac{\frac{5}{2}R}{\frac{3}{2}R}=1.67$$

The value γ is found to be true experimentally for monoatomic gases like argon and helium etc.

Diatomic gases. For diatomic gases a single molecule has in all five degrees of freedom, three translational and two rotational.

The total energy E of one mole of a diatomic gas therefore is given by

$$E=\frac{5}{2}N_AkT$$

$$C_V=\frac{dE}{dT}=\frac{d}{dT}\left(\tfrac{5}{2}N_AkT\right)=\tfrac{5}{2}N_Ak=\tfrac{5}{2}R$$

$$C_P=C_V+R=\tfrac{5}{2}R+R=\tfrac{7}{2}R$$

$$\gamma=\frac{C_P}{C_V}=\frac{7}{5}=1.40$$

The theoretical results for γ agree with the experimental values for large number of diatomic gases such as hydrogen, oxygen, nitrogen etc.

From above discussions we see that kinetic theory of gases has explained the behaviour of the gases. within a reasonable range of temperature and pressure. However it gives no idea about the velocity distribution i.e. the number of molecules moving with different velocities. Besides, this theory is applicable only to perfect gases i.e. gases obeying Boyle's law. The theory therefore needs improvements regarding the size of the molecule and the forces between them. These considerations have been done and they lead to the Van derwaal's equation of state and Maxwell's Boltzman distribution law both of which are beyond the scope of this book.

Example 9.5. Calculate the ratio of the specific heat of nitrogen given that the specific heat at constant pressure is 0.235 and density at NTP is 1.234 gm/litre.

Solution. The difference between two specific heats of a gas when unit mass of it is to be considered is given by

$$C_P-C_V=\frac{r}{J}$$

where r is the gas constant for unit mass of the gas.

we know for a gas. $PV=rT$

or
$$r=\frac{PV}{T}$$

Here V is the volume of the unit mass of gas, and $T=0°C=273k$ and normal pressure $P=76\times13.6\times981$ dynes/cm²

Hence,
$$r=\frac{PV}{T}=\frac{76\times13.6\times981}{273\times0.001234}$$

$$\left(\text{As } V=\frac{\text{mass}}{\text{density}}=\frac{1\text{ gm}}{0.001234\text{ gm/cm}^3}=\frac{1}{0.001234}\text{ cm}^3\right)$$

$$=3.0\times10^6\text{ erg gm}^{-1}\,°\text{C}^{-1}$$

$$\therefore \quad C_p - C_v = \frac{r}{J} = (3.0\times10^6)/4.18\times10^7$$

$$= \frac{0.072 \text{ cal}}{(\text{gm})(°\text{C})}$$

$$(\text{As } J = 4.18\times10^7 \text{ ergs/cal})$$

$$\therefore \quad C_p = 0.235 \text{ cal (gm)}^{-1} (°\text{C})^{-1}$$

$$\therefore \quad 0.235 - C_v = 0.072$$

or $$C_v = 0.163 \text{ cal/(gm) } (°\text{C})$$

$$C_p / C_v = \frac{0.235}{0.163} = 1.4$$

QUESTIONS

1. The temperature of a gas kept at constant volume is increased. Does the number of collisions per second of gas molecules with walls change? Does the average change of momentum per collision increased?
2. If the speed of every molecule of a gas were doubled, what would happen to the temperature ? explain.
3. Why may there be greater danger associated with the failure of storage tank for gas at 30 MPA than with the failure of a water tank at the same pressure ?
4. State basic assumptions of kinetic theory of gases and derive the expression for the pressure of a gas on the basis of kinetic theory.
5. What do you mean by r.m.s. velocity of gas ? How does it differ from mean velocity of a gas ? Derive the relationship between pressure and r.m.s. velocity of a gas.
6. Derive important gas laws from kinetic theory.
7. Deduce the relation between C_{rms} and the temperature.What is the kinetic interpretation of temperature ?
8. What do you mean by the degrees of freedom of a system ? Calculate the number of degrees of freedom possessed by a molecule of (*a*) perfect gas, (*b*) mono atomic gas, (*c*) diatomic gas.
9. What do you mean by equipartition of energy ? With example show that principle of equipartition of energy is appropriate.
10. Why do you talk of two specific heats in case of gases. Establish a relationship between molar specific heat at constant volume and molar specific heat at constant pressure.
11. Write short notes on: (*a*) Brownian motion, (*b*) Evaporation, and (*c*) Mean free path.

PROBLEMS

1. Ten molecules have speeds as follows in arbitrary units 1, 3, 0, 2, 2, 4, 3, 6, 1, 2. Calculate (a) Average speed and the r.m.s. speed. [*Ans.* 2.4, **2.9**]
2. Calculate the root mean square velocity of **nitrogon molecules at STP**

The atomic weight of ntirogen is 14 and the molecule is diatomic.

[*Ans.* $4.9 \times 10^2\ ms^{-1}$]

3. Under what pressure is a gas if the mean square velocity of its molecule is 580 m/s and its density is 9×10 gm/cm^3 [*Ans.* $1.1 \times 10^6\ N/m^2$]
4. The density of helium at STP is 0.356 kg/cm^3. Determine the r.m.s. speed of helium molecules under these conditions. Compare its value with that at 473 K. [*Ans.* $9.29 \times 10^2\ m/s$]
5. Calculate the value of universal gas constant R using a perfect gas equation. (one mole of a gas at STP occupies a volume of 22.9×10^{-3}).

   [*Ans.* 8.3 *J*)/(*mole* K)]
6. Calculate the mean kinetic energy of one mole of helium gas at 300 K. Given that the gas constant $R=8.3$ *Joule*/(*mole* K)
7. If the root mean square speed of nydrogen molecule at 0°C is 1840 ms^{-1}. What is the r.m.s. speed at 100°C. [*Ans.* 21, 50.96 ms^{-1}]
8. The mean kinetic energy per molecule of hydrogen at 273K is $5{\cdot}64 \times 10^{-21}$ Joules and molar gas constant R is 8·32 J (mole-K). Calculate Avogadro's number. [*Ans.* 6.04×10^{23}]
9. Calculate the specific heat of air at constant volume given that specific heat at constant pressure is 0.23 kcal/(kg)(C°), density at NTP is 1.293 gm/litre and J=4.2 joules/cal? [*Ans.* 0.1617 *kcal*/(*kg*)(C°)]

10

THERMODYNAMICS

Thermodynamics mainly deals with the conversion of heat energy to mechanical energy and the reverse process i.e. the mechanical energy into heat energy. However in some situations it is also concerned with conversion of heat energy into some other form of energy. Its scope is very wide and covers all branches of science in which heat or something depending on heat plays a role. Expansion of gases, change of state of matter on heating, exothermic and endothermic chemical reactions, thermo-electricity, nuclear trans formations and many biological process, are few examples which come under its domain. It is an important branch of science and has innumerable applications. But the detailed and extensive discussions of them are beyond the scope of this unit. However we will try to discuss some general principles and their applications to some familiar situations.

10.1. System

An object or a group of objects we are investigating is called a system. Anything outside the system which has got some bearing on the behaviour of the system is called surronding. To investigate a system there are two kinds of approaches which we can follow. In one approach we can go into the detailed internal structure of the system i.e. to take into account the properties of atoms or molecules constituting the system. This is called the microscopic approach and the properties associated with atoms or molecules which we consider, are called microscopic properties. One such approach is the kinetic theory, where bulk behaviour of a system is explained in terms of microscopic properties. The other approach is called macroscopic approach. Here we take into consideration the properties of the system as a whole without any reference to internal structure. These properties are called macroscopic, large scale or bulk properties. Macroscopic properties are measurable quantities. The volume, the pressure and the temperature of the system—are examples of macroscopic or large scale properties. Thermodynamics in fact deals with the large scale or gross property of the system and it does not pay any attention to the nature of the material of the system and its internal structure. As said earlier, macroscopic properties are measurable quantities and thermodynamics deals with them. Thermodynamics is therefore essentially an experimental or emperical science.

10.2. Thermodynamic System and its State

Any system which is subjected to thermodynamic study is called thermodynamic system. We need three properties of the system, such as Temperature (T), Volume (V) and Pressure (P) to describe it. These three quantities i.e. T, V and P are called thermodynamic co-ordinates or state variables. A particular set of such values specify a particular state of the system and another set of values takes the system into some other state. The process by which the system goes from one thermodynamic state to the other is called thermodynamic process. When such process takes place thermodynamics co-ordinate changes from one set of values to another set of values. Heating a gas contained in a cylinder with a piston or compressing the gas in it are familiar examples of thermodynamic process. It is to be noted here that change of state is different from change of phase. The change of phase refers to a change in form such as from liquid to vapour. But change of state means only change in the thermodynamic co-ordinates.

10.3. Thermal Equilibrium and Temperature

We distinguish a hot body from a cold body through our sense of touch. We say that the body which feels hotter is at higher temperature and the body which feels colder is at lower temperature. This procedure of determining the temperature of a body is a subjective one and certainly not very useful for purpose of science. We can however develop a more exact concept of what temperature means by discussing what happens when bodies are placed in thermal contact. (If two systems are placed in contact in such way as to allow thermal interaction, they are said to be in thermal contact).

Consider two systems A and B. Let the system A be such that it feels cold to the hand and the system B be such that it feels hot to the hand. Let the two systems be kept in thermal contact with each other. After a sufficient length of time, A and B give rise to same temperature sensation. Then we say A and B are in thermal equilibrium with each other.

This idea of thermal equilibrium forms the basis of the concept of temperature. If some systems are in thermal equilibrium they possess a physical property which is same for all of them This physical property is known as temperature. Temperature therefore determines, whether a system is in thermal equilibrium with an other system or not. This means if two systems are in thermal equilibrium; their temperature will be same otherwise their temperature will be different.

The above concept of thermal equilibrium has been embodied in a law called Zeroeth Law of thermodynamics. (So called because the nead for it, was not recognised until after the first law had been established). The law states that if two systems are separately in thermal equilibrium with respect to a third one then they must also be in thermal equilibrium with each other. According to this law if *A* and *B* are two systems which are separately in thermal equilibrium with another system *C*, then *A* and *B* are also in thermal equilibrium with each other. Since we are free to choose anything we like for the third system, it follows that the condition for thermal equilibrium cannot depend upon the nature of the system concerned. All systems in thermal equilibrium have the same temperature irrespective of their nature.

10.4. Thermodynamical Equilibrium

We know when there is no unbalanced force in the interior of the system or between the system and its surroundings, the system is said to be in mechanical equilibrium. Similarly if the temperature in all parts of the system is same and is also equal to the temperature of the surroundings, the system is said to be in thermal equilibrium. Further, if the net rate of any chemical reaction or change of internal structure in a system is zero, the system is said to be in chemical equilibrium. A system which is simultaneously in mechanical, thermal and chemical equilibrium, is said to be in thermodynamical equilibrium. Such a system is represented by specifying definite value to temperature, pressure and volume, i.e., by specifying definite value of thermodynamic co-ordinates P, V, T The subject of thermodynamics chiefly deals with systems which are in thermodynamic equilibrium.

10.5. Heat and Work

(*a*) **Heat.** When two bodies at different temperatures are brought in contact with each other, the temperature of one falls and the temperature of the other rises. This process will continue till both of them attain a common temperature.

To explain the facts stated above we suppose that something has flowed from the hot body to the cold body to equalise the temperature. This 'something' we call heat. Heat is a form of energy and flows from a body at higher temperature to a body at lower temperature in thermal contact.

It is to be noted that although heat is a form of energy, it is only evident when it flows from one body to another by virtue of temperature difference. If there is no temperature difference between two bodies in contact, no flow of heat takes place. Conventionally the quantity of heat energy transferred to a body is said to be positive and that transferred from the body is said to be negative. Heat is a form of energy and should be expressed in terms of units of energy. Recently when SI unit was introduced it was agreed upon that all types of energy should be expressed in same unit i.e. in Joule. So the unit of heat is Joule in SI units. The earlier unit is calorie and one calorie is equal to 4.186 Joules.

(*b*) **Work**. Our concepts of work and energy treated in mechanics are further developed in thermodynamics. An essential condition for the performance of work by a body (system) is the displacement of body when acting forces are present. It is measured by the product of the displacement and the component of the force in the direction of displacement. If the system as a whole exerts a force on its surroundings and a displacement takes place, the work that is done either by or on the system is called external work. For example when a gas contained in a cylinder expands and pushes out the piston, external work is done by the gas on the piston.

When the work is done by one part of the system on another part of the same system then the work is called internal work. For example the molecules of an actual gas attract one another. Therefore when a gas expands work is done against the mutual attraction between its molecules. The work done in this case is called internal work.

Internal work has no place in thermodynamics. Only the external work which involves the interaction between a system and its surroundings is important in thermodynamics. Therefore we speak of work only when the macroscopic state of a body (or a system) changes i.e. only the external work. In thermodynamics the work done by the system is taken as positive and work done on the system is taken as negative. For example if in a process the volume of a system decreases, then the work is done on the system and this work is said to be negative. On the other hand if the volume of the system increases then work is done by the system and work is considered to be positive.

10.6. Work in Change of Volume (Positive and Negative Work)

Whenever a gas expands against some external force, it does work on the external agency. Conversely whenever a gas is compressed by the action of some outside force, work is done on the gas. To calculate the work done when a gas expands, consider an ideal gas enclosed in a cylinder, equipped with movable piston as shown in Fig. 10.1. If the pressure exerted by the gas be P and the area of cross section of the cylinder be A then the force exerted by the gas on the piston will be $P.A$. Suppose the piston moves out a small distance Δx, the work ΔW done due to this force is given by

$$\Delta W = P.A\Delta x$$

But $A\Delta x = \Delta V$ where ΔV is the change in volume.

$$\therefore \quad \Delta W = P.\Delta V \qquad \ldots(10.1)$$

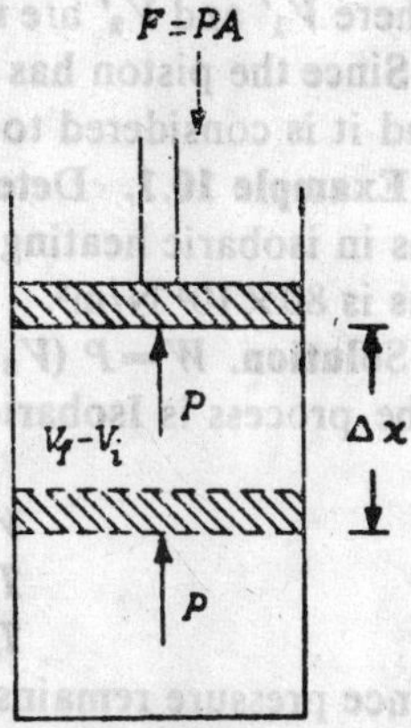

Fig. 10.1. Expansion of a gas at constant pressure (Isobaric process).

If the pressure is constant i.e. the process is isobaric* and the gas expands from its initial volume V_1 to final volume V then the total work W is equal

$$\text{to,} \quad W = \sum \Delta W = \sum_{V_1}^{V_2} P\Delta V \qquad \ldots(10.2a)$$

Changing the above summation to integration we have,

$$W = \int_{V_1}^{V_2} PdV = P(V_2 - V_1) \qquad \ldots(10.2b)$$

According to convention, this work is taken to be positive. Since the volume increases in the process.

Let us now consider the converse process. Suppose the gas is compressed at constant pressure by adding weights on the piston. The work done on the gas by the piston,

*When the volume of the gas changes at constant pressure, the process is known as Isobaric Process.

$$W' = \int_{V_1'}^{V_2'} PdV = P(V_2' - V_1') = -P(V_1' - V_2') \quad \ldots(10\ 2c)$$

where V_1' and V_2' are initial and final volumes of the gas.

Since the piston has pushed the gas, the work is done on the gas and it is considered to be negative.

Example 10.1. Determine the work of expansion of 20 litres of gas in isobaric heating from 300 K to 393 K. The pressure of the gas is 80×10^3 N/m².

Solution. $W = P(V_2 - V_1)$

The process is Isobaric and the constant pressure is

$$= 80 \times 10^3 \text{ N/m}^2$$

$$V_1 = 20 \text{ litre} = 20 \times 10^{-3} \text{ m}^3.$$

$$T_1 = 300 \text{ K},$$

$$T_2 = 393 \text{ K}.$$

Since pressure remains constant, from gas laws

$$\frac{PV_1}{T_1} = \frac{PV_2}{T_2} \quad \ldots(i)$$

From (i), we get

$$\frac{V_1}{T_1} = \frac{V_2}{T_2}$$

$$\therefore \quad V_2 = \frac{V_1 T_2}{T_1}$$

Hence,

$$W = PV_1\left(\frac{T_2}{T_1} - 1\right)$$

$$= 80 \times 10^3 \times 20 \times 10^{-3}\left(\frac{393}{300} - 1\right)$$

$$= 496 \text{ J}$$

Example 10.2. Six moles of an ideal gas kept at a constant temperature of 300 K and compressed from a volume of 16 litres to a volume of 4 litres. Calculate the work done in the process.

Solution.

$$W = \int_{V_1}^{V_2} PdV$$

Since for ideal gas $PV = nRT$.

We get,

$$W = \int_{V_1}^{V_2} \frac{nRT}{V} dV$$

$$= nRT \log_e \frac{V_2}{V_1}$$

Since $n=6$ moles, $T=300$ degree Kelvin. $V_1=16$ litre. $V_2=4$ litre. $R=8.31$ J/mole-degree and $\log_e x=2.3 \log_{10} \frac{1}{4}$

We get,

$$W=6\times8.31\times300\times2.3 \log_{10} \frac{1}{4}$$
$$=-20640\text{J}.$$

The negative sign indicates that work is done on the system.

10.7. Indicator Diagram

The work done by or on a gas enclosed in a container can be found by means of a graph. This graph is called the indicator diagram or *P-V* diagram. In the indicator diagram the pressure is plotted as ordinate and the volume as abscissa. The points thus obtained are joined to obtain a smooth curve. The area under this smooth curve extending down to *V*-axis, represents the work done.

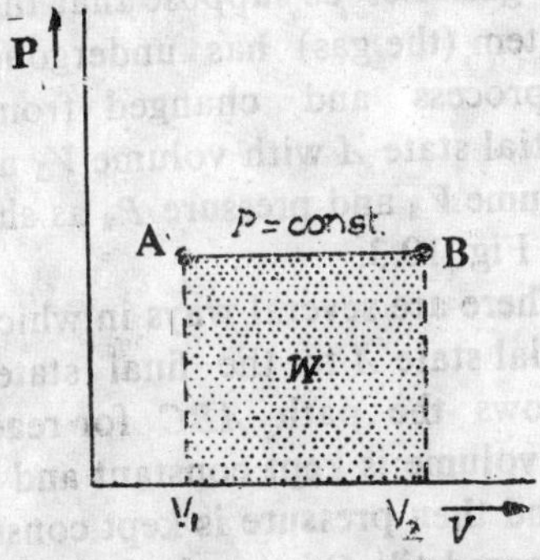

Fig. 10.2. Graphical representation of work done by a gas expanding at constant pressure (Indicator diagram).

The Fig. 10.2 represents the indicator diagram for the expansion of gas at constant pressure. Here the constant pressure is represented by line the *AB*, which is parallel to the *V*-axis. The work done by the gas in increasing volume from V_1 to V_2 is represented by the area under this line i.e. by the area of the rectangle ABV_2V_1.

If the gas is compressed, at constant pressure, the work done will still be represented by the same area but it is to be considered negative since the volume decreases in the process.

In the above example we have described the indicator diagram for a gas, at constant pressure. However indicator diagram for a gas, which expands or contracts under variable pressure can be drawn. Such a diagram is particularly very helpful to evaluate the work done, where exact relation ship between pressure and the volume is not known. It plays very important role in the theory of heat engines.

10.8. Dependence of Work and Heat Transfer on the Path of the Process

(*a*) **Dependence of work done, on the path.** We have seen that a system changes from one state to another when work is performed on it or by it. But the amount of work done depends upon not only on the initial and final state of the system but also on the intermediate states that is on the path of the process. To show this, let us consider a system consisting a gas enclosed in a cylinder, in which a piston operates to change the volume of the gas. Let us suppose that the system (the gas) has undergone a process and changed from initial state A with volume V_1 and pressure P_1 to final state C with volume V_2 and pressure P_2 as shown in the indicator diagram in the Fig. 10.3.

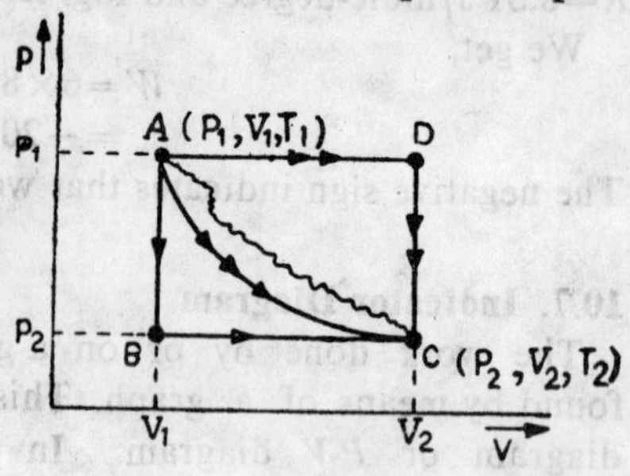

Fig. 10.3. work depends upon path.

There are several ways in which the system can be taken from an initial state A to the final state C. For example if the system follows the path ABC for reaching the final state C, then at first the volume is kept constant and the pressure is reduced from A to B and then pressure is kept constant from B to C and the volume is increased till C is reached. No work is done in the change from A to B as the volume remains constant. So the work done in expanding along the path ABC is given by the area under the curve BC i.e. area BCV_2V_1. The final state C can also be reached by following the path ADC. Here the pressure is kept constant from A to D and the volume is allowed to expand from A to D. Then the volume is kept constant and the pressure is reduced from D to C to reach the final State C. The work done by the gas in this case is the area ADV_2V_1. The third possibility is the path from A to C and in the case the work done is given by the area ACV_2V_1. The fourth possibility is the zigzag path AC in which both pressure and volume change along the path. The work done is equal to the area under the curve zigzag path AC. Thus there are number of paths along which the change of state from A to C (initial state to final state) can be accomplished. We can see now that the work done for all the paths are not same but different for different paths.

Thus we can conclude that work done by a system depends not only on the initial and final states but also on the intermediate states that is, on the path followed in the process.

(b) Dependence of heat transfer on path. Similarly the amount of heat Q flows into the system or out of the system, when it passes from initial state A at temperature T_1 to final state C at T_2, also depends on how the system is heated.

For example in Fig. 10.3, we can heat the gas at constant pressure P_1 until the temperature rises to T_2 and then lower the pressure to P_2 keeping the temperature constant, at T_2 (Path *ADC*). Alternatively we can first lower the pressure to P_2 and then heat it to the temperature T_2 keeping the pressure constant at P_2 (Path *ABC*). Each process would give a different value of Q. Thus we can see that heat flowing into (or out of) a system depends not only on the initial and final states but also on the intermediate state i.e. on the path of the process.

10.9. Internal Energy

When a body is in motion it possesses kinetic energy and by virtue of this kinetic energy it can do work. Similarly a body raised to a height is said to possess an energy called potential energy, with which the body can do some work. The kind of energy, mentioned above, which a body possesses due to its motion as a whole or due to its position or configuration is called external energy. In addition to this energy, all bodies possess energy which is not externally apparent. This energy is called internal energy. The internal energy is related to the microscopic aspect of matter. We know that all matter consists of atoms or molecules which are in perpetual motion. So they possess kinetic energy of motion. Again there exists a force called intermolecular force between molecules of matter. Due to this force the molecules possess potential energy. All the molecules or atoms of the matter possess such kind of energy. The sum of kinetic and potential energy which are related to molecules or atoms of matter are called internal energy. From thermodynamical point of view internal energy is the property of the state of matter and it has definite value for definite thermodynamic state. When matter changes its state its internal energy also changes.

10.10. Work, Heat and Internal Energy

We are now in a position to throw more light on work, heat, internal energy and their inter relationships. The following two experiments will be illustrative for that purpose.

The Fig. 10.4 (*a*) depicts the first experiment. It shows a weight P connected to a paddle is immersed in water in a container. When it is let fall a certain distance, the paddle rotates and the water is churned. The temperature of the water is found to rise as a result.

The second experiment is depicted in the Fig. 10.4(*b*). It shows a container, containing some water in contact with a Bunsen flame. The water in the container gets heated up by the Bunsen flame and the temperature of the water is found to rise.

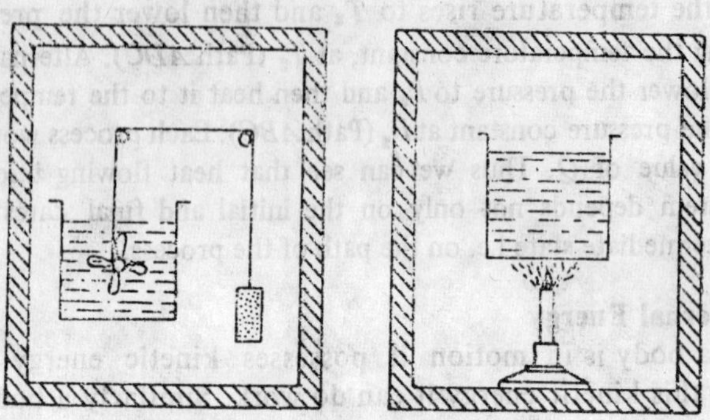

Fig. 10.4. (*a*) Work, (*b*) Heat.

In the first experiment the work done by falling weight is responsible for the rise in the temperature of water, because no heat has been supplied to the system from outside. We can say that it is the mechanical work which has brought about the change in the state of the system by raising its temperature. In the second case (Fig. 10.4*b*) the temperature of the water in the container has risen because of heat supplied by the burning flame of the Bunsen burner which is at a higher temperature. Here the heat energy has brought about the change in the state of the system by raising its temperature. Thus we see that the temperature of a system can be raised either by doing external work on it, or putting it in contact with another body which is at higher temperature. Therefore if one is given a hot body it is not possible on his part to tell in which of the above two ways, the body has been heated. Therefore so long as energy transfer continues, one can talk of either work or heat but it is meaning less to talk of heat or work in a body when energy transfer stops.

Thus we conclude that work and heat are energies in transit. The work is the form of energy transfer of an ordered motion as it is associated with ordered motion of a body as a whole like falling of weight in the last example. But heat is the form of energy transfer of disordered motion as it is associated with disordered motion of the bodies like atoms and molecules. Further the former can take place by any means but the latter takes place only by temperature difference.

We have seen that heat and work have some meaning so long as energy transfer into (out of) a system continues. But when the transfer of energy is stopped, we cannot identify the energy contained in the system either with work or with heat. The energy contained in the system is then called the internal energy. The work and the heat energy imparted to a system are stored as internal energy and thereby internal energy increases The internal energy is purely a function of state of the system. If the state of the system changes the internal energy changes along with it, but it does not depend upon the path taken by the system for the change of state.

10.11. Measuremeut of Internal Energy

The internal energy is a microscopic property of matter. So it cannot be directly measured. However, its value can be indirectly measured by formulating a macroscopic definition of it (internal energy). This definition would provide a means to measure the internal energy. Let a thermodynamic system interact with its surroundings and pass from an initial state i to the final state f through a certain process (path). Let Q be the heat absorbed by the system and W be the work done by the system during the process. The quantity $(Q—W)$ which is left in the system can be computed. It is experimentally found that if the system be carried from an initial state i to a final state f through different paths, the quantity $(Q—W)$ is found to be same although Q and W are individually different for different paths. Thus when a thermodynamic system passes from initial state i to final state f the quantity $Q—W$ depends only upon the initial and final state and is independent of the path taken by the system between the states. This quantity $Q—W$ is defined as the change in the internal energy of the system. Thus if U_i and U_f are the internal energies of the system in the initial and final states respectively, we have then,

$$U_f - U_i = \Delta U = Q - W \qquad \ldots(10.4)$$

We conclude therefore that every state of a system is associated with definite amount of energy called internal energy. When the system changes from one state to the other the internal energy also changes. But the difference in internal energy between two states is definite, in whatever manner the change is brought about.

The equation 10.4 is used to measure the change in the internal energy of the system. However, if some arbitrary value is assumed for the internal energy in some standard reference state, the value of internal energy in any other state can be found out from the above equation. But in practice, however, only the change in internal energy is important.

10.12. The Mechanical Equivalent of Heat

It is well known that when two bodies are rubbed against each other, heat is produced. It is produced at the expense of the work done. Similarly, when a weight, falling from a height strikes against the ground, it loses its kinetic energy acquired during the fall, which is converted into heat. Conversely heat energy is transformed into work in case of steam engine, diesel engine and jet propulsion engine etc. In all these engines fuel (coal, diesel etc.) is burned to produce heat and by expanding gases, the heat is converted into mechanical work.

The above examples illustrate that heat and work can be converted into each other. Now question arises how much of heat is converted to get a definite amount of work or vice versa? That means what is the exact relationship that exists between the amount of heat spent and the mechanical work obtained? The answer to the above questions was furnished by Dr Joule, who experimentally established the rate of exchange between these two forms of energy. In his experimental study, he proceeded as follows.

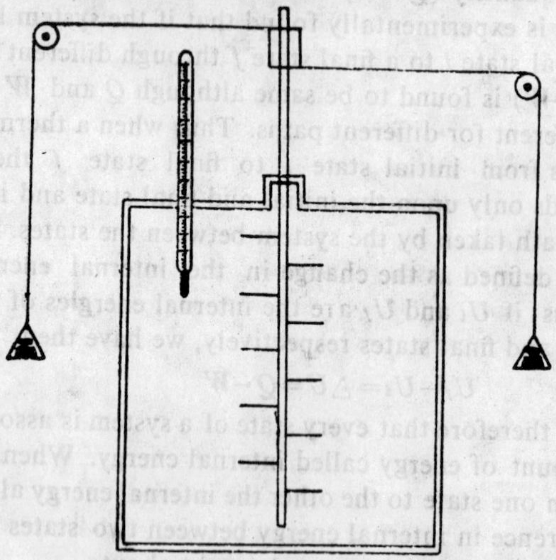

Fig. 10.5. Apparatus for Joule's experiment.

A vertical rod having some paddles attached to it, was mounted in a cylindrical vessel containing water. The rod could be revovled by a thread and p lley arrangement as shown in Fig. 10.5. When the weights fell, the rod rotated and water got churned up and heated thereby. The amount of heat gained by the water was calculated by noting the temperature rise of water.

The quantity of work done in revolving the rod was calculated by noting the distance through which the weights fell. After making necessary corrections for the frictional losses and the heat energy lost due to radiation and moreover taking number of observations, Joule found that a definite relationship exists between the work done and the heat produced. Mathematically he expressed the relation as,

$$W=JQ$$

$$\text{or} \quad J=\frac{W}{Q} \qquad \ldots(10.5)$$

where W is the work done in Joule and Q is the heat in calorie and J is a constant. J is called the mechanical equivalent of heat and from his experiment Joule found that its value is 4.18 joules/calorie.

So we conclude that whenever work is converted into heat or vice versa, there is a fixed ratio between the quantities of work and heat thus converted.

10.13. First Law of Thermodynamics

The first law of thermodynamic is a particular case of conservation of energy where heat energy is specifically involved. To understand this law let us consider a gas contained in a cylinder with a movable piston. Let all the walls except the bottom be thermally insulated. Let the bottom be brought in contact with a Bunsen Burner (Fig. 10.6) so that, it absorbs a quantity of heat Q. Due to this absorption the temperature of the gas will rise and it will expand. In this process the piston will be pushed upward and some work (W) will be done by the gas, and the internal energy of the gas will increase from u_i to u_f.

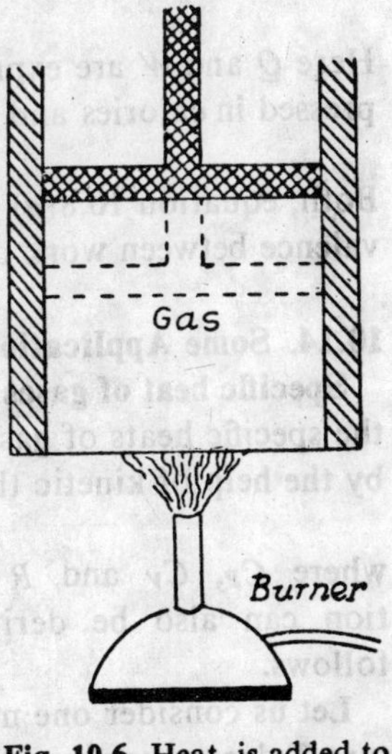

Fig. 10.6. Heat is added to a gas enclosed in a cylinder.

Thus the amount of heat (Q) energy absorbed by the gas is used in doing external work (W) and increasing the internal energy ($u_f - u_i$). Since the total energy has to be conserved, we can write,

$$Q=W+(u_f-u_i) \quad \ldots(10.6)$$

The above equation mathematically represents what is called the first law of thermodynamics. It can be stated in words as follows. If heat will be supplied to a system which is capable of doing work then quantity of heat supplied to the system will be equal to sum of the external work done by the system and the increase in internal energy.

In applying, the above equation it should be remembered that,

1. all the quantities must be expressed in same units.
2. Q is taken to be positive when it goes to the system.
3. W is taken positive when work is done by the system.

The form of the equation 10.6 expressing the first law, applies to a process in which the change in the internal energy and work done are finite. However, if the internal energy and work done are infinitesimally small, when infinitesimal amount of heat is absorbed by the system, the equation 10.6 can be written as

$$dQ=dW+dU \quad \ldots(10.7)$$

where dQ is small amount of heat absorbed by the system and dW is the small amount of external work done and du is the small change in the internal energy of the system. The equation 10.7 is known as the differential form of first law of thermodynamics:

If in a process internal energy of the system does not change then $(u_f-u_i)=0$.

Then the equation 10.6 reduces to

$$Q=W \quad \ldots 10.8(a)$$

Here Q and W are expressed in same units i.e. Joule. If Q is expressed in calories and W is in joule then equation 10.8 is written as:

$$W=JQ \quad \ldots 10.8(b)$$

Both, equation 10.8(a) and 10.8(b) represents Joule's law of equivalence between work done and the heat developed.

10.14. Some Applications of First Law of Thermodynamics

Specific heat of gases. We have derived the relationship between the specific heats of gases at constant pressure and constant volume by the help of kinetic theory of gas. The relationship is given as,

$$C_P-C_V=R$$

where C_P, C_V and R are having usual meaning. The same equation can also be derived by thermodynamic considerations as follows.

Let us consider one mole of an ideal gas at pressure P, temperature T degree kelvin and Volume V. Let the gas be heated at constant volume so that its temperature is raised by an infitesimal amount dT. The heat supplied will be $C_V dT$, where C_V is the specific heat at constant volume. As the volume remains constant, the external work which is given by PdV is zero ($\because dV$ is zero). In the differential form of the first law i.e. $dQ=du+dW$,

Putting $dQ=C_V dT$ and $dW=0$

we get,

$$C_V dT=dU \quad \ldots(10.9)$$

Let the same gas be heated at constant pressure P until the temperature is raised by the same amount dT. The heat supplied will be PdT. Now the gas would expand and external work against pressure P, would be done. If dV is the change in volume of the gas, the external work would be PdV. Thus for this process,

$$dQ=C_P dT \text{ and } dW=PdV$$

Hence from the first law of thermodynamics we obtain

$$C_P dT=dU+PdV \quad \ldots(10.10)$$

We have seen in the topic on kinetic theory of gases that the internal energy of an ideal gas depends only upon the temperature. So in the above process, since the change in temperature i.e. dT is same, the change in internal energy i.e. du will also be same in the above cases. Hence we can write,

$$C_P dT=C_V dT+PdV \quad \ldots(10.11)$$

or

$$C_P dT-C_V dT=PdV \quad \ldots(10.12)$$

We know for a mole of gas,

$$PV=RT$$

Therefore, we can write,

$$PdV=RdT \quad \ldots(10.13)$$

Then we have $(C_P-C_V)dT=RdT$

or

$$C_P-C_V=R \quad \ldots(10.14)$$

This relation was obtained by Mayer and is, therefore, called Mayer's relation.

(*b*) **Boiling process.** When a liquid is heated it absorbs heat and its temperature rises. After some time a stage is reached when the liquid starts boiling and changes its phase from liquid to vapour. During the process of change, the temperature remains constant although heat is continuously supplied from outside. The constant temperature is known as boiling point and it depends upon pressure. The boiling point of water is 373.15 K at atmospheric pressure.

Due to the change of phase from liquid to vapour, the volume increases by a large factor, and work is being done thereby. As the process involves heat and work, we can apply first law of thermodynamics to it.

Let us consider the vaporisation (boiling) of mass M of a liquid at constant temperature and pressure P. Let V_A be the volume of the liquid and V_B be the volume of vapour formed. The work W done by the liquid in expanding will be given by,

$$W=P(V_B-V_A) \qquad \ldots(10.15a)$$

Let L be the latent heat of vaporisation. It represents heat per unit mass to convert from liquid to vapour phase at constant pressure and temperature. Thus heat absorbed by the liquid of mass M, for conversion to its vapour will be given by

$$Q=ML \qquad \ldots(10.15b)$$

By applying first law of thermodynamics to this change of phase, we get,

$$Q=U_B-U_A+W \qquad \ldots(10.15c)$$

where U_B and U_A are internal energies in liquid and vapour stage respectively.

$$\text{or} \quad ML=U_B-U_A+P(V_B-V_A) \qquad \ldots(10.16)$$

$$U_B-U_A=\Delta U=ML-P(V_B-V_A) \qquad \ldots(10.17)$$

Thus we see by knowing M, L, P, V_B and V_A we can easily evaluate the increase in internal energy from equation 10.17. It is to be noted that as pressure remains constant at boiling, the boiling process is considered as an isobaric process.

Example. 10.3 At atmospheric pressure 1.00 gram of water having volume 1.00 cm^3 becomes 1671 cm^3 of steam when vaporised. The latent heat of vaporisation is 540 cal/gm. Calculate the external work done and increase in internal energy.

Solution. The atmospheric pressure is

$$P=76 \text{ cm of mercury.}$$
$$=76\times13.6\times9.8 \text{ dynes/cm}^2.$$
$$=1.013\times10^6 \text{ dynes/cm}^2.$$

As pressure remains constant $W=P\ (V_B-V_A)$

$$=1.013\times10^6 \text{ dynes/cm}^2\times(1671-1) \text{ cm}^3.$$
$$=1.013\times10^6 \text{ dynes}\times1670 \text{ cm.}$$
$$=1.013\times10^6\times1670 \text{ dynes/cm.}$$
$$=1.013\times10^6\times1670 \text{ ergs.}$$
$$=169.2\text{J}=40.5 \text{ cal.}$$

Heat absorbed $Q=ML=540$ calories.

Substituting the value of Q and W in the equation,

$$Q=U_B-U_A+ML$$

We get $\quad U_B-U_A=540-40.5$

$$=499.5 \text{ cal.}$$
$$=2087.9 \text{ J.}$$

(*c*) **Isothermal process.** When a process takes place in such a manner that the volume and the pressure undergo change but the temperature remains constant, then the process is known as Isothermal process.

Let us consider an isothermal process like expansion of a gas when its temperature remains constant during the expansion. Let ΔQ be the amount of heat supplied to the gas and ΔW be the amount of work done. Here, as the temperature remains constant, the change in internal energy

$$\Delta U=0$$

Now from first law of thermodynamics,

$$\Delta Q=\Delta U+\Delta W$$

But $\Delta U=C_V\Delta T=0$ as the temperature rise, $\Delta T=0$.

Therefore $\Delta Q=\Delta W$

Thus for an isothermal process whole of heat supplied, is used in doing external work in expansion. For a perfect gas isothermal change is also represented by:

$PV=$constant, which is Boyle's law.

(*d*) **External work done by an ideal gas in an isothermal expansion.** We know that when a gas undergoes an expansion in its volume by an amount dV at pressure P, then work done by the gas is given by

$$dW=PdV$$

If the volume changes from V_1 to V_2 then total work done W is given by:

$$W=\int_{V_1}^{V_2} PdV \quad \text{...(10.18)}$$

Case I. If the pressure remains constant during expansion then equation 10.18 can be written as:

$$W=P(V_2-V_1)$$

Case: II. When the pressure changes during isothermal expansion, then we can use the perfect gas equation

$PV=nRT$ to evaluate the work.

We have

$$P=\frac{nRT}{V} \quad \text{...(10.19)}$$

Putting the value of pressure in equation 10.18 we get,

$$W = nRT\int_{V_1}^{V_2} \frac{dV}{V}$$

as n, R, T are constants

$$W = nRT \log_e \frac{V_2}{V_1} \qquad \ldots(10.20)$$

If we consider 1 mole of gases then $n=1$ and the equation 10.20 reduces to:

$$W = RT \log_e \frac{V_2}{V_1} \qquad \ldots(10.21)$$

(*e*) **Adiabatic process.** When a process takes place in such a manner that no amount of heat either enters or goes away from the system, then that process is called adiabatic process.

Let us consider one mole of gas which is undergoing adiabatic expansion. In this case no heat enters or leaves the gas.

So, $dQ=0$.

From first law of thermodynamics we have

$$dQ = dU + dW$$

Putting $dQ=0$ in this equation, we get,

$$0 = dU + dW$$

or $$dW = -dU \qquad \ldots(10.22a)$$

Thus in an adiabatic process all the external work done, is at the expense of the internal energy of the system. The $-$ve sign in equation 10.22(*a*) shows that the internal energy decreases in these process. For an adiabatic expansion of one mole of perfect gas, the perfect gas equation is represented by,

$$PV^{\gamma} = RT$$

where $\gamma = \frac{Cp}{Cv}$ is the, ratio of two specific heats of gas.

To derive an adiabatic relation between temperature, pressure and volume of an ideal gas, let us consider one mole of gas having a volume V, pressure P and temperature T°K Suppose the gas undergoes a small adiabatic expansion. In doing so it does the necessary external work at the expense of its own internal energy which therefore decreases and hence the temperature falls.

Let dV, be the small change in the volume of the gas at pressure P. Then the external work done by the gas in its expansion will be

$$dW = PdV \qquad \ldots(i)$$

If dT be the fall of temperature of the gas in the process the heat lost by it will be $C_V dT$, where C_V is the molar specific heat at constant volume.

As no heat is transferred from or added to the system, then this amount of heat is equal to the change in internal energy of the system.

Therefore $dU = C_V dT$ . . .(ii)

Now from first law of thermodynamics:

$$dQ = dU + dW$$

In adiabatic process no heat is supplied from outside. This means

$$dQ = 0.$$

Hence $dU + dW = 0$ . . .(iii)

Substituting the values of dW and dU from equation (i) and (ii) in equation (iii) above we get,

$$C_V dT + PdV = 0 \qquad \text{. . .(iv)}$$

We know that the equation of state for one mole (one gram molecule) of an ideal gas is $PV = RT$.
where R is the universal gas constant.

This on differentiation gives

$$PdV + VdP = RdT$$

$$\therefore\ dT = \frac{PdV + VdP}{R}$$

Substituting this value in equation (iv) we get,

$$C_V \left(\frac{PdV + VdP}{R} \right) + PdV = 0$$

$$C_V (PdV + VdP) + RPdV = 0$$

But $R = C_P - C_V$

$$\therefore \qquad C_V(PdV + VdP) + (C_P - C_V) PdV = 0$$

or $$C_V = VdP + C_P PdV = 0$$

Dividing by $C_V PV$ we get,

$$\frac{dP}{P} + \frac{C_P}{C_V} \frac{dV}{V} = 0$$

$$\frac{dP}{P} + \gamma \frac{dV}{V} = 0 \qquad \text{. . .(v)}$$

where $\gamma = C_P / C_V$. Integrating the equation (v) we get

$$\log P + \gamma \log V = \text{constant.}$$

or $\log PV^{\gamma} = \text{constant.}$

$\therefore \qquad PV^{\gamma} = \text{constant.}$

(*f*) **External work done by an ideal gas in adiabatic expansion.** In adiabatic expansion no heat enters and leaves the system that is $dQ = 0$.

It does not mean that work is not being done by the gas. In fact there is change in temperature and work is done at the cost of its internal energy.

Let us consider one mole of an ideal gas. If it expands adiabatically from an initial volume V_1 to final volume V_2 the external work done is given by

$$W=\int_{V_1}^{V_2} PdV \quad \ldots (i)$$

where P is the instantaneous pressure of the gas when it undergoes a small expansion dV.

Since the expansion is adiabatic,

$PV^{\gamma}=K$ (a constant).

$$P=\frac{K}{V^{\gamma}}$$

$$W=\int_{V_1}^{V_2} \frac{K}{V^{\gamma}}dV$$

$$=K\int_{V_1}^{V_2} \frac{dV}{V^{\gamma}}$$

$$\text{or } W=K\left[\frac{V^{-\gamma+1}}{-\gamma+1}\right]_{V_1}^{V_2}=\frac{K}{1-\gamma}\left[V_2^{1-\gamma}-V_1^{1-\gamma}\right]$$

$$\text{or } W=\frac{1}{1-\gamma}\left[KV_2^{1-\gamma}-KV_1^{1-\gamma}\right] \quad \ldots (ii)$$

But the adiabatic process implies

$$P_1V_1^{\gamma}=P_2V_2^{\gamma}=K$$

Substituting the values of K in the bracket of equation (*ii*) we get,

$$W=\frac{1}{1-\gamma}\left[P_2V_2^{\gamma}V_2^{1-\gamma}-P_1V_1^{\gamma}V_1^{1-\gamma}\right]$$

$$\text{or } W=\frac{1}{1-\gamma}\left[P_2V_2-P_1V_1\right] \quad \ldots (iii)$$

Let T_1 and T_2 are the temperatures of the gas before and after the expansion respectively, then,

$$P_1V_1=RT_1 \text{ and } P_2V_2=RT_2$$

$$\therefore W=\frac{R}{\gamma-1}\left[T_2-T_1\right] \quad \ldots (10.23)$$

(*g*) **Cyclic process.** A thermodynamic process in which a system after undergoing a series of changes returns back to original state is called cyclic process. Since the internal energy is fixed for a particular state, the change in internal energy $dU=0$ as the system returns back to its original state. So from first law of thermodynamics it is obvious that $dQ=dW$.

We can find out the work done in a cyclic process by considering the *P-V* diagram. Suppose a gas undergoes a series of changes in its volume and pressure. Let its initial state be respresented by the point *A* (Fig. 10.7). Let it go along the path *ABC* to reach its final state *C*. Then let it return to its origional state *A* along the path *CDA* shownin Fig. 10.7. During the first process path (*ABC*) the gas expands. The work done by the gas is equal to the area *ABCLMA*. This work is taken as positive. During the second process (*CDA*) the gas is compressed and the work done on the gas

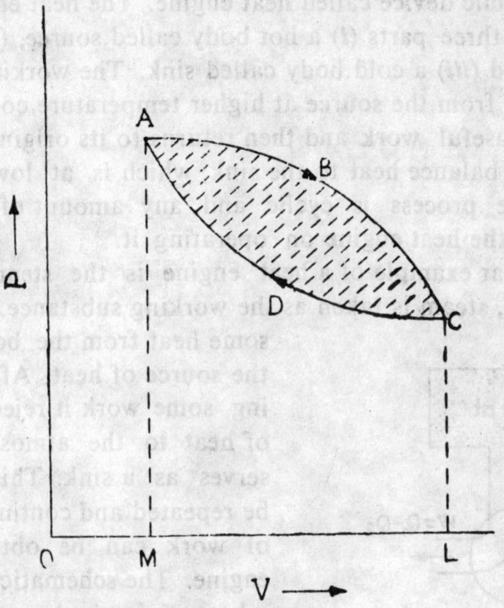

Fig. 10 7. Cyclic process.

is equal to the area *CLMADC* This work is taken as negative. So the net work done by the gas in the cycle is equal to the difference between two areas, that is, equal to the area enclosed by the curve (shown shaded). As in a cyclic process change in internal energy is zero, we can conclude that heat absorbed = work done = the area enclosed by the cycle of changes It is obvious from Fig. 10.7 that when the *P-V* cycle is traced clockwise, the net work done by the system is taken to be positive. When the cycle is traced in anticlockwise direction then, work is done on the system and is said to be negative.

(*h*) **Isochoric process.** A process taking place at constant volume ($\Delta V=0$) is called isochoric process. In such a process work done by the system is equal to zero. Hence from first law of thermodynamics,

$$\Delta U=\Delta Q \text{ as } \Delta W=0.$$

Thus in an isochoric process the heat added to (or taken from) the system goes entirely to increase (or decrease) the internal energy of the system.

10.15. Conversion of Heat into Work (Heat Engine)

The conversion of heat into mechanical work always takes place through a cyclic device called heat engine. The heat engine mainly consists of three parts (*i*) a hot body called source, (*ii*) a working substance and (*iii*) a cold body called sink. The working substance takes in heat from the source at higher temperature, converts a part of it into useful work and then returns to its original state after rejecting the balance heat to the sink, which is at lower temperature. So the process is cyclic and any amount of work can be obtained by the heat engine on operating it.

The familiar example of a heat engine is the steam engine. In steam engine, steam is taken as the working substance. It absorbes some heat from the boiler which is the source of heat. After performing some work it rejects the excess of heat to the atmosphere which serves as a sink. This process can be repeated and continuous amount of work can be obtained by this engine. The schematic diagram of a heat engine is shown in Fig. 10.8. This diagram is sometimes called flow diagram.

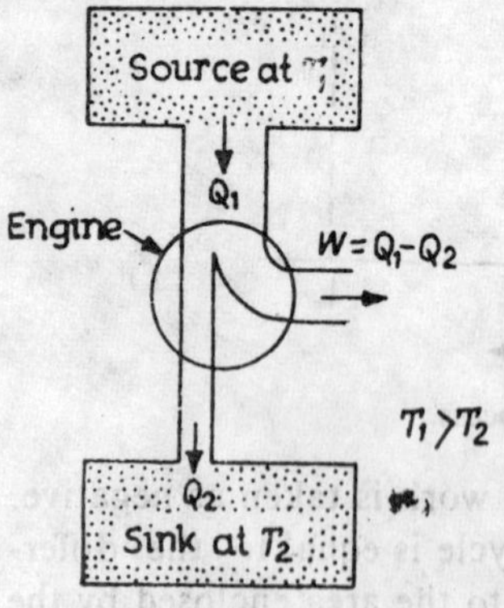

Fig. 10.8. Schematic flow diagram of the operation of a heat engine.

Thermal efficiency of a heat engine The thermal efficiency η of a heat engine is defined as the ratio of the net work done by the engine during one cycle to the quantity of heat absorbed from the source during that cycle. Suppose the working substance takes in an amount of heat Q_1 from the source and delivers an amount Q_2 to the sink. Suppose W is the amount of work obtained there in. The net amount of heat absorbed is Q_1-Q_2. Since the process is cyclic, there will be no change in the internal energy and according to the first law of thermodynamics

$$Q_1-Q_2=W$$

$$\text{Hence thermal efficiency } \eta=\frac{W}{Q_1}$$

$$=\frac{Q_1-Q_2}{Q_1}$$

$$\text{or } \eta=1-\frac{Q_2}{Q_1} \quad \ldots(10.24)$$

This equation shows that the efficiency of the heat engine will be unity (100%) when $Q_2=0$. Such a perfect or ideal engine will convert the whole of the heat extracted from the source into mechanical energy and will reject nothing to the source. This is, however, not possible in practice. Thus we conclude that a practical engine cannot convert entire amount of heat taken in from the source into work.

10.16. Second Law of Thermodynamics

The first law of thermodynamics tells us that energy is conserved in any thermodynamical process. It does not, however, tell us, whether that thermodynamical process is actually possible or not. For example if a hot body and a cold body are brought in contact the first law is not violated whether heat flows spontaneously from the hot body to the cold body or vice versa. (That is because in both the processes total energy remains constant). But by experience, we know heat never flows spontaneously from a cold body to a hot body. Similarly if we propose to extract a certain amount of heat from a body and convert it completely into work, the first law would not be violated. But in practice this is found to be impossible. If it were possible, we would drive ships across the ocean by extracting heat from water of the ocean. There would be no necessity of fuelling the ship. Moreover we know that heat engine converts heat into work. If the entire heat extracted from the hot reservoir is converted into work and the engine becomes 100% efficient, first law is not violated. But in practice no engine is 100% efficient and complete conversion of heat into work is impossible.

These facts were practically recognised and they infact led to the formulation of second law of thermodynamics. Second law of thermodynamics in fact deals with the question whether a process assumed to be consistent with first law is practically possible or not in nature.

The law can be stated in many forms although all of them are equivalent. However, we prefer to give one, which combines the statement of Kelvin and Planck and is known as Kelvin-Planck statement. It can be stated as follows.

It is impossible to construct an engine that operating in a cycle will produce no effect other than the exraction of heat from a reservoir and performing an equivalent amount of work. The above law implies that working substance working in a cycle cannot convert all the heat extracted, into work. It has to reject some amount of heat into the sink. So in order to convert heat into work it is necessary to have both a source and a sink. Since all heat extracted is never converted into work, no matter how good the engine design, it follows that the efficiency of an engine is never unity. It can be seen further that the efficiency of a heat engine can never be more than one. If so, then Q_2 will be negative and the engine will extract heat, both from the source and the sink and rejects no heat, which we know by second law, is not feasible.

10.17. Refrigerator

We have seen that in a heat engine, working substance working in a cycle, takes in same quantity of heat from a hot reservoir, converts a part of it into work and rejects the rest into the sink at lower temperature. In practice it is possible to construct a device which operates exactly in the reverse way. In this, working substance takes heat from a sink at low temperature and on some work being done on it, a larger amount of heat is rejected to the reservoir at higher temperature. Such a device is called refrigerator and its working substance is called refrigerant. The working principle of a refrigerator can be explained

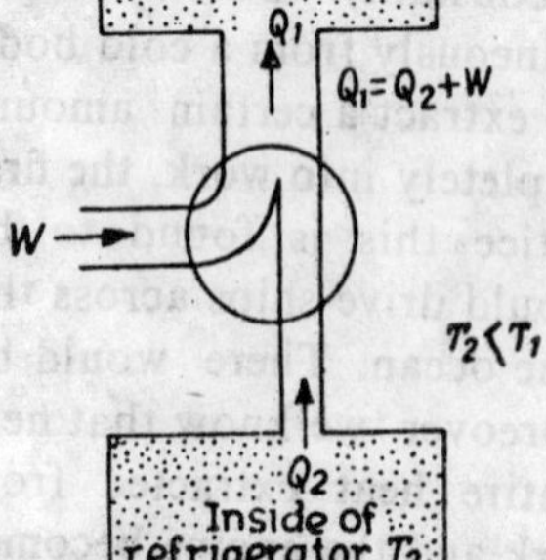

Fig. 10.9(*a*). Schematic flow diagram of the operation of a refrigerator.

by means of flow diagram shown in Fig. 10.9(*a*).

Let the quantity of heat absorbed by the refrigerant from the sink at lower temperature be Q_2 (Fig. 10.9*a*). The work done on the refrigerant be W. The heat rejected by the refrigerant to the hot reservoir be Q. The substance is working in a cycle, so there is no change in the internal energy. Applying the first law of thermodynamics.

$$dQ = dU + dW$$

we have

$$Q_2 - Q_1 = -W$$

$$Q_1 = Q_2 + W \qquad \ldots(10.25)$$

W is taken as negative as work is done, on the working substance.

The equation 10.25 indicates that it is always necessary to do work on the substance in order to transfer heat from a reservoir at low temperature to a reservoir at high temperature. In house hold refrigerator the work is done by the electric motor. The working substance commonly used in it is feron (CCl_2F_2). It absorbs heat from the materials say food, ice cubes etc. kept inside the refrigerator and rejects it to the surrounding air which is at higher temperature. However, the working of refrigerator does not violate the universal principle that heat does not flow from a body at lower temperature to a body at higher temperature because here heat does not flow spontaneously;it is forced to flow with the help of the motor.

The principle of a household electric refrigerator is shown in Fig. 10.2(*b*). In it, the liquid feron (refrigerant) is pumped into the cooling coils, where the pressure is reduced. As a consequence the liquid vaporises and the gas expands. Both the processes remove heat from the surroundings, which are cooled. The gas is pumped out of the cooling chamber and in an external set of coils it is compressed and liquified. In both these processes large amount of heat are given up.

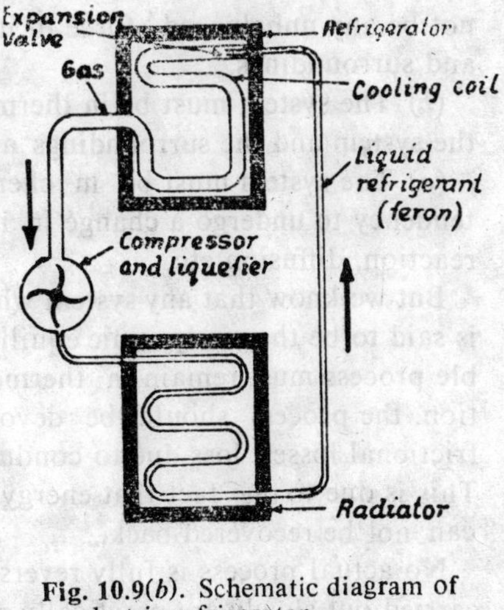

Fig. 10.9(*b*). Schematic diagram of an electric refrigerator.

The heat is removed by water cooling in large systems or air cooling in small systems. The liquid is then ready to repeat the cycle. The work required to transfer heat from lower to higher temperature is supplied by electric motor.

Co-efficient of performance. From economic point of view the best refrigerator is one that removes the greatest amount of heat Q_2 from the refrigerator, for the least expenditure of mechanical work W. We therefore define the co-efficient of performance of a refrigerator as the ratio Q_2/W and since $W=Q_1-Q_2$,

Co-efficient of performance

$$=\frac{Q_2}{Q_1-Q_2} \quad \ldots (10.26)$$

Thus if Q_2 is large the co-efficient of performance is higher.

10.18. Reversible and Irreversible Process

Reversible process. From thermodynamical point of view a process is said to be reversible if it can be retraced in the opposite direction such that the system and the surroundings pass through exactly the same states at each stage as in the direct process. After the end of the process, the systems and the surroundings taking part in the process are restored to their initial state without producing any change in either of them. A reversible process has to be done very slowly such that it satisfies the following requirements at every stage of the process.

(*a*) The system must be in mechanical equilibrium i.e. there must not be any unbalanced force in its interior or between the system and surroundings.

(*b*) The system must be in thermal equilibrium i.e. all parts of the system and the surroundings are at the same temperature.

(*c*) The system must be in chemical equilibrium i.e. it has no tendency to undergo a change in internal structure due to chemical reaction, diffusion etc.

But we know that any system which satisfies the above conditions is said to be thermodynamic equilibrium. So we infer that a reversible process must remain in thermodynamic equilibrium. In addition, the process should be devoid of dissipative effect such as frictional losses, loss due to conduction, radiation and inelasticity. This is due to the fact that energy losses due to dissipative effects can not be recovered back.

No actual process is fully reversible, but many processes when carried out slowly are practically reversible. For example, the slow compression of spring is practically reversible process. Because if the compressing force is slightly decreased, it expands and performs the work, equal to the work done in compressing it. Similarly the slow evaporation of a substance in an insulated container is practically reversible for if the temperature is slightly lowered condensation can be made to occur, returning energy to the heater until both it and the substance are in their original condition. Therefore, it can be concluded that all slow process are treated to be reversible process.

10.17: Irreversible Process

Any process which is not reversible is irreversible. This means if we retrace back we cannot exactly pass through changes, which take place during the direct process. It is obvious that a process in which changes occur suddenly is an irreversible process. During an irreversible process heat energy is always used to overcome friction. Energy is also dissipated in the form of conduction and radiation. This energy loss always takes place whether we carry on the process in one direction or reverse direction. For example, if an electric current is passed in the wire, the wire gets heated and resistance increases. There also occurs loss of heat. Now if we reverse the currents similar heat loss and increase in resistance also occur. So process of passing current is an example of irreversible process. All chemical reactions are irreversible. In general all natural process are irreversible.

10.19. Carnot's Engine

According to the second law of thermondynamics the efficiency of heat engine cannot be 100%. This leads us to question as to what can be the maximum efficiency of an engine when it works with a heat source and a sink maintained at two given temperatures, and how to operate an engine so as to achieve this maximum efficiency out of it. Sadi Carnot, a French engineer developed the principle to obtain maximum efficiency out of an engine. Any engine, which operates adopting this principle is known as Carnot's engine.

Carnot engine is an ideal heat engine, which acting with a source and sink achieves the maximum efficiency. It consists of the following components: (*a*) A cylinder containing working substance and fitted with a perfectly non-conducting and frictionless piston.

(*b*) A hot reservoir of infinite capacity maintained at a constant temperature T_1 to serve as a source.

(*c*) A similar cold reservoir of infinite heat capacity maintained at a lower temperature T_2 to serve as a sink.

(*d*) A perfectly non-conducting pad which serves as a stand for the cylinder.

For providing continuous work the working substance is taken through series of operations which constitute a cycle known as Carnot cycle.

Carnot's cycle. *Step I.* Initially the cylinder is put on the source maintained at temperature T_1 Fig. 10.10(*a*). Weights were added till the piston remains in a fixed position. The state of the working

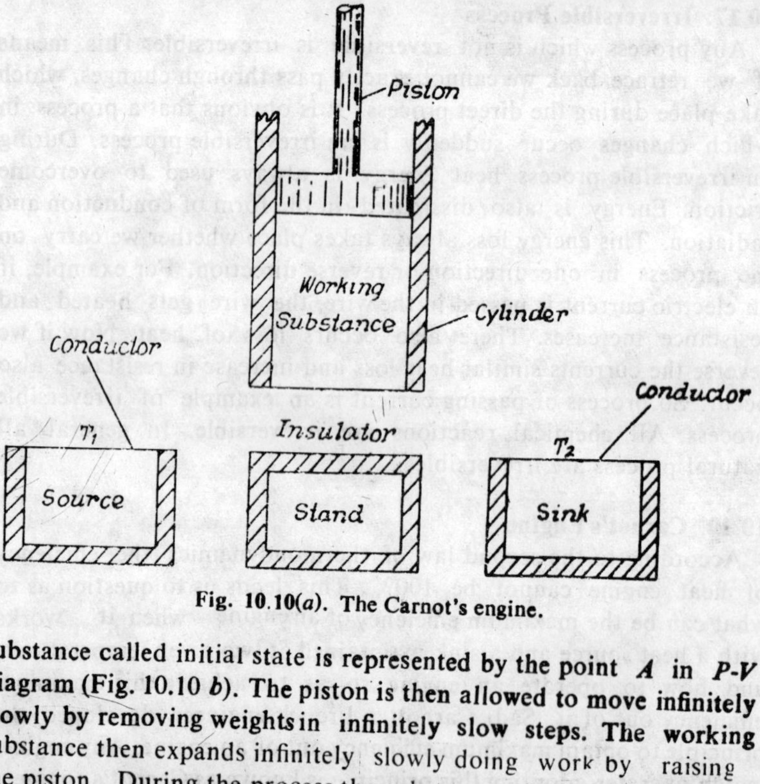

Fig. 10.10(*a*). The Carnot's engine.

substance called initial state is represented by the point A in P-V diagram (Fig. 10.10 *b*). The piston is then allowed to move infinitely slowly by removing weights in infinitely slow steps. The working substance then expands infinitely slowly doing work by raising the piston. During the process the substance takes in heat from the source by conduction through the base. Hence the temperature of the working substance remains constant. Therefore, the expansion is isothermal. It is continued until the state represented by the point B is reached. The curve AB represents the isothermal expansion of the substance at the constant temperature T_1 (of the source).

Step II. The cylinder is next placed on the heat insulating stand and by further removing weights on the piston, the working substance is further expanded. The expansion is adiabatic because no heat can leave or enter the working substance through the cylinder. The working substance does work in raising the piston and its temperature falls. The expansion is continued until the temperature

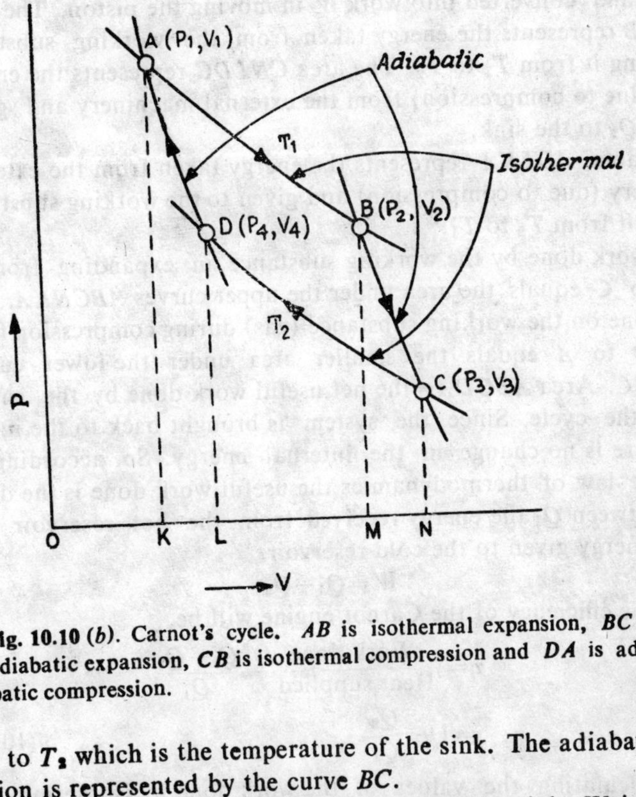

Fig. 10.10 (*b*). Carnot's cycle. *AB* is isothermal expansion, *BC* is adiabatic expansion, *CB* is isothermal compression and *DA* is adiabatic compression.

falls to T_2 which is the temperature of the sink. The adiabatic expansion is represented by the curve *BC*.

Step III. The cylinder is then placed on the sink. Weights are now placed on the piston in infinitely small steps. The working substance is thus compressed isothermally until state represented by *D* is reached. The curve *CD* represents the isothermal compression at the constant temperature T_2 (of the sink). During this process work is done on the working substance by the piston and a certain quantity of heat is given out to the sink.

Step IV. The cylinder is placed back on the insulating stand and by further placing the weights on the piston, the working substance is compressed adiabatically. A further amount of work is done on the working substance and its temperature rises. The process is continued until the temperature rises once more to T_1 and the state A is reached back. The adiabatic compression is represented by the curve DA.

The curve $ABCD$ so obtained in the Fig. 10.10(*b*) is called Carnot cycle diagram. Now let us examine the diagram.

The area $ABMKA$ represents the energy Q_1 taken from the hot source and converted into work W in moving the piston. The area $BCNMB$ represents the energy taken from the working substance in cooling it from T_1 to T_2. The area $CNLDC$ represents the energy taken (due to compression) from the external machinery and given as heat Q_2 to the sink.

The area $ADLKA$ represents the energy taken from the external machinery (due to compression) and given to the working substance heating it from T_2 to T_1.

The work done by the working substance in expanding from A to B to C equals the area under the upper curves $ABCNKA$. The work done on the working substance (gas) during compression from C to D to A equals the smaller area under the lower curves $CDAKNC$. Area $ABCDA$=the net useful work done by the engine during the cycle. Since the system is brought back to the initial state there is no change in the internal energy. So according to the first law of thermodynamics the useful work done is the difference between Q_1 the energy received from the hot reservoir and Q_2 the energy given to the cold reservoir.

$$W=Q_1-Q_2$$

Hence the efficiency of the Carnot engine will be,

$$\eta=\frac{\text{Work done}}{\text{Heat supplied}}=\frac{Q_1-Q_2}{Q_1}$$

$$=1-\frac{Q_2}{Q_1} \qquad \ldots (10.27)$$

By calculating the values of the work done by isothermal and adiabatic process as done in article 10.14, it can be shown that,

$$\frac{Q_2}{Q_1}=\frac{T_2}{T_1}$$ where T_1 and T_2 are the temperature on the perfect gas scale. So we can write,

$$\eta=1-\frac{T_2}{T_1} \qquad \ldots (10.28)$$

It is quite clear from the equation 10.28 that the η is always less than unity and efficiency depends only on the temperature of the source and the sink. As, $T_1>T_2$, smaller the value of T_2 greater will be efficiency. If T_2 is zero that is if it is possible to have sink at absolute zero temperature, then efficiency would be

100%. But absolute zero cannot be attained in actual practice. This is the consequence of another law called third law of thermodynamics. Thus we find that even the ideal Carnot's engine cannot have 100% efficiency. Moreover, it may be noted that Carnot's engine is an ideal engine and cannot be achieved in practice.

10.20. Reversibility of Carnot's Engine

Carnot's refrigerator. We have considered four reversible operations *AB* *BC*, *CD* and *DA* while describing the Carnot's engine. Let us now consider the reverse process *AD*, *DC*, *CB*, and *BA*. To start with, let the working subtance whose state is represented by the point *A* be allowed to undergo expansion along the adiabatic *AD*. During this reversible process its temperature will fall from T_1 to T_2. When representative point is at *D*, the cylinder is put on the sink at temperature T_2. The working substance is now allowed to undergo reversible isothermal expansion along the path *DC*. The working substance absorbs a quantity Q_2 and maintains its temperature T_2.

When the representative point reaches *C*, the cylinder is put on the insulating stand and is allowed to undergo adiabatic compression along the path *CB*. During this reversible process the temperature of the gas will rise from T_2 to T_1.

When the representative point reaches *B*, the cylinder is put on the source at temperature T_1 and is further compressed along the reversible path *BA*. In this reversible process, the gas rejects a quantity Q_1 to the source at higher temperature. Thus the process is performed in the reverse direction.

In the above set of processes the system and the surrounding have come to the orginal state *A*. As the amount of heat Q_1 is rejected into the source at higher temperature and the amount of heat Q_2 is absorbed from the sink at lower temperature, the engine behaves as a refrigerator and the work (*W*) done on the working substance is equal to Q_1-Q_2. This amount of work $W=Q_1-Q_2$ also represents the work done by the working substance in the direct process. So Carnot's cycle is a reversible cycle and the Carnot's engine is a reversible engine.

It can also be shown that no engine is more efficient that Carnot's engine and its efficiency is maximum. All reversible engines working between the same two temperatures have their efficiency same and is equal to that of Carnot's engine. In the above process we have considered perfect gas as the working substance. However, it can be shown that efficiency of reversible engine is independent of the nature of the working substance. This fact led to the formulation of the thermodynamic scale of temperature by Lord Kelvin.

All actual engines such as steam engine, petrol engine and diesel engine are irreversible and so they have less efficiency, than Carnot's engine.

Example 10.4. A Carnot's engine working as a refrigerator between 260 K and 300 K receives 500 calories of heat from the reservoir at lower temperature. Calculate the amount of heat rejected to the reservoir at higher temperature. Calculate also the amount of work done in joule in each cycle to operate the refrigerator.

Solution. For Carnot's refrigerator,

$$\frac{Q_1}{Q_2}=\frac{T_1}{T_2}$$

where Q_2 is the heat taken from the cold body at temperature T_2 and Q_1 is the heat delivered to the hot body at temperature T_1.

Here $Q_2=500$ cal., $T_1=300$ K.
$T_2=260$ K

Thus $$\frac{Q_1}{500 \text{ cal}}=\frac{300 \text{ K}}{260 \text{ K}}$$

$$\therefore Q_1=\frac{500\times 300}{260}=577 \text{ cal.}$$

Thus the excess of heat rejected to the reservoir at higher temperature is $577-500=77$ cal. and this must come from the work done by the motor running the refrigerator. Thus, $W=77$ cal$=77\times 4.18=322$ joules.

Examples 10.5. Find the thermal (maximum) efficiency of a reversible engine working between temperature 0°C and 100°C.

Solution. If Q_1 be the heat taken in by the working substance from the source at Kelvin temperature T_1 and Q_2 be the heat given up to the sink at Kelvin temperature T_2, then the efficiency of the engine is given by,

$$\eta=1-\frac{Q_2}{Q_1}$$

For Carnot's (reversible) engine,

$$\frac{Q_2}{Q_1}=\frac{T_2}{T_1}$$

$$\therefore \eta=1-\frac{T_2}{T_1}$$

Here $T_1=100+273=373$ K.
$T_2=0+272 \quad =273$ K.

$$\therefore \text{ Efficiency}=\eta=1-\frac{273}{373}=0.27$$

$$27\%$$

Example 10.6. A Carnot's engine has an efficiency of 50% when its sink temperature is 27°C. What must be the change in its source temperature for making efficiency 60%?

Solution. Let T_1 be the initial temperature of the source. The temperature of the sink is $T_2=27°C=300$ K and efficiency $\eta=50\%$ $=0.5$.

$$\text{Now, } \eta=1-\frac{T_2}{T_1}=1-\frac{300}{T_1}=0.5$$

$$\therefore T_1=600 \text{ K}$$

Now let the temperature of the source be raised to $(600+T)$ K, when the efficiency becomes 60% i.e. 0.6. Thus,

$$\eta=1-\frac{300}{600+T}=0.6$$

$$\therefore T=150 \text{ K}$$

The temperature of the source should be raised by 150 K.

10.21. Radiation

Conduction, convention and radiation are the three methods by which heat is transferred from one place to another place. In conduction and convention heat is transferred only through material medium. This material medium should fill the space between source and receiver. On the other hand in radiation no material medium is necessary, and heat can be transferred even in vacuum. The heat transferred from the sun to the earth is an example of heat radiation. We know that there is almost no material medium between the sun and the earth still earth receives heat from the sun through radiation.

When radiation takes place, the energy is transferred in the form of electromagnetic waves having wave length from 8×10^{-5} cm to 0.04 cm. The energy associated with the wave is called radiant energy. In fact it is the radiant energy but not the heat energy that travels in space in radiation. When such radiant energy falls on an object, the object absorbs it and converts it into heat energy. The so called radiant energy is sometimes called thermal radiations.

10.22. Properties and Nature of Heat Radiation

All bodies emit radiant energy irrespective of their temperature. When such energy falls on an object a part is reflected, a part is transmitted and the rest is absorbed. In all respects it behaves like light energy except it does not produce sensation of sight. The wave length of radiant energy as stated earlier ranges from 8×10^{-5} cm to 0.04 cm, whereas light has the wave length range from 4×10^{-5} cm to 8×10^{-5} cm. Since wave length of thermal radiation fall in the region beyond the red colour in the electromagnetic spectrum, they are called infra-red radiation.

Some of the common properties of radiant energy and light energy.

1. Radiant energy and light energy both travel in straight lines.
2. Both can travel in vacuum with velocity 3×10^8 m/sec.
3. Both of them obey the inverse square law i.e. intensity is inversely proportional to the square of the distance.
4. Both of them obey the laws of reflection and refraction,
5. Both of them show the phenomena of interference, diffraction and polarisation.

10.23. Detection of Thermal Radiation

Since thermal radiation carries energy the simplest way of detecting it is to absorb it and measure the resulting temperature rise.

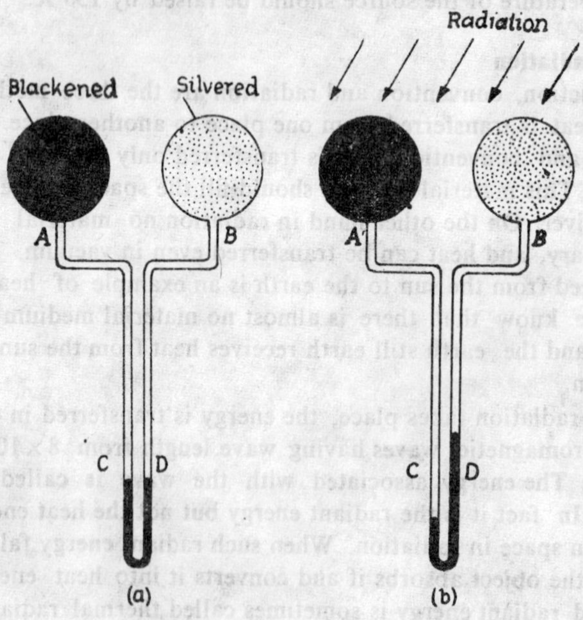

Fig. 10.11. A differential air thermometer exposed to radiation.

The principle is illustrated by a simple device called differential air thermometer as shown in Fig. 10.11 *a*. A differential air thermometer consists of two glass bulbs *A* and *B* of equal size, which contain air. The blubs *A* is coated with lamp black and the other, *B*, is silvered. Two blubs are connected by a narrow glass U-tube which contains sulphuric acid. If the blubs are at the same temperature the levels of sulphuric acid in two arms *C* and *D* are same.

When this thermometer is exposed to thermal radiation, the blackened bulb *A* absorbs radiant energy whereas silvered one reflects. As a result, the blackened bulb *A* warms up more than the silvered one *B*. The air inside the blackened bulb *A* expands and pushed down the level of sulphuric acid in *C*. Consequently the level of sulphuric acid in arm *D* rises up as shown in the Fig. 10.11*b*, and a difference in the level of sulphuric acid in two arms occurs. This instrument is sensitive and detects even feeble intensity of radiation.

10.24. Thermopile

This instrument is based on thermoelectric effect and can detect and measure the thermal radiation. It consists of a number of thermocouples (Ref–Art. 11.23) connected in series as shown in the Fig. 10.12. The thermocouples are made out of antimony and bismuth rods. The junctions on one side are blackened and are exposed to the thermal radiations. The junctions on the other side are kept cooled. The free ends of the thermopile are connected to a sensitive Galvanometer.

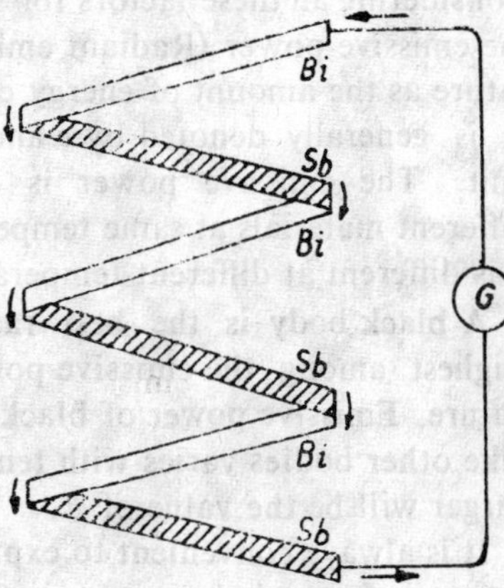

Fig. 10.12. A thermopile : A number of thermocouples is connected in series and arranged so that the radiation falls on one set of junction.

When the junctions are at the same temperature, no deflection in the Galvanometer is seen. This indicates that no current flows through the Galvanometer But when the blackened junctions are exposed to the radiation they absorb radiant energy and their temperature rises. This brings about a temperature difference between exposed junctions and unexposed junctions. As a result a current flows in the circuit due to the thermo-electric effect. This current causes the Galvanometer to deflect. The amount of deflection depends upon the intensity of radiation. Greater the intensity larger will be the deflection. Therefore the Galvanometer can be calibrated to measure the intensity of radiation also.

There are other more sensitive instruments like Bolometer, Boy's radiometer etc. which are frequently used for detection and measurement of thermal radiation. They are not discussed here and the interested students may refer to books on advanced text books, on heat to know about them.

10.25. Emission and Absorbtion of Radiation

All bodies at all temperatures emit rediation. The amount of radiation emitted by a body depends upon the nature and the temperature of its surface. A dull black surface emits more radiation than a polished surface under identical condition. Moreover a body at higher temperature is seen to emits more radiation than when it is at lower temperature. It is further observed that the amount of radiation from a body is directly proportional to its surface area. This means that different bodies of same material with different surface area emit different amount radiation even at same temperature. Considering all these factors for study of energy emission, we define the emissive power (Radiant emittance) of a body at a given temperature as the amount of energy emitted per unit area per second. It is generally denoted by e and has unit joule/m^{-2}s or Jm^{2}s^{-1} in SI unit. The emissive power is obviously different for bodies of different materials at same temperature and even for the same body, it is different at different temperatures.

A black body is the best radiator and its emissive power is highest among the emissive powers of all bodies at a given temperature. Emissive power of black body is denoted by E and its value like other bodies varies with temperature. Greater the temperature larger will be the value of E.

It is always convenient to express emissive power of a body relative to the emissive power of a black body. For that we define emissivity or relative emittance as the ratio of emissive power the body to the emissive power E of the black body at the same temperature. It is denoted by ϵ

Thus $$\epsilon = \frac{e}{E}$$

$$\therefore \quad e = \epsilon E \quad \ldots(10.29)$$

The emissivity of a body is a dimensionless quantity and its value is less than one for all real bodies at all temperature.

The radiant energy emitted by a hot body at a given temperature does not correspond to a single wave length but are spread over a range of wave length of electromagnetic spectrum. We are sometimes interested in the emissive power of a body in small band of radiation between say λ and $\lambda+\Delta\lambda$. We define a quantity called spectral emissive power e_λ as the amount of energy radiated per second per unit area between the wave length λ and $\lambda+\Delta\lambda$ divided by $\Delta\lambda$. The energy radiated per unit area per second between wave length λ and $\lambda+\Delta\lambda$ is then given by $e_\lambda\Delta\lambda$.

It is noted that e_λ is not a constant but a variable quantity. It varies with temperature of the body and wave length.

We are also sometimes interested in the ability of a body to absorb radiation of a given wave length. We may then define the spectral absorbtive power a_λ for a given temperature and a wave length λ as the ratio of the radiation absorbed to the radiation incident on it between λ and $\lambda+\Delta\lambda$.

10.26. A Perfectly Black Body*

A body which absorbs all the radiations (corresponding to all wave lengths) incident upon it, is called a perfectly black body or simply a black body. A black body does not transmit or reflect any radiation. It is known that whenever radiation falls on a body, a fraction of it, is absorbed, a fraction is reflected and another fraction is transmitted if the body is transparent to the incident radiation. If a represents the fraction of the incident radiation absorbed, r the fraction reflected and t the fraction transmitted then $a+r+t=1$. In the case of black body $r=0=t$. Hence $a=1$. This means the absorbtive power of a black body is equal to one. When such body is heated, it emits radiations of all wave lengths. Therefore radiations of black body are called full radiation or complete radiation. No actual black surface satisfies the above conditions completely. Some may approximate these conditions only better than others. Lamp black absorbs 95% of incident radiation whereas platinum black absorbs 98%. However a body showing close approximation to perfectly black body can be artificially produced. This is discussed below

A hollow sphere is taken and coated with lamp black on its inner surface. A fine hole is made on it to allow radiations to enter in. When the radiations enter the hole they suffer multiple reflections and are completely absorbed Fig. 10.13*b*. To eliminate direct reflection of radiation from the interior surface just opposite the hole

*Definition of black body has been given later in this Chapter.

a conical projection P is made. So this body acts as a black body absorber. When this body is placed in a bath at fixed temperature

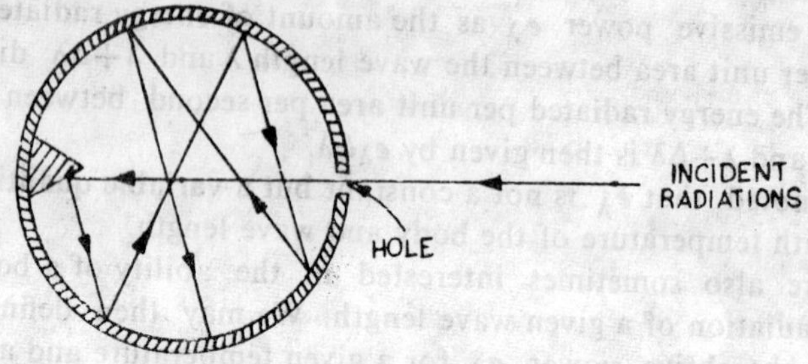

Fig. 10.13(*a*). Black body absorber.

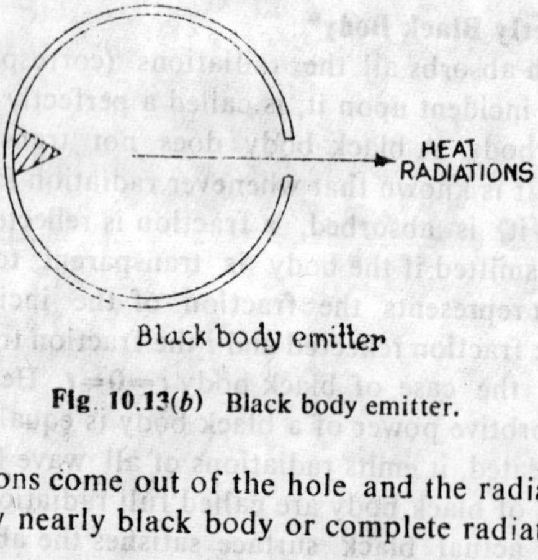

Fig. 10.13(*b*) Black body emitter.

the heat radiations come out of the hole and the radiation is considered to be very nearly black body or complete radiation.

10.27. Prevost's Theory of Exchange

Prevost in 1972 enunciated a theory called theory of exchange to account for the phenomena of radiation. According to this theory every body at all temperature above absolute zero continually emits radiant energy in all directions. The rate of emission per unit area depends only on the nature of its own surface and its own temperature. While emitting radiations it also absorbs radiant energy simultaneously from all surrounding bodies at a rate depending on the nature of its surface and the temperature of the sorrounding. Let us apply this theory to an enclosure in which two bodies at different temperature are placed. Each will emit radiations according to its own temperature independent of the other and will receive heat being placed in the field of radiations of the second. The one initially hotter of the two, gives out more heat than it receives, while the colder one gives out less heat than it receives. As a result of exchange of heat the hotter body falls in temperature and colder gains in temperature until a common temperature is attained by both. We then say that a thermal equilibrium has been reached. Even when the temperature is equalised the exchange does not cease, each will continue to radiate as much as heat as it receives from the other. So this thermal equilibrium is dynamic and not satic. This theory applies to any number of bodies at a time.

The sensation of warmth of a man before a hot body and the sensation of cold before a block of ice are consequences of heat exchange. In the case of man and the hot body as a result the differential effect of the exchange, the man receives more heat than he loses. Therefore, he feels the sensation of warmth. On the other hand, in the case of a man and the block of ice, the man loses more heat by exchange process than he receives from the ice. So he feels the sensation of cold, moreover the ice receives more heat than it loses and consequently it melts down.

10.28. Kirchoff's Law

It states that the ratio of the emissive power to the absorbtive power for radiations of a given wave length is the same for all bodies at the same temperature. Moreover it is equal to the emissive power of perfectly black body.

To prove the above statement let a body be placed in a uniformly heated enclosure at a constant temperature. Let an amount of heat radiation dQ between the wave length λ and $\lambda+\Delta\lambda$ be incident per second on unit surface area of the body. Let a_λ be the absorptive power of the body for the wave length λ. Then an amount of energy $a_\lambda dQ$ will be absorbed per second by unit surface area of the body. The rest amount $(dQ-dQ\times a_\lambda)$ i.e. $dQ(1-a_\lambda)$ of the incident radiation will be either reflected or transmitted.

Let e_λ be the emissive power of the body at wave length λ. Then an amount of radiation $e_\lambda d\lambda$ between the wave length λ and $\lambda+d\lambda$ will be emitted per second by unit surface area of the body due to its temperature.

Thus the total energy sent out by unit area of the body per second is $(1-a_\lambda)\,dQ+e_\lambda dQ$ . . .(10.30)

For equilibrium the amount of heat received must be equal to the amount sent out. Therefore,

$$(1-a_\lambda)\, dQ + e_\lambda d\lambda = dQ.$$

or
$$e_\lambda d\lambda = .a_\lambda dQ \qquad \ldots(10.31)$$

For the perfectly black body the absorptive power$=a_\lambda=1$. Therefore if E_λ be the emissive power of a black body we shall have,

$$E_\lambda d\lambda = dQ$$

Substituting this value of dQ in equation 10.31 we get,

$$e_\lambda d\lambda = a_\lambda E_\lambda d\lambda$$

$$\frac{e_\lambda}{a_\lambda} = E_\lambda \qquad \ldots(10.32)$$

Since E_λ is constant at a given temperature, it follows that,

$$\frac{e_\lambda}{a_\lambda} = \text{constant} \qquad \ldots(10.33)$$

for all substances at a given temperature. This is Kirchoff's law. This law is applicable to each wave length separately if the radiation covers a wide range.

From the equation 10.33 it is evident that if a_λ is large, e_λ must be proportionately large to maintain the ratio constant. This indicates that a good absorber, is also a good emitter. This means if a body absorbs radiation of a particular wave length strongly, it also emits the radiation of the same wave length equally strongly.

10.29. Illustrations and Applications of Kirchoff's Law

When a polished metal ball having a black spot on its surface is heated to a very high temperature and then suddenly transferred to a dark room, the black spot appears more bright than polished surface. The reason is that the black spot which absorbs light more strongly than the polished surface at ordinary temperature emits more strongly at high temperature. Similarly a piece of red glass which absorbs green light strongly at ordinary temperature glows with green light when strongly heated. In other words a body which emits strongly a particular radiation when hot, absorbs the same radiation strongly when cold.

Kirchoff's law is of great help to study the atmosphere of the sun. Its role is illustrated by the following experiment. When the Sodium vapour is heated to a temperature high enough to emit visible radiations its emission spectrum consists of two yellow lines. On the other hand when white light is passed through a cooler sodium vapour and then seen through a spectroscope, the continuous spectrum of the white light is found to consist of two dark lines in exactly the same position as the yellow lines in the emission spectrum. Thus sodium vapour which emits two yellow lines strongly is also a good absorber of light of these two wave lengths. These phenomenon is used to explain the dark lines in the sun's spectrum.

The sun consists of a central glowing mass which emits a continuous spectrum without any dark lines. It is surrounded by a comparatively cooler atmosphere which contains various elements like hydrogen, nitrogen, sodium, calcium, copper etc, in the gaseous state. When the radiation from the central mass passes through the surrounding atmosphere the various elements absorb those wave lengths which they can emit at higher temperature. As a result those wave lengths are missing from the spectrum and we see dark lines. These dark lines were first studied by Fraunhoffer and therefore known as Fraunhoffer lines and are characteristic of the different elements present in sun's atmosphere. By studying emission spectra of different elements in the laboratory and comparing them with sun's spectrum, the elements, in the sun can be identified.

Kirchoff's law is responsible for the development of science of spectrum analysis. It is established that atoms of each element give a spectrum which is the charateristic of that element alone. Hence if the characteristic spectral lines of an element are seen in a spectrum, it is sure that the element is present in the substance of which the spectrum is being observed.

10.30. Stefan's Law

The total energy of all wave lengths radiated from a black body depends upon its temperature only. The exact relationship between rate of emission per unit area of a black body and its temperature was experimentally established by Stefan. This relationship is known as Stefan's law. The law states that the total radiant energy E (called radiant emittance) emitted per second from unit surface area of a black body is proportional to the fourth power of its absolute temperature T.

Thus $E \propto T^4$

or $$E = \sigma T^4 \quad \ldots (10.34)$$

where σ (sigma) is a constant called Stefan's constant and has a value 5.67×10^{-8} watt/m^2 (1°C)^4. Boltzman established this law theoretetically. Hence it is also called Stefan-Boltzman law.

If the body is not perfectly black and has relative emittance ϵ

then $$E = \epsilon \sigma T^4 \quad \ldots (10.35)$$

Here ϵ varies from zero to one depending upon the nature of the surface. For a perfectly black body $\epsilon = 1$, so that equation 10.35 reduces to equation 10.34. The law is not only true for emission, but also for absorption of radiant energy.

Now a black body at absolute temperature T surrounded by another black body at temperature T_0 will lose an amount of energy σT^4 per second per unit area and gain from the surroundings an amount σT_0^4 per second per unit area, according to Prevost's theory of heat exchange. Hence the net loss of energy per second per unit area of the black body will be given by,

$$E_{net} = \sigma(T^4 - T_0^4) \qquad \ldots (10.36)$$

This equation is used to estimate energy loss or gain of a black body when surrounded by another black body at different temperature. For nonblack bodies the equation 10.36 is modified and is written as, $E_{net} = \sigma\epsilon(T^4 - T_0^4)$ . . . (10.37)

Example 10.7. Two large closed spaced concentric spheres both of which are black body radiators are maintained at temperature 27° C and 73°C, respectively. The space between the two spheres are evacuated. Calculate the net rate of heat transfer per metre square of the body at higher temperature. Given that Stefan constant $\sigma = 5.672 \times 10^{-8}$ SI units.

Solution. The net heat transtfer per unit area i.e.

$$E_{net} = \sigma(T^4 - T_0^4)$$

where T is the temperrture of the body at higher temperature and T_0 is the temperature of the body at lower temperature.

Here $T = 27° + 273°K = 300°K$. $T_0 = -73 + 273 = 200°K$

$= 5.672 \times 10^{-8}$ watt/(m)^2 (°K)^4

$E_{net} = 5.672 \times 10^{-8}$ watt/m^2 $-$(°K)^4$\{(300)^4 - (200)^4\}$(°K)^4

$E_{net} = 5.672 \times 10^{-8}$ watt/m$^2 \times 10^8 \times 65$

or $E_{net} = 5.672 \times 65$ watt/m^2

or 368.68 J m^{-2} s^{-1}

Example 10.8. What energy per second is radiated from a tungston filament 20 cm long and 0.010 mm in diameter ? When the filament is kept at 2500 K in an evacuated bulb ? The tungston radiates at 30 per cent of the rate of a black body at the same temperature. Neglect the conduction losses.

Solution. If P is the energy radiated then $P = EA$ where E is the energy radiated per second per unit area and A is the area of the body.

According to the equation 10.35,

$$P = EA = \sigma\varepsilon AT \qquad \ldots (i)$$

Here $A = \pi(0.0010 \text{ cm}) (20 \text{ cm})$

$= 0.0628 \text{ cm}^2$, $T = 2500°K$

$T^4 = (2500°K)^4 = 39.1 \times 10^{12}$ (°K)^4

$$\epsilon = \frac{30}{100} = 0.30$$

$\therefore \quad P = 0{\cdot}30\sigma AT^4 = (0.30)\ [5.672\times10^{-12}\ \text{watt/cm}^2(°\text{K})^4]$
$\times\{0.0628\ \text{cm}^2\}\ [39.1\times10^{12}(°\text{K})^4]$
$=4.2$ watts.

Example 10.9. An aluminium foil of relative emittance 0.1 is placed in between two concentrate spheres at temperature 300 °K and 200 °K respectively. Calculate the temperature of the foil after the steady state is reached. Assume that the spheres are perfect black body radiators. Also calculate the energy transfer between one of the spheres and to foil ($\sigma = 5.672\times10^{-8}$ SI unit)

Solution. Here $T = 300$ °K
$T_0 = 200$ °K
$\epsilon = 0.1$
$\sigma = 5.672\times10^{-8}$ SI units.

(*i*) Let x be the temperature of the foil. After the steady state is reached ra 'iant absorbance of the aluminium foil is equal to radiant emittance of it. Therefore

$$\epsilon\sigma(T^4 - x^4) = \epsilon\sigma(x^4 - T_0^4)$$

or $$[(300)^4 - x^4] = [x^4 - (200)^4]$$

or $$x^4 = 48.5\times10^8$$

$\therefore$ $$x = 263.8°\text{K}.$$

(*ii*) $$E_{\text{net}} = \epsilon\sigma(T_1^4 - x^4)$$
$$= 0.1\times5.672\times10^{-8}\ \text{watt/m}^2 - (°\text{K})^4 \times[(300)^4 - (263.8)^4\ (°\text{K})^4]$$
$$= 18.5\ \text{watts/m}^2.$$

10.31. Newton's Laws of Cooling

It states that if temperature difference is small, the rate of loss of heat of a body is directly proportional to the difference of temperatures T of the body and that of surroundings at T_0. This law can be derived from Stefan's law as follows.

According to Equation 10.37, the net loss of heat by a body is given by,

$$E_{\text{net}} = \epsilon\sigma(T^4 - T_0^4)$$
$$= \epsilon\sigma(T - T_0))(T^3 + T^2T_0 + TT_0^2 + T_0^3) \quad \ldots(10.38a)$$

If $(T - T_0)$ is very small T can be taken approximately equal to T_0. Then above Equation 10.37 can be written approximately as

$$E_{\text{net}} = \sigma\epsilon 4T_0^3(T - T_0) = 4T_0^3\sigma\epsilon(T - T_0)$$
$$= K(T - T_0) \quad \ldots(10.38\,b)$$

If the surroundings is maintained at T_0 then $K = 4\epsilon T_0^3\sigma$ is a constant. So $E \propto (T - T_0)$. The above relation is Newton's law of cooling.

10.32. Experimental Study of the Distribution of Energy in the Spectrum of Black Body

Lummer and Pringheim studied experimentally the distribution of energy amongst various wave lengths of the thermal spectrum of a black body radiation. They used an electrically heated chamber with a small aperture as the black body and measured its temperature by a thermocouple. The whole arrangement is shown in Fig. 10.14. The radiation from the black body O is made to fall on a slits S by means of a concave mirror m_1. The slit S is in the focal plane of a second concave mirror m_2, which reflects the radiations as a parallel beam which is incident on a flour spar prism P which is transparent to the heat radiation. The beam then gets dispersed and after emerging from the prism falls on a concave mirror m_3 which directs it to a platinum resistance thermometer called bolometer. On rotating the mirror m_3 about a vertical axis, the radiations of different wave lengths are made to

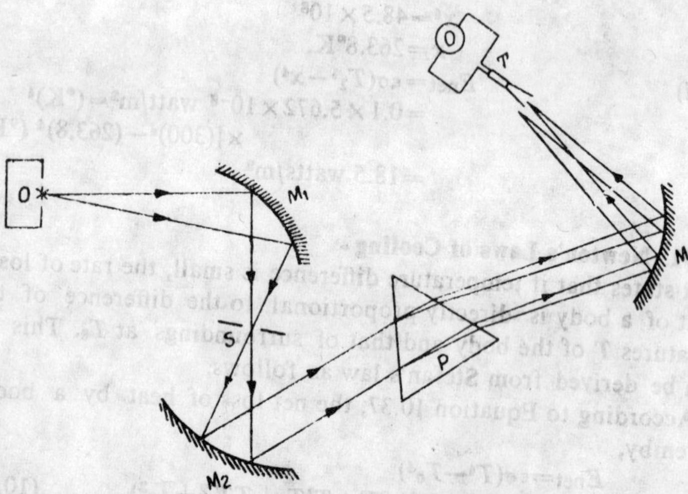

Fig. 10.14. Study of spectral distribution of energy in black body radiation.

fall on the bolometer one after the other. As a result of this, radiations are absorbed and the resistance of the resistor in the bolometer changes due to rise of temperature. Measuring the change in resistance it is possible to calculate the spectral emmissive power E_λ of the source between the wave length λ and $\lambda+\Delta\lambda$. It is to be noted that the wave length λ can be measured by a diffe-racting grating and a graph can be plotted between $E\lambda$ and λ to study the spectral distribution. The distribution of energy in the spectrum for different temperatures in the range of 904°K to 1646°K are shown in the Fig. 10.15. Here the wave length are plotted along *x*-axis and the spectral emissive power $E\lambda$ along the *y*-axis. The curves obtained shows the following important features.

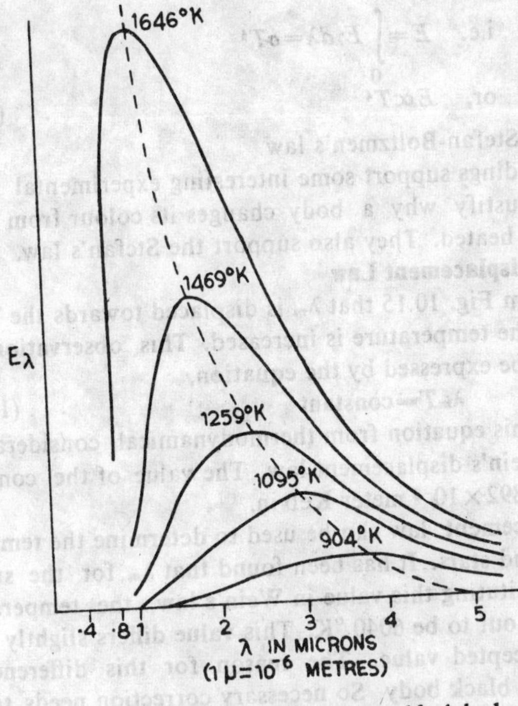

Fig. 10.15. Energy distribution in a black body spectrum.

1. At a given temperature the energy is not uniformly distributed in the radiation spectrum of hot body.

2. At a given temperature the spectral emissive power $E\lambda$ increases with increase in wave lengths and at particular wave length its value is maximum. With further increase in wave length the spectral (radiancy) emissive power decreases.

3. With increase of temperature λm the wave length at which maximum emission of energy takes place decreases. The points on the dotted lines represent λ_m at various temperatures.

4. For all wave lengths an increase in temperature cause an increase in the energy emission.

5. The area under each curve represents the total energy emitted per second per unit area for the complete spectrum at particular temperature. This area increases with the increase in temperature of the body. It is found that this area is directly proportional to the fourth power of the temperature of the body

$$\text{i.e.}\quad E=\int_{0}^{\infty} E_\lambda d\lambda=\sigma T^4$$

$$\text{or,}\quad E\propto T^4 \qquad \ldots (10.39)$$

This represents Stefan-Boltzmen's law.

The above findings support some interesting experimental observations. They justify why a body changes its colour from red to white, on being heated. They also support the Stefan's law.

10.33. Wein's Displacement Law

It is seen from Fig. 10.15 that λ_m is displaced towards the lower wave length as the temperature is increased. This observation can mathematically be expressed by the equation,

$$\lambda_m T=\text{constant} \qquad \ldots (10.40)$$

Wein derived this equation from thermodynamical consideration. So it is called Wein's displacement law. The value of the constant is found to be 2.892×10^{-3} meter-Kelvin.

Wein's displacement law can be used to determine the temperature of the sun and stars. It has been found that λ_m for the sun is 4753 Å. On substituting this value in Wein's law, the temperature of the sun comes out to be 6040 °K. This value differs slightly from the currently accepted value. The reason for this difference is that, sun is not a black body. So necessary correction needs to be made.

10.34. Measurement of High Temperature

Radiation pyrometry. The art of measuring high temperature is called pyrometry and the instruments used for the purpose are called pyrometers. Thermal radiation can be used to measure high temperature, since the radiation from black body depends solely on its temperature. When radiation is used for high temperature measurements it is called radiation pyrometry and the instrument is called radiation pyrometer. These instruments differ from thermometer, thermocouple etc. in that they are not to be in contact with the hot body.

The pyrometers are capable of measuring temperature howsoever high. But their lower limit is about 900 °K. The principle of measurement of high temperature by radiation pyrometers is based on the laws of radiation such as Stefan-Boldzmann and Wein's law.

Since these laws are applicable strictly to black bodies, and no actual body is black, the result obtained for temperature of actual hot body by using pyrometers only will be approximate

A typical radiation pyrometer called total radiation pyrometer is shown schematically in Fig. 10.16. It consists of a concave mirror *M* of copper and plated with nickel in front surface. The mirror can be moved forward and backward by rack and pinion *R*. *E* is an eye piece fitted at the pole of the mirror through an opening *P*. *D* is a diaphragm placed in front of the mirror and placed at its focus. Immediately behind *D*, a metallic strip *S* is arranged. The front surface *S* facing *D* is blackened while its back surface is connected to one junction of a thermopile *T*. The other junction of *T* as well as *S* are protected from the direct rediation by suitable screen *Q*. The leads from the thermocouple are connected to a millivoltmeter *mV* which records the thermo e.m.f.

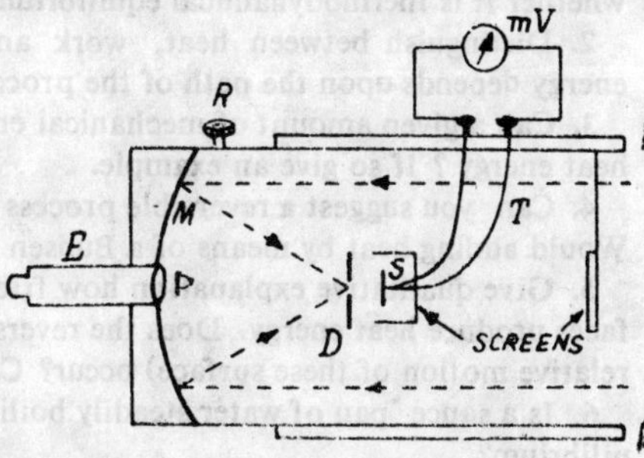

Fig. 10.16. A radiation pyrometer.

The beam of radiation from the hot body enters the instrument through an aperture *AA'* and falls on the concave mirror *M*. The mirror converges the beam to the diaphragm *D* which is at its focus. The metallic plate *S* receives the focussed radiation, and its temperature rises. As the thermocouple arrangement is connected to the plate, a thermo e.m.f. would be developed in the thermocouple. The millivoltmeter will record the thermo e.m.f. so developed.

The e.m.f. developed is proportional to the temperature difference of hot and cold junction. If T and T_0 are the temperature of the hot body and the receivers (receiving junction of thermocouple) respectively then e.m.f. developed should be proportional to $(T^4-T_0^4)$. Since T is very large compared to T_0, T_0^4 can be neglected in comparison with T^4 and e.m.f. should be proportional to T^4.

But in actual measurement it is found that e.m.f. is not exactly proportional to the fourth power of the absolute temperature of the source. The power of T varies from 3.8 to 4.2.

Some of the reasons of this discrepancy in the observations are (*i*) the source is not a black body, (*ii*) T_0^4 is not zero. (*iii*) Stray radiations and conduction along the wires raises the temperature of the cold junctions. So before use the pyrometer has to be suitably calibrated.

QUESTIONS

1. What requirements should a system meet in order to be in thermodynamical equilibrium ? If a system is in thermal equilibrium can you say, whether it is thermodynamical equilibrium or not?

2. Distinguish between heat, work and internal energy. Does internal energy depends upon the path of the process?

3. Can a given amount of mechanical energy be converted completely into heat energy ? If so give an example.

4. Can you suggest a reversible process whereby heat is added to a system? Would adding heat by means of a Bunsen burner be reversible process?

5. Give qualitative explanation how frictional forces between moving surfaces produce heat energy. Does the reverse process (heat energy producing relative motion of these surface) occur? Can you give plausible explanation?

6. Is a sauce pan of water steadily boiling on a stove a case of thermal equilibrium?

7. What are the purpose of second law of thermodynamics? State the limitation of first law with some examples?

8. Can a kitchen be cooled by leaving the door of the electric refrigerator open?

9. Discuss whether following phenomena are reversible.
   (*a*) Waterfall.
   (*b*) Rusting of iron.
   (*c*) Electrolysis.

10. Why are the efficiencies of heat engines so low? Comment on the statement a heat engine converts disordered mechanical motion into organised mechanical motion?

11. What you mean by isothermal and adiabatic process? Is any work done by adiabatic process ? If so where does the energy come?

12. Is there any difference between heat and radiant energy? Justify your answer with suitable example?

13. What are the reasons for Fraunhoffer lines in sun's spectrum? Can you produce such lines in the laboratory?

14. Can ice radiate?

15. Why does a piece of green glass plate heated to 1000 K and taken out glow with red light?

16. What do you mean by black body ? Why is it so important in the study of radiation?

17 Why does a body become first red hot and then white hot, when strongly heated?

18. Explain the following:
(*a*) For cooking purpose, vessels should, be preferably black.
(*b*) Black clothing is preferred in winter but white clothing is preferred in summer.

(c) Clear night is colder than cloudy night.

19. You are to paint the walls and the roofs of your house. Between white and black paint, which one you will prefer ? Why?

PROBLEMS

1. In a Joule experiment two weights 10 kg each fall through a height of 3 m and rotate the paddle wheel which stires 0.1 kg water. What is the change in the temperature of the water? *[Ans. 1.4 K]*

2. A lead bullet strikes a target with a velocity of 630 m/s and the bullet falls dead. Calculate the rise in temperature of the bullet assuming that 25% of the heat produced is used in heating the bullet. (Specific heat of lead 0.03). *[Ans. 393.75 K]*

3. Determine the work of expansion of 20 litres of gas in isobaric heating from 300 K to 393 K. The pressure of the gas is 80×10^3 N/m^2. *[Ans. 496 J]*

4. 2 kg of water at 100°C is converted into steam at the same temperature. The volume of 1 cm^3 of water changes to 1671 cm^3 on boiling. Calculate the change in the internal energy of the steam if the latent heat of vaporisation of water is 540 cal/gm. *[Ans. 4175.8 Joule]*

5. A quantity of oxygen (γ=1.4) is compressed to occupy a volume of 3 litres at 10 atm. pressure and 27°C. If it is allowed to expand adiabatically to 1 atmosphere pressure find the new volume and temperature. *[Ans. 15.5 litres, 155 K]*

6. Carnot engine takes 100 cal of heat from the source of temperature at 400 K and gives up 80 cal to the sink. What is the temperature of the sink and thermal efficiency of the engine? *[Ans. 320 K, 70%].*

7. The surface and deep water temperature of an ocean are 25°C and 10°C respectively. What will be the maximum efficiency of an engine operating upon this temperature difference? *[Ans. 60%]*

8. If it were possible to make an ideal refrigerator utilising carnot cycle, how many joules of mechanical energy would be required to remove 10 KJ from cold compartment at −13°C and deliver it to the outside air at 37°C? How much work would be required to remove this same 10 KJ from a cold compartment at —63°C and reject it at 37° C? *[Ans. 1.92 KJ, 4.76 KJ]*

9. The efficiency of carnot engine is 1/6. On decreasing the temperature of the sink by 65°C the efficiency increases to 1/3. Calculate the temperature of the source and the sink? *[Ans. 390 K, 325 K]*

10. A carnot engine has an efficiency 30% when the temperature of the sink is 27°C. What must be the change in the temperature of the source for making efficiency 50%. *[Ans. raise by 171 K]*

11. A reversible heat engine takes in heat from a reservoir at 527°C and gives up heat to the sink at 127°C. How many calories per second must it take from the reservoir in order to produce useful mechanical work at the rate of 750 watts. *[Ans. 358.8 cal/sec]*

12. In a refrigerator heat from inside at 277 K is transferred to a room at 300 K. How many joules of heat will be delivered to the room for each joule of electric energy consumed ideally ? *[Ans. 13 Joules]*

13. Wave length of corresponding to E_{max} for the moon is 14 microns. Estimate the temperature of the moon by taking the value of σ given in the book. *[Ans. 200 K]*

14. To what temperature must a black body be raised in order to double its total radiation if its original temperature is 1000 K ? *[Ans. 1190.5 K]*

15. The temperature of a tungsten filament in an incandescent lamp is 3000 K. Find the surface area of the filament of 100 W lamp. The emissivity of the filament is 0.30. $\sigma = 5.735 \times 10^{-8}$ Wm^{-2} K^{-4} *[Ans. 7.175×10^{-5} m^2]*

11

ELECTRICITY

11.1 Statical Electricity

(*a*) **Electrification by friction**. Many substances such as glass ebonite, sealing wax, resin etc., when rubbed with silk, flannel, catskin or other suitable materials, acquire the property of attracting light bodies like bits of paper, piece of pith,hair, wool etc The substances in such a state are said to be electrified or to possess electric charges, or they are simply called the *charged bodies*.

The electricity so produced by friction (rubbing) called frictional electricity. It is called ***statical electricity***, when it does not move from one place to another in the substance, in which it is produced. The study of the property of electrified bodies, on which the electric charge is at rest forms the subject matter of a branch of Physics called electrostatics.

(*b*) **Charge positive and negative**. Experimental studies have revealed that there exists only two kinds of charges namely positive and negative. The charge produced on the glass rod being rubbed with silk is arbitrarily called positive, while other type of charge is called negative. The negative charge is produced on ebonite rod on being rubbed with animal fur. The process of rubbing or friction does not electrify always the glass positively or ebonite negatively, but the kind of electric charge or electricity developed in the glass, ebonite etc. depends upon the nature of the substances used for rubbing. In the following series known as electrostatic series, the substances are arranged in such way that if any of them is rubbed with any other, the substance higher up in the series will be charged positively and the lower one will be charged negatively.

Electrostatic series. Fur, flannel, shellac, sealing wax, glass, paper, silk, human body, wood, metals, India rubber, resin, amber, sulphur, ebonite and guttaparcha.

For example, if glass rod is rubbed with flannel, it becomes negatively charged, but, when rubbed with silk it is charged positively. Therefore the substance used for rubbing is responsible for producing positive or negative charge, in the same substance. Thus any given substance can be charged positively or negatively as desired, by suitable choice of the substance for rubbing. It is to be remembered that bodies charged with same kind of electricity repel each other, while bodies charged with different kinds of electricity attract each other. In other words like charges repel each other while unlike charges attract.

11.2. Electronic Structure of Matter and Explanation of Electrification by Friction

Matter may be considered to be made up of atoms. Each atom contains a number of three kinds of subatomic particles called neutron, proton, and electron. Neutrons and the protons are situated at the centre called the nucleus and the electrons constantly move round the nucleus in different orbits.

The neutron is electrically neutral where as proton has positive charge and electron has negative charge. The amount of positive charge contained in a proton is equal to the amount of negative charge contained in an electron. Ordinarily in an atom the total number of protons is equal to the total number of electrons and hence the net charge contained in an atom is zero. This suggests that ordinarily an atom as well as substances, which are also composed of atoms are electrically neutral. A body is charged positively if some of the electrons are removed from it and charged negatively if some of the electrons are attached to it. In fact, when two bodies are rubbed with each other, some of the electrons of one body are transferred to the other. The body from which the electrons are removed is charged positively, while the other body into which, electrons get transferred is charged negatively. For example if a glass rod is rubbed with silk some of the electrons get detached from the glass rod and get attached to the silk. Hence glass rod is charged positively and the silk is charged negatively.

The above discussions reveal that the process of rubbing does not create any charge. It merely transfer electrons from one body to the other body, when they are rubbed. Besides it tells us that by friction equal and opposite charges are produced simultaneously.

If instead of electrons protons are removed from or added to a body, body can become electrified. But protons are fixed in the nucleus and cannot be easily removed. Therefore protons do not take part in the mechanism of electrification.

11.3. Conductors and Insulators

When a glass rod is rubbed with silk, it exhibits the property of electrification only at the portion rubbed. The other portion of the rod does not show any property of electrification. This implies that charged developed in glass on electrification, remains locally, where it is produced. Such materials (like glass), in which charges, are localised, when electrified are called insulators. The example of such materials are glass, fur, ebonite, silk etc.

On the other hand, when a brass rod, fitted with glass or ebonite handle, is rubbed with silk or fur, charges are found to spread, to the portion which is not rubbed. This implies that charge developed on elecrification in brass rod, does not remain locally, where it is produced but spread throughout the brass rod. Such materials like brass in which charges get spread through out the body, when charged are called conductors. The example of conductors are all metals and their alloys.

There is no sharp boundary between materials, which are insulators and those which are conductors. However a rough classification can be made on the basis of atomic structure of matter. As stated earlier, all substances, consist of atoms and each atom consists of equal number of protons and electrons. But under the influence of nearby atoms some of the electrons lose their association, with particular atoms and become free electrons. The number of these electrons in a substance, determines, whether a substance is insulator or conductor. If the number of these free electrons is large in a body, the body behaves as a conductor. On the other hand if these free electrons are few in a body, the body behaves as an insulator.

Electrostatic induction. It is a known fact that like charges repel and unlike charges attract. So when a conductor is brought near a positively charge body, the free electrons being negatively charged will be dragged towards the end of the conductor nearer to the charged body. As a result the nearer end has got excess of electrons *i.e.* negative charges and the farther end has got excess of positive charges. If the conductor is earthed, electrons from the earth moves to the body and neutralise the positive charges at the farther end. The electrons at the nearer end being bound by the positive charges of the charging body, would remain as such. On removing the charging body and disconnecting the earthing it would be found that the conductor is negatively charged. The electrons are spread throughout the body. On the other hand if a conductor is placed near a negatively charged body the electrons being negatively charged would be driven away from the nearer end to the farther end of the conductor. As a result the nearer end will have excess of positive charge and farther end will have excess of negative charge, on connecting to earth momentarily and then disconnecting, it would be found that the conductor has become positively charged. The above process of charging a conductor is known as charging by induction. And also the charges developed by such process are known as induced charges. It is obvious that in the process of charging by induction, opposite charges are developed on induced body.

Charges are measurable quantities and are expressed in coulombs.

11.4. Coulomb's Law

It was found experimentally that there is a force of attraction between two oppositely charged bodies and a force of repulsion between two similarly charged bodies. Coulomb in 1887 gave a law for the force of attraction or repulsion between two electrically charged bodies separated from each other by a definite distance. The law is known as Coulomb's law after his name and is given below.

The force of attraction or repulsion between two point charges separated from each other is directly proportional to the product of the magnitudes of the two charges and inversely proportional to the square of the distance between them. The force between the two charges always acts along a straight line joining the point charges.

Thus if q_1 and q_2 are two point charges, separated by a distance r (Fig. 11.1), then the magnitude of the force

$$F \propto q_1 q_2 \quad \ldots(11.1)$$

$$F \propto \frac{1}{r^2} \quad \ldots(11.2)$$

Combining Equations 11.1 and 11.2 we get,

$$F \propto \frac{q_1 q_2}{r^2}$$

$$F = K\frac{q_1 q_2}{r^2} \quad \ldots(11.3)$$

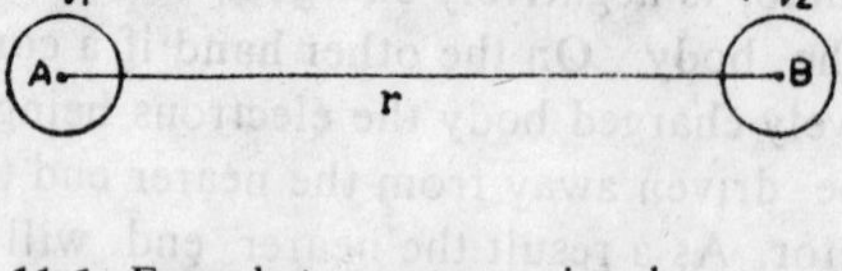

Fig. 11 1. Force between two point charges q_1 and q_2.

where K is a constant whose value depends upon the nature of the medium and units in which charges are expressed. In SI system for convenience K is taken to be equal to $\frac{1}{4\pi\epsilon}$ where ϵ (epsilon) is the permittivity of the medium. The permittivity of the medium may be considered as a measure of the property of the medium sorrounding the charge which determines the force between two charges.

Hence, in SI system,

$$F=\frac{1}{4\pi\epsilon}\frac{q_1q_2}{r^2} \qquad \ldots(11.4)$$

If the force between two charges is considered when charges are kept in empty space then force F is given by,

$$F=\frac{1}{4\pi\epsilon_0}\frac{q_1q_2}{r^2} \qquad \ldots(11.5)$$

where ϵ_0 is the permittivity of the empty space. The value of ϵ_0 has been computed and found that it is approximately equal to 8.85×10^{-12} coul^2/newton-m^2. The factor $\frac{1}{4\pi\varepsilon_0}$ has value approximately equal to 9.0×10^9 newton-m^2/(coul)^2.

11.5. Concept of Electric Field

We are familiar with interactions, which involves, direct physical contact of the interacting bodies. Here the bodies in contact exert forces on each other. Pushing a car, pulling a rope are such familiar examples. But there are interactions in which the inter acting bodies do not remain in contact still they exert forces on one another. To explain how this happens we develop a concept called concept of field. Here we imagine that one of the interacting bodies, sets up a field in the space around itself. This field acts on the other body and produces the observed force. Since the force is mutual we may regard any one of the interacting bodies to be the source of the field, which acts upon the other. The field thus plays an intermediary role in producing the forces between two bodies.

When the interacting bodies are two static charges separated by a distance, the field that is set up is called electrostatic fieids or simply electric field, field. Here one of the charges sets up electrostatic field which acts on the other charge to produce the force between them.

The field concept is very useful to explain the interactions between objects, which are not in contact. The gravitational and magnetic interactions can easily be explained by this concept.

11.6. The Electric Field Intensity

Consider a point a charge q situated at some point in space. Let us call this as the source charge. This charge sets up an electric field around itself. If another positive point charge q_0, which we call the test charge (conventionally positive point charge is taken to be test charge) is placed at some point in the field, it experiences a force, which is given by,

$$\vec{F}=q_0\vec{E} \qquad \ldots(11.6)$$

where $\vec{E}$ is the relectric field strength or intensity of the electric field at that point. From Equation 11.6 we get,

$$\vec{E}=\frac{\vec{E}}{+q_0} \qquad \ldots(11.7)$$

Hence the electric field intensity $\vec{E}$ at any point is defined as the force experienced by a unit positive charge placed at that point. It is a vector quantity and its direction at any point is same as that of the direction of the force on a test charge $+q_0$ at that point.

The magnitude of the electric field intensity can be obtained from expression for the force on charge $+q_0$ due to the change q. From Coulomb's law we can write this force as,

$$F=\frac{1}{4\pi\varepsilon_0}\frac{qq_0}{r^2}$$

Hence the magnitude of the electric field intensity E is given by

$$E=\frac{F}{+q_0}=\frac{1}{4\pi\varepsilon_0}\frac{q}{r^2} \qquad \ldots(11.8)$$

The unit of electric field intensity at a point can be found out from the expression for E. In MKS system, force is expressed in Newton and charge in Coulomb. Therefore unit of electric field intensity is newton per coulomb (N/c).

In the above discussion the source of the field is taken to be stationary charge. But it is observed that a moving charge can also produce on electrostatic field. However if the velocity is small, the field, produced, by a moving charge is exactly similar to the field, produced, when it is at rest. Therefore in this case the force experienced by a test charge $+q_0$, is given by the same expression

$$\vec{F}=q_0\,\vec{E},$$

where $\vec{E}$ is the electric field intensity due to the moving charge at the test charge, at the time of observation. Moreover it should be borne in mind that a moving test charge would experience the same force that experienced by the test charge when at rest.

Thus we conclude that a test charge whether at rest or in motion experiences the same force at any point in a static electric field.

11.7. Lines of Force and Electric Field Intensity

The electric field in the vicinity of one or more charged bodies is frequently represented by drawing lines of force to help us

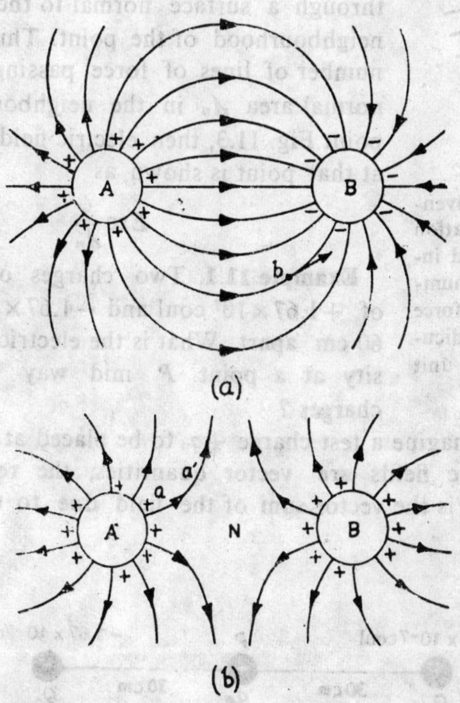

Fig. 11.2. (*a*) Lines of force for a pair of equal and opposite charges. (*b*) Lines of force for two identical positive charges. At the neutral point *N* field intensity is zero.

visualise the field. An electric line of force is a line so drawn that a tangent to it at any point shows the direction of the electric field intensity at that point. Some of the lines of force associated with a pair of equal and opposite charges are shown in Fig. 11.2(*a*) while some due to two identical positive charges are indicated in Fig. 11.2 (*b*). It is to be noted that lines of force always start from positive charges and terminate on negative charges.

One may think from the definition of line of force that line of force can give only the direction of the electric field intensity at a point but not its magnitude. However the magnitude of the field intensity at a point can also be obtained from the number of lines of force passing near that point. This is done by the coventional agreement that the intensity of the electric filed at any point in space will be directly proportional to the number of lines of force per unit area through a surface normal to the field in the neighbourhood of the point. Thus if ψ is the number of lines of force passing through a normal area A_n in the neighbourhood of a point Fig. 11.3, then electric field intensity E at that point is shown as

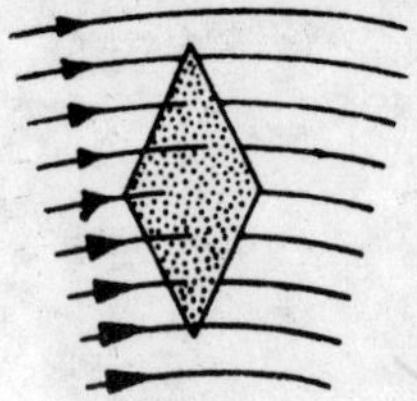

Fig. 11.3. Conventional representation of electric field intensity by the number of lines of force passing perpendicularly through a unit area.

$$E \propto \frac{\psi}{A_n} \quad \ldots (11.9)$$

Example 11.1. Two charges of magnitude of $+1.67 \times 10^7$ coul and -1.67×10^7 coul are 60 cm apart. What is the electric field intensity at a point P mid way between the charges ?

Solution. Imagine a test charge $+q_0$ to be placed at P, Fig. 11.4. As electrostatic fields are vector quantities, the resultant field intensity at P is the vector sum of the field due to two charges, Q_1 and Q_2

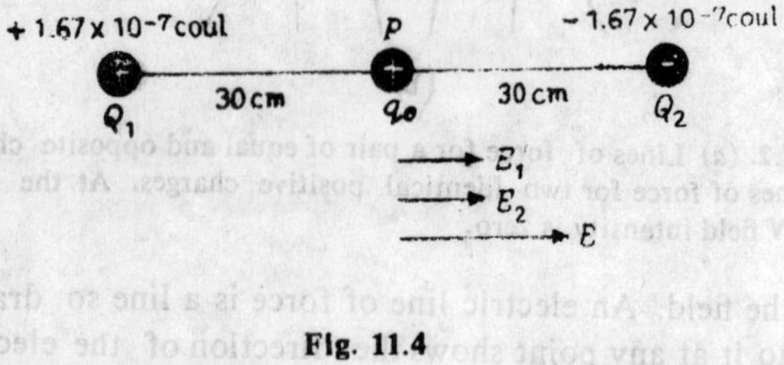

Fig. 11.4

Here both the fields are in the same direction. Therefore if E is the resultant field,

$$E=\frac{1}{4\pi\varepsilon_0}\ \frac{Q_1}{r_1^2}+\frac{1}{4\pi\varepsilon_0}\ \frac{Q_2}{r_2^2}$$

where r_1 and r_2 are the distances of P from Q_1 and Q_2 respectively.

$$\therefore \qquad E=9.0\times10^9 \text{N-m}^2/(\text{coul})^2\times\frac{1.67\times10^{-7}\text{ coul}}{(0.30\text{m})^2}$$

$$+9.0\times10^0\text{N-m}^2(\text{coul})^2\times\frac{1.67\times10^7\text{ coul}}{(0.30)^2}$$

$$=3.33\times10^4\text{newton/coul}$$

The resultant field is directed along the line from Q_1 to Q_2, Fig. 11.4.

11.8. Eelectrostatic Potential

Electrostatic potential at a point is defined as the amount of work done in taking a unit positive charge from infinity to that point.

Suppose a point A is at a distance from a conductor having charge q. The electric field intensity at A is given by

$$\frac{1}{4\pi\varepsilon}\ \frac{q}{r^2}$$

Let us find out the potential at A

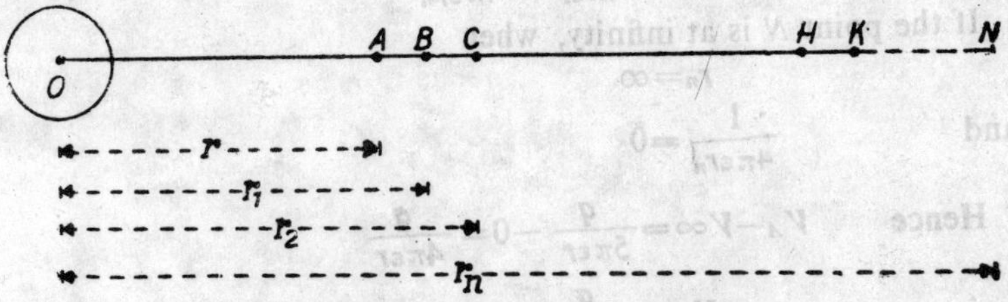

Fig. 11.5. Potential due toa charge q.

Consider close points $B, C \ldots N$ at a distance $r_1, r_2 \ldots r_n$ in Fig. 11.5. The electric intensity at B is given by,

$$\frac{1}{4\pi\varepsilon}\ \frac{2}{r_1^2}$$

If the points A and B are very close to each other, average intensity between A and B may with good approximation be taken as

$$\sqrt{\frac{q}{4\pi\epsilon r^2}\times\frac{q}{4\pi\epsilon r_1^2}}=\frac{q}{4\pi\epsilon r r_1} \qquad \ldots (11.10)$$

We have the distance between B and A is $=r_1-r_2$. The work done in bringing a unit positive charge from B to A=Electric field intensity × distance.

$$=\frac{1}{4\pi\varepsilon}\times\frac{t}{rr_1}(r-r_1)=\frac{q}{4\pi\varepsilon r}-\frac{q}{4\pi\varepsilon r_1} \quad \ldots (11.11)$$

The work done is bringing a unit positive charge from B to A is equal to the potential difference between A and B.

The petential difference between A and B *i.e.*

$$V_A-V_B=\frac{q}{4\pi\varepsilon r,}-\frac{q}{4\pi\varepsilon r_2} \quad \ldots, (1)$$

Similarly we can find that potential difference between B and C and D and so on. Then we hove,

$$V_B-V_C=\frac{q}{4\pi\varepsilon r_1}-\frac{q}{4\pi\varepsilon r_2} \quad \ldots, (2)$$

$$V_C-V_D=\frac{q}{4\pi\varepsilon r_2}-\frac{q}{4\pi\varepsilon r_3} \quad \ldots (3)$$

....

....

$$V_{N-1}-V_N=\frac{q}{4\pi\varepsilon r_{n-1}}-\frac{q}{4\pi\varepsilon r_n} \quad \ldots (n)$$

Adding all the equation, from (1) to (n), we have

$$V_A-V_N=\frac{q}{4\pi\varepsilon r}-\frac{q}{4\pi\varepsilon r_n} \quad \ldots (11.12)$$

If the point N is at infinity, when

$$r_n=\infty$$

and $$\frac{1}{4\pi\varepsilon r_n}=0$$

Hence $$V_A-V_\infty=\frac{q}{5\pi\varepsilon r}-0=\frac{q}{4\pi\varepsilon r}$$

$\therefore$ $$V_A=\frac{q}{4\pi\epsilon r} \quad \ldots (11.13)$$

where V_A is the potential at point A due to charge q.

Potential at a point is a scalar quantity. It is positive due to positive charge and negative due to negative charge. The potential at a point due to number of charges is equal to the algebraic sum of the individual potentials.

11.9. Potential Gradient and Electric Field Intensity

Consider two close points A and B in Fig. 11.5, which are $r_1-r=\Delta r$ apart and the field between them is E. Let A be at a higher potential then B. Now $V_A-V_B=\Delta V$=work done in moving a unit positive charge against the forces=−(force on a unit positive × distance) in the direction of the force

$\therefore$ $$\Delta V=-E\Delta r$$

$$\vec{E} = -\frac{\Delta V}{\Delta r} \quad \ldots (11.14(a)$$

when $\Delta r \to 0$ the quantity $\frac{\Delta V}{\Delta r}$ is called potential gradient. So the electric field intensity at a point in an electric field is equal to the negative of the potential gradient at that point. If however the field is uniform over a large distance between two points then electric field intensity at any point is found by dividing potential difference with distance of separation between two points. Therefore if V is the potential difference between any two points d distance apart, then the magnitude of the electric field intensity between them is given by,

$$E = \frac{V}{d} \quad \ldots (11.4b)$$

11.10. Capacity of a Conductor

If the charge on a conductor is gradually increased its potential also increases and at any instant, the charge given to a conductor is directly proportional to its potential. If a coductor has a charge q and its potential is V, then,

$$q \propto V$$

or

$$q = CV \quad \ldots (11.15)$$

where C is called capacity of a conductor.

The capacity of a conductor is defined as the amount of charge that, has to be given to raise its potential by unity. In SI system the unit in which capacity is expressed is farad. As farad is a big unit, it is usually expressed in microfarad, μF ($1 \mu F = 10^{-6}F$).

11.11. Condenser or Capacitance

A condenser consists of two conductors, one charged and the other is usually earth connected. The principle of a condenser is to increase the capacity of a conductor. The condenser is also called capacitor.

Let A be the charged conductor and B the earth connected conductor Fig. 11.6.

In the absence of B let the charge on A be $+q$ and the potential be V. The capacity of the conductor

$$= \frac{q}{V}$$

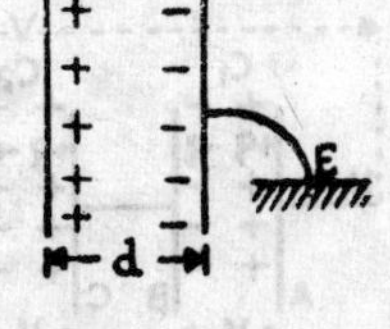

Fig. 11.6. Condenser (capacitor).

when B is kept near A, electrostatic induction takes place. As a result negative charges equal in magnitude to $+q$ will appear on the near side and positive charges equal in magnitude to $+q$ will appear on the other side of B. If B is connected to the earth, negative charges will flow from the earth to B and neutralise the positive charges on it. So only the negative charges equal in magnitude to the positive charges $+q$ remain on the conductor B. Now due to the presence of negative charges on B, the potential of A will decrease and this will result in an increase in the capacity of the conductor A. This is because, with the presence of B the amount of work done in bringing a unit positive charge from infinity decreases as there will be a force of repulsion due to A and a force of attraction due to B on the unit positive charge. The resultant force of repulsion on a unit positive charge is therefore reduced. So the amount of work done is less and the potential of A decreases. Thus the capacity of the conductor increases by bringing earth connected conductor near to it.

There are many kinds of capacitors but the parallel plate capacitor is the simplest of them. It consists of two parallel plates of any metal separated by a distance and the space between them is being filled with a dielectric. Its capacity C is given by

$$C=\frac{\varepsilon A}{d} \qquad \ldots(11.16)$$

where ε is the permittivity of the dielectric, A is the area of the conductor and d is the distance of seperation between two plates. The capacity of a capacitor is also called the capacitance of a capacitor.

11.12. Grouping of Capacitors

If a number of capacitors is given we can obtain a desired value of capacitance by joining the capacitors, either in series or in parallel or in both series and parallel. Such different combinations are discussed below.

(*a*) **Series combination.** In this arrangement the second plate of the first capacitor is joined to the first plate of the second capacitor and so on, Fig. 11.7. Here all the plates are insulated except the last one, which is earthed. If charge $+q$ is given to the plate A of the first capacitor, it induces $-q$ on the inner side of the other plate B and $+q$ goes to the first plate of

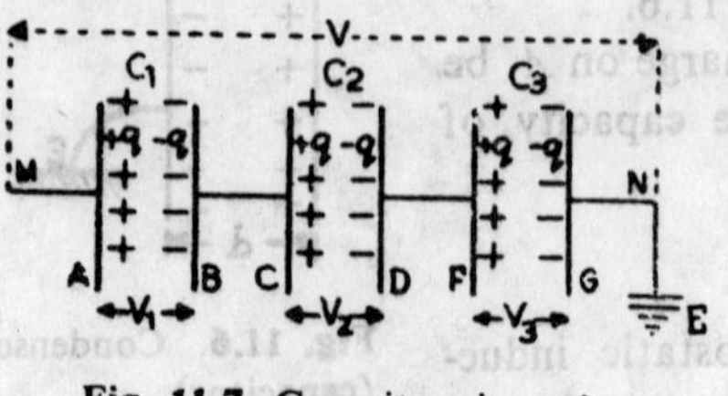

Fig. 11.7. Capacitors in series.

the capacitor C of the next capacitor. This is repeated. So each capacitor acquires $+q$ charge on one plate and $-q$ charge on the other. If V is the potential difference of the series and V_1, V_2 and V_3 the potential difference between the plates of the separate capacitors, we have,

$$V = V_1 + V_2 + V_3 \qquad \ldots(11.17)$$

Let C be the combined capacity of the system and C_1, C_2, C_3 be their individual capacitances, then $V = q/C$,

$$V_1 = q/C_1 \quad V_2 = \frac{q}{C_2}, V_3 = \frac{q}{C_3} \text{ and so on}$$

Hence from Equation 11.17,

$$\frac{q}{C} = \frac{q}{C_1} + \frac{q}{C_2} + \frac{q}{C_3} + \ldots$$

$$\frac{1}{C} = \frac{1}{C_1} + \frac{1}{C_2} + \frac{1}{C_3} \ldots \qquad \ldots(11.18)$$

Thus the reciprocal of the combined capacitance of a number of capacitors in series is the sum of the receprocals of the capacitance of the separate capacitors.

From the above result it is clear that is series combination the resultant capacitance is always less than that of individual capacitor. Such a grouping is made, when a small capacitor is sought to be made out of some large units available.

(*b*) **Parallel combination**. In this arrangement, the insulated plates are joined to common terminal M, which is connected with the source of potential and similarly, other plates are joined to another common terminal N which is earthed, Fig. 11.8. It is clear that all the capacitors being directly connected to the source and the Earth have the same potential difference V. When a charge is given at M, the charge is distributed to the capacitors according to their capacitances. If q_1, q_2, q_3 be the charges of the capacitors, the total charges,

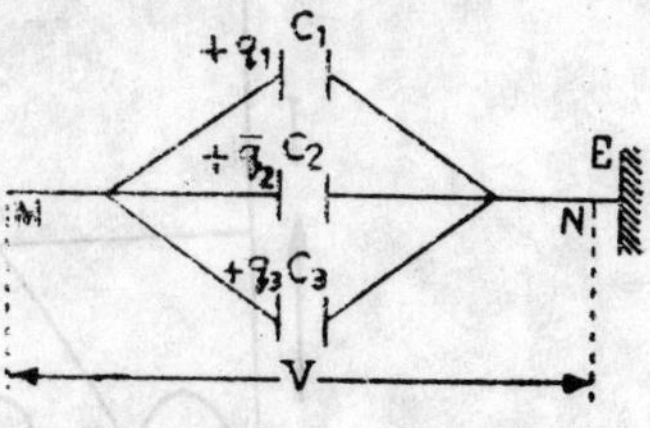

Fig. 11.8. Capacitors in parallel.

$$q = q_1 + q_2 + q_3 + \ldots \qquad \ldots(11.19)$$

If the C_1, C_2, C_3 be the individual capacitances and C is the combined capacitance we have

$$q = CV,\ q_1 = C_1V,\ q_2 = C_2V,\ q_3 = C_3V$$

From Equation 11.19,

$$VC = VC_1 + VC_2 + VC_3 = V(C_1 + C_2 + C_3)$$

Hence $$C = C_1 + C_2 + C_3 + \ldots \qquad \ldots(11.20)$$

Thus the combined capacitance of a number of capacitors in parallel is the sum of the separate capacitances.

It should be noted that this arrangement is used when a large capacitance is required to be built up out of a number of small units available.

Example 11.2. Three capacitors have capacitance 0.20, 0.30, 0.50 μf. What is their joint capacitance. When arranged to give (*a*) a minimum capacitance (*b*) a maximum capacitance.

Solution. (*a*) For minimum the combination should be in series.

If C is the minimum capacitance then,

$$\frac{1}{C}=\frac{1}{0.20\mu f}+\frac{1}{0.30\mu f}+\frac{1}{0.50\mu f}$$

$$\therefore \quad C=0.097\mu f$$

(*b*) For maximum capacitance the combination should be in parallel.

If C is the maximum capacitance,

$$C=(0.20+0.30+0.50)\mu f=1.00\mu f.$$

11.13. Current Electricity

(*a*) **Electric current.** Motion of charges constitute an electric current. This motion may be that of electrons or that of charged atoms or molecules called ions. In solid conducters current is only

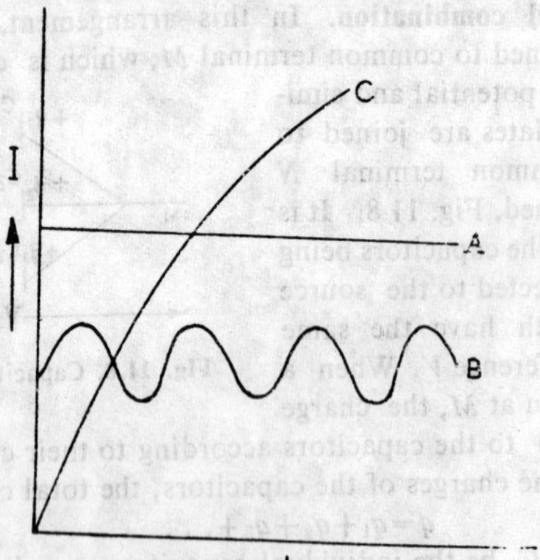

Fig. 11.9. Examples of steady and varying current.

due to the motion of electrons but in liquids it is due to positive or negative ions, where as in gases motions of both ions and electron constitute the electric current. Current is a measurable quantity and measured as the number of charges crossing any cross section of a medium in unit time. Mathematically it is expressed by

$$I=\frac{q}{t} \qquad \ldots(11.21)$$

where as I is the current, q is the quantity of charges in coulomb that flows in any cross section in time t. It is a scalar quantity and its unit is ampere in SI units. One ampere is one coulomb per second.

Current can be of two types, such as steady and varying. In steady current the rate of flow of charge remains constant with time where as in varying current the rate of flow charge does not remain constant but varies with time. To show the nature of different types of currents mentioned above graphs may be drawn between current and time, taking current in y-axis and time in x-axis. For steady current, the time-current graph is a straight line parallel to time axis, as shown in Fig. 11.9(A), whereas in varying current the graph will be curved lines, two of which are shown in Fig. 11.9 (B,C).

(*b*) **Direction of current**. Although motions of positive and negative charges constitute an electric current, the direction of motion of positive charge is conventially taken as the direction of current. However a negative charge moving in one direction is considered to be equivalent to an equal positive charge moving in the opposite direction. Therefore direction of the current is taken to be opposite to the direction of motion of negative charge. The direction of current can be considered as follows. If the positive charges move from

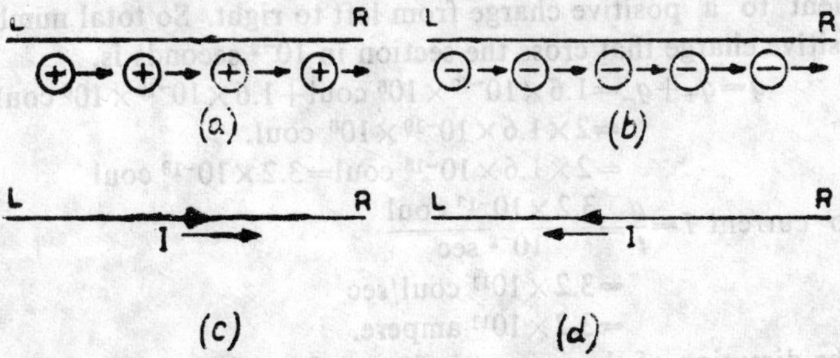

Fig. 11.10. (*a*) Positive charges in motion, (*b*) Negative charges in motion, (*c*) Current is from L to R, (*d*) Current is from R to L.

left to right, then the direction of the current is from left to right, whereas if the negative charge moves from left to right, the direction of the current is from right to left. The above facts have been shown in the Fig. 11.10.

Example 11.3. One million electrons pass from a point P towards another point Q in 10^{-3} seconds. What is the current in ampere? What is its direction.

Solution. Charge on one electron $=1.6\times10^{-19}$ coul. Charge an one million electron

$$q=1.6\times10^{-19}\times10^{6}=1.6\times10^{-13}\text{ coulomb}$$

$$t=10^{-3}\text{ second}$$

$$I=\frac{q}{t}=\frac{1.6\times10^{-13}\text{ coul}}{10^{-3}\text{ sec}}$$

$$=1.6\times10^{-10}\text{ coul/sec}$$

$$=1.6\times10^{-10}\text{ ampere}$$

Since the electrons are negatively charged, the direction of the current is from Q to P.

Example 11.4. Calculate the current in a conductor, when 1 million single charged positive ions cross a section from left to right and one million singly charged negative ion cross the same section from right to left simultaneously in 10^{-2} seconds.

Solution. A singly charged positive ion means, it has positive charge equal to the charge of an electron. If q_+ is the total positive charge of 1 million singly charged positive ion then

$$q_+=10^{6}\times1.6\times10^{-19}\text{ coul}$$

A singly negatively charged ion has a charge equal to the charge of an electron and it is negative. So if q_- is the charge of negatively charged ions then

$$q_-=1.6\times10^{-19}\times10^{6}\text{ coul}$$

We know motion of a negative charge from right to left is equivalent to a positive charge from left to right. So total number of positive charge that cross the section in 10^{-2} seconds is,

$$q=q_++q_-=1.6\times10^{-19}\times10^{6}\text{ coul}+1.6\times10^{-19}\times10^{6}\text{ coul.}$$

$$=2\times1.6\times10^{-19}\times10^{6}\text{ coul.}$$

$$=2\times1.6\times10^{-13}\text{ coul}=3.2\times10^{-13}\text{ coul}$$

So current $I=\dfrac{q}{t}=\dfrac{3.2\times10^{-13}\text{ coul}}{10^{-2}\text{ sec}}$

$$=3.2\times10^{11}\text{ coul/sec}$$

$$=3.2\times10^{11}\text{ ampere.}$$

The direction of the current is from left to right.

Example 11.5. The electron in the hydrogen atom, circles around the proton with a speed of 2.18×10^{6} m/s in an orbit of radius 5.3×10^{-11} m. Find the equivalent current.

Solution. Radius of the orbit $r=5.3\times10^{-11}$m

Circumstance of the orbit$=2\pi r=2\pi\times5.3\times10^{-11}$m.

Speed of the electron $v=2.18\times10^{6}\text{ms}^{-1}$

In one second the electron will go round the proton n times, where

$$n=\frac{v}{2\pi r}=\frac{2.18\times10^6\text{m/s}}{2\pi\times5.3\times10^{-11}\text{m}}=\frac{2.18\times10^6\text{m/s}}{2\times3.14\times5.3\times10^{-11}\text{m}}$$

The electric current is defined as the quantity of charge that flows past a fixed point in one second. Athough it is the same electron going round and round the proton, the quantity of charge that passes across a fixed point on the orbit in one second is given by ne where is the electric charge of an electron.

Hence the equivalent current

$$I=ne=\frac{2.18\times10^6\text{m/s}\times1.6\times10^{-19}\text{coul}}{2\times3.14\times5.3\times10^{-11}\text{m}}$$

$$=1.05\times10^{-3}\text{ coul/sec}=1.05\times10^{-3}\text{ ampere.}$$

11.14. Potential Difference and Flow of Charge

We know that the flow of charges in a medium is current. But charges can move from one point to other only when there exists a potential difference between the points in the medium. This is because potential difference gives rise to an electric field, which acts upon the charge to make it move. On the other hand, if there exists no potential differences between any two points in the medium, no current is possible, for, here, there exists no electric field to make the charges move. Thus we conclude that to have a current in a medium, a potential difference is absolutely necessary. Further it can he told that if current is to be made continuous, between two points a steady potential difference has to be maintained between these points, by a battery or any other device and the medium must be continuous. The medium through which charges can move is called conductor. This may be a metal, may be a liquid called electrolyte, a gas under certain condition or vacuum.

11.15. Electromotive Force (e.m.f.)

It has been told earlier that to maintain constant current in a conductor, it is necessary to maintain a steady potential difference across the conductor. This potential difference can be supplied only if some device converts some other form of energy into electric energy. Such a device is called *a source of electromotive force* (abbreviated e.m.f.).

There are many kinds of e.m.fs. In batteries chemical reactions occur transforming chemical energy into electric energy. In generators, mechanical energy is converted into electric energy. In a thermo couple, it is heat energy while in the photoelectric cell, it is radiant energy which is transformed.

When a charge q receives an energy W in passing through bettery or some other source of electric energy, the e.m.f. is given by,

$$\mathcal{E}=\frac{W}{q} \quad \ldots (11.22)$$

when W is in joules, q in coulombs then e.m.f. is in joule per coulomb, or volts.

When a charge of one coulomb receives one joule of energy upon passing through a source, the source is said to have an electromotive force of one volt. A 12 volt battery delivers 121 J of energy to each coulomb which passes through it. Electromotive force and potential difference are measured in the same units. An e.m.f. is a particular kind of potential difference namely one which arises through the transformation of some other form of energy into electric energy. In contrast, the potential difference between any two points in an electrostatic field is not an electromotive force and cannot be used to maintain steady current in a conductor connecting these points.

11.16. Ohm's Law

In 1826 George Simon Ohm discovered that for all metallic conductors the potential difference between the ends of the conductor is directly proportional to the current provided other quantities like pressure, temperature and tension etc. are kept constant.

Mathematically, $V \propto I$

or $\quad V=RI \quad \ldots (11.23)$

where V is the potential difference between the two ends of the conductor and I is the current through conductor and R is the proportionality constant called resistance. If I is expressed in ampere, potential difference in volts then R is expressed in ohms. It has been found that Ohm's law is valid for a wide range of currents in metallic conductor. However it is not valid for currentts in all conducting medium. For example in some conductors like electronic tube, arcs, ionic conductors, there is not a direct proportion between V and I and I may actually decrease as V is increased.

Some of these cases are depricted in Fig. 11.11 as plotting graph between potential difference and current.

From above graphs, it is clear that for all metallic conductors V-I curve is a straight line *i.e.*, resistance of the conductor is independent of potential difference. In this case, we say, the conductors obey Ohm's law. All such conductors for which V-I curve is a straight line is called ohmic resistors. Others for which *V*-curve is not a straight line, are called non-ohmic resistors.

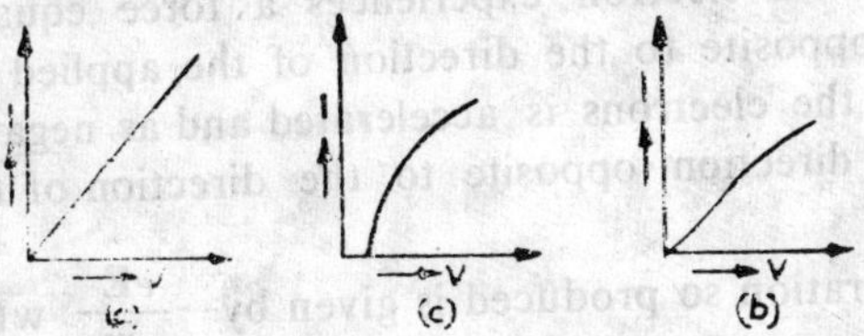

Fig. 11.11. *V-I* plots for, (*a*) Metallic conductor, (*b*) Vacuum tube, (*c*) Electrotype.

11.17. Current in Metallic Conductor

Metals are the best conductors of electricity and they obey Ohm's law perfectly well. Therefore it is desirable to discuss the mechanism of currents in them as to understand Ohm's law and furnish an explanation to the observed electrical effects such as cause of resistance, heating effects, etc.

Metals consist of atoms, which are very closely packed. These atoms are arranged in fixed points called lattice points and they exert electrical influence on each other through the electrons orbiting in them. Due to the action of nearby atoms, some electrons lose their association with particular atom and become free electrons as a result of this the atoms fall short of electrons and become positively charged called positive ions. The number of electrons so freed are enormously large and approximately equal to 10^{28} per cubic meter. They possess thermal energy, hence thermal motion. They move about within lattice, with average speed equal to $10^{-5} ms^{-1}$, at room temperature. While in motion they collide constantly with ionic core of the conductor, that is, they interact with lattice often suffering sudden changes in speed and direction, much like gas molecules in a container. So their velocities are not ordered rather randomly distributed in all directions. The average velocity of electrons is given by,

$$\vec{U} = \frac{\vec{U}_1+\vec{U}_2+\vec{U}_3+\ldots+\vec{U}_N}{N}$$

$$=0 \qquad \ldots (11.24)$$

where $\vec{U}_1, \vec{U}_2, \vec{U}_3 \ldots \vec{U}_N$ are volocities of individual electrons and N the total number of free electrons. Thus we see there is no net flow of charge in any direction and hence no current although in the metal the electrons move with non-zero speed.

When a potential difference is applied across the conductors an electric field $\vec{E}$ is established inside it. This field acts on the electrons and each electron experiences a force equal to $-e\vec{E}$ in a direction opposite to the direction of the applied field. Hence the motion of the electrons is accelerated and as negative charges they move in a direction opposite to the direction of the applied field.

The acceleration so produced is given by—$\frac{e\vec{E}}{m}$ where e and m are charge and mass of the electron respectively. The positive ions also experience this force. But they cannot move as they are heavy and tightly fixed to the lattice.

Thus we see that the motion of the electron is modified by the application of electric field and between two successive collision electron acquires an additional velocity in addition to the thermal velocity in a direction opposite to the direction of the field. However this velocity is very small, and is lost in the next collision with ionic core at lattice points.

If we consider the motions of the electrons at an orbitrary instant, we could see that electron with thermal velocity $\vec{U_1}$ is moving with a velocity equal to $\vec{U_1}+\vec{a}t_1$ where $\vec{a}t_1$ is the velocity aquired due to the applied electric field $\vec{E}$, $\vec{a}$, being the acceleration and t being the time that has elapsed since the last collision. Similarly the motion of all other electrons will change and their velocities are given by $\vec{U_2}+\vec{a}t_2$, $\vec{U_3}+\vec{a}t_3$ . . . . . . $\vec{U_n}+\vec{a}t_n$, respectively. Where t_2, t_3 . . . . . . t_n are time since last collision with ionic core. The average velocity of all the electrons at this instant is given by

$$\vec{v}=\frac{(\vec{U_1}+\vec{a}t_1)+(\vec{U_2}+\vec{a}t_2)+(\vec{U_3}+\vec{a}t_3)+\ldots+(\vec{U_N}+\vec{a}t_N)}{N}$$

$$=\frac{\vec{U_1}+\vec{U_2}+\vec{U_3}+\ldots+\vec{U_N}}{N}+\frac{\vec{a}(t_1+t_2+t_3+\ldots+t_n)}{N}$$

$$=0+\vec{a}\tau, \text{ where } \tau=\frac{t_1+t_2+t_3+\ldots t_n}{N} \qquad \ldots (11.25)$$

Where τ represents the average time elapsed since each electron suffered its last collision. τ is called the relaxation time and its value is order of 10^{-14} seconds for pure metals. It is to be noted here that relaxation time varies from metal to metal and is also temperature dependent.

Now putting the value of acceleration $\vec{a}$ in the Equation 11.25, we have,

$$\vec{v} = -\frac{e\vec{E}\tau}{m} \quad \ldots(11.26)$$

Thus we see that before the application of the field the average velocity of the electrons is zero and there is no net transportation of charge. But after the application of the field electrons acquire a constant average velocity equal to $\frac{e\vec{E}\tau}{m}$ in a direction opposite to $\vec{E}$ and there is a net transport of negative charge in that direction. This constant average velocity is called drift velocity ($\vec{v}_d$) and a typical value of it is 1 mm/sec, which is very small compared to the random thermal velocity which is equal to 10^5 m/sec. It should be noted that although the drift velocity is small, the number of electrons that cross any cross section of the conductor are so large, that we get a sizeable amount of current, even with a small potential difference.

11.18. Speed of Propagation of Electrical Action

Whenever we switched on an electric current an electric bulb at a distance say 100 meters, away lights up almost in no time. Hence the question arises, if the electrons drift so slowly then how is it that, an electric bulb at such a large distance, lights up so soon? In order to answer this question let us first consider an example, of long water filled tube, to one end of which a pressure is applied. As soon as the pressure is applied pressure wave or the disturbance travels rapidly along the tube and when it reaches the other end, the flow of water starts. The water inside the tube also starts moving forward but the speed at which water moves through the tube is much lower than the speed of pressure waves. Somewhat similar phenomena occur in case of electrical conductor carrying electric current. In a conductor free electrons are present every where, when a potential difference (by switching on) is applied to a circuit, an electric field gets established throughout the circuit almost with the speed of light, equal to 3×10^8 m/sec. As soon as the electric field is established, the elctrons begin to drift every where under the influence of the electric field and current is established. It is for this reason, it takes almost no time for an electric bulb to glew when a switch in an electric circuit is switched on. Thus we conclude that it is the electric impulse (field) that travels with a velocity, which approaches velocity of light but not the electrons. However the electrons move slowly which is the order of 1 mm/sec called drift velocity.

11.19(a). Resistance of a Conductor

Consider a conductor PQ of length l and of uniform cross sectional area A, across which a potential difference V is applied Fig.

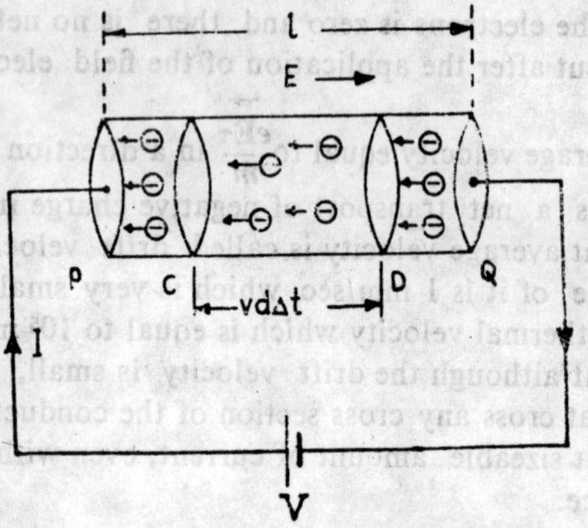

Fig. 11.12. Flow of current in a conductor.

11.12. As a result, an electric field $\vec{E}$ is established towards right and electrons start drifting towards left. Suppose there are n electrons per unit volume, all moving with the drift velocity v_d. In a small time interval Δt each advances a distance $v_d\Delta t$.

The number of electrons crossing any section such as C in Δt is equal to the number contained in a portion of the conductor of length $v_d\Delta t$ and the volume $v_dA\Delta t$. This number is $nv_dA\Delta t$ and the total charge crossing the section is

$$\Delta q = nev_dA\Delta t \qquad \ldots(11.27)$$

where e is the charge of an electron.

The magnitude of current is therefore given by,

$$I = \frac{\Delta q}{\Delta t} = neAv_d \qquad \ldots(11.28)$$

Using Equation 11.26 for the value of v_d, we get,

$$I = enA\,\frac{eE\tau}{m} \qquad \ldots(11.29)$$

or $$I = \left(\frac{e^2nA\tau}{m}\right)E = \frac{e^2nA\tau}{m}\,\frac{V}{l} \text{ as } E = \frac{V}{l} \text{ (the potential gradient)}$$

or $$I = \frac{e^2nA\tau}{ml}\,V \qquad \ldots(11.30)$$

Hence resistance of the conductor $$R = \frac{V}{I} = \frac{ml}{e^2nA\tau} \qquad \ldots(11.31)$$

On the right of Equation 11.31 all the quantities for a given conductor of length l and area of cross section A, are constants. Therefore R is constant. This explains Ohm's law. We can write, Equation 11.31 as

$$R=\rho\frac{l}{A} \quad \ldots(11.32)$$

where $\rho=m/e^2n\tau$ is called resistivity or specific resistance of the material of the conductor. At a given temperature, pressure and tension etc., the resistivity is a constant of the material of the conductor and its units are ohm- meter.

11.19(b). Conductivity

The inverse of resistivity is known as conductivity and is usually denoted, σ

Thus $$\sigma=\frac{l}{\rho}=\frac{e^2n\tau}{m} \quad \ldots(11.33)$$

In SI units, the conductivity is expressed in $ohm^{-1}\ m^{-1}$.

Example 11.6. Calculate the drift velocity of free electrons in a copper conductor of cross sectional area $10^{-4}\ m^2$ and in which there is a current 200 amperes. Assume that there are 8.5×10^{28} free electrons per cubic meter of copper.

Solution. $I=enAv_d$

The charge of electron,

$$e=1.6\times10^{-19}\ \text{coul},\ n=8.5\times10^{28}/m^3$$

$$A=10^{-4}m^2,\ I=200\ \text{ampere}.$$

$$v_d=\frac{I}{enA}=\frac{200\ \text{coul/sec}}{1.6\times10^{-19}\ \text{coul}\times8.5\times10^{28}/m^3\times10^{-4}m^2}$$

$$=1.5\times10^{-4}m/sec.$$

Example 11.7. Resistance of a copper wire of length 1 m is 0.082 ohms at 20°C. If the diameter of the wire is 0.511 mm. What is the resistivity of copper at this temperature?

Solution. The resistivity $\rho=\frac{RA}{l}$

$$=\frac{0.082\ \text{ohms}\times\pi\left(\frac{0.511}{2}\right)^2\times10^{-6}m^2}{lm}$$

$$=\pi\,8.42\times\frac{(0.511)^2}{4}\times10^{-8}\Omega\text{-m}$$

$$=1.73\times10^{-8}\Omega\text{-m}.$$

11.20. Heating Effect of Electric Current and Joule's Law of Heating

Whenever there is an electric current in a conductor, the conductor gets heated up. This fact can be explained on the basis of microscopic picture of the structure of the conductor. We have seen that when a potential difference is applied across a conductor the free electrons under the action of the field move from one end to the other end. During their motion they frequently collide with ionic cores at lattice points and their motion is frequently hampered. This is considered to be the basic cause of resistance in a conductor. At the time of collision the electrons give up to the ionic core whatever kinetic energy they again on motion due to the applied electric field. Thus the ionic core which is already in a state vibration due to thermal agitation acquires more energy and its amplitude of vibration increases. This leads to the increase in the average internal energy of vibration of ionic core. As we know the internal energy of vibration is directly proportional to the temperature, the temperature of the conductor rises up.

Let us now derive an expression for the rate of heat development due to a current in a conductor. Consider the portion *AB* of a circuit shown in Fig. 11.13 in which there is a conventional current *I* from *A* to *B*. Let V_A and V_B be the potentials at points *A* and *B* and $V_A > V_B$. Suppose, the charge Δq passing from *A* to *B* in time Δt constitute the current. The charge Δq entering at *A*, acquires potential energy equal to $\Delta q V_A$ and leaves the end *B* with potential energy $\Delta q V_B$. But we know that V_A is greater than V_B. So there is a loss of potential energy equal to $(\Delta q V_A - \Delta q V_B)$ in the passage from *A* to *B*. This loss in potential energy infact appears as heat energy in a conductor. If ΔW is the potential energy lost by the charge Δq in moving from *A* and *B*. Then we have,

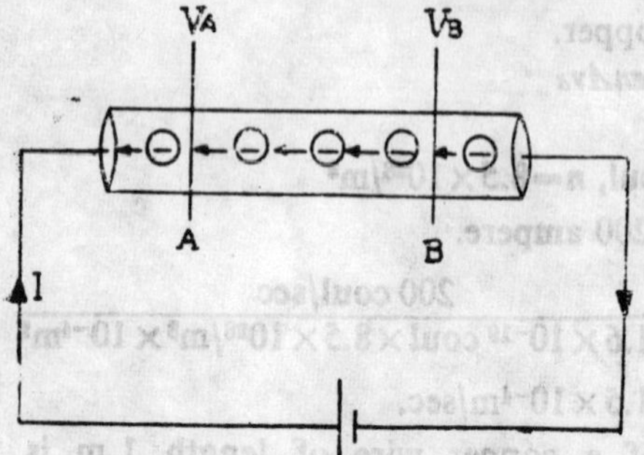

Fig. 11.13. For calculation of power dissipation in a conductor.

$$\Delta W = (\Delta q V_A - \Delta q V_B)$$
$$= \Delta q (V_A - V_B) \quad \ldots (11.34)$$

Therefore the rate at which the energy is converted into heat *i.e.* power *P* is given by

$$P=\frac{\Delta W}{\Delta t}=\frac{\Delta q}{\Delta t}(V_A-V_B)$$
$$=I(V_A-V_B)=IV_{AB} \quad \ldots (11.35)$$

where $\qquad V_{AB}=V_A-V_B.$

Since all metals obey Ohm's law, $V_{AB}=IR$
where R is the resistance of the AB portion of the conductor. Then we have,

$$P=I.IR=I^2R \quad \ldots (11.36)$$

This is known as Joule's law of heating and was discovered by Joule in course of his determination of mechanical equivalent of heat.

11.21. Units of Electric Power and Energy

(*a*) If current is expressed in ampere or coul/sec, and potential difference is in volts or Joule/coul the power is expressed in Joule/sec or watts. Since power=current × voltage,

$$1 \text{ amp} \times 1 \text{ volt}=\frac{1 \text{ coul}}{\text{sec}}\times\frac{1 \text{ Joul}}{\text{coul}}=\frac{1 \text{ Joul}}{\text{sec}}=1 \text{ watt}$$

(*b*) Heat energy Q developed in time t is given by $Q=Pt=I^2Rt$ Joules.

$$=\frac{I^2Rt}{4.2} \text{ calories.}$$

Since 1 calorie=4.2 Joules.

(*c*) **Unit of electric energy—the killowatt hour.** If P is the power in watt and t is the time in seconds, the energy consumed is

$$W=P \text{ watt} \times t \text{ sec.}$$
$$=Pt \text{ watt} \times \text{sec}=Pt \text{ Joule.}$$

For calculation of consumption of electrical energy Joule is considered to be too small a unit. Therefore calculation monthly electricity bill killowatt hour, a bigger unit is introduced. This is defined by the equations $W=Pt$.

Since $\qquad 1 \text{ kw}=100 \text{ watts and}$

$$1 \text{ hour}=3600 \text{ seconds}$$
$$1 \text{ killowatt. hour}=1000 \text{ watt}\times 3600 \text{ sec}$$
$$=1000\times 3600 \text{ watt}\times\text{sec}$$
$$=3.6\times 10^6 \text{ Joules.}$$

(*d*) **Cost of electric energy.** The cost of electric energy is given by the equation as,

$$\text{cost}=P\times t\times \text{unit cost}$$
$$=\frac{VIt\times \text{cost per kw-hr}}{1000 \text{ watts/kw}}$$

where V is expressed in volts, I in amperes and t in hours.

Example 11.8. What is the cost of operation for 24 hour a lamp requiring 1.0 amp on a 100 volt line if the cost of energy is Rs 0.50/kw.hr.

Solution.
$$\text{Cost}=\frac{100\ \text{volt}\times 1\ \text{amp}\times 24\ \text{hr}\times \text{Rs}\ 0.50/\text{kw-hr}}{1000\ \text{watts/kw}}$$
$$=\text{Rs}\ 1.20.$$

Example 11.9. An electric bulb is marked 200 watt, 220 volt. Find (*a*) current it draws, (*b*) its resistance.

Solution. Power of the bulb=200 watt

Voltage at which it gives this power=220 volt

$$\text{Power}=V\times I$$
$$\text{Current}=\frac{P}{V}=\frac{200\ \text{watt}}{220\ \text{volt}}=\frac{20}{22}\text{A}=0.909\ \text{amp}.$$
$$\text{Resistance}=\frac{V}{I}=\frac{220\ \text{volts}}{0.909\ \text{A}}=202.02\ \text{ohms}.$$

11.22. Variation of Resistance with Temperature

The electrical resistance of a metal increases when it is heated. This phenomenon can easily be explained on the basis of microscopic picture of the structure of the metals. We have considered that the atoms in ionic state occupy fixed positions in metals. They however are not at rest but vibrate to and fro about their fixed positions. When the metal is heated the atoms in it absorb energy and their kinetic energy increases. As a result, the atoms vibrates more vigorously with larger amplitude. Further at higher temperature the speed of the electrons becomes more also. Both of these facts lead to the frequent collision of electrons with the atoms and a consequent decrease in relaxation time τ. The decrease in relaxation time always results in the increase in the resistance. Therefore resistance increases.

Experiments have shown that for a moderate temperature range, the change of resistance with temperature of metallic conductors can be approximately represented by the equation.

$$R_t=R_o+R_o\alpha\, dt=R_o(1+\alpha t) \qquad \ldots(11.35)$$

where R_t is the resistance at temperature t°C, R is the resistance at 0°C and α is a quantity characteristic of the substance known as that temperature co-efficient of resistance. Moreover as resistivity changes with temperature resistivity at any temperature t is also given by

$$\rho_t=\rho_o(1+\alpha t) \qquad \ldots(11.36)$$

From the Equation 11.35,

$$\alpha=\frac{R_t-R_o}{R_o t} \qquad \ldots(11.37)$$

Since (R_t-R_o) and R_o have the same units, their units will cancel in the fraction of Equation 11.37. Hence the unit of α depends upon the unit of temperature and as such given by $°C^{-1}$. While computing the value α, the base temperature is referred to 0°C. If the base temperature is referred to some other value, than the value of α will be different. Therefore while computing the value of α at any temperature the value of R_o at 0°C has to be always considered.

The temperature co-efficient of resistance is positive for all metals. For most alloys α is much smaller than for metals. Indeed for some alloys e.g. constantan and manganin, the temperature co-efficients of resistance are very near zero. For this reason such alloys are often used for resistors, when it is desirable that the resistance be independent of temperature. Carbon and other semi-conductors give a negative co-efficients *i.e.* their resistance decreases with rise of temperature. However it is beyond the scope of this book to find out an explanation for such negative value of α.

The fact that the resistance increases with temperature is used to measure the temperature of a heat bath. The material used for this purpose is platinum and the instrument by which temperature is measured is called platinum resistance thermometer. When such thermometer is properly callibrated, it gives a high precision in the measurement of temperature over a wider range.

Example 11.10. A tungsten filament has a resistance of 133 ohms at 150°C. If $\alpha=0.00450/C°$, what is the resistance of the filament at 500°C?

Solution. The resistance at temperature t is given by

$$R_t=R_o\ (1+\alpha_t)$$

$$\text{or}\quad R_o=\frac{R_t}{1+\alpha t}=\frac{133 \text{ ohms}}{(1+0.00450/°C\times 150°C)}$$

$$=79.4 \text{ ohms}$$

$$R_{500}=R_o\ (1+\alpha t_{500})$$

$$=79.4\ (1+0.00450/°C\times 500°C)$$

$$=258 \text{ ohms.}$$

11.23. Thermoelectric Effect

If two wires of dissimilar metals are joined at the ends to make a closed circuit and one of the junctions thus formed is maintained at a temperature different from that of the other, an electric current is established in the circuit Fig. 11.14. The current thus established is called the thermoelectric current and the phenomenon is called the thermoelectric effect. This phenomena is also called, seeback effect after the name of seeback, who discovered it first.

Seeback investigated this effect for a number of metals in combination and found that the direction of the current at hot or cold junction depends upon the nature of metals used. He then arranged the metals in a series such that when any two of them form a circuit, current is established at cold junction, from the metal occurring earlier in the series to the one occurring later. The series is known as seeback series and some of them are given below.

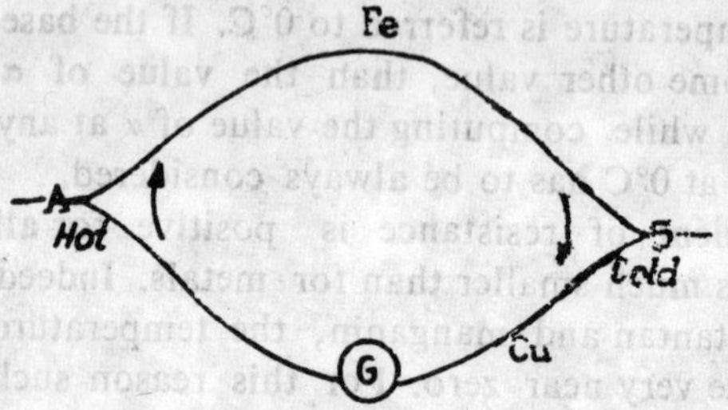

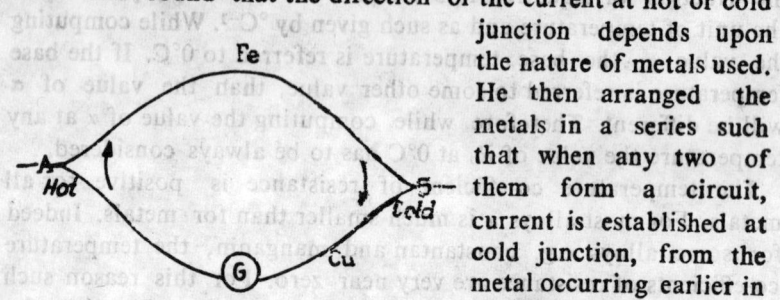

Fig. 11.14. A thermo couple of Cu-Fe. A is the hot junction and *B* is the cold junction. The direction of flow of current is given by arrow shown.

1. Antimony, 2. Iron, 3. Silver, 4. Gold, 5. Lead, 6. Copper. 7. Platinum, 8. Nickel, 9. Bismuth.

Thus from the above list, it can be seen that in the iron-copper combination, current is established from iron to copper at cold junction and from copper to iron at hot junction, Fig. 11.14.

The combination of two different metals to get thermoelectric current is called thermoelectric couple or simply thermocouple. In the thermocouple, current is established due to an e.m.f., which arises on account of temperature difference, between ends of it. This e.m.f. is known as thermo e.m.f. The thermo e.m.f. can be accurately measured by potentiometer.

Variation of thermo e.m.f. with the temperature. It has been found that the thermo e.m.f. varies, when the temperature of the junctions are changed. To study such variations, usually one of the junctions is kept at 0°C and the other one is heated. The thermo e.m.f. generated is measured by a potentiometer. A graph is then plotted between the temperature difference and thermo e.m.f. measured. The nature of the graph so obtained is shown in Fig. 11.15.

As the temperature increases, the thermo e.m.f. goes on increasing. At a temperature t_1°C, it reaches a maximum value and on further increases temperature e.m.f. starts decreasing and attains a zero value at temperature t_2°C. On further heating it is found that the e.m.f. reverses in direction and similar plot is obtained in the reverse direction.

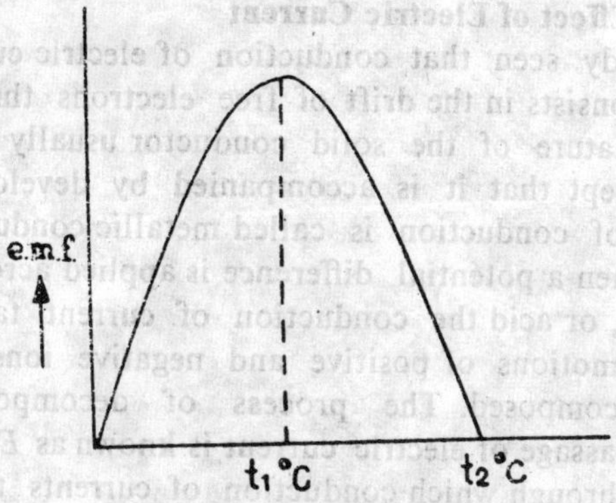

Fig. 11.15. Variation therme e.m.f. with temperature

The temperature at which the maximum e.m.f. is obtained is known as neutral temperature and the temperature at which e.m.f. reverses in direction is called inversion temperature. The inversion temperature is not constant but depends upon the temperature of the cold junction. In the above graph t_1°C represents the neutral temperature and t_2°C represents the inversion temperature.

11.24. Thermo Couple and Temperature Measurement

Thermo couples are very convenient for measuring temperature, since the e.m.f. depends on the temperature difference between the junctions in a known way. Ordinarily one junction is kept a constant temperature often in an ice bath at 0°C and the other, on the body, whose temperature is to be measured. Due to smallness, the junction can be installed at some inaccessible points, where it would be impossible to place and read a standard mercury thermometer. A galvanometer may be included in the circuit to read the temperature directly if it is calibrated suitably. The thermo couples can be designed to measure wide range of temperature of the small region in a *cavity* where other conventional methods of measurement fail. The e.m.f. involved in the thermocouples are usually small. To obtain higher e.m.f., we can connect a number of thermocouples in series thus producing a thermopile. Thermopiles are often used for measuring radiant energy (Thermopile has been discussed in radiation chapter *i.e.* in Art. 10.24).

11.25. Chemical Effect of Electric Current

We have already seen that conduction of electric current in a solid conductor consists in the drift of free electrons through the conductor. The nature of the solid conductor usually undergoes a little change except that it is accompanied by development of heat. This type of conduction is called metallic conduction. On the other hand when a potential difference is applied across a solution of salts, base, or acid the conduction of current takes place by simultaneous motions of positive and negative ions, and the solution gets decomposed. The process of decomposition of liquids, by the passage of electric current is known as *Electrolysis* and the solution through which conduction of currents take place is known as *Electrolyte*. The conductors by which the current enters or leaves the electrolyte are called electrodes, through which enters being the *anode* or positive electrode and that by which it leaves, the *cathode* or negative electrode. The apparatus containing the electrolytic solution is called a *voltameter* or electrolytic cell. The products of the decomposition are liberated at the electrodes. That part which is liberated at the cathode is called *cation* and that liberated at the anode is called *anion*.

11.26. Electrolysis

When two copper electrodes are inserted into a beaker filled with pure water and connected to the terminals of a battery an *ammeter* in the circuit shows no current. But if a little copper sulphate ($CuSO_4$) is dropped into the beaker there is a current through the solution. This indicates that pure water is not basically conductor of electricity, but when some salt is dissolved in it, it becomes conducting. The mechanism of this phenomenon can be understood as follows. When copper sulphate crystal is dropped into water, it breaks up into two charged particles called ions. One is a copper atom from which two electrodes are missing and second is a sulphate ion (SO_4^{--}) with two excess electrons. The net charge in atomic units carried by an ion is called valence of an ion. Thus the valence of Cu^{++} ion is +2 while that of sulphate ion is —2. In the solution the positively charged copper ion migrates to the cathode (negatively charged electrode) and negatively charged sulphate ion into the anode (positively charged electrode). Copper ions on reaching the cathode takes two electrons there and is deposited as neutral copper atom. On the other hand when sulphate ion (SO_4^{--}) reaches the anode it gives two of its electrons to the anode. The anode being positive readily accepts them. The sulphate ions (SO_4^{--}) after losing two excess electrons becomes SO_4 radical, which cannot have independent existence. Hence it joins with an atom of copper to form $CuSO_4$ molecule and goes into solution, keeping the amount of $CuSO_4$ constant. The net effect is gain of copper by the cathode and loss of copper by the anode. Thus we see there has not only been transfer of electricity through the solution but copper has been carried from one plate to another.

Similarly water can also be made conducting by adding dilute acids to it. During the conduction process water breaks up into two of constituents *i.g.* hydrogen and oxygen. The volume of hydrogen released in the process is twice that of oxygen. A simplified explanation of the process by which water is decomposed can be given as follows. A sulphuric acid molecule in solution splits up into one H^+ and one HSO^-_4 ion. The hydrogen ions is attracted to the cathode where it receives an electron to form a neutral hydrogen atom. Then two of such atoms promptly form a hydrogen molecule which rises up and can be collected. An HSO^-_4 ion gives an electron at the anode and then unites with an atom of hydrogen to form sulphuric acid. These atoms of hydrogen are taken from the water and oxygen is set free. Two atoms of oxygen unite to form a molecule of oxygen gas.

The sulphuric acid formed at the anode goes into solution. The amount of sulphuric acid in the water does not change. The net result of passing current through the liquid is to decompose water into hydrogen and oxygen as shown below.

$$2H_2O \longrightarrow 2H_2 + O_2$$

From above discussion it is clear that when potential difference is applied across a conduction solution, not only electricity is transferred but also the solution chemically decomposes into its constituents. The products of decomposition are deposited at the electrodes.

11.27. Some Terms Used in Electrolysis

(*a*) In electrolysis it is found that metals and hydrogen always travel with the current and they are therefore called **electropositive,** where acid radicals and non-metals travel against the current and go to the anode and are called ***electronegative.***

(*b*) **Valency.** Valency of an element is given by number of hydrogen atoms which will combine with or replaced by one atom of the element.

(*c*) **The chemical equivalent.** Chemical equivalent is numerically equal to its atomic weight divided by valency.

(*d*) **Kilogram equivalent.** Kilogram equivalent of a substance is the quantity of the substance equal to its chemical equivalent expressed in kilogram.

(*e*) **Gram equivalent.** Chemical equivalent expressed in gram is called gram equivalent. The atomic weight of silver is 108 and its valency is 1. So the chemical equivalent of silver is 108/1, *i.e.* 108 and the gram equivalent is 108 grams. Kilogram equivalent of silver is 108 kilograms.

11.28. Faraday's Laws of Electrolysis

Faraday conducted a series of experiments on electrolysis and found a striking relationship between the electrolytic behaviour and chemical properties of various substance. From his experimental findings, he announced the following two laws.

(*a*) **First law.** The mass of any substance liberated in electrolysis is proportional to the quantity of electricity that passes. Hence the mass liberated is proportional to the product of the current and time.

Let m kg of substance is liberated due to the passage of I amperes of currents that flows for time t seconds. If Q is the quantity of charge then,

$$Q=It$$

According to first law of Faraday,

$$m\propto Q$$

$$m=ZQ \qquad \ldots(11.38)$$

where Z is a constant and is called the electrochemical equivalent (e.c.e.) of the substance, under consideration. In the Equation 11.38 if Q is one coulomb, then m is numerically equal to Z. Therefore we say that electrochemical equivalent of a substance is the mass of the substance liberated when one coulomb of charge passes through the electrolyte. The value of e.c.e. of copper is 0.329×10^{-6} kg/(coul)$^{-1}$ and that of silver is 1.118×10^{-6}kg (coul)$^{-1}$.

As $Q=It$, then Equation (11.38) can be written as

$$m=ZIt \qquad \ldots(11.39)$$

Both the Equations 11.38 and 11.39 are the mathematical form of Faraday's first law of electrolysis.

(*b*) **Second law.** The masses of different substances liberated by a given charge are proportional to their chemical equivalents. (Chemical equivalent=Atomic weight divided by valency). Mathematically,

$$\frac{m_1}{m_2}=\frac{C_1}{C_2} \qquad \ldots(11.40)$$

where m_1 and m_2 are the masses of two substances liberated by the passage of same quantity of electric charge and C_1 and C_2 are their chemical equivalents. If m_a and m_b be the masses of two elements *A* and *B* of e.c.e. Z_a, Z_b and chemical equivalents C_a, C_b liberated by the same current in same time, according to first law,

$$m_a=Z_aIt \qquad \ldots(i)$$

$$m_b=Z_bIt \qquad \ldots(ii)$$

Dividing quantity in equation (*i*) by the quantity in equation (*ii*), we get,

$$\frac{m_a}{m_b}=\frac{Z_a}{Z_b} \qquad \ldots(iii)$$

But by second law, we have,

$$\frac{m_a}{m_b}=\frac{C_a}{C_b} \qquad \ldots(11.41)$$

Therefore

$$\frac{Z_a}{Z_b}=\frac{C_a}{C_b} \qquad \ldots(11.41)$$

That is the ratio of electrochemical equivalents is equal to the ratio of chemical equivalents of the substances.

Example 11.11. How long will it take to electroplate 6.00 gram of silver on to a brass costing by the use of steady current of 15 amp? The electrochemical equivalent of silver is 1.1180×10^{-6} kg/coul.

Solution. $m=ZIt$

$$6.00\times10^{-3}\text{kg}=(1.1180\times10^{-6}\text{ kg/coul})\times(15.0\text{ amp})\times t$$

$$t=358\text{ seconds.}$$

Example 11.12. A circuit consists of a solution of silver salt and a coil of wire resistance 20 ohms immersed in an oil bath in series. Constant current flows for 10 seconds and deposits 0.0279 grams of silver. Calculate how much heat energy is developed in the oil bath (e.c.e.) of silver is 1.1183×10^{-6} kg/coul.

Solution. From relation,

$$m=ZIt, \text{ we have, } I=\frac{0.0279\times10^{-3}\text{ kg}}{1.1183\times10^{-6}\text{ kg/coul}\times10\text{ sec}}$$

$$=\frac{0.0279\times10^{3}\text{ kg}}{1.1183\text{ kg/coul}\times10\text{ sec}}$$

$$=2.49\text{ amperes.}$$

The amount of heat energy developed $=I^2Rt$ Joules.

$$=(2.49 \text{ amp})^2 \times 20 \text{ ohm} \times 10 \text{ sec}$$
$$=(2.49)^2 \times 20 \times 10 \times (\text{amp})^2 \times \text{ohm} \times \text{sec}$$
$$=(2.49)^2 \times 20 \times 10 \times \text{amp} \times \text{ohm} \times \text{sec} \times \text{amp}$$
$$=295.2 \text{ coul} \times \text{volt}$$
$$=295.2 \text{ Jouls.}$$

Example 11.13. Calculate the charge necessary to deposit one kilogram equivalent of silver.

Solution. The mass m deposited by charge Q is given by

$$m=QZ$$

$$Q=\frac{m}{Z}$$

The kilogram equivalent of silver is 108 kg.

The electrochemical equivalent silver $=1.1180 \times 10^{-6}$ kg/coul. From the definition of electrochemical equivalent,

$\therefore$ 1.1180×10^{-6} kg of Ag is deposited by 1 coul charge

$\therefore$ 1 „ „ „ $\dfrac{1}{1.1180 \times 10^{-6}}$ coul

$\therefore$ 108 „ „ „ $\dfrac{1 \times 108}{1.1180 \times 10^{-6}}$ coul

$$=9.6487 \times 10^7 \text{ coul.}$$

Thus we see 9.6487×10^7 coul is necessary to deposit one kilogram equivalent of Ag. But from second law of Faraday this amount of charge, when passed through any electrolyte, will deposite the mass in equal to chemical equivalent. Therefore it can be concluded that 9.6487×10^7 coul is necessary to deposite one kilogram equivalent of any substance.

11.29. Faraday

From above it is seen that 9.6487×10^7 coul of charge is necessary to deposit a kilogram equivalent of any substance. However if we consider one gram equivalent or 1/1000th of a kilogram equivalent, then charge necessary to deposit this amount of any substance will be equal to 9.6487×10^4 coul. This amount of charge *i.e.* 9.6487×10^4 coul is called one Faraday. Thus we define Faraday as, the charge in coulomb, which is necessary to deposit a mass equal to 1/1000th one kilogram equivalent.

11.30. Faraday Constant

It has been found the electrochemical equivalent is proportional to chemical equivalent, *i.e.*

$$Z \propto C$$

where C is the chemical equivalent

$$\therefore \quad Z=\frac{1}{F_c}\,C \qquad \ldots(11.42)$$

where F_c is a constant, called Faraday constant.

$$F_c=\frac{C}{Z} \qquad \ldots(11.43)$$

Putting the value of Z from Equation 11.38 in the Equation 11.43 we get,

$$F_c=\frac{C}{\frac{m}{Q}}=\frac{QC}{m} \qquad \ldots(11.44)$$

Now if m is taken to be 1 kilogram equivalent then $m=C$ kg and $Q=9.6487\times10^7$ coul.

$$F_c=\frac{9.6487\times10^7 \text{ coul}\times C}{C \text{ kg}}$$

$$F_c=9.6487\times10^7 \text{ coul/kg} \qquad \ldots(11.45)$$

Thus value of Faraday constant $=9.6487\times10^7$ coul/kg.

11.31. Magnetic Action of Currents

When a compass needle is brought near a magnet, the compass needle gets deflected. But when it is brought near a charged body at rest it shows no deflection. This happens, because, a magnet has a magnetic field around it. The compass needle is being a magnet is acted upon by it and gets deflected. But on the other hand, a charged body at rest has an electric field around it. The electric field has no action on a magnet. Therefore compass needle experiences, no force near a charged body and hence does not deflect.

Let us consider a compass needle which is put near a current carrying conductor connected to a battery. Soon after the circuit is closed, a current is established and the compass needle shows deflection. When the current is switched off, the compass needle returns to its original position and shows no deflection. This phenomenon is not only seen with currents in a conductor but also with charges moving freely in space. Therefore, it is believed that charges, whether moving in a conductor or moving freely in space, set up magnetic field in space and possess magnetic action in addition to the ever present electrostatic action. For this reason, the compass needle shows deflection on being acted upon by the magnetic field of the current which is nothing but the charges in motion.

11.32. Magnetic Field and the Moving Charges

Experiments have shown that charges in motion experience a force in a magnetic field. This fact has been used to define a magnetic field. We define the magnetic field is a region in which a moving charge experiences a force in addition to the electrostatic force. This force depends upon the quantity of charge, the speed of the moving charge, the direction of motion and some property of the magnetic field. This property of the field is called the magnetic induction or magnetic field and is usually denoted by $\vec{\mathbf{B}}$. It is a vector quantity and its magnitude and direction at any point can be found out from the force acting on a moving charge at that point.

The force $\vec{\mathbf{F}}$ on a charge q moving with a velocity $\vec{\mathbf{v}}$ in a magnetic field of induction $\vec{\mathbf{B}}$ is given by

$$\vec{\mathbf{F}}=q(\vec{\mathbf{v}}\times\vec{\mathbf{B}}) \quad \text{...(11.46)}$$

This equation gives the value of magnetic induction $\vec{\mathbf{B}}$, both in direction and magnitude. From the above equation the magnitude of the force is given by

$$F=qvB\sin\theta \quad (11.47a)$$

where θ is the angle between the direction of the moving charge and the direction of magnetic induction. The Equation 11.47 indicates that the force acting on a moving charge for a given value of v and B depends upon the direction of the motion of the moving charge. If θ is zero, that is, the direction of motion coincides with that of the direction of the magnetic induction, then force on the moving charge is equal to zero. This direction along which a moving charge experiences no force is taken to be the direction of the magnetic induction $\vec{\mathbf{B}}$. From the Equation 11.47, the magnitude of the magnetic field or induction at any point is given by

$$B=\frac{F}{qv\sin\theta} \quad (\text{...}11.47b)$$

Up to this point, we have defined the magnitude of the magnetic induction and its direction but not its sense. The definition of magnetic induction now be completed by specifying that when the moving charge is positive, the normal component of velocity vector $\vec{V}$ *i.e.* vector $\vec{V_L}$ magnetic induction $\vec{B}$ and the force $\vec{F}$ on the moving charge form a mutually perpendicular set of axes as shown in Fig. 11.16. It is to be noted if the directions of any two quanties said above are known the direction of the third can be found out from this figure.

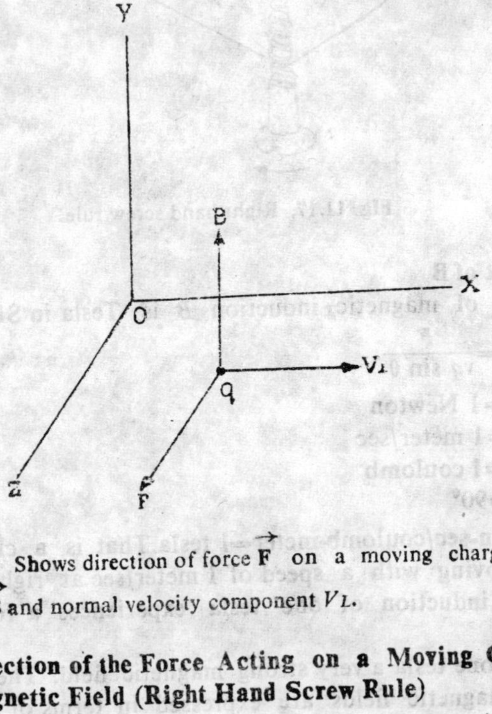

Fig. 11.16. Shows direction of force $\vec{F}$ on a moving charge, magnetic induction $\vec{B}$ and normal velocity component V_L.

11.33. Direction of the Force Acting on a Moving Charge in a Magnetic Field (Right Hand Screw Rule)

The force on a moving charge is given by $\vec{F}=q(\vec{v}\times\vec{B})$ where the symbols have usual meaning. If the charge is positive the force has the same direction as that or the vector $(\vec{v}\times\vec{B})$ and is given by the direction of the advance of the right hand screw when rotated from $\vec{v}$ to $\vec{B}$, Fig. 11.17. It is to be noted that this direction is opposite to that of $\vec{v}\times\vec{B}$ if the charges are negative *i.e.* for electrons and negative ion.

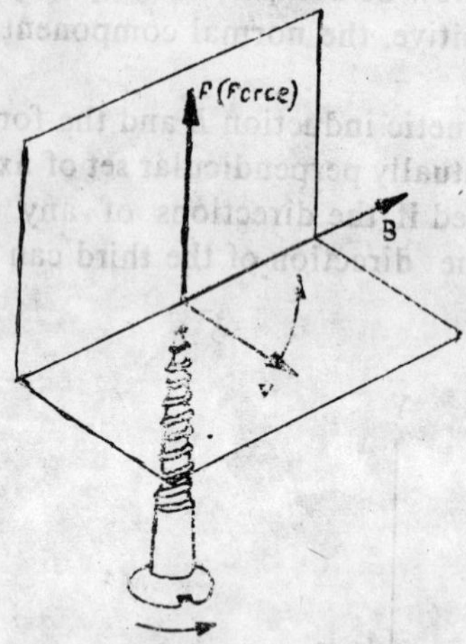

Fig. 11.17. Right hand screw rule.

11.34. Unit of B

The unit of magnetic induction B is Tesla in SI unit. In the equation $B = \frac{F}{vq \sin \theta}$

If $F = 1$ Newton
$v = 1$ meter/sec
$q = 1$ coulomb
$\theta = 90°$

$B = 1$ Newton-sec/coulomb-meter = 1 tesla. That is a charge of one coulomb moving with a speed of 1 meter/sec at right angles to a magnetic induction of one tesla experiences a force of one Newton (N).

A field of one tesla a very strong magnetic field. Therefore very often the magnetic fields are expressed in terms of smaller unit called Gauss.

$$1 \text{ Gauss} = 10^{-4} \text{ tesla.}$$

Example 11.14. An electron is projected into a magnetic field of magnetic induction $B = 10$N/(coul-m)/sec) with a velocity of 3×10^7 m/sec in a direction at right angles to the field. Compute the magnetic force on the electron and compare with the weight of the electron. Given mass of electron is 9.1×10^{-3} kg.

Solution. The magnitude of the force $\vec{F}_m$ on a charge in a magnetic field is

$$F = |\vec{F}_m| = |q(\vec{v} \times \vec{B})| = qvB \sin\theta$$
$$= 1.6 \times 10^{-19} \text{ coul } \times 3 \times 10^7 \text{ m/sec} \times 10\text{N} \times 1 \text{ (coul—m)/sec}$$
$$= 4.8 \times 10^{-11} \text{ N}$$

The gravitational force or the weight of the electron is

$$Fg = |\vec{F}_g| = \text{mg} = 9.1 \times 10^{-31} \text{ kg} \times 9.8 \text{ m/sec}^2$$
$$= 8.8 \times 10^{-31} \text{ N}$$

The gravitational force is therefore negligible in comparison with the magnetic force.

11.35. Magnetic Field and Lines of Induction

Like electric field, magnetic fiəld can be graphically represented by lines called lines of induction. A line of induction in a magnetic field is a line so drawn that tangent to it at any point, shows the direction of the magnetic induction at that point. These lines can be used to obtain the magnitude of the magnetic induction at a point. This is done by the conventional agreement that magnetic induction at a point in space will be represented by the number of lines of Induction per unit area through a surface normal to the field in the neighbourhood of the point. The unit of magnetic induction is chosen to be one weber per square meter. Therefore one weber would represent one line of induction.

Mathematically, magnetic induction is given by

$$B = \frac{\phi}{A_n} \qquad \ldots(11.48)$$

where ϕ is the total number of lines induction of called magnetic flux and A_n is the normal area. Magnetic induction is some times called magnetic flux density and expressed in tesla. Therefore 1 tesla is equal to one weber/meter^2 (wb/m^2).

11.36. Pictorial Representation of Uniform Magnetic Field

The uniform magnetic field is pictorially represented by lines of induction, Fig. 11.18(*a*) which are straight lines with arrow heads and uniformally spaced over paper. The direction of the field is known from arrow head. If the field is parallel to the plane of the paper (or black board), the lines are shown to lie on the plane of the paper. However it is often necessary to represent pictorially a uniform magnetic field perpendicular to the plane of the paper (or black board) directed either towards or away from the reader. This is usually done with number of dots or crosses. The dots may be thought of as the magnetic field directed towards the reader, Fig. 11.18.(*b*) and crosses as the magnetic field directed away from the reader *i.e.* into the paper (or black board), Fig. 11.18(*c*).

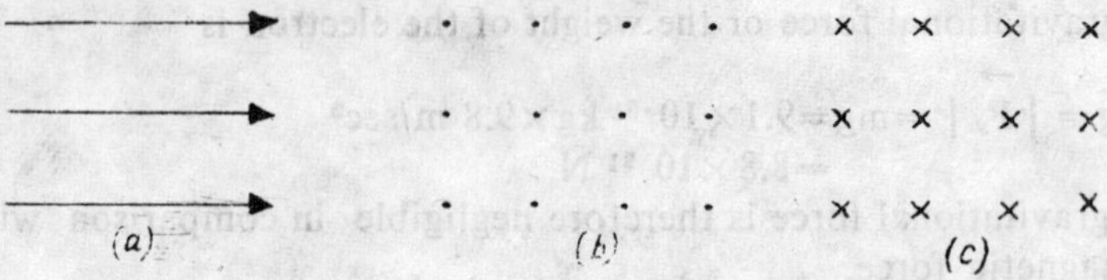

Fig. 11.18. (*a*) Uniform magnetic field parallel to the plane of the paper, (*b*) Uniform perpendicular magnetic field directed out of the paper, (*c*) Uniform perpendicular magnetic field directed into the paper.

11.37. Magnetic Induction due to Current in a Conductor

Current in a conductor sets up magnetic field around the conductor. The magnetic induction $\vec{B}$, at any point of this field depends upon the shape of the conductor, direction of the current, the distance of the point from the conductor and the medium surrounding the conductor. In order to find out the magnetic induction at any point due to current in a conductor, the conductor is divided in imgination into short elements of length *dl* called the current element, one of which is shown in Fig. 11.19. Then the field intensity or the field $\vec{dB}$ due to each of the elements at that point is found out. The resultant of all these fields gives the field due to the entire conductor.

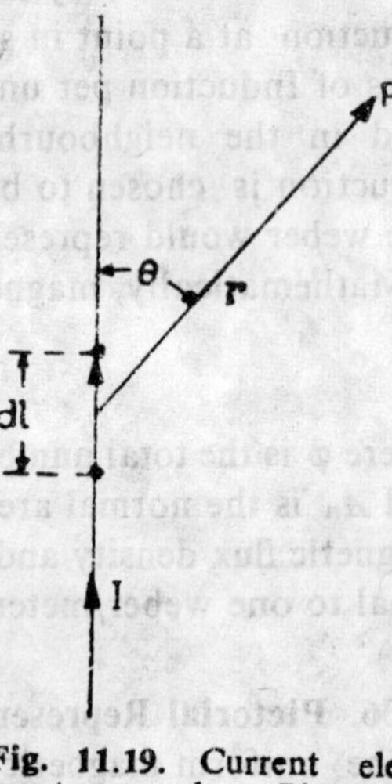

Fig. 11.19. Current element and Biot Savart law.

11.38. Biot-Savart Law

Biot-Savart experimentally arrived at a formula for the magnetic field due to a current element. The formula is known as Biot-Savart law. This formula is used to compute the magnetic field due to the current of any shape.

According to this law the magnetic field $\vec{B}$ due to current element

$\vec{dl}$ of a conductor carrying current I, at a point P, at a distance r (Fig. 11.19) is given by.

$$\vec{dB}=\frac{\mu_0}{4\pi}\,I\left(\frac{\vec{dl}\times\vec{r}}{r^3}\right) \qquad \ldots(11.49)$$

where μ_0 is a constant called permeability of the vacuum and its value is $4\pi\times10^{-7}$ weber per ampere meter. The vector expression in Equation 11.49 gives complete information regarding the magnitude as well direction of $\vec{dB}$ at P. The magnitude of the field $\vec{dB}$ is given by,

$$dB=\frac{\mu_0 I dl \sin\theta}{4\pi r^2} \qquad \ldots(11.50)$$

where θ is the angle between the vector $\vec{dl}$ and $\vec{r}$, $\vec{r}$ being the position vector of the point P relative to the current element. The direction of $\vec{B}$ is same as the direction of the vector $(\vec{dl}\times\vec{r})$. In Fig. 11.19 for example $\vec{dB}$ at the point P is directed into the page and at right angles to the plane of the figure.

11.39. Magnetic Induction at the Centre of a Circular Current (Coil)

Consider a circular coil of one turn of radius r carrying a current I as shown in Fig. 11.20. When the coil lies in the plane of the

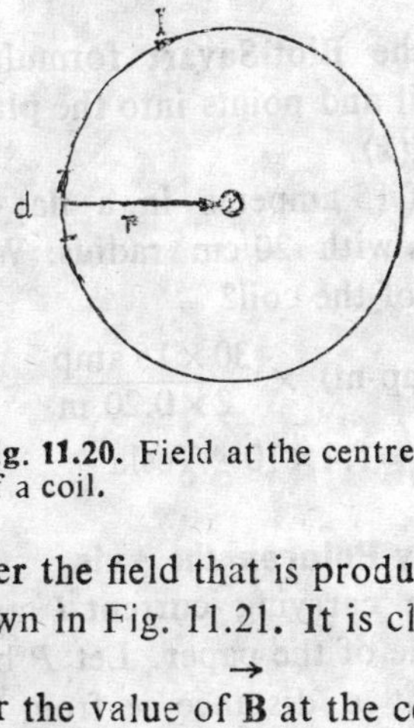

Fig. 11.20. Field at the centre of a coil.

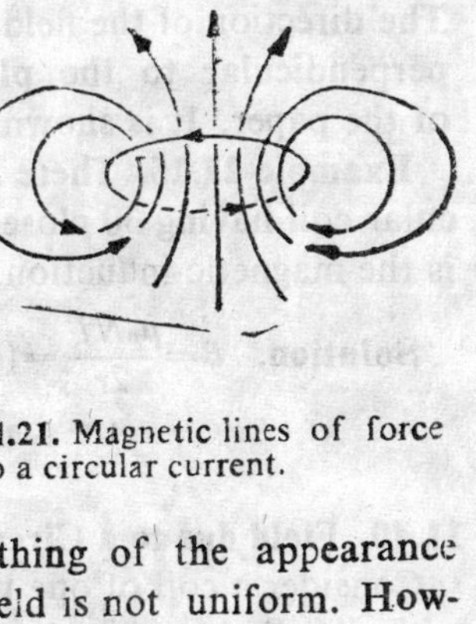

Fig. 11.21. Magnetic lines of force due to a circular current.

paper the field that is produced has some thing of the appearance shown in Fig. 11.21. It is clear that the field is not uniform. However the value of $\vec{B}$ at the centre of the coil can be obtained from Biot-Savart law as follows.

Consider a small element $\vec{dl}$ of the coil. The magnetic induction $\vec{dB}$ at the centre of the coil due to this element is given by

$$\vec{dB}=\frac{\mu_o}{4\pi}\,I\left(\frac{\vec{dl}\times\vec{r}}{r^3}\right)$$

Since the angle between $\vec{dl}$ and $\vec{r}$ is 90°, sin $\theta=1$ and the magnitude of $\vec{dB}$ using above equation is given by

$$dB=\frac{\mu_o I dl}{4\pi r^2} \qquad \ldots(11.51)$$

If we devide the coil into large number of such elements then the field due to all these elements is obtained by integrating the expression 11.51 over the whole length of the coil,

$$\therefore \quad B=\oint\frac{\mu_o I dl}{4\pi r^2} \qquad \ldots(11.52)$$

Since I and r are constants,

$$B=\frac{\mu_o}{4\pi r^2}I\oint dl$$

$$=\frac{\mu_o I}{4\pi r^2}.2\pi r=\frac{\mu_o I}{2r} \qquad \ldots[11.53(a)]$$

If instead of a coil of one turn there is a coil of N turns that are closely wound, then

$$B=\frac{\mu_o NI}{2r} \qquad \ldots[11.53(b)]$$

The direction of the field as seen from the Biot-Savart formula is perpendicular to the plane of the coil and points into the plance of the paper. It is shown by the symbol(x).

Example 11.15. There is a current of 15 amperes in a flat circular coil having 30 closely wound turns with 20 cm radius. What is the magnetic induction, at the centre of the coil?

Solution. $B=\frac{\mu_o NI}{2r}=(4\pi\times10^{-7}\text{wb/amp-m})\times\frac{30\times15\text{ amp}}{2\times0.20\text{ m}}$

$=1.4\times10^{-3}\text{wb/m}^2=1.4\times10^{-3}$ Tesla

11.40. Field due to a Circular Coil at any Point on the Axis

Consider a coil of one turn of radius a carrying current I amp, with its plane perpendicular to the plane of the paper. Let P be a point on the axis of this circular coil at a distance x from the centre O. We have to calculate the magnetic induction at P. Consider a small element of the coil length dl at a point A. The magnetic induction $\vec{\mathbf{dB}}$ at the point P due to this element is given by

$$\vec{\mathbf{dB}}=\frac{\mu_0 I}{4\pi}\cdot\frac{\vec{\mathbf{dl}}\times\vec{\mathbf{r}}}{r^3} \qquad \ldots(i)$$

$\vec{\mathbf{dB}}$ is perpendicular to the plane containing $\vec{\mathbf{dl}}$ and $\vec{\mathbf{r}}$ and its direction is found out by right hand screw rule. As the angle between $\vec{\mathbf{dl}}$ and $\vec{\mathbf{r}}$ is 90°, the magnitude of the magnetic field $\vec{\mathbf{dB}}$ is given by

$$dB=\frac{\mu_0}{4\pi}\frac{Idl}{r^2} \qquad \ldots(ii)$$

In the Fig. 11.22. $\vec{\mathbf{dB}}$ is represented by $\vec{\mathbf{PR}}$. The vector $\vec{\mathbf{PR}}$ can be resolved into two components namely $\vec{\mathbf{PQ}}$ and $\vec{\mathbf{PS}}$ along and perpendicular to the axis respectively and their magnitudes are $\vec{\mathbf{PQ}}$ $=dB \sin\phi$ and $\vec{\mathbf{PS}}=dB\cos\phi$ [as shown in the figure]

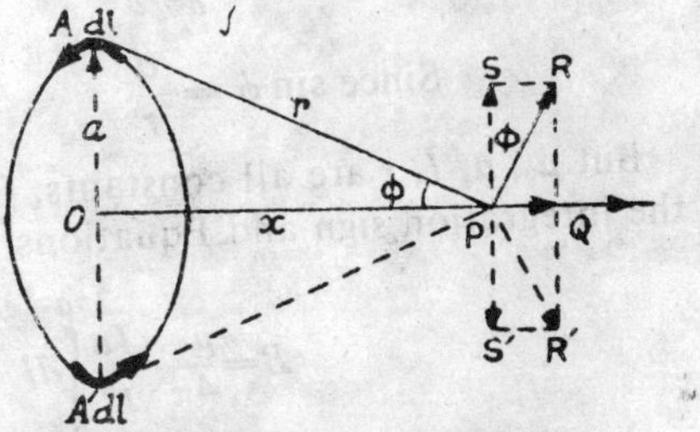

Fig. 11.22. Magnetic field due to a circular current at any point P along the axis.

Now consider another current element of same length at $\vec{\mathbf{dl}}$ at A' in the lower half of the coil. This element is taken to be diametrically opposite to the current element at A. The contribution to the magnetic induction at P due to this element has been respresented by vector $\vec{\mathbf{PR}}$ whose magnitude is $\frac{\mu_0}{4\pi}\cdot\frac{Idl}{r^2}$. $\vec{\mathbf{PR'}}$ can be resolved into two components namely $\vec{\mathbf{PQ'}}$ and $\vec{\mathbf{PS'}}$ along and perpendicular to the axis respectively and their magnitudes are given by,

$$PQ=dB\sin\phi \text{ and } PS'=dB\cos\phi$$

It is clear from the figure that $\vec{\mathbf{PS}}$ and $\vec{\mathbf{PS'}}$ are equal in magnitude but opposite in direction. Therefore they cancel out. On the other hand the axial components are found to be parallel and hence they add up. Dividing the entire coil into such small sigments and considering two diametrically opposite segments together, it can be seen that sum of the perpendicular components equals to zero. Therefore only parallel components contribute to the net magnetic induction at P.

Hence we can write the total magnetic induction as

$$B = \Sigma dB \sin \phi$$

Changing the summation into integration, we have

$$B = \int_0^{2\pi a} \frac{\mu_0}{4\pi} \cdot \frac{dl}{r^2} \sin \phi = \int_0^{2\pi a} \frac{\mu_0 I dl}{4\pi r^2} \frac{a}{r}$$

$$= \int_0^{2\pi a} \frac{\mu_0 I a}{4\pi r^3} dl \qquad \ldots(11.54)$$

Since $\sin \phi = \frac{a}{r}$

But μ_0, a, I, r are all constants. They can be taken out side of the integration sign and Equations 11.54 can be written as,

$$B = \frac{\mu_0}{4\pi} \cdot \frac{Ia}{r^3} \int_0^{2\pi a} dl$$

$$= \frac{\mu_0}{4\pi} \cdot \frac{Ia}{r^3} \cdot 2\pi a$$

$$= \frac{\mu_0}{2} \cdot \frac{Ia^2}{r^3} \qquad \ldots(11.55)$$

Substituting $r = (x^2 + a^2)^{\frac{1}{2}}$ in Equation 11.55 we get,

$$B = \frac{\mu_0 I}{2} \cdot \frac{a^2}{(x^2 + a^2)^{3/2}} \qquad \ldots(11.56)$$

This is the expression for the magnetic induction for a coil of single turn. If the number of turns in a coil is N, and are closely wound, then,

$$B = \frac{\mu_0}{2} NI \frac{a^2}{(x^2 + a^2)^{3/2}} \qquad \ldots(11.57)$$

If $x = 0$, then P is at the centre of the coil and B at the centre of the coil is given by,

$$\frac{\mu_0 NI}{2a}$$

which is same as the Equation 11.53.

Example 11.16. Calculate the magnetic induction due to a circular coil of 100 turns, and radius 0.1 m carrying a current of 5*A* at a point on the axis of the coil. The distance or the point being 0.2 m from the centre or the coil.

Solution. Magnetic induction at any point on the axis of a coil is given by

$$B=\frac{\mu_0 N I a^2}{2(x^2+a^2)^{3/2}}$$

where *a* is the radius and *x* is the distance of the point from the centre.

$$\therefore \quad B=\frac{4\pi\times10^{-7}\text{wb/amp-m}\times100\times(0.1\text{m})^2\times5\text{ amp}}{2\times[(0.10\text{ m})^2+(0.02\text{ m})^2]^{3/2}}$$

$$=2.81\times10^{-4}\text{wb/m}^2=2.81\times10^{-4}\text{ tesla}$$

11.41. Magnetic Induction near a Long Straight Conductor

Let *AB* in the plane of paper be a straight wire carrying a current *I* ampere and *P* be the point at a distance *a* from the wire, at which induction is to be determined Fig. 11.23(*a*). Consider a small element *QR* of length *dl* of the conductor, which is at a distance *r* from the point *P* such that angle between *r* and the conductor is θ.

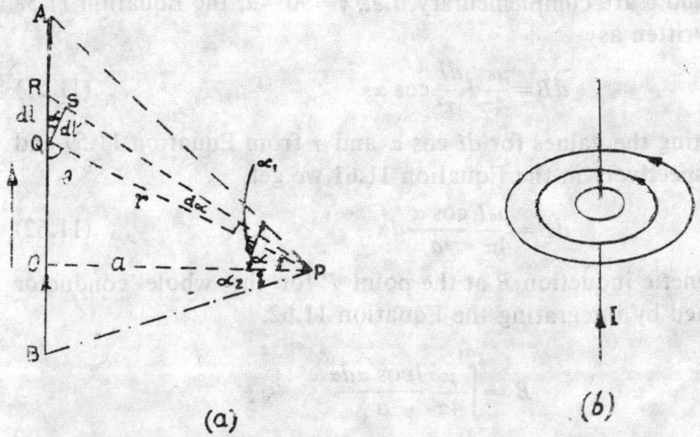

Fig. 11.23. (*a*) Field due to a straight current, (*b*) Magnetic lines of force due to a straight current.

The contribution to the magetic induction $\vec{\mathbf{dB}}$ at point *P* from this current element $\vec{\mathbf{dl}}$ is given by Biot-Savart law as,

$$\vec{\mathbf{dB}}=\frac{\mu_0}{4\pi}I\frac{\vec{\mathbf{dl}}\times\vec{\mathbf{r}}}{r^3} \qquad \ldots(i)$$

Here $\vec{dB}$ is perpendicular to the plane of the paper and its direction is given by right hand screw rule. The magnitudes of $\vec{dB}$ is given by,

$$dB = \frac{\mu_0}{4\pi} I \frac{dl}{r^2} \sin\theta \qquad \ldots(11.58)$$

(where θ is the angle between current element $\vec{dl}$ and radius vector $\vec{r}$).

If we drop the perpendicular QS on PR from the point Q then from the right angled triangle RQS, $QS = QR \cos(90-\theta) = QR \sin\theta = dl \sin\theta$. Let $OPQ = \alpha$ and also let element QR subtend an angle $d\alpha$ at P.

$$\therefore \qquad QS = r d\alpha = dl \cos\alpha \qquad \ldots(11.59)$$

From the right angled triangle QOP

$$\cos\alpha = \frac{a}{r}, \text{ or } r = \frac{a}{\cos\alpha} \qquad \ldots(11.60)$$

Since α and θ are complementary, *i.e.*, $\theta = 90 - \alpha$, the Equation 11.58 can be written as

$$dB = \frac{\mu_0}{4\pi} I \frac{dl}{r^2} \cos\alpha \qquad \ldots(11.61)$$

Substituting the values for $dl \cos\alpha$ and r from Equation 11.59 and 11.60 respectively in the Equation 11.61 we get,

$$dB = \frac{\mu_0 I}{4\pi} \frac{\cos\alpha}{a} d\alpha \qquad \ldots(11.62)$$

The magnetic induction B at the point P for the whole conductor is obtained by integrating the Equation 11.62.

$$\therefore \qquad B = \int_{-\alpha_2}^{\alpha_1} \frac{\mu_0}{4\pi} \frac{I \cos\alpha \, d\alpha}{a}$$

$$= \frac{\mu_0}{4\pi a} I \int_{-\alpha_2}^{+\alpha_1} \cos\alpha \, d\alpha$$

$$= \frac{\mu_0 I}{4\pi a} \Big[\sin\alpha \Big]_{-\alpha_2}^{\alpha_1}$$

$$= \frac{\mu_0 I}{4\pi a} \Big[\sin\alpha_1 - \sin(-\alpha_2) \Big]$$

$$= \frac{\mu_o I}{4\pi a}\left[\sin \alpha_1 + \sin \alpha_2\right] \quad \ldots (11.63)$$

For an infinitely long conductor,

$$\alpha_1 = \alpha_2 = 90°$$

Therefore for infinitely long conductor

$$B = \frac{\mu_o}{4\pi a} I\,[1+1]$$

$$= \frac{\mu_o}{4\pi a} I \times 2 = \frac{\mu_o I}{2\pi a} \quad \ldots (11.64)$$

This equation although is valid for infinitely long conductor, it also applies satifactorily to shorter conductor if the distance from the conductor is not too great.

Example 11.17. There is a current of 50 amperes in a long straight wire. What is the flux density at a point 6 cm from the wire.

Solution. $B = \frac{\mu_o I}{2\pi a} = \frac{4\pi \times 10^{-7} \text{ wb/amp-m} \times 50 \text{ amp}}{2\pi \times 0.0600 \text{ m}}$

$$= 1.67 \times 10^{-4} \text{ wb/m}^2$$

11.42. Magnetic Induction due to a Solenoid

A long cylindrical coil of wire is called a solenoid. It has a magnetic field similar to a bar magnet, whenever there is a current in it. The poles and lines of induction due to a solenoid is shown in Fig. 11.24.

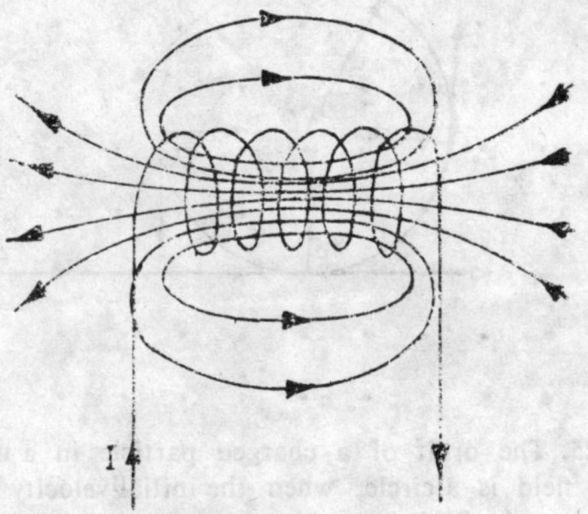

Fig. 11.24. Current in a solenoid.

If solenoid is very long, then the field is fairly uniform inside the interior of the coil and is given by,

$$B = \mu_o n I \quad \ldots (11.65)$$

where I is the current and n is the number of turns per unit length of the solenoid.

11.43. Rule to Find out the Direction of the Magnetic Induction due to Currents

The direction of the magnetic induction due to different kinds of currents can conveniently be found out with help of the following rule:

(*i*) **For circular current.** Curl the fingers of the right hand in the direction of the current, the stretched thumb then points in the direction of the field.

(*ii*) **For straight current.** Grasp the wire in the right hand with thumb pointing in the direction of the current, the field lines then follow the direction of the curled fingers.

11.44. Orbit of a Charged Particle in Uniform Magnetic Field

Consider a charged particle carrying a positive charge q to be moving with a velocity v at right angles to a uniform mag netic field of magnetic induction B. The field is taken to be perpendicular to the plane of the paper and is directed towards the reader as shown by dotes in the Fig. 11.25. The charge will then experience a force

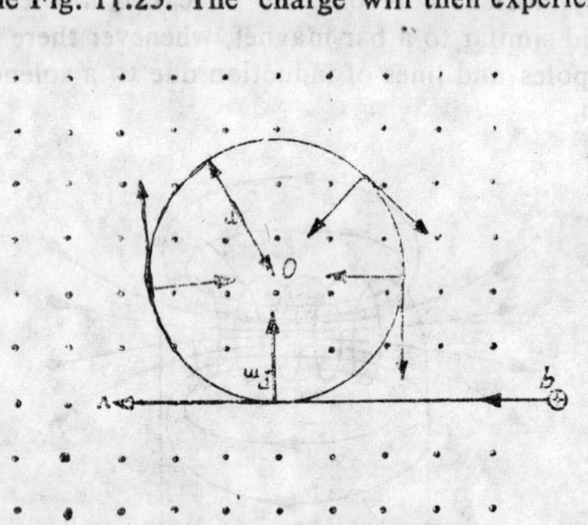

Fig. 11.25. The orbit of a charged particle in a uniform magnetic field is a circle, when the initial velocity is perpendicular to the field.

$\vec{F}=q(\vec{v}\times\vec{B})$, which lies in the plane of the paper and is perpendicular to both $\vec{v}$ and $\vec{B}$. The magnitude of the force is given by qvB as $\vec{B}$

and $\vec{v}$ are at right angles to each other. Since the force is normal to the direction of the velocity, it will not affect the magnitude of the velocity but only change its direction. This indicates that the magnitude of the force which is given by qvB always remains constant during the motion. The particle therefore moves under the influence of the force whose magnitude is constant but whose direction is always at right angles to the velocity, of the particles. This motion is seen to be similar to the motion of a particle which moves with constant speed in a circular path. Therefore the path of the charged particle in a uniform magnetic field is a circle described with a constant tangential velocity with the force $\vec{\mathbf{F}}$ being the centripetal force.

It is known that if a particle of mass m moves with a constant speed v along a circular path of radius r, the magnitude of the centripetal force is given by,

$$\frac{mv^2}{r}$$

Hence we can write,

$$qvB = \frac{mv^2}{r}$$

$$\text{or} \quad r = \frac{mv}{qB} \qquad \ldots (11.66)$$

It is seen from the above equation that the faster the particle moves, the larger is the radius and larger is the value of magnetic induction smaller is the radius of the circular path.

The angular velocity ω of the charged particle is determined from the Equation 11.66. *i.e,*

$$\omega = \frac{v}{r} = \frac{qB}{m} \qquad \ldots (11.67)$$

The frequency ν measured say in rev/sec is given by,

$$\nu = \frac{\omega}{2\pi} = \frac{qB}{2\pi m} \qquad \ldots (11.68)$$

It is to be noted that the angular velocity and frequency do not depend upon the speed of the charged particle. The frequency given above is sometimes known as cyclofron frequency.

Example 11.18. A 10 ev electron is circulating in a plane at right angles to a uniform magnetic field of induction 1.0×10^{-4} weber/meter^2 (=1.0 gauss).

(*a*) What is its radius?

(*b*) What is its cyclofron frequency?

(*c*) What is the period of revolution?

Given charge of an electron is $=1.6\times10^{-19}$ coul and mass of the electron 9.1×10^{-31} kg.

Solution. 1 ev(1 electron volt) means energy carried by electrons in moving through a potential difference 1 volt. This is equal to charge of electron in coulomb x volt.

$$=1.6\times10^{-19} \text{ coul x 1 volt.}$$
$$=1.6\times10^{-19} \text{ joule.}$$

Therefore energy carried by 10 ev electron.

$$=10\times1.6\times10^{-19} \text{ joule}$$

Let the mass of the electron be m and its velocity v. The kinetic energy of the electron $=\frac{1}{2}mv^2$

$$\therefore \quad \tfrac{1}{2}mv^2=10\times1.6\times10^{-19} \text{ J}$$

$$v^2=\frac{2\times10\times1.6\times10^{-19} \text{ J}}{m}$$

$$v=\sqrt{\frac{2\times10\times1.6\times10^{-19} \text{ joule}}{9.1\times10^{-31} \text{ kg}}}$$

$$=1.9\times10^6 \text{ meter/sec.}$$

(*a*) from Equation 11.66.

$$r=\frac{mv}{qB}=\frac{(9.1\times10^{-31} \text{ kg})\ (1.9\times10^6 \text{ meter/sec})}{(1.6\times10^{-19} \text{ coul})\ (1.0\times10^{-4} \text{wb/m}^2)}$$

$$=0.11 \text{ meter}=11 \text{ cm}$$

(*b*) From Equation 11.68.

$$\nu=\frac{qB}{2\pi m}=\frac{(1.6\times10^{-19} \text{ coul})\ (1.0\times10^{-4} \text{ wb/m}^2)}{2\pi\ (9.1\times10^{-31} \text{ kg})}$$

$$=2.8\times10^6 \text{ rev/sec.}$$

(*c*) Period of revolution T is given by,

$$T=\frac{1}{\nu}=\frac{1}{2.8\times10^6 \text{ rev/sec}}=3.6\times10^7 \text{ seconds.}$$

11.45. e/m by Thomson's Method

Sir J.J. Thomson, a scientist from Cavendish laboratory at Cambridge discovered the electron and measured its charge to mass ratio (e/m) by performing an experiment. He was awarded the nobel prize of the year 1906 for this famous work.

A modern version of the apparatus used in his experiment is shown schematically in the diagram 11.26. It consists of a highly evacuated glass tube, the narrow portion of which contains a metal filament F. The filament is heated by means of a low voltage source

V_f, so as to emit electrons. His method of determination of e/m is briefly described below. The electrons after being emitted from the hot filament get accelerated towards a plate A called anode, which is maintained at a potential V_a, higher than that of the filament. Most of the electrons strike the anode A and are stopped but there is a small hole in A through which some of them pass. The electrons

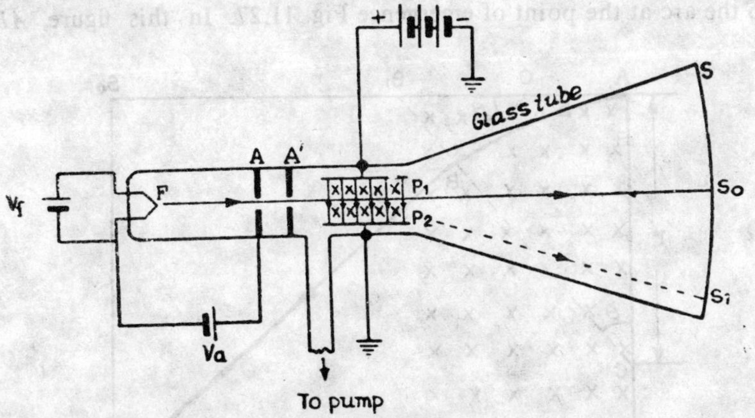

Fig. 11.26. Thomson's apparatus for e/m of electrons.

are further restricted by an electrode A' in which there is another small hole and which is maintained at the potential of the anode A. Thus a narrow beam of electrons is obtained. This narrow beam then enters into a region of uniform electric and magnetic field, which are perpendicular at each other and are also perpendicular to the direction of the velocity of the undeflected beam. The electric field is obtained by applying d.c. voltage from outside to two metal parallel plates P_1 and P_2 and the magnetic field is obtained by means of two co-axial parallel current carrying coils called Helmhotz coil. After passing the field region the electron beam finally hits the wider end of tube, which is called the screen S. The inside of the screen is coated with fluorescent materials. When electrons hit the screen, a luminous glow is seen, thus making the electron impact visible.

(*a*) **Theory and working.** (*i*) First of all the magnetic and electric fields are switched off and position of the undeflected beam spot is noted on the screen.

(*ii*) Then a uniform magnetic field of magnetic induction B is applied, in the region enclosed by the plate P_1 and P_2. The magnetic field is arranged in a manner, so that it is perpendicular to the direction of motion of the electrons and also perpendicular to the plane of the paper being directed away from the reader. As explained in section 11.44 the electron experiences a force of magnitude Bev in downward direction, where e is the charge of the electron and v its velocity. Due to this force the electron will traverse an arc of a circle of radius r in the region of the magnetic field and then emerge out of the field in a straight line, tangential to the arc at the point of emergence Fig. 11.27. In this figure, AB

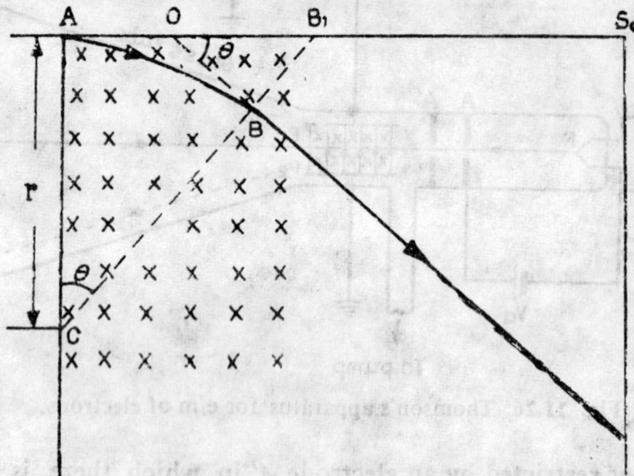

Fig. 11.27. Motion of an electron in the magnetic field.

is the region of the magnetic field, O the centre of the magnetic field. S_0 the undeflcted spot, S_1 the deflected spot, AB the circular arc of radius r along which the electron moves in the magnetic field and BS_1 is the tangential path followed by the electron after its emergence from the magnetic field, C is the centre of the circular path of radius r. The force on the electron perpendicular to the direction of motion is Bev and this is equal to the centripetal force $\frac{mv^2}{r}$, where m is the mass of the electron.

$$Bev = \frac{mv^2}{r}$$

or
$$\frac{e}{m} = \frac{v}{rB} \quad \ldots (11.69)$$

Here, B is known. If v and r are found out, then e/m can be calculated.

(*b*) **Calculation for r.** The triangle ABC and OS_0S_1 (Fig. 11.27) are similar. (This is because CA and CB are two radii and AS_0 and

OS_1 are the two tangents at A and B respectively to the circular arc. Therefore,

$$\underline{|ACD}=\theta=\underline{|SOS_1} \text{ and } \underline{|B_1AC}=90^\circ=\underline{|OS_0S_1})$$

$$\therefore \quad \frac{AB_1}{AC}=\frac{S_0S_1}{OS_0} \quad \ldots(11.70)$$

or

$$\frac{AB_1}{r}=\frac{S_0S_1}{OS_0}$$

or

$$r=\frac{AB_1\times OS_0}{S_0S_1} \quad \ldots(11.71)$$

Where OS_0 is the distance between the centre of the magnetic field and the centre of the screen, AB_1 is the length of the magnetic field and S_0S_1 is the observed shift of the fluorescent spot. AB_1 and OS_0 can be determined, which depends upon the geometry of the apparatus and is noted experimentally. Hence in the Equation 11.71 the quantities on the right side are known and therefore r can be easily determined.

(*c*) **Calculation for v.** To determine the velocity of the electron in the presence of the magnetic field, the electric field is switched on. The strength of the field was so adjusted that, the fluorescent spot is obtained at the centre of the screen. Under this conditions forces experienced by the electrons due to the electric and magnetic field are equal and opposite (Both the forces are in the plane of the paper and perpendicular to the direction of the motion of the electron), Fig. 11.28. If E is the strength of the electric field then magnitude of the force experienced by the electron is eE.

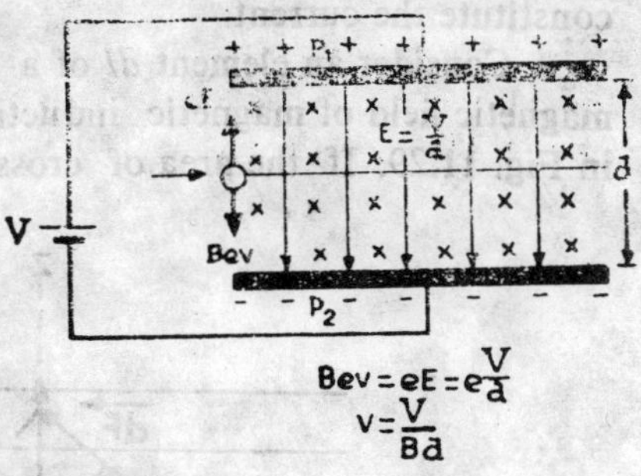

Fig. 11.28. Motion of the electron in simultaneous electric and magnetic field normal to each other.

Hence we can write

$$Bev=eE$$

or

$$v=\frac{E}{B} \quad \ldots(11.72)$$

But if V is the potential difference, between the plates and d is the distance between them, electric field E is given by the Equation 11.14 (*b*) as,

$$E=\frac{V}{d}$$

$$\therefore \qquad v=\frac{V}{Bd} \qquad \ldots(11.73)$$

By substituting the value v and r as given above in the Equations 11.71 and 11.73, we have

$$\frac{e}{m}=\frac{V}{B^2 d}\times\frac{S_0S_1}{OS_0\times AB} \qquad \ldots(11.74)$$

All the quantities in the right side of the Equation 11.74 are known. Hence e/m can be found out.

The ratio e/m is called the specific charge of an electron and its value is 1.759×10^{11} coul/kg.

11.46. Force on a Current Carrying Conductor

A current carrying conductor when placed in a magnetic field experiences a force. This force arises from the action of magnetic field, on the moving electrons, that constitute the current. Therefore the determination of force on a current carrying conductor means the determination of force that acts on the electrons, which constitute the current.

Consider an element dl of a conductor PQ that lies in the magnetic field of magnetic induction B and carries a current I as in Fig. 11.29. If the area of cross section of the conductor be A,

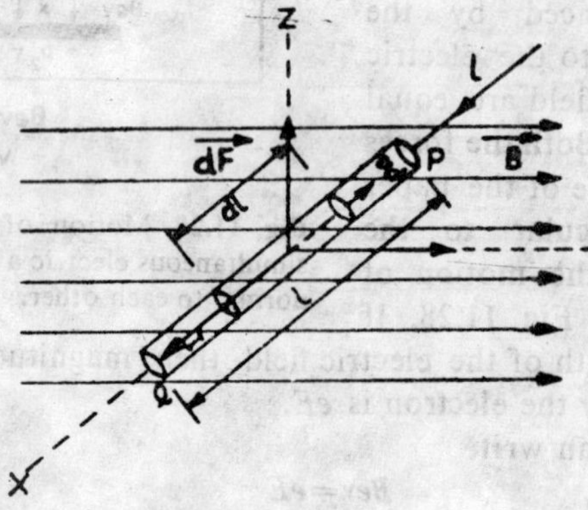

Fig. 11.29. Force on a current carrying conductor placed inclined to the uniform magnetic fied.

drift velocity of the electrons be v_d electronic charge be e and the number of free electrons per unit volume of the conductor be n, then the magnitude of the current is given by $I = enAv_d$ and the total number of electrons in the length dl is given by $nAdl$.

Now each of the electrons experiences a force equal to $-e(\vec{v}_d \times \vec{B})$. Therefore the total force on the electrons in the length dl is given by,

$$\overrightarrow{dF} = -neAdl\,(\vec{v}_d \times \vec{B}) \qquad \ldots(11.75)$$

If we represent the vector length in the direction of the current, then obviously $v_d = -\frac{\vec{dl}}{dt}$, $-$ve sign is attached because $\vec{v}_d$ and $\vec{dl}$ are in opposite direction.

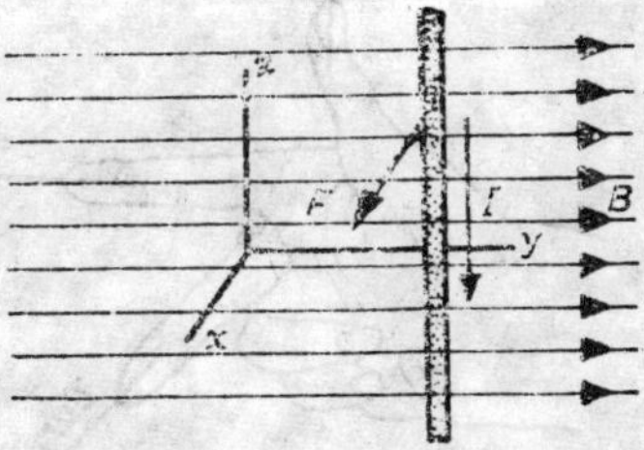

Fig. 11.30. Force on a current carrying conductor placed perpendicular to the uniform magnetic field.

Substituting this value of $\vec{v}_d$ in the Equation 11.75, we get,

$$\overrightarrow{dF} = -neAdl\left(-\frac{\vec{dl}}{dt} \times \vec{B}\right)$$

$$= -neA\frac{dl}{dt}\left(-\vec{dl} \times \vec{B}\right)$$

$$= +neAv_d\,(\vec{dl} \times \vec{B})$$

$$= I\,(\vec{dl} \times \vec{B}) \qquad \ldots(11.76)$$

If B is uniform over a whole length of the conductor of length l,

then total force $\vec{F} = I(\vec{l} \times \vec{B})$ $\qquad \ldots(11.77)$

The magnitude of the force is given by $IlB \sin\theta$ where θ is the angle between the direction of the magnetic induction and the direction of the current. The direction of the force is given by the vector $\vec{l} \times \vec{B}$ where $\vec{l}$ is the vector taken in the direction of the current. If the conductor is held at right angle to field as in Fig. 11.30, the magnitude of the force is given by,

$$F = IlB \text{ newtons} \qquad \ldots(11.78)$$

where I is in ampere, l in meters and B is in tesla. The force is perpendicular to both the conductor and the magnetic field. However if the conductor is along the direction of the magnetic induction, then the force acting on the conductor is zero.

11.47. Fleming's Left Hand Rule (or Left Hand Rule)

The direction of the force $\vec{F}$ on a current carrying conductor in a magnetic field can conveniently be found by Fleming's left hand rule which is given as follows. Hold the thumb, the first finger, and the middle finger of the left hand mutually at right angles to each other. Point the fore finger in the direction of the magnetic field, and the middle finger in the direction of the current, then thumb indicates the direction of the force on the conductor, Fig. 11.31. The conductor obviously tends to move in the direction of the force.

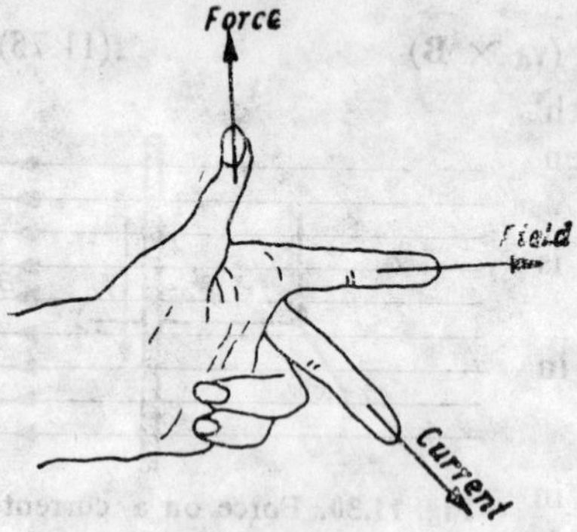

Fig. 11.31. Fleming's left hand rule.

11.48. Force between Two Parallel Conductor Carrying Current

When two parallel current carrying conductors are adjacent, each exerts a force on the other. We may think of one of the currents produces a magnetic field and this magnetic field, exerts a force on the other current. If the currents are in the same direction, the force is found to be attractive, but if the currents are in opposite direction, the force is found to be repulsive. The magnitude as well as direction of this force can be found as follows.

Let *AB* and *CD* be two very long parallel straight conductors carrying currents I_1, and I_2 ampere respectively in the same direction and placed at a distance r meter apart Fig. 11.32 (*a*). For convenience let the conductors be in the plane of the paper. By Equation 11.64, the magnetic field B at any point P, on the conductor *CD*, due to current in *AB* is given as by

$$B=\frac{\mu_0 I_1}{2\pi r} \text{ tesla}$$

The field B is directed into the paper ⊗ and is at right angle to *CD*. The force per dl length of the conductor *CD* at P, according to Equation 11.78 is

$$F=BI_2 dl=\frac{\mu_0 I_1 I_2 dl}{2\pi r} \text{ newton-meter} \quad \ldots(11.79)$$

This force is at right angle to the *CD* and is directed towards

AB in the plane of paper. Hence the force is attractive. If we consider $dl=1$ meter, then the force on *CD* per unit length is given by

$$F=\frac{\mu_0 I_1 I_2}{2\pi r} \quad \ldots(11.80)$$

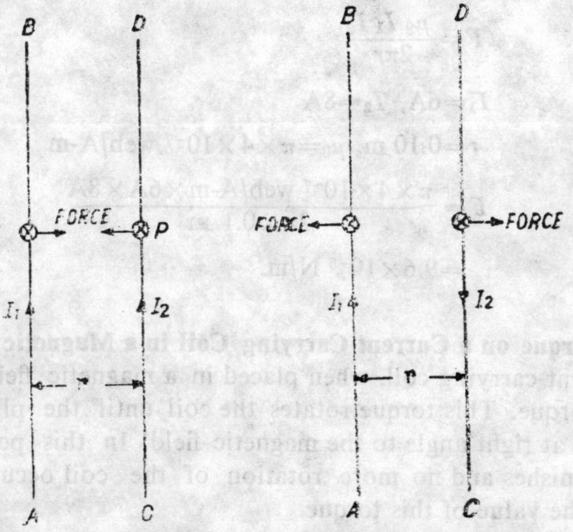

Fig. 11.32. (*a*) Force between two like parallel currents. Force is attractive. (*b*) Force between two unlike parallel currents. Force is repulsive.

If we consider the currents in opposite direction, Fig. 11.32 (*b*), the magnitude of the force will be same as Equation 11.79 but the force will be directed away from the conductor *CD*. Therefore the force is repulsive. It is to be noted that same result will be obtained by considering the conductor, *CD* first in place of *AB* and then finding out that force per unit length of the conductor *AB*.

11.49. Definition of Ampere

The relation expressed in Equation 11.79 can be used to define the MKS unit of current *i.e.* the ampere. In Equation 11.79 if we substitute $I_1=I_2=1$ ampere, $r=1$ meter, then $F=\frac{\mu_0}{2\pi}=2\times10^{-7}$ newton. So the ampere is a current that when it is maintained in two infinitely long conductors parrellel to each other and 1 meter apart in vacuum, will cause a force on each other of exactly 2×10^{-7} newton per meter length.

Example. 11.19. Two parallel wires carry currents of 6A and 8A respectively. If the separation between them is 0.10 meter, what are the magnitude and sense of the force per unit length between two currents.

Solution. $F=\dfrac{\mu_0 I_1 I_2}{2\pi r}$

Here $I_1=6A,\ I_2=8A$

$r=0.10$ m, $\mu_0=\pi\times 4\times 10^{-7}$ web/A-m

Hence $F=\dfrac{\pi\times 4\times 10^{-7}\text{ web/A-m}\times 6A\times 8A}{2\pi\times 0.1\text{ m}}$

$=9.6\times 10^{-5}$ N/m.

11.50. Torque on a Current Carrying Coil in a Magnetic Field

A current carrying coil, when placed in a magnetic field experiences a torque. This torque rotates the coil until the plane of the coil is set at right angle to the magnetic field. In this position the torque vanishes and no more rotation of the coil occurs. Let us find out the value of this torque.

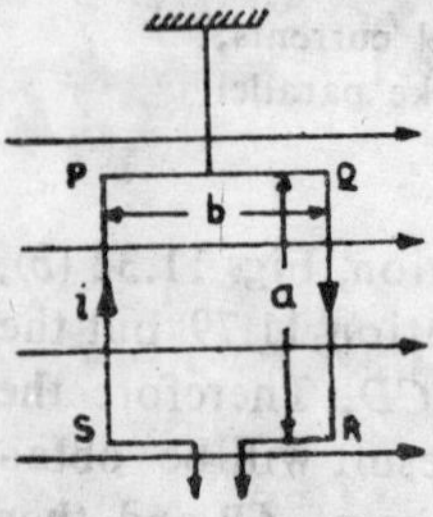

Fig. 11.33. Torque on a suspended coil in a uniform magnetic field.

Consider rectangular coil *PQRS* of one turn having the length *a* and the breadth *b*, Fig. 11.33. The coil is freely suspended in a uniform horizontal magnetic field of magnetic induction $\vec{\mathbf{B}}$. Let the plane of the coil be parallel to the magnetic field. In this position if a current *I* is passed as indicated in the diagram, each of the vertical sides *PS* and *QR* will experience a force but *PQ* and *RS* will experience no force as the currents in them are in the direction of the magnetic field. Since *PS* and *QR* are normal to the magnetic field, the force *F* experienced by them is given by

$$F=aBI\sin 90^\circ=IBa$$

By left hand rule, the force on *PS* is directed into the plane of the paper and the force on *QR* is directed out of the plane of the paper. This means *PS* will move downward and *QR* will move upward.

The top view of the transverse section of the coil is shown in Fig. 11.34 where sign ⊙ indicates the cross section of the *PS* with current coming out of the paper and sign ⊗ indicates the cross section of the wire QR with current going into the paper. The small arrows show the direction of the forces acting on the wires of the coil.

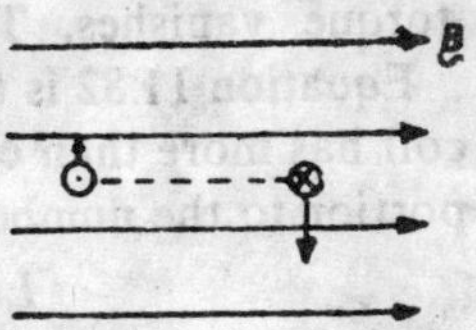

Fig. 11.34. Torque on a freely suspended coil, when the plane of the coil is parallel to the magnetic field (Top view).

These forces on PS and QR are equal in magnitude but opposite in direction. Therefore they constitute a couple, which exerts a torque on the coil to rotate it. The magnitude of this torque L, which is the product of the one of the forces and the perpendicular distance between them is given by

$$L = F \times b = IBa \times b = IB \times (a \times b)$$

where b is the perpendicular distance between two forces. But b is the breadth of the coil and a is the length of the coil. Therefore,

$$L = IBA \qquad \ldots (11.81)$$

where as $a \times b = A$, the area of the coil.

As the coil rotates, the value of the torque changes although the forces on its sides remain the same. When the plane of the coil makes an angle α with the direction of the magnetic field, the torque is given by

$$L = IAB \cos \alpha \qquad \ldots (11.82)$$

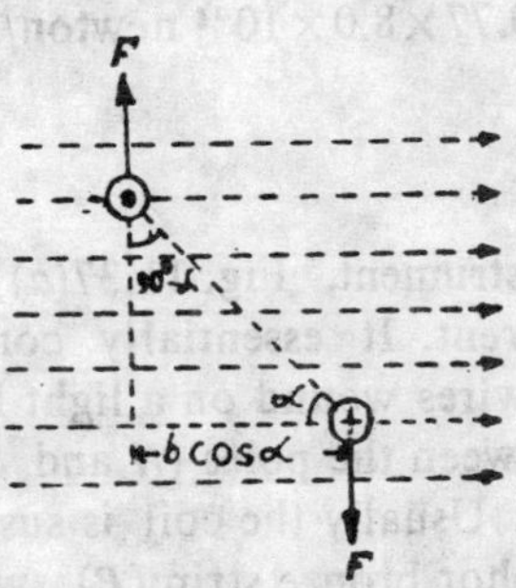

Fig. 11.35. Torque on a freely suspended coil when the coil makes an angle α with the magnetic field (Top view).

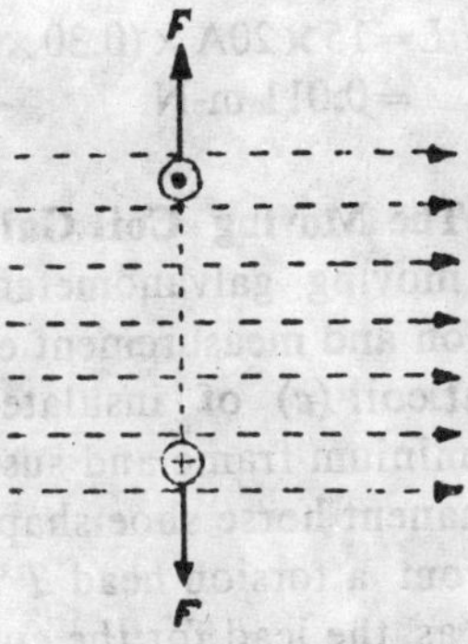

Fig. 11.36. Torque when the plane of the coil is normal to the magnetic field (Top view).

The top view of the transverse section of the coil in this condition is shown in Fig. 11.35.

In general a coil if free will tend to rotate in a field until its plane is perpendicular to the direction of the field. In this condition the forces on PS and QR remain along the same line and the torque vanishes. This has been shown in Fig. 11.36.

Equation 11.82 is the basic equation of electric motors. If the coil has more than one turn, the torque will be increased in proportion to the number of turns N. The general equation is then

$$L = NIAB \cos \alpha \qquad \ldots (11.83)$$

The Equation 11.83 has been derived for the case of the rectangular coil. However it can be shown that the same equation applies to a coil of any shape.

Example 11.20. A rectangular coil of 30 cm long and 10 cm wide is mounted in a uniform field of magnetic induction 8.0×10^{-4} tesla. There is a current of 20 amp in the coil, which has 15 turns. When the plane of the coil makes an angle of 40° with the direction of the field, what is the torque tending to rotate the coil.

Solution. 1 tesla = 1 newton-sec/coul-meter.

$= 1$ newton/(coul/sec) × meter.

$= 1$ newton/ampere-meter. (1N/A-m)

$\therefore \quad 8.0 \times 10^{-14}$ tesla $= 8.0 \times 10^{-4}$ newton/A-m

We have torque $L = NIAB \cos \alpha$

Here $\quad N = 15$, $B = 8.0 \times 10^{-4}$ newton/A-m, $I = 20$A,

Area $\quad A = (0.30 \times 0.10)$ m^2, $\cos 40° = 0.77$,

Hence $L = 15 \times 20\text{A} \times (0.30 \times 0.10)\text{m}^2 \times 0.77 \times 8.0 \times 10^{-4}$ newton/A-m

$= 0.011$ m-N.

11.51. The Moving Coil Galvanometer

The moving galvanometer is an instrument, Fig. 11.37(a)) for detection and measurement electric current. It essentially consists of a flat coil (c) of insulated copper wires wound on a light brass or aluminium frame and suspended between the poles (N and S) of a permanent horse shoe shaped magnet. Usually the coil is suspended from a torsion head T by a phosphor bronze strip (P), which serves as the lead for the current to the coil and is finally connected to a terminal T_1 at the base or side of the instrument. The other end of the coil is connected to one end of a very light spring S_p, which serves as the second lead but which exerts a negligible control on the coil. The other end of the spring is then connected to another terminal T_2. A mirror M is fixed along the phosphor-bronze strip by means of which the deflection of the coil is measured with the lamp scale arrangement.

A soft iron cylinder A is generally placed mid way, between poles of the magnet and within the frame of the coil but quite detached from it. Due to this arrangement the magnetic field gets concentrated in the vicinity of the coil and the coil is subjected to a strong magnetic field.

When a current is passed through the coil, (by connecting T_1 and T_2 to the current source) the coil experiences a torque due to the action of the magnetic field. This torque is called deflecting torque which tends to rotate the coil to set it, at right angles to the magnetic field. However this is prevented by an opposing torque called controlling torque which is set up by the twist of the phosphor-bronze strip. As a result, the coil takes up an intermediate equilibrium position making an angle θ with the field direction.

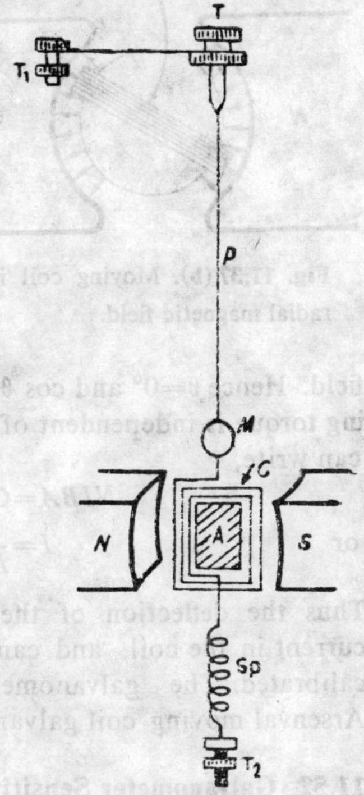

Fig. 11.37(*a*) Moving coil galvanometer.

The deflecting torque, in this position when the plane of the coil makes an angle θ with the field direction is given by

$$L = IBNA \cos \theta,$$

according to the Equation 11.83. But the restoring torque in this position is given by $C\theta$ where C is the restoring torque per unit twist between the ends of the suspension. In equilibrium position the deflecting couple is equal to the controlling couple. Therefore we can write,

$$IBNA \cos \theta = C\theta \qquad \ldots (11.84)$$

or $$I = \frac{C}{BAN \cos \theta} \theta = K \frac{\theta}{\cos \theta} \qquad \ldots (11.85)$$

where $K = \frac{C}{NAB}$ is a constant for the particular galvanometer.

Thus in the above case the deflection θ is not proportional to the current and the instrument will not have a linear scale, i.e. equal increments of current will not be represented by equally spaced marks on the scale. To over come this difficulty the poles N and S are made concave and soft iron cylinder is placed inside the coil but detached from it, Fig. 11.37(*b*). By this modification the field becomes radial and whatever be the position of the coil, the plane of the coil remains parallel to the magnetic field. Hence $\theta=0°$ and $\cos\theta=1$. Under this condition the deflecting torque is independent of the position of the coil. Therefore we can write,

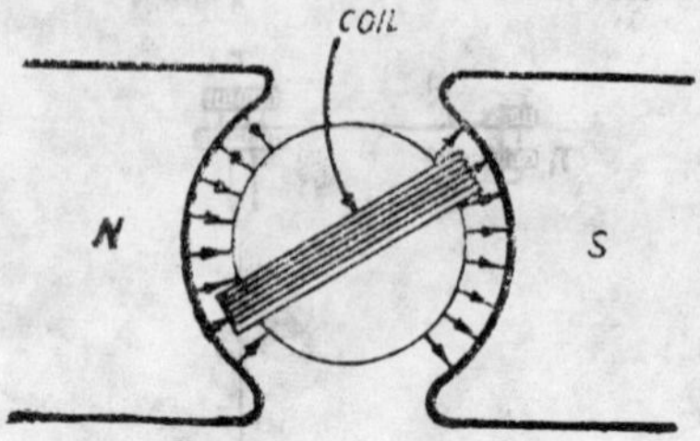

Fig. 11.37 (b). Moving coil in a radial magnetic field.

$$NIBA=C\theta \quad \ldots (11.86)$$

or

$$I=\frac{C}{NAB}\theta=K\theta \quad \ldots (11.87)$$

Thus the deflection of the coil is directly proportional to the current in the coil, and can be made direct reading if suitably calibrated. The galvanometer described above is known as de-Arsenval moving coil galvanometer.

11.52. Galvanometer Sensitivity

The deflection caused by a given current in a galvanometer depends upon the design of the instrument. This characteristic is known as current sensitivity of the galvanometer. It is defined as the amount of currents necessary to produce a deflection of 1 mm on a scale placed at a distance of 1m from the galvanometer. Thus from the Equation 11.87,

$$\frac{I}{\theta}=\frac{C}{NAB}=K,$$ where K is a constant is called sensitivity of a galvanometer. Thus sensitiveness is reciprocal of sensi tity.

11.53. Pivoted Coil Galvanometer

The suspended coil galvanometer is a very delicate instrument and needs careful handling. It is also not an easily portable one. Therefore to make it rugged and portable, the coil instead of being suspended, is pivoted at its two ends between two jewel bearings. Then a light alluminium pointer is attached to it, which moves on a graduated scale to give the deflection of the coil, Fig. 11.38. Usually two phosphor-bronze springs are attached to the two ends of the coil to provide necessary restoring torque and to serve as leads to the currents into and out of the coil. Such a modified moving coil galvanometer is called pivoted coil galvanometer and is also called Weston type of galvanometer as the modification was suggested by Weston.

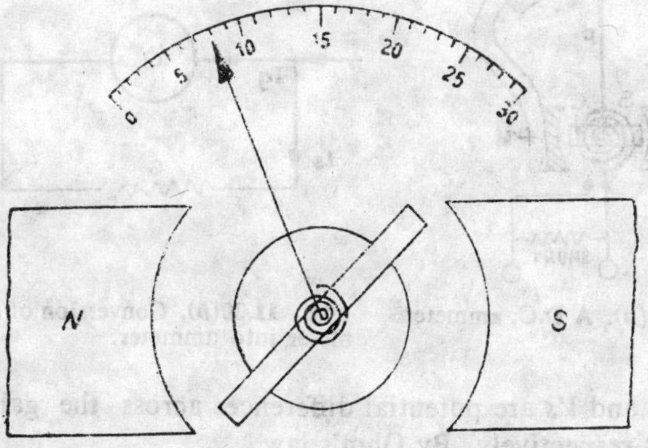

Fig. 11.38. Pivoted type of galvanometer.

A pivoted coil galvanometer is very small in size and can conveniently be used for laboratory work. When it is suitably modified it serves as a current measuring device called ammeter and potential difference measuring device called voltmeter.

11.54. Ammeter

A basic galvanometer cannot be used to measure large currents. This is because its range is small and a small current in it causes full scale deflection. However by placing a suitable low resistance resistor in parallel with it a large current of any magnitude can be measured. The low resistance resistor placed in parallel with the galvanometer is called shunt and a shunted galvanometer is called ammeter, Fig. 11.39(*a*). It is to be noted that the ammeter always used in series in the circuit.

Let us compute the value of shunt resistor needed to convert a galvanometer into an ammeter for the measurement of current I. To do so consider a galvanometer shown in Fig. 11.39(*b*), whose resistance is Rg and which produces a full scale deflection with a current Ig. Let the galvanometer be shunted with a low resistance

R_s. The value of the shunt resistance Rs is so adjusted that only Ig amount of current passes through the galvanometer and rest of the current $I - I_g = I_s$ passes through the shunt. Since the potential differences across shunt as well as galvanometer terminals is same (as they are joined in parallel) then

$$V_g = V_s \quad \ldots(11.88)$$

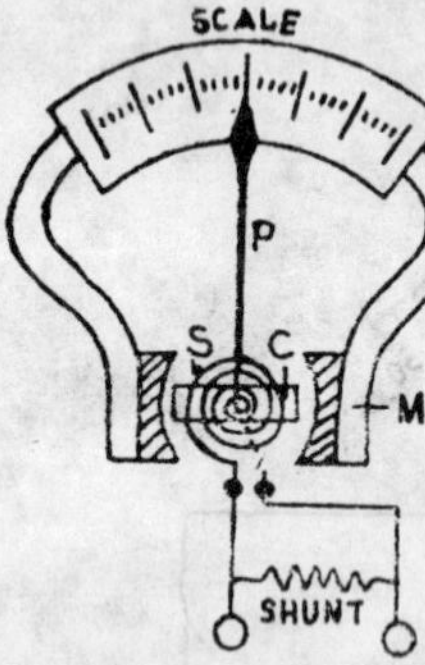

Fig. 11.39(*a*). A D.C. ammeter.

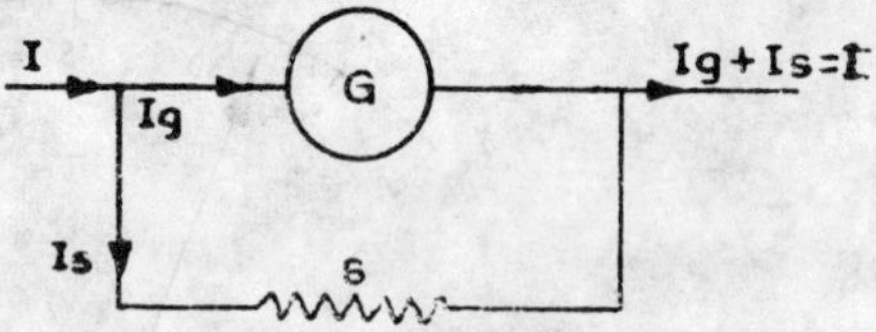

Fig. 11.39(*b*). Conversion of a galvanometer into ammeter.

where Vg and Vs are potential differences across the galvanometer and shunt respectively. By Ohm's law,

$$I_g \times R_g = I_s \times R_s = (I - I_g) \times R_s$$

$$\text{or } R_s = \frac{I_g \times R_g}{I - I_g} \quad \ldots(11.89)$$

The Equation 11.89 gives value of shunt to be used for a current I. As Ig and Rg are fixed for a galvanometer the Equation 11.89 shows that the value of the shunt depends upon the value of the current to be measured. This means greater is the current smaller will be the value of shunt resister and viceversa.

Example 11.21. A galvanometer requires 0.00015 A to produce a full scale deflection. The coil has a resistance of 60 Ω. What shunt resistance is needed to convert this galvanometer into an ammeter, reading 2A full scale?

Solution. Given $I = 2\text{A}$

$Ig = 0.00015$ A

$Rg = 60$ Ohm

we have $Rs = \dfrac{Ig \times Rg}{I - Ig}$

$$\therefore Rs = \frac{0.00015 \text{ A}}{(2.00 - 0.00015) \text{ A}} \times 60 \text{ ohms}$$

$$= 0.0045 \text{ ohms.}$$

11.55. Voltmeter

The voltage across the usual galvanometer is very low. Therefore it cannot be used for the measurement of high voltage. For instance consider the galvanometer mentioned in the example 11.47. The maximum potential difference that a galvanometer can measure is that, which produces full scale deflection in it and is given by $V_g = R_g I_g$. Thus in the above example 11.21 the maximum potential difference which this galvanometer can measure is equal $V_g = I_g \times R_g = 0.00015 \text{ A} \times 60 = 0.009$ volts only. This value is indeed very small. Therefore a galvanometer cannot be used to measure high potential difference. However, it can be made to measure high potential difference if suitably modified. For this, it is necessary to connect a high resistance R in series with the galvanometer. As a result when a high potential difference $V(>Vg)$ is applied across the combination, Fig. 11.40(*b*). The current passing through the galvanometer can be limited to a value, which will not exceed full scale deflection. Hence for full scale deflection the relation between potential difference V to be measured and high resistance R to be used is given by

$$V=(R_g+R)I_g \quad \text{...(11.90)}$$

$$R=\frac{V}{I_g}-R_g \quad \text{...(11.91)}$$

This high resistance R connected in series with galvanometer is called multiplier and can be seen from Equation 11.91 that its value depends upon the range of the potential difference which is

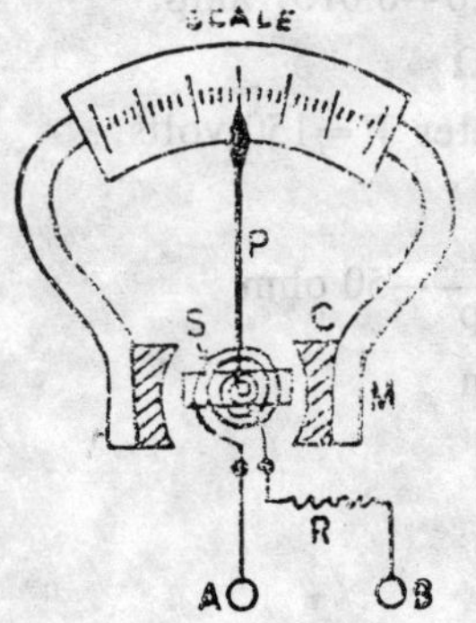

Fig. 11.40(*a*). A D.C. voltmeter.

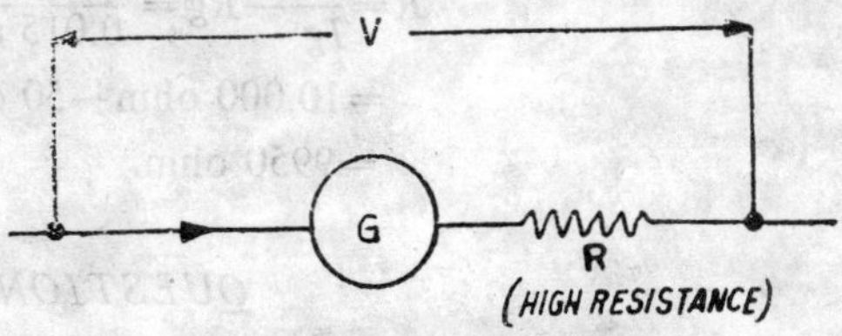

Fig. 11.40(*b*). Conversion of a galvanometer into a voltmeter.

to be measured. It is to be noted that the voltmeter is always connected in parallel with the element, across whose terminals, potential difference is to be measured.

Example 11.22. A galvanometer of resistance 60 Ω requires 0.00015 A for full scale deflection. Calculate the resistance that must be placed in series with it, so that it can measure a maximum potential difference 60 volts.

Solution. Given $V = 60$ volts

$Ig = 0.00015$ A

$Rg = 60\ \Omega$

we have $R = \frac{V}{Ig} - Rg = \frac{60 \text{ volt}}{0.00015\text{A}} - 60\ \Omega$

$= 40{,}000\ \Omega - 60\ \Omega$

$= 39{,}940\ \Omega$

Example 11.23. It is required to convert a galvanometer of current range 15 milliamperes and voltage range 750 millivolts into (*a*) an ammeter of range 25 amperes and (*b*) a voltmeter of range 150 v. Calculate the value of the necessary resistance for the purpose.

Solution. Current for full scale deflection = 15 milliamperes.

= 0.015 amp.

Potential difference for full scale deflection = 750 millivolts.

= 0.75 volts.

Resistance of the galvanometer,

$$Rg = \frac{0.75 \text{ volt}}{0.015 \text{ A}} = 50\ \Omega$$

(*a*) Range of the conversion into an ammeter = 25 amp.

Shunt resistance $Rs = \frac{Rg Ig}{I - Ig} = \frac{50 \text{ ohm} \times 0.015 \text{ amp}}{(25.0 - 0.015) \text{ amp}}$

$= 0.03\ \Omega$

(*b*) Range of the conversion into voltmeter $V = 150$ volts

Series resistance,

$$R = \frac{V}{Ig} - Rg = \frac{150 \text{ volt}}{0.015 \text{ amp}} - 50 \text{ ohm}$$

$= 10{,}000$ ohm $- 50$ ohm

$= 9950$ ohm.

QUESTIONS

1. Explain why it is so much easier to remove an electron from an atom of large atomic weight than it is to remove a proton.
2. Can two lines of force intersect? Explain your answer.
3. What is the capacity of a short circuited capacitor?

4. What are the drift velocity and relaxation time? Are they different for different materials?

5. What is non-ohmic resistance? Can $V=IR$ be applied to non-ohmic resistance?

6. What characteristic must (*a*) heating wire and (*b*) fuse wire have?

7. The expression $P=I^2R$ seems to suggest that the rate of joule heating in a resistor is reduced if the resistance is made less. But the expression for $P=\frac{V^2}{R}$ seems suggest just opposite. How do you reconcile with this paradox?

8. Is the resistance of 100 watt bulb greater or less than that of 60 watt bulb? Remember that the voltage applied across either is 220 volt.

9. A potential difference V is applied to a copper wire of diameter d and length l. What is the effect of electron drift velocity of (*a*) doubling v (*b*) doubling d (*c*) doubling l.

10. Discuss the advantages of a thermocouple as temperature measuring device.

11. A pair of platinum plates is inserted into a solution of copper sulphate and connected to a battery. Describe the electro chemical actions, which take place. Will this process continue indefinitely? Why?

12. A cable carrying a direct current is buried in a wall that stands in north-south plane. On the west side of the wall a horizontal compass needle points south instead of north. What are (*a*) the position of the cable, (*b*) the direction of the current in the cable.

13. Of the three vectors in the equation $\vec{F}=q(\vec{v}\times\vec{B})$ which pairs are always at right angles? Which may have any angle between them?

14. If an electron is not deflected in passing through a region of spaces. Can we be sure that there is no magnetic field in that region?

15. Is any work done by a magnetic field on a moving charge?

16. Assuming that earth's magnetic field is due to a large circular loop of current in the interior of the earth. What is the plane of the loop and what is the direction of the current around it?

17. A stream of electrons is moving towards the west. If the stream passes through a uniform magnetic field directed upward, in what direction is the electron beam deflected?

18. How would you reverse the direction of motion of a beam of electrons without changing the speed of the electrons at any time?

19. Two parallel wires carrying current in the same direction attract each other, while two beams of electrons travelling in the same direction repel each other. Explain, why?

20. A current bearing wire has no net static charge. Why can a magnetic field exert a force on the wire in spite of its uncharged condition?

21. Imagine that the room in which you are seated is fitted with magnetic field $\vec{B}$ pointing vertically upward. A circular loop of wire has its plane horizontal. For what direction of the current in the loop as viewed from above will the loop be in stable equilibrium with respect to forces and torques of magnetic origin.

22. An ammeter and voltmeter of suitable range are to be used to measure

the current and voltage of an electric lamp. If a mistake were made and the meters inter changed. What would happen?

PROBLEMS

1. Two small gilded pith balls carrying charges $+0.00250\ \mu$ coul and $-0.00600\ \mu$ coul are 250 mm apart in air. Compute the force between them. *(Ans. 2.16×10^{-6} newton)*

2. Calculate the position of the point in the neighbour of two point charges of $1.67\ \mu$ coul and $-0.600\ \mu$ coul situated 400 mm apart, where a third charge would experience no force. *(Ans. 600 mm from the second charge)*

3. Three point A, B and C are on the corners of an equilateral triangle. At A there is a point charge $+0.100\ \mu$ coul. What is the magnitude of the electric field intensity at a point midway between B and C if BC is 10.0 cm? *(Ans. 1.20×10^5 newton/coul)*

4. Show how one might connect a group of 1 μf capacitors so as to obtain the following capacitance 0.5, 1, 1.5, 2 and 2.25 μf respectively.

5. How many electrons passes through a lamp in one minute if the current is 300 A. *(Ans. 1.12×10 electrons)*

6. The resistance of a copper wire 2,500 cm long nnd 0.090 cm in diameter is 0.67 ohms in 20°C. What is the resistivity of copper at this temperature? *(Ans. 1.7×10^{-6} ohm-m)*

7. A 60 ohm electric lamp is left connetced to a 240 volt line for 3 minutes. How much energy is taken from the line. *(Ans. 1.72×10^4 Joule)*

8. A silver wire has a resistance of 1.25 ohm at 0°C and a temperature co-efficient of resistance of 0.00375 per °C. To what temperature must the wire be raised to double the resistance. *(Ans. 266°C)*

9. A copper coil of resistance 100 Ω at 40°C is heated to 100°C. Given the temperature co-efficient of resistance is $0.00427°C^{-1}$. Calculate the resistance when hot. *(Ans. 121.8 Ω)*

10. A copper wire of diameter 0.559 mm carries a current of 1 A. Calculate the drift velocity of the electrons assuming that one Cu atom contributes one free electron. *(Ans. 0.30 mm/sec)*

11. The heater element of an electric kettle has a constant resistance of 100 Ω and the applied voltage is 250 volts. Calculate the time taken to raise the temperature of 1 litre of water from 15°C to 90°C assuming that 85% of the power input to the kettle is usually employed. *(Ans. 9 min 50 seconds)*

12. A copper plate of 107.392 gm mass is placed in an electroplating bath. A steady current is sent through the bath for 30 min and the mass of plate is inereased to 109.074 gm. What total charge passed through the bath and what was the current in ampere assuming copper is divalent. *(Ans. 5.108 coul, 2.84 A)*

13. A current of 2.25 A flows for 27 min through a series of cells containing nickel nitrate, copper sulphate and silver nitrate. Find the masses of nickel (valence=2), copper (valence=2) and silver (valence=1) deposited. *(Ans. 1.11 gm., 1.20 gm., 4.07 gm.)*

14. The electron in the hydrogen atom circles around the proton with a speed of 2.18×10^6 m/s in an orbit of radius 5.3×10^{-11}m. What is the

strength of the magnetic induction does it produce at the proton?

(*Ans. 1.25 tesla*)

15. At what distance from a long straight wire carrying a current of 12 amperes will the magnetic field be equal to 3×10^{-5} weber/meter^2.

(*Ans.* $8 \times 10^2 m$)

16. A long straight conductor when carrying a certain current produces a magnetic field B at a point 7 cm from the axis of the wire. If a circular coil of single turn carrying the same current is to produce the same field B at its centre, what must be its radius? (*Ans. 0.22m*)

17. Calculate the magnetic induction B at a point P distance 2 meters from the centre of a circular coil of radius 1 mm carrying a current of 0.5 amp. The point is situated on the axis of the coil. (*Ans.* 39.3×10^{-14} *wb/m*^{2})

18. What will be the magnetic field at a point on the axis of a long solenoid containing 5 turns per cm length when a current of 0.8 amp flows through it.

(*Ans.* 5.027×10^{-8} *tesla*)

19. Two long straight parallel wires 5 cm apart each carries a current of 25 A. Calculate the magnetic induction at a point between the wires 2 cm from one and 3 cm from the other, when the currents are (*a*) in the same direction, (*b*) in the opposite direction.

[*Ans.* (*a*) 0.833×10^{-4} *wb/m*^{2} (*b*) 4.17×10^{-4} *wb/m*^{2}]

20. Two long straight wires and carrying a current of 5 A each are placed parallel to each other 5 cm apart. Find the force each wire exerts on the other? (*Ans.* 1×10^{-4} *N/m*)

21. A horizontal wire of 0.10 m long carries a current of 5 A. Find the magnitude and the direction of the magnetic field, which can support the weight of the wire assuming its mass is 3.0×10^{-3} kg/m? (*Ans.* 5.88×10^{-2} *tesla*)

22. A long straight wire carries a current of 2A. An electron travels with a velocity of 4.0×10^4 m/sec parallel to the wire, 0.1 m from it and in a direction opposite to the current. What force does the magnetic field of the current exert on the moving electron? (*Ans.* $2.56 \times 10^{-2} N$)

23. What is the velocity of a beam of electrons when the simultaneous influence of the electric field of intensity 34×10^4 volts/m and a magnetic field of flux density 2×10^{-3} tesla both fields being normal to the beam and to each other produces no deflection of the electron? What is the radius of curvature of the path of the electron beam when the electric field is removed?

(*Ans.* 1.7×10^8 *m/s*, $47.8 \times 10^{-3} m$)

24. A galvanometer having a resistance of 2000 ohm shows a full scale reading when a current of 200 mA flows through it. What shunt resistance across this instrument will enable it to be used as an ammeter with 2A as a full scale reading? (*Ans.* 222 Ω)

25. A voltmeter with a resistance of 180 Ω shows a full scale reading when 5 volts are applied to its terminals. What resistance connected to this instrument will give it a full scale reading when 25 volts are applied.

(*Ans.* 720 Ω *in series*)

12

ELECTRO MAGNETIC INDUCTION

We have seen that an electric current in a conductor produces magnetic field. This phonomenon was discovered by Hans Christian Oersted in 1820. Soon after his discovery a number of scientists engaged themselves to find, whether it could be made possible to produce electric current with the help of magnetic field. Among them, Michel Faraday of England and Joseph Henry of America were successful in their efforts. They independently demonstrated that when magnetic flux linked with a conductor is made to change an e.m.f. is induced in it, and a current is established thereby. The e.m.f. thus set up is called induced e.m.f. and the current is called induced current. The phenomenon of production of e.m.f. and current is called electro magnetic induction. Faraday published his findings first and is usually credited with the discovery.

12.2. Magnetic Flux

A change in magnetic flux linked with a circuit gives rise to electro magnetic induction. Therefore, to understand the pheno-

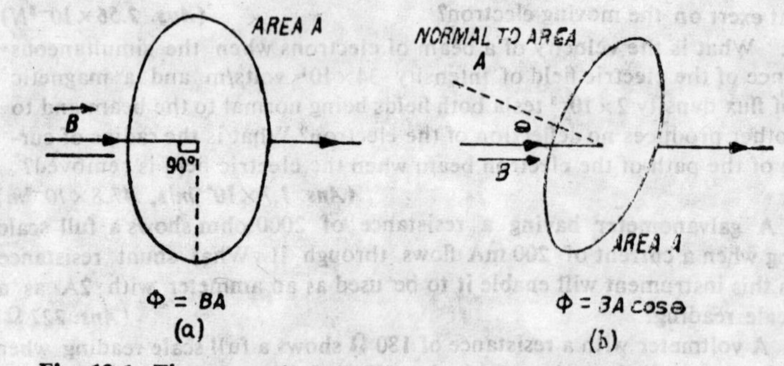

Fig. 12.1. The magnetic flux linking an area is the product of the area and the component of the magnetic intensity perpendicular to the area (*a*) Flux normal to the area $\phi = AB$, (*b*) magnetic intensity inclined to the area $\phi = AB$ cos.

menon of electromagnetic induction it is necessary to become familiar with the meaning of the magnetic flux.

Consider an area *A*, Fig. 12.1 (*a*) with its plane perpendicular to a uniform magnetic field of intensity $\vec{B}$. The magnetic flux ϕ through this area is defined as the scalar product of $\vec{B}$ and $\vec{A}$.

$$\text{i.e.} \quad \phi = B.A = BA \quad \ldots (12.1)$$

If $\vec{B}$ is net perpendicular to the plane of the area through which the flux is desired then $\phi = BA\cos\theta$. ... (12.2)

where θ is the angle between $\vec{B}$ and the normal to the area Fig. 12.1(*b*).

In SI system unit of flux is weber. A *uniform* magnetic field of 1 tesla normal to an area of 1 m² would give rise to a flux of 1 weber.

12.2. Positive and Negative Flux

A normal can be drawn to a plane in two ways. The flux is taken as positive when the normal points out in the direction of the field i.e. $\theta = 0$. When the normal points in opposite direction i.e. $\theta = 180°$ then the flux is taken as negative. Thus for negative flux.

$$\phi = BA \cos\theta = BA \cos 180° \\ = -BA \quad \ldots (12.3)$$

12.3. Faradays Experiments on Electromagnetic Induction

In one of his experiments Faraday took one closed circuit consisting of a coil and a galvanometer. When the bar magnet was moved in the neighbourhood of this coil following observations were made.

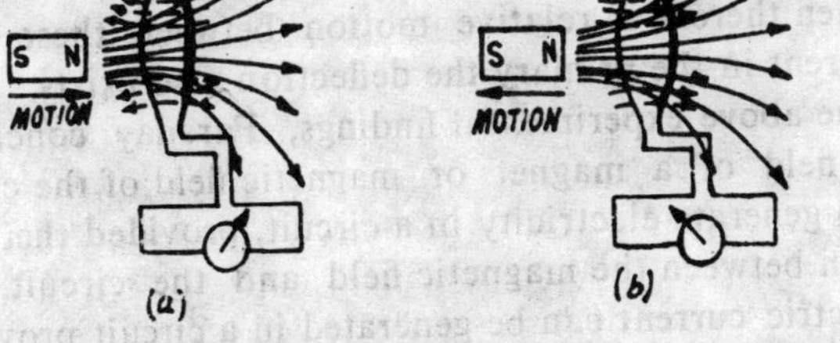

Fig. 12.2. Induced e.m.f. on change of flux due to motion of a magnet.

(*i*) If the magnet is moved towards the coil, with its north pole facing the coil the galvanometer deflects, the electric current is thus produced in the circuit, Fig. 12.2 (*a*).

(*ii*) If the magnet is moved away from the coil, the galvanometer again deflects but in opposite direction, Fig. 12.2 (*b*).

(*iii*) On the other hand, if the south pole end of the magnet is towards the coil, the deflections observed in (*i*) and (*ii*) are reversed.

(*iv*) The galvanometer also shows deflection if the magnet is kept stationery and the coil is moved.

(*v*) The deflection in the galvanometer is found to be proportional to the relative speed of the magnet and the coil.

(*vi*) The galvanometer shows no deflection when the magnet is held stationary with respect to the coil.

In another experiment two coils were placed close to each other, Fig. 12.3. One coil was connected to a galvanometer and the other to a battery with a plug key. When the battery circuit known as a primary circuit was closed by putting the plug with the key there was a sudden deflection in the galvanometer circuit, known as secondary circuit. Following results were observed.

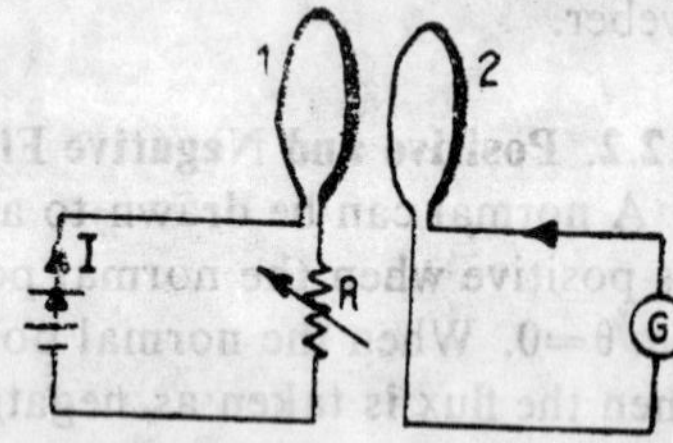

Fig. 12.3. Induced e.m.f. due to the motion of a current carrying circuit.

(*i*) When the primary circuit is closed or broken, the galvanometer needle deflects momentarily. The directions of deflection these cases are opposite to each other.

(*ii*) The galvanometer also shows deflection if the current in the primary circuit is varied. The deflection depends upon the time rate of change of current and not upon the value of the current.

(*iii*) The current produced in the secondary circuit in (*i*) and (*ii*) is momentary and does not axist when the current in the primary reaches a steady value.

(*iv*) When there is a relative motion between these coils with steady current in the primary the deflection also exists.

From the above experimental findings, Faraday concluded that magnetic field of a magnet or magnetic field of the current in a circuit can generate electricity in a circuit, provided there is a relative motion between the magnetic field and the circuit. In other words electric current can be generated in a circuit provided there is a change of flux linked with the circuit.

12.4. Faraday's Laws of Electro Magnetic Induction

From his experimental results, Faraday formulated two laws as given below.

(*i*) An induced e.m.f. is established in a circuit whenever magnetic flux linked with the circuit is changed.

(*ii*) The magnitude of the induced e.m.f. is equal to the time rate of change of the magnetic flux linked with the circuits, stated mathematically,

$$|\mathcal{E}| = \frac{d\phi}{dt} \quad \ldots (12.4)$$

where $\mathcal{E}$ is the indvced *e.m.f.*

Let us consider a coil of one turn, placed in a magnetic field. If the flux through the coil changes from ϕ_1 to ϕ_2 during a time interval t an e.m.f. $\mathcal{E}$ is produced in the coil.

where

$$|\mathcal{E}| = \frac{\phi_2 - \phi_1}{t} \quad \ldots (12.5)$$

If a coil consists of N turns, the same flux change is linked with all these turns. Therefore e.m.f. generated in the coil is given by,

$$|\mathcal{E}| = \frac{N(\phi_2 - \phi_1)}{t} \quad \ldots (12.6)$$

12.5. Lenz's Law

Faraday's laws stated above give the magnitude of the induced e.m.f. but not its direction. The direction of the induced e.m.f. is determined by Lenz's law which is stated as follows.

The direction of the induced e.m f. is always such that any current it produces opposes through its magnetic effect, the change producing the e.m.f.

To illustrate the application of Lenz's law, consider a north pole, which is being pushed towards a coil, Fig. 12.4. The current induced in the coil is in such a direction that its magnetic field opposes the approach of the magnet towards the coil. Hence the face of the coil towards the magnet would behave as a north pole. This means, the magnetic field of the coil is directed towards the north pole of the magnet. This requires that the current in the coil must be anticlockwise. When the north pole is withdrawn, the direction of the induced e.m.f. and the current reverse. The nearer face of the coil behave as a south pole, which opposes, the withdrawal of the north pole of the magnet.

Similar phenomenon is observed, when a south pole is used instead of a north pole. The direction of the current in this case is clockwise, when the south pole is pushed towards the coil and anticlockwise, when the south pole is taken away from the coil.

Taking into consideration the direction of the induced e.m.f. as given by Lenz's Law the Equation 12.4 is rewritten as,

$$\mathcal{E} = -\frac{d\phi}{dt} \qquad \ldots (12.7)$$

This equation is for a coil containing one turn only. However if the coil consists of N turns then, the induced e.m.f. is given by,

$$\mathcal{E} = -N\frac{d\phi}{dt} \qquad \ldots (12.8)$$

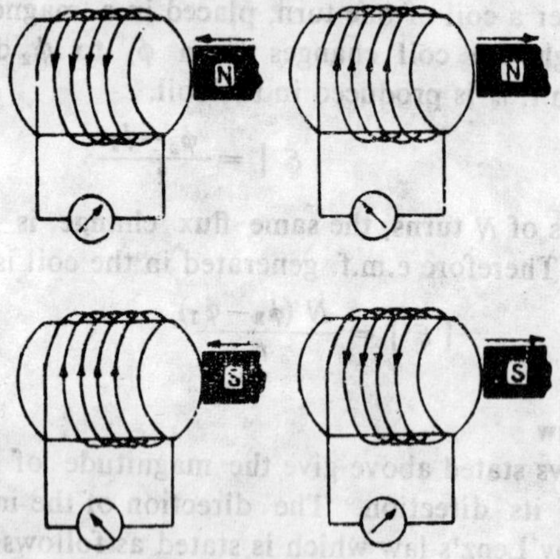

Fig. 12.4. The current arising from an induced e.m.f. always opposes through its magnetic effects, the change inducing the e.m.f. whether a bar magnet is pushed into the coil or pulled out of it, the magnetic field of the induced current opposes the motion of the magnet (Lenz's law).

12.6. Lenz's Law and Principle of Conservation of Energy

Lenz's law is a particular example of the principle of conservation of energy. When the north pole of the magnet is pushed towards the coil, the face of the coil towards the magnet acquires north polarity. The north pole thus formed opposes the motion of the magnet. So mechanical work has to be done to overcome the force of repulsion, to push it further. This mechanical work gets transformed into electrical energy which appears as induced current in the coil. If the magnet is moved more rapidly, more work is done thereby and a larger current is induced in the coil.

Similarly when the north pole is withdrawn, a south pole is induced on the very face of the coil and work has to be done against the force of attraction. This mechanical work gets transformed to electrical energy and hence an electric current is established in the coil. On the other hand when a change in current in a primary coil induces an e.m.f. in a neighbouring secondary coil, the current in the secondary will be in such direction as to require expenditure of additional energy in the primary to maintain the current in it.

Example 12.1. A uniform magnetic field of magnetic flux density 0.5 web/m² extends over a plane circuit of area 0.5 m² and is normal to the plane of the circuit. How quickly must the field be reduced to zero if e.m.f. of 500 volts is to be induced in the circuit.

Solution. $\phi = BA$

and an induced e.m.f.

$$\mathcal{E} = -\left(\frac{0 - BA}{t}\right)$$

The negative sign merely indicates that the induced e.m.f. ($\mathcal{E}$) is in opposition to the change causing it but has no other significance.

Hence,

$$t = \frac{BA}{\mathcal{E}} = \frac{0.5 \text{ web/m}^2 \times 0.5 \text{ m}^2}{500 \text{ volts}}$$

$$= 5 \times 10^{-4} \text{ seconds.}$$

Example 12.2. A 200 turn closed packed coil with cross sectional area of 0.16 m² is placed with its plane perpendicular to a uniform magnetic field. The field value then varies at a uniform rate from 0.10 web/m² to 0.50 web/m² in 2×10^{-2} seconds. Find the e.m.f. induced in the coil.

Solution. $$\mathcal{E} = -\frac{N(\phi_2 - \phi_1)}{t}$$

Here $N = 200$

$$\phi_2 - \phi_1 = A(B_2 - B_1)$$

$$= 0.16 \text{ m}^2 \ (0.50 - 0.10) \text{ web/m}^2$$

$$= 0.16(0.50 - 0.10) \text{ web}$$

and $t = 2 \times 10^{-2}$ seconds

$\therefore$ $$\mathcal{E} = -\frac{200 \times 0.16(0.50 - 0.10) \text{ web}}{2 \times 10^{-2} \text{ sec}}$$

$$= -640 \text{ web/sec} = -640 \text{ volts.}$$

Negative sign indicates that the e.m.f. acts in such a direction as to send a current, which would oppose increase of flux.

12.7. Methods of Producing Induced e.m.f.

An induced e.m.f. is produced when the magnetic flux passing through the circuits is made to change. The change may be brought about by any of the following ways in view of the Equation 12.2.

(*i*) Changing the magnetic field $\vec{B}$.

(*ii*) Changing the area of the circuit.

(*iii*) Changing the relative orientation (*i.e.* changing the angle θ)

(*i*) **Induced e.m.f. by changing flux $\vec{B}$.** When there is a relative motion between the magnetic field of any origin and a closed circuit, the magnetic flux ϕ passing through the circuit gets changed. Thereby an induced e.m.f. is established which causes an induced current in the circuit. This has already been discussed in article 12.3 and 12.5.

(*ii*) **Induced e.m.f. by changing A.** Consider a conductor *OR* of length *l*. It is sliding with a velocity *v* towards right, on an U-shaped conducting rails placed in a uniform magnetic field of flux density $\vec{B}$. The field is normal to the plane of the paper and is directed into it, Fig. 12.5.

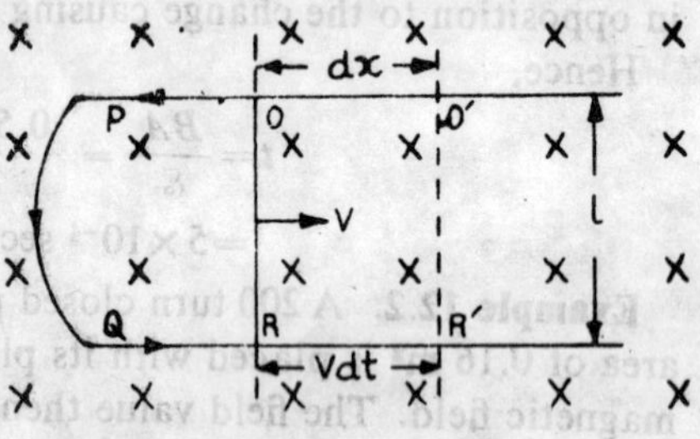

Fig. 12.5. Induced e.m.f. caused by change in flux due to change in area of the circuit.

Let the conductor cover a distance dx in time dt and occupy new position $O'R'$. As a result of this, the area of the circuit changes from $OPQR$ to $O'PQR'$. This causes an increase in flux given by

$$d\phi = B \times (\text{area } O'PQR' - \text{area } OPQR)$$
$$= B \times \text{area } OO'RR'$$
$$= B \times l \times dx = Blvdt \quad \ldots (12.9)$$
$$(\because \; dx = vdt)$$

Induced e.m.f.

$$|\mathcal{E}| = \frac{d\phi}{dt} = Blv \quad \ldots (12.10)$$

The induced e.m.f. thus produced is called motional e.m.f.

12.8. Fleming's Righthand Rule

The direction of the induced e.m.f. and the current in general are given by Lenz's law. But it is more convenient to use Fleming's Righthand rule to determine them. The rule is given as follows.

Stretch the thumb, the forefinger and the central finger of the righthand mutually perpendicular to one another as shown in,

Fig. 12.6. If the thumb represents the direction of the motion of the conductor, the forefinger represents the direction of the magnetic field, then the central finger points in the direction in which the current is induced in the circuit.

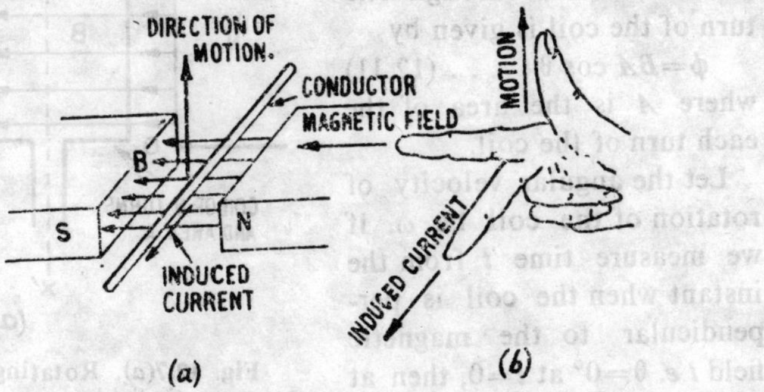

Fig. 12.6. Fleming's righthand rule.

Example 12.3. An aeroplane with a wing span of 50 meters flies at a horizontal speed of 200 meter/sec, in a region where the vertical component of the magnetic field due to earth is 5×10^{-5} web/m². What is the potential difference between the tips of the wing?

Solution. The metal between the wing tips can be considered to be a single conductor, which is moving at right angles to the magnetic field. The induced e m f. creates a potential difference between the tips. Its value is given by

$$|\mathcal{E}| = Blv$$

Here $B = 5.0 \times 10^{-5}$ web/m²

$l = 50$ m

$v = 200$ m/sec

Hence $\mathcal{E} = 5 \times 10^{-5}$ web/m² $\times 50$ m $\times 200$ m/sec

$= 5 \times 10^{-5} \times 10^{4}$ web/sec

$= 5 \times 0^{-1}$ volts $= 0.5$ volts.

12.9. Induced e.m.f. by Changing Relative Orientation of the Coil and the Magnetic Field (Rotating Coil in a Magnetic Field)

Consider a rectangular coil *OPQR* which is free to rotate about an axis XX^1 in a magnetic field of flux density $\vec{\mathbf{B}}$. The axis

XX' lies in the plane of the coil but perpendicular to the magnetic field as shown in Fig. 12.7 (*a*). At any instant of time, when the normal to the coil makes an angle θ, with the field, the flux through one turn of the coil is given by

$$\phi = BA \cos\theta \quad \ldots (12.11)$$

where A is the area of the each turn of the coil.

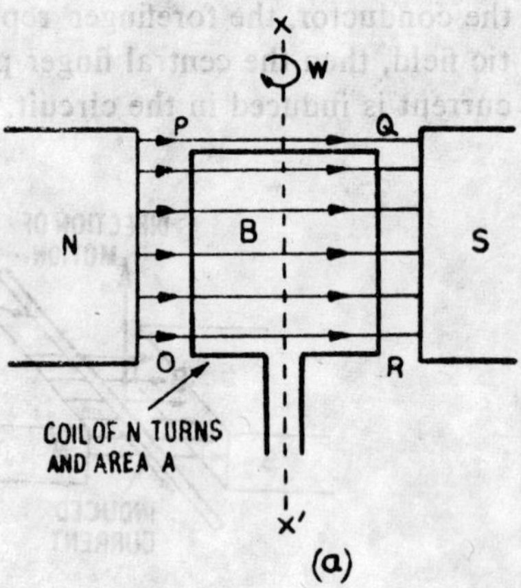

Fig. 12.7(*a*). Rotating coil in a magnetic field.

Let the angular velocity of rotation of the coil be ω. If we measure time t from the instant when the coil is perpendicular to the magnetic field *i.e.* $\theta = 0°$ at $t = 0$, then at any instant of time $t, \theta = \omega t$ and Equation 12.11 can be written as:

$$\phi = BA \cos \omega t \quad \ldots (12.12)$$

The induced e.m.f. for one turn at any time is

$$-\frac{d\phi}{dt} = -\frac{d}{dt}\,(BA \cos \omega t)$$

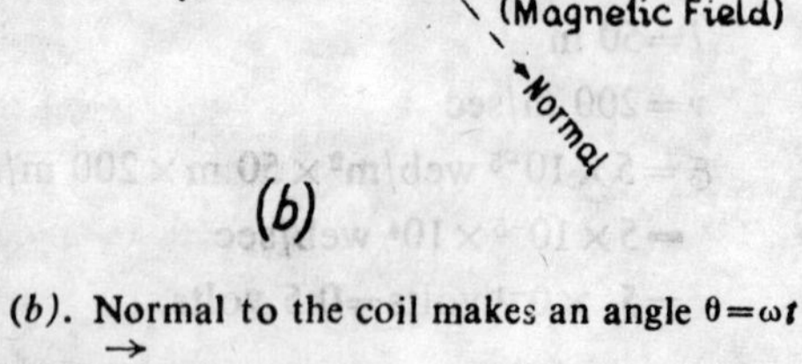

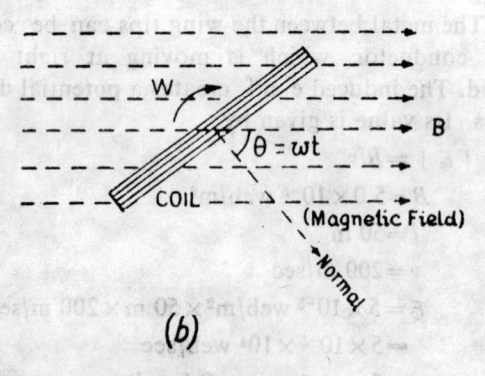

Fig. 12.7 (*b*). Normal to the coil makes an angle $\theta = \omega t$ with the magnetic field $\vec{\mathbf{B}}$.

If the coil consists of N turns, the induced e.m.f. is given by,

$$\mathcal{E} = -N\frac{d\phi}{dt} = -N\frac{d}{dt}\,(BA \cos \omega t)$$

$$= NBA\,\omega \sin \omega t \quad \ldots (12.13)$$

The Equation 12.13 indicates that the induced e.m.f. ($\mathcal{E}$) varies with time. Since maximum value of sin $\omega t = +1$, then we have the maximum value of $\mathcal{E}$ as

$$\mathcal{E}_{max} = \mathcal{E}_0 = NBA\omega \quad \ldots (12.24)$$

$\mathcal{E}_0$ is called the maximum or peak value of induced e.m.f. The value of this obviously depends upon the:

(*i*) Number of turns in the coil.
(*ii*) Strength of the magnetic field.
(*iii*) The area of the coil.
(*iv*) Speed of rotation of the coil.

We can write the Equation 12.13 in view of Equation 12.14 as

$$\mathcal{E} = \mathcal{E}_0 \sin \omega t \quad \ldots (12.15)$$

The Equation 12.15 gives the variation of e.m.f. with respect to time. If $\mathcal{E}$ is plotted against t, the graph is a sine curve as shown

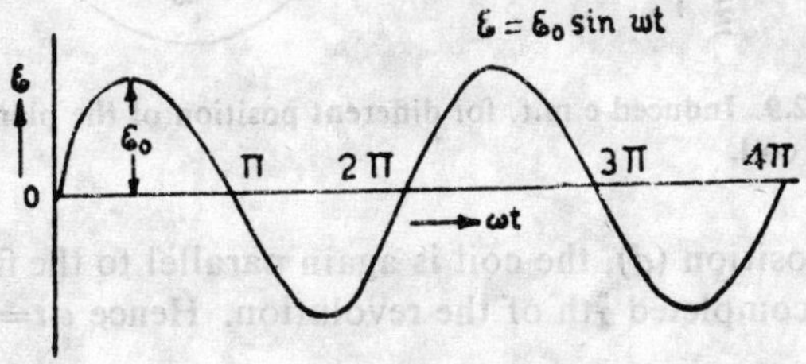

Fig. 12.8. Wave form of alternating e.m.f. $\mathcal{E} = \mathcal{E}_0 \sin \omega t$.

in Fig. 12.8. Such e.m.f. is called sinusoidal or alternating e.m.f. The e.m.f. will appear at the terminal of the rotating coil and may be transferred to an external circuit by suitable means.

The variation of e.m.f. with respect to time may also be diagramatically shown by considering the different positions of the coil. This has been shown in Fig. 12.9.

1. In the position (*a*) the coil is perpendicular to the field

$$\omega t = 0 \text{ and } t = 0$$

Thus induced e.m.f. is given by

$$\mathcal{E} = \mathcal{E}_0 \sin \omega t = \mathcal{E}_0 \times 0 = 0$$

2. In the position (*b*), the coil is parallel to the magnetic field and it has completed a quarter revolution. Hence $\omega t = \pi/2$ and $t = T/4$ if T is the time for one complete revolution.

Then induced e.m.f. is given by

$$\mathcal{E} = \mathcal{E}_0 \sin \omega t = \mathcal{E}_0 \sin \pi/2 = \mathcal{E}_0$$

3. In the position (*c*), the coil again perpendicular to the magnetic

field and has completed 1/2 revolution. Hence $\omega t=\pi$ and $t=T/2$. The induced e.m.f. in the position is given by,

$$\mathcal{E}=\mathcal{E}_0 \sin \omega t=\mathcal{E}_0 \sin \pi=0$$

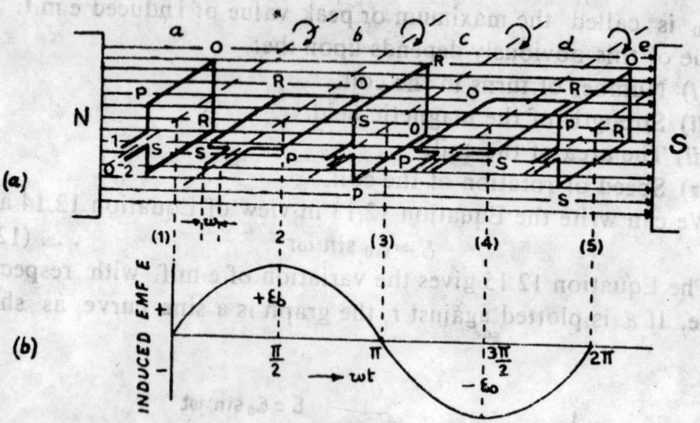

Fig. 12.9. Induced e.m.f. for different position of the plane of the coil.

4 In the position (*d*), the coil is again parallel to the field direction and has completed ¾th of the revolution. Hence $\omega t=3\pi/2$ and $t=3T/4$.

The induced e.m.f. is given by

$$\mathcal{E}=\mathcal{E}_0 \sin \omega t=\mathcal{E}_0 \sin 3\pi/2=-\mathcal{E}_0$$

In this position the e.m.f. attains its peak value but in opposite direction and the face of coil is reversed.

5. In the position (*e*), the coil is again perpendicular to the field direction and completed one full revolution. Hence, $\omega t=2\pi$ and $t=T$ in this position.

The induced e.m.f is given by,

$$\mathcal{E}=\mathcal{E}_0 \sin \omega t=\mathcal{E}_0 \sin 2\pi=0$$

Now if the coil keeps on rotating, the above operations from (1) to (5) gets repeated and we get a sinusoidal or alternating e.m.f. continuously.

It is to be noted that had we measured the time since the instant, when the coil was parallel to the field i.e. at $t=0$ and $\theta=\pi/2$ then at time t, we would have $\phi=NBA \cos (\omega t+\pi/2)$ and $\mathcal{E}$ would have been given as,

$$\mathcal{E}=\mathcal{E}_0 \cos \omega t \qquad \ldots(12.16a)$$

The curve of e.m.f. would have been shown as in Fig. 12.10. Except for the matter of choosing zero of time, there is no essential

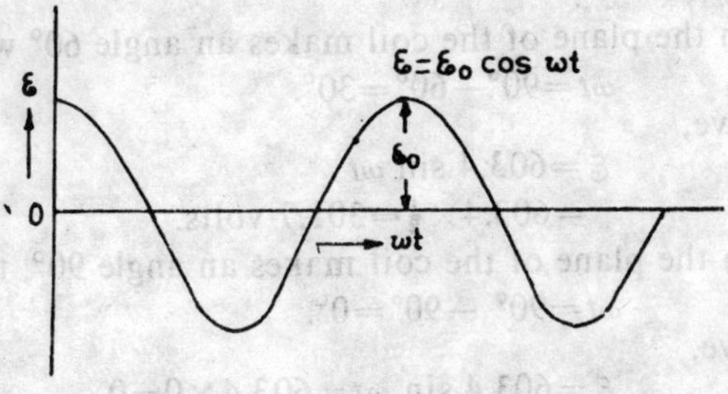

Fig. 12.10. Wave form of alternating e.m.f. $\mathcal{E}=\mathcal{E}_0 \cos \omega t$.

difference between the sine and cosine forms. Hence both curves are called sinosoidal curve. Further if the frequency of rotation of the coil is f i.e. the number of rotation the coil makes per second is f, then Equation 12.15 can be written as,

$$\mathcal{E}=\mathcal{E}_0 \sin 2\pi f t \qquad \ldots(12.16b)$$

as $$f=\frac{1}{T} \text{ and } \omega=2\pi f$$

Example 12.4. A rectangular coil of dimension 0.3m × 0.4m consisting of 200 turns rotates about an axis parallel to its long side, making 3000 revolutions per minute in a magnetic field of 0.08 tesla. What are the instantaneous values of induced e.m.f., when the plane of coil makes angle (*a*) 0°, (*b*) 50° and 90° with the field direction.

Solution. Given f=3000 revolutions per minute.

=50 revolutions per second

$\omega=2\pi\times 50=100\pi$

N=200 turns

A=0.3m × 0.4m=0.12m²

B=0.08 tesla=0.08 web/m²

The formula is to be used is,

$$\mathcal{E}=\mathcal{E}_0 \sin \omega t=NA\omega B \sin \omega t \qquad \ldots(i)$$

Here the magnitude of $\mathcal{E}_0$ is

$$\mathcal{E}_0=NA\omega B=200\times 0.12\times 0.08\times 100\pi$$

$$=192\pi=192\times\frac{22}{7}=603.4$$

So the formula (*i*) becomes,

$$\mathcal{E}=603.4 \sin \omega t$$

(*a*) When the plane of the coil makes angle 0° with the field, the normal to it makes an angle 90° or $\pi/2$.

$\therefore$ $\omega t=\pi/2$

Hence $\mathcal{E}=603.4 \sin \omega t=603.4\times1=603.4$ volts

(*b*) When the plane of the coil makes an angle 60° with $\vec{\mathbf{B}}$,

$$\omega t=90°-60°=30°.$$

So we have,

$$\mathcal{E}=603.4 \sin \omega t$$
$$=603.4\times\tfrac{1}{2}=301.7 \text{ volts.}$$

(*c*) When the plane of the coil makes an angle 90°, then

$$\omega t=90°-90°=0°.$$

So we have,

$$\mathcal{E}=603.4 \sin \omega t=603.4\times0=0.$$

Example 12.5. A loop of 0.04 m² in area having 100 turns rotates in a magnetic field of 0.01 *T*. The period of revolution of the loop is 0.1 sec Find the maximum value of the induced e.m.f. in the loop. The axis of rotation is perpendicular to the magnetic field lines.

Solution. Given, $A=0.04$ m², $N=100$

$B=0.01$ tesla.

$$\mathcal{E}_{max}=\mathcal{E}_0=\omega BNA=\frac{2\pi}{T}BNA$$

$$=\frac{2\times3.14\times0.01\times0.04\times100}{0.1}$$

$$\approx 2 \text{ volts.}$$

12.10. The Generator or Dynamo

The principle discussed above has the practical application in a generator or dynamo.

The generator is an instrument by which we get electrical energy at the expense of mechanical energy. If the current set up by a generator alternates i.e. changes its direction at regular interval of time, in the outer part of the circuit, it is called a.c. generator. If the current generated always flows in the same direction in the outer part of the circuit, it is called d.c. generator.

(*a*) **A.C. generator**. A schematic diagram of a.c. generator is shown in Fig. 12.11. It essentially consists of the following parts. 1. Field magnet, 2. Armature, 3. Slip ring, 4. Brushes, 5. Source of mechanical energy.

1. Field magnet. The field magnet provides the magnetic field within the space where the armature rotates. It is usually a permanent horse shoe shaped magnet for small dynamo and a strong electromagnet for a generator producing large e.m.f. and current.

2. Armature. The coil *OPQR* shown in Fig. 12.11 is called armature. It consists of a laminated core of soft iron on which a number of turns of insulated copper wire is wound. The iron core concentrates the magnetic lines of force to create a strong magnetic field. The core is laminated to reduce the induced currents in it to the

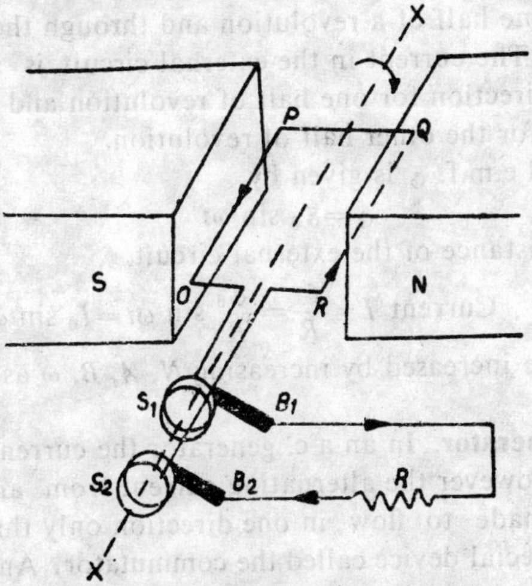

Fig. 12.11. A.C. generator (dynamo).

minimum. The armature can be rotated about an axis XX^1, which is perpendicular to the direction of the field.

3. The slip ring. There are two metal rings *SS'*, Fig. 12.11 to which the ends of the armature coil are connected. These rings are co-axial with the axis of rotations and fixed rigidly to the same shaft, which is used for rotating the coil. They are insulated from the shaft and rotate with the armature, which is also properly insulated from the shaft.

4. Brushes. The brushes B_1 and B_2 Fig. 12.11 are made of carbon rods and are kept lightly pressed against the slip rings by means of springs and have leads to the external circuit.

5. **Source of mechanical energy**. The armature is rotated about the axis *XX'* by means of steam engine, oil engine, steam turbine or water turbine. The mechanical energy involved in the rotation of the armature by means of turbine, gets converted into electrical energy.

Principle. The basic principle of operation of an a.c. generator is the production of induced e.m.f. across the ends of the coil rotating in a magnetic field. This has been discussed in the article 12.10.

As the coil *OPQR* rotates in the magnetic field, Fig. 12.11 an induced e.m.f. develops in the coil, which causes induced current in it. This current is collected by means of brushes, bearing on the two slip rings and is conveyed to the external circuit.

The output e.m.f. of the generator is alternating as shown in Fig. 12.8. As the armature rotates the current flows out through the brush B_1 for one half of a revolution and through the brush B_2 for the next half. The current in the external circuit is alternating *i.e.* flows in one direction for one half of revolution and flows in opposite direction for the other half of revolution.

The induced e.m.f. $\mathcal{E}$ is given by,

$$\mathcal{E}=\mathcal{E}_0 \sin \omega t$$

If R is the resistance of the external circuit,

$$\text{Current } I=\frac{\mathcal{E}}{R}=\frac{\mathcal{E}_6}{R} \sin \omega t=I_0 \sin \omega t \quad \ldots(12.17)$$

$\mathcal{E}$ and I can be increased by increasing N, A, B, ω as explained in article 12.9

(b) **D.C. generator**. In an a.c. generator the current is essentially alternating. However the alternating current from an a.c. generator can be made to flow in one direction only through external circuit by a special device called the commutator. An a.c. generator modified in this way called direct current (D.C.) generator.

For this modification the two ends of the armature are connected to two halves C_1, C_2 of a metal cylinder, called the split ring commutator instead of slip rings Fig. 12.12. These two half cylinders are fixed round the main shaft but insulated from each other. They are also, so arranged that during a half revolution of the coil each half cylinder C_1 or C_2 makes contact with a particular fixed brush and when the current is reversed during the other half revolution, the same half cylinder (C_1 or C_2) is in contact with other fixed brush. Referring to Fig. 12.12 we see that during the first half of the revolution (say anticlockwise as shown by arrow) the current is from C_2

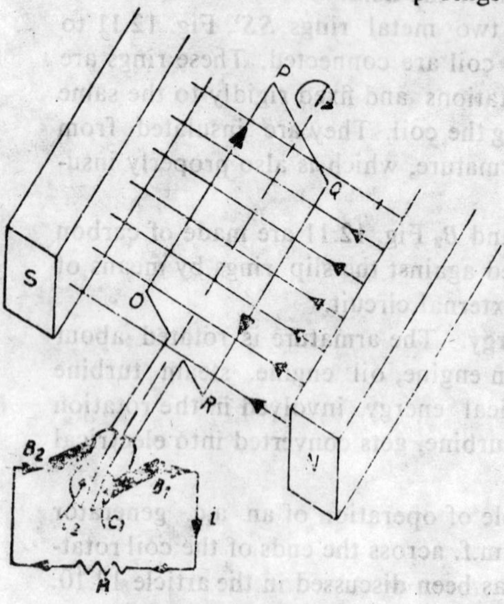

Fig. 12.12. D.C. generator (dynamo).

to C_1 in the coil *OPQR* and from positive brush B_1 to the negative brush B_2 through the external circuits. During the other half revolution, when the direction of the current is reversed the current is from C_1 to C_2 in the coil but during this time the position of C_1 to C_2 are also interchanged. Therefore C_2 makes contact with the positive brush and C_1 with the negative brush. Here in the external circuit, the direction of the current remains always from positive to negative brush

The e.m.f. generated by a single coil, connected to two split rings and rotating in a magnetic field is in fact unidirectional but not constant. Its value fluctuates or pulsates from maximum to zero value as shown in Fig. 12.13(*a*). To make such pulsating e.m.f. nearly constant, a large number of coils in series are spaced uniformly over the armature. As a result, the maximum value of e.m.f.

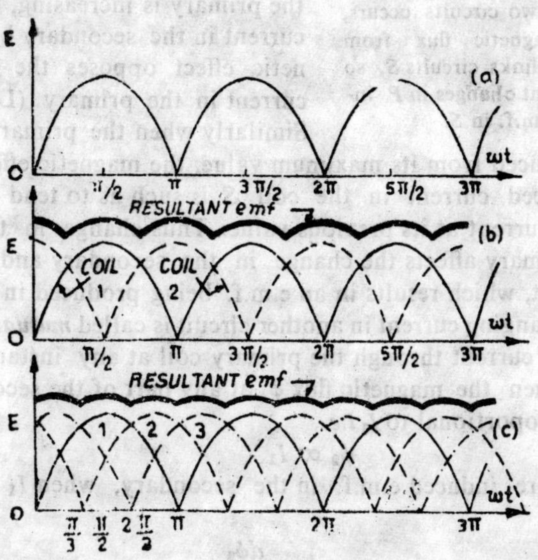

Fig. 12.13. e.m.f. of a D.C. generator. (*a*) Single coil, (*b*) Two coils, (*c*) Three coils.

occurs in each coil at different instant and the net effect is an almost constant unidirectional e.m.f. The resultant e.m.f. for two and three such coils have been shown in Fig. 12.13(*b*) and (*c*) respectively.

12.11. Mutual Inductance

Consider the neighbouring coils of Fig. 12.14. The coil P is connected to a cell B through a key K and is called the primary. The coil S is connected to a sensitive galvanometer G and is called the secondary. When the key is pressed, the current in the coil P rises and the associated magnetic lines of flux produces a change in the magnetic flux linking the coil S. As a result, there is an induced e.m.f. and a resulting induced current in the coil S. The current in the coil S lasts only, as long as the current in P is changing.

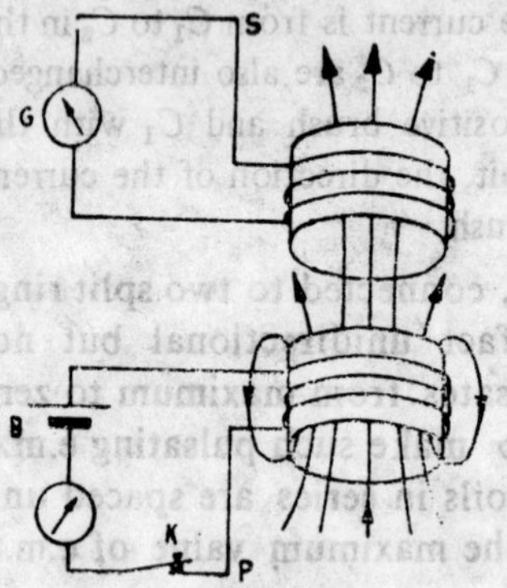

Fig. 12.14. Mutual inductance between two circuits occurs, when magnetic flux from current P links circuits S, so that current changes in P induce an e.m.f. in S.

There is another effect taking place simultaneously. When the current in the primary is increasing, the induced current in the secondary by its magnetic effect opposes the rise of the current in the primary (Lenz's law). Similarly when the primary current is being reduced from its maximum value, the magnetic effect due to the induced current in the coil S is such as to tend to keep the primary current at its previous value. Thus change in the current in the primary affects the change in the secondary and viceversa. This effect, which results in an e.m.f. being produced in one circuit due to changing current in another circuit is called ***mutual induction.***

Let the current through the primary coil at any instant of time be I_1. Then the magnetic flux ϕ_2 at any part of the secondary coil will be proportional to I_1 *i.e.*

$$\phi_2 \propto I_1$$

Therefore induced e.m.f. in the secondary, when I_1 changes is given by

$$\mathcal{E}_2 = -\frac{d\phi_2}{dt}$$

i.e.

$$\mathcal{E}_2 \propto -\frac{dI_1}{dt}$$

or

$$\mathcal{E}_2 = -\frac{M dI_1}{dt} \quad \ldots (12.18)$$

The proportionality constant M is called the co-efficient of mutual inductance or simply mutual inductance. Mutual inductance is the property of a pair of circuits. Therefore if the current in the secondary changes at the rate of $\frac{dI_2}{dt}$ then the e.m.f. in the primary is also given by

$$\mathcal{E}_1 = -M \frac{dI_2}{dt} \quad \ldots (12.19)$$

The proportionality constant M is the same in both the Equations 12.18 and 12.19.

In the Equation 12.18 if $\frac{dI_1}{dt}=1$, then induced e.m.f. $\mathcal{E}_2$ is numerically equal to M. Hence M is defined as the e.m.f. induced in the secondary by unit rate of change of current in the primary, when $\mathcal{E}$ is expressed in volts and dI/dt in ampere/second, M is expressed in volt-sec/amp. This unit is also called henry (H). The mutual inductance between the circuits is 1 henry (H) if a rate of change of current of 1 amp/second in the primary produces an induced e.m.f. of 1 volt in the secondary.

The value of M depends upon (*i*) the number of turns in the coils (*ii*) their geometrical shape, (*iii*) relative orientation of the coils and their separation. It is maximum when entire flux of the primary links with the secondary. Moreover the mutual inductance can be increased by a factor μ if the coils are wound over an iron core. Here μ is the permeability of iron.

Example 12.6. The mutual inductance between two circuits is 0.4 H. Find the e.m.f. induced in the secondary at an instant, when the current is changing at the rate of 90 amp/second in the primary.

Solution. $M = 0.4\text{H} = 0.4$ volt-sec/amp.

$$\mathcal{E}_2 = M \frac{dI_1}{dt} = 0.4 \text{ volt-sec/amp} \times 90 \text{ amp/sec.}$$

$$= 36 \text{ volts.}$$

12.12. Self Inductance

When there is a current in a circuit, it gives rise to a magnetic field, which itself links with the circuit. If the current in the circuit varies the magnetic flux links with it also varies. As a result an induced e.m.f. is set up in it, which opposes the change in the current. Such an induced e.m.f. is called self induced electromotive force and the phenomenon is known as Self Induction.

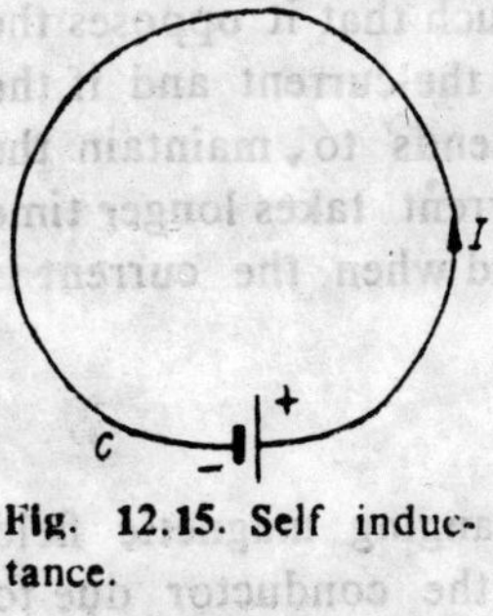

Fig. 12.15. Self inductance.

Let C be a circuit, Fig. 12.15 in which a current I is flowing. This current, sets up a magnetic field and hence a magnetic flux ϕ is linked with it. As the magnetic field strength is proportional to the current I flowing through a circuit, the magnetic flux, which is proportional to the magnetic field is proportional to the current I in the circuit. Hence we can write,

$$\phi \propto I$$

Therefore, induced e.m.f. $\mathcal{E}$ is given by

$$\mathcal{E} = -\frac{d\phi}{dt}$$

Hence $$\mathcal{E} \propto -\frac{dI}{dt}$$

or $$\mathcal{E} = -L\frac{dI}{dt} \quad \ldots (12.19)$$

where L is a constant called co-efficient of self inductance or simply inductance. In the Equation 12.19 if, $dI/dt=1$ then $\mathcal{E}$ is numerically equal to L. Therefore self inductance may be regarded as the self induced e.m.f. per unit rate of change of current. Like M, L is expressed in henry. A circuit has a self inductance of one henry if an e.m.f of one volt is induced in the circuit, when the current in the circuit changes at the rate of one ampere per second.

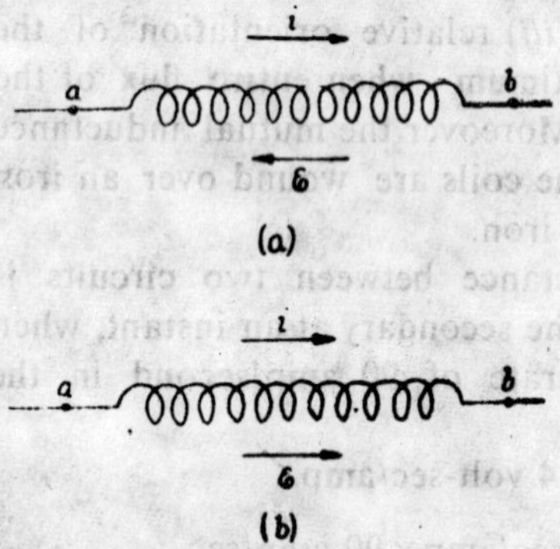

Fig. 12.16. Direction of a self induced e.m.f. (*a*) *i* increasing, $\mathcal{E}$ is opposite to *i*, point *a* is at higher potential than *b*. (*b*) *i* decreasing $\mathcal{E}$ and *i* in the same direction *b* is at higher potential than *a* ($R=0$)

This direction of the self induced e.m.f. is found from Lenz's law. If the current is increasing the direction of the induced e.m.f. is such that it opposes the growth of the current and if the current, is decreasing the induced e.m.f. tends to maintain the current Fig. 12.16. Due to this cause, a current takes longer time to attain its full value, when switched on and when the current is switched off, it does not step suddenly.

12.13. Eddy Currents (or Foucault Current)

When a solid conductor is put in a changing magnetic field currents are found to set up in the body of the conductor due to change of flux linked with it. The currents are found to be circular and are in a direction perpendicular to the magnetic flux. These currents are called eddy currents as they look like eddies or whirl pool and also sometimes called Faucault current as it was first discovered by Foucault. The sense of these currents are found by applying Lenz's law. Fig. 12.17 shows some of eddy currents in a metal sheet placed in an increasing magnetic field pointing into the plane of the paper. They are circular and are found to be in anticlockwise direction.

Like any other currents eddy currents heat up the conductor in which they flow. The heating effect is quite large in a bulk of metal since the currents are quite large due to its low resistance. There-

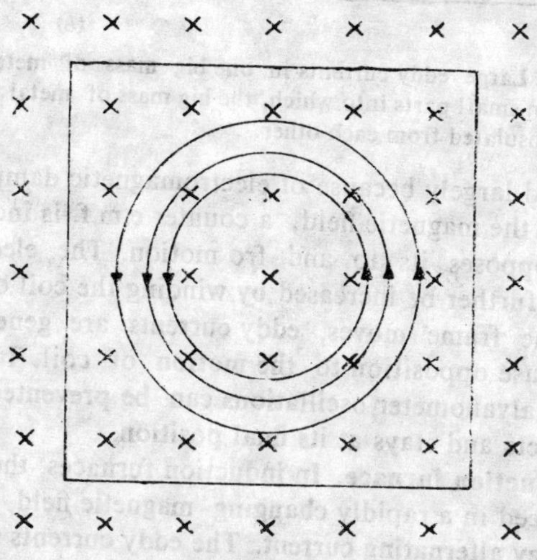

Fig. 12.17. Eddy currents.

fore eddy currents are considered undesirable in many electrical appliances and machineries as they cause unnecessary heating and wastage of energy derived from the source. Hence to reduce these currents, the metals used in these machines, are not taken in one solid piece but are made in parts or laminations insulated from each other by a coat of varnish. In this way eddy currents are restricted to the individual laminations instead of flowing across the whole metal, Fig. 12.18 (*a*) and (*b*). As a result of this arrangement, the resulting length of the path of the current is also greately increased with consequent increase in resistance. Hence the currents and their heating effects are minimised. For this reason, the iron core, in motor, transformer etc are made in laminations to reduce eddy current effect.

Practical uses of eddy currents. (*i*) **Electromagnetic damping.** When a current is passed through a galvanometer, the galvanometer coil usually suffers a few to and fro oscillations before settling down to its proper deflected position. The motion of the

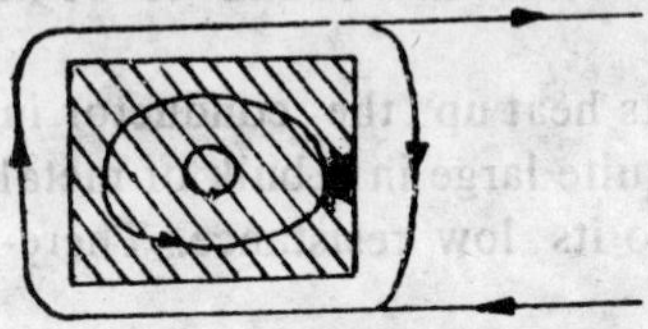

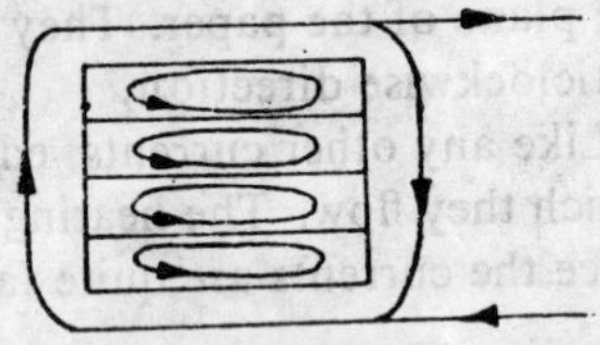

Fig. 12. 18. (*a*) Large eddy currents in one big mass of metal. (*b*) Small eddy currents in small parts into which, the big mass of metal is broken up and the parts insulated from each other.

coil is damped largely because of electromagnetic damping. As the coil moves in the magnetic field, a counter e.m.f. is induced in the coil, which opposes its to and fro motion. The electromagnetic damping can further be increased by winding the coil on a metallic frame. As the frame moves, eddy currents are generated in the frame and cause opposition to the motion of coil. In a properly constructed galvanometer oscillations can be prevented completely; the coil deflects and stays at its final position.

(*ii*) **Induction furnace.** In induction furnaces, the metal to be heated is placed in a rapidly changing magnetic field, provided by high frequency alternating current. The eddy currents set up in the metal produce so much of heat that the metal melts. The process is used in extracting the metal from ores.

(*iii*) **Electric brakes.** When strong stationary magnetic field is suddenly applied to a rotating drum, the eddy currents set up in the drum exert, torque which stops the motion of the drum. This principle is used in stopping electric trains.

12.14. Motors

An electric motor is a device used for converting electric energy into mechanical energy. So an electric motor can be called a reversed dynamo. There are two types of motors (*i*) A.C. motor that utilises a.c. (*ii*) D.C. motor which utilises d.c. for conversion into mechanical energy. A.C. motors have wide industrial applications but their principle of action is complicated. Therefore we will consider only the principle of d.c. motor as it can be easily under-

D.C. Motors. In construction a d.c. motor is similar to a d.c. dynamo but reversed in function (Fig. 12.19). Its chief parts are:

(*a*) *The armature.* It consists of a soft iron laminated core and a coil of insulated copper wire wound over the core. It is capable or rotation about an axis.

(*b*) *Commutator* C_1 *and* C_2. It consists of two segments made of copper, separated by mica insulation (segments are formed by cutting a copper cylinder into two pieces).

(*c*) *Brushes* B_1 *and* B_2. These are made of copper or carbon and remain in contact with the segments.

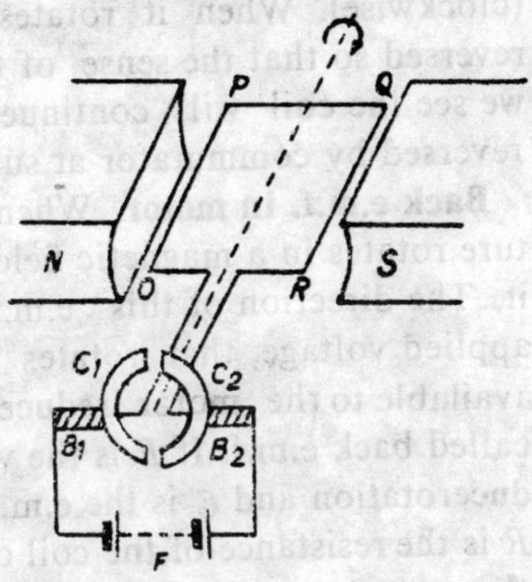

Fig. 12.19. D.C. motor.

(*d*) *Field magnet*. The field magnet *N-S* provides the magnetic field within the space, where the armature rotates. It is usually permanent horse shoe shaped magnet or an electromagnet.

(*e*) *The source of electric energy.* The electric energy in provided by a battery or a d.c. source. The source is connected to the brushes.

Principle. A current carrying coil experiences a torque in a magnetic field. This forms the basis for the operation of a d.c. motor.

Working of a D.C. motor. Consider the armature coil *OPQR* placed between the pole pieces *N* and *S* of a permanent magnet as shown in Fig. 12.19. Suppose the plane of the coil is parallel to the magnetic field. If the current from the battery is passed through the coil along the path *RQPO* (with segment C_2 in contact with the brush B_2)the coil will experience a torque due to which the coil will start rotating in clockwise direction. This means *PO* will move up and *RQ* will move down. But when the plane of the coil becomes perpendicular to the magnetic field, the torque vanishes. However due to inertia, the coil will over shoot its position which means *PO* will move towards right and *RQ* towards left. If the direction current in the coil is not changed, then a new torque will set up, which will bring the coil back to its previous position, normal to the field. But the arrangement is made such, that during the over shooting, the position of the segments get inter changed, that is C_1, will be in contact with the brush B_2 and C_2 will be in contact with the brush B_1. As a result, the direction of the current will reverse and will follow the path $OPQR$. In this condition, the torque which wili result, will rotate the coil in the same direction (clockwise). When it rotates through 180°, the current is again reversed so that the sense of the torque remains unchanged. Thus we see the coil will continue to rotate if the current is suitably reversed by commutator at suitable points.

Back e.m.f. in motor. When an electric motor works, the armature rotates in a magnetic field and therefore an e.m.f. is induced in it. The direction of this e.m.f. is opposite to the direction of the applied voltage, that rotates the motor. Thus the total voltage available to the motor reduces. The induced e.m.f. so produced is called back e.m.f. If E is the voltage applied to the motor to produce rotation and $\mathcal{E}$ is the e.m.f. induced in opposite direction and R is the resistance of the coil of the armature, then the current in the motor at any instant of time is given by,

$$I=\frac{E-\mathcal{E}}{R} \qquad \ldots(12.20)$$

But at the time of starting the motor is at rest. There fore $\mathcal{E}=0$ and the current in the coil is E/R. This current is very large and therefore likely to burn the motor. To avoid this risk, a variable resistance is placed in series with the armature coil to resist the large current. This resistance is known as the 'starter'. As the motor picks up the speed, the e.m.f. induced, grows up and the above said extra resistance is gradually cut off, by suitable arrangement.

12.15. Alternating Current or A.C.

We have seen that when a coil of wire is made to rotate in a uniform magnetic field an e.m.f. is induced in the coil, which at any instant of time is given by $\mathcal{E}=\mathcal{E}_0 \sin \omega t$. This e.m.f. is alternating and denoted by a symbol ~. Let us apply an alternating e.m.f. to a circuit containing a resistance R, Fig. 12.20(a). We shall denote this e.m.f. by $E=E_0 \sin \omega t$. At any instant of time, the current in the circuit would be given by Ohm's law as,

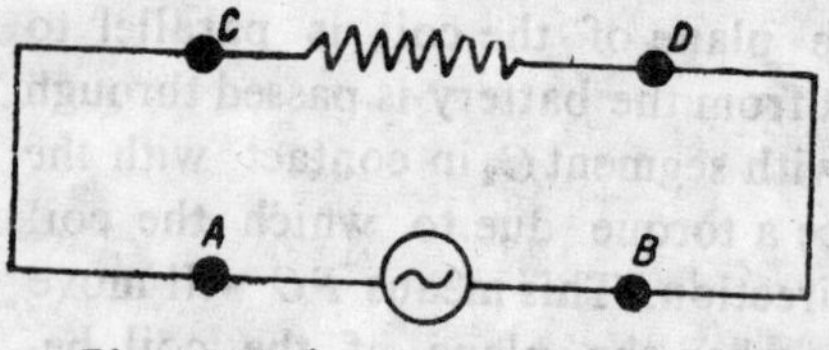

Fig. 12.20(a). A.C. source with resistance.

$$I=\frac{E}{R}=\frac{E_0 \sin \omega t}{R}=I_0 \sin \omega t$$

$$=I_0 \sin 2\pi ft \qquad \ldots(12.21)$$

where $I_0 = \frac{E_0}{R}$ is the maximum or peak value of current, f is the frequency, t is the time and ωt is the phase angle. This current is sinusoidal and known as alternating current or A.C.

This current varies with time and its variation with time is shown in Fig 12.20 (*b*). The time scale is chosen for commonly used 50 c/s alternating current *i.e.* $T = \frac{1}{50}$ seconds where T is the time for one complete cycle. It is seen from the figure that for half cycle, the current flows in one direction and for the other half in opposite direction.

12.16. Mean or Average Value of A.C. Over one Complete Cycle

Let us find the mean or average values of current considered over

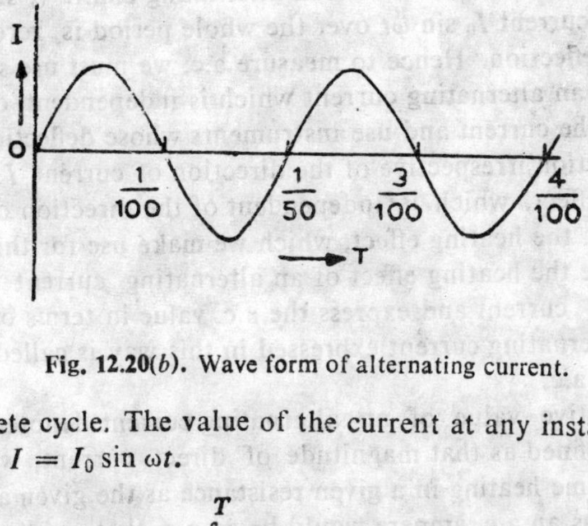

Fig. 12.20(*b*). Wave form of alternating current.

one complete cycle. The value of the current at any instant of time is given by $I = I_0 \sin \omega t$.

$$I_{\text{average}} = \frac{\int_0^T I_0 \sin \omega t \, dt}{\int_0^T dt} = -\frac{I_0}{\omega} \frac{\left[\cos \omega t\right]_0^T}{T}$$

$$= -\frac{I_0}{\omega T}\left[\cos \frac{2\pi}{T} t\right]_0^T \left(\because \omega = \frac{2\pi}{T}\right)$$

$$= -\frac{I_0}{\omega T}\left[\cos 2\pi - \cos 0\right] = -\frac{I_0}{T}\left[1 - 1\right] = 0 \quad \ldots (12.22)$$

Thus average value of alternating current over one complete cycle is zero. The same is also true of the alternating e.m.f. considerd over the whole cycle.

12.17. Measurement of an A.C.

A D.C. is measured in ampere. An ampere is that current, which would deposit a definite amount of silver per second from a standard silver nitrate solution in a silver voltmeter. But if an a.c. is passed through such a voltmeter, no silver is deposited, because the polarities of electrode, change periodically. Hence alternating current cannot be measured by this method. Also a moving coil ammeter or a voltmeter cannot be used for measuring alternating current or voltage for they will indicate the mean value of the quantity and the mean value of an alternating e.m.f. $\mathcal{E}$ sin ωt or an alternating current I_0 sin ωt over the whole period is zero. So they show no deflection. Hence to measure a.c. we must use some other property of an alternating current which is independent of the direction of the current and use instruments whose deflections are all in one direction irrespective of the direction of current I. There is only one effect, which is independent of the direction of the electric current, the heating effect, which we make use for this purpose. We compare the heating effect of an alternating current with that of a direct current and express the a.c. value in terms of d.c. The value of alternating current expressed in this way is called effective value of an a.c.

The effective value of an alternating current denoted by I_{eff} is therefore defined as that magnitude of direct current, which produces the same heating in a given resistance as the given alternating current. Thus an a.c. ampere would be an a.c. that produces a heating effect equal to that of one d.c. ampere. The a.c. ampere is sometimes called virtual ampere.

We know when a current is passed through a conductor, heat produced is proportional to the square of the current. Therefore for an a.c., $I=I_0 \sin \omega t$, the heat produced is proportional to $I_0^2 \sin^2 \omega t$. Over the complete cycle, on integration, it is found that heat produced is proportional to $I_0^2/2$. This is same as, the heat produced by d.c. of value $I_0/\sqrt{2}$ So the effective value of an a.c. *i.e.*

$$I_{\text{eff}} = \frac{I_0}{\sqrt{2}} = 0.707\ I_0 \quad \ldots (12.23)$$

This I_{eff} is also called root mean square or r.m.s. current for the fact that this value is obtained by taking the square root of the mean of the squares of instantaneous current over a cycle.

The effective or r.m.s. value of an alternating voltage is defined in an exactly similar way and we have

$$E_{eff} = \frac{E_0}{\sqrt{2}} = 0.707\ E_0 \qquad \ldots (12.24)$$

It is to be noted that usually the value of an alternating current or voltage is expressed in terms of its effective or r.m.s. value. The a.c. voltmeters or ammeters, also record the r.m.s. value. Our domestic supply is 220 v(r.m.s.). This means the effective or r.m.s. value is 220 volts. But the peak value of the voltage is $220\times\sqrt{2}=310$ volts. Similarly an a.c. of 3 amperes means a current, whose effective value is 3 amperes. The peak value of the current *i.e.* I_0 is $3\sqrt{2}=4.24$ amperes and the instantaneous value I may be anything from 0 to 4.24.

Example 12.7. What is the effective or r.m.s. value of an a.c. having a peak value of 7.1 amperes? What will be the reading shown for this current by (*i*) a.c. ammeter, (*ii*) an ordinary moving cell galvanometer.

Solution.

$$I_{eff} = 0.707\ I_0$$

Here $I_0 = 7.1$ A

∴ $I_{eff} = 0.707\times 7.1 = 5$ A

(*i*) An a.c. ammeter will read 5 A.

(*ii*) An ordinary galvanometer will read zero.

12.18. A.C. Circuit containing Resistance only

Consider an a.c. circuit (Fig. 12.21) in which a resistance R is connected in series with an a.c. source e.m.f. $E=E_0 \sin \omega t$

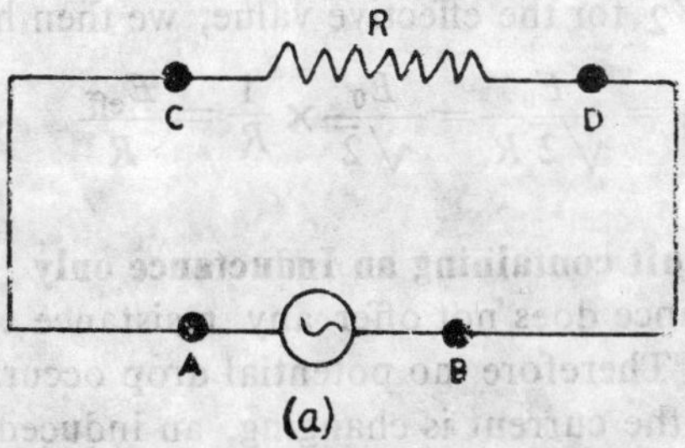

Fig. 12.21. A.C. circuit containing resistance only.

At any instant, the potential difference across the resistance R between the points C and D is given by

$$V = IR \qquad \ldots (12.25)$$

Where I is the instantaneous value of the current. At every instant of time the potential difference between C and D must be equal to the potential difference across the source terminals A and B. Thus

$$IR = E_0 \sin \omega t$$

or
$$I = \frac{E_0 \sin \omega t}{R} = I_0 \sin \omega t \qquad \ldots (12.26)$$

where $I_0 = \dfrac{E_0}{R}$ is the amplitude or peak value of the current.

Graph for the instantaneous values of E and I against time are shown in (Fig. 12.22). Both the current and the voltage start at

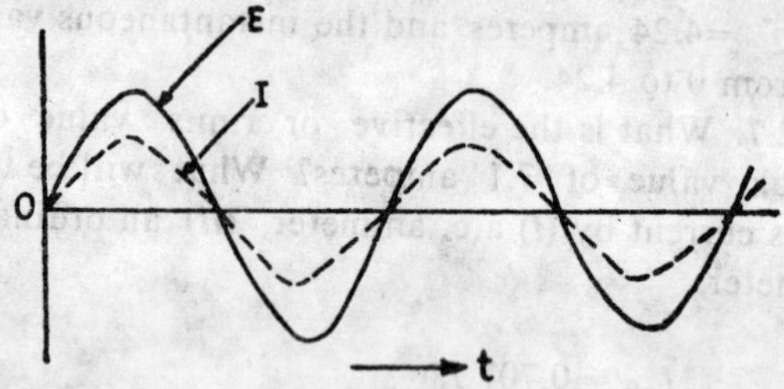

Fig. 12.22. Current in phase with voltage in a resistive a.c. circuit.

zero at the same time reach maximum at the same time and have the same sinusoidal shape. The voltage and current under these conditions are said to be in phase.

A voltmeter connected across the resistor would read effective voltage, whose value is $E_0/\sqrt{2}$. Similarly ammeter in the circuit would read $I_0/\sqrt{2}$ for the effective value, we then have

$$I_{\text{eff}} = \frac{E_0}{\sqrt{2}\,R} = \frac{E_0}{\sqrt{2}} \times \frac{1}{R} = \frac{E_{\text{eff}}}{R} \qquad \ldots (12.27)$$

12.19. A.C. Circuit containing an Inductance only

A pure inductance does not offer any resistance to the flow of d.c. through it. Therefore no potential drop occurs across the inductance. But if the current is changing, an induced e.m.f. is produced, which opposes change in the current. This opposing e.m.f has to be overcome in order to maintain the current. Therefore an e.m.f. is to be applied across the inductance, which must be equal and opposite to the induced e.m.f., for the current to flow.

Consider a circuit of Fig. 12.23 in which a pure inductance CD

is connected with an a.c. source. The induced e.m.f. $\mathcal{E}_L$ in this circuit is given by $\mathcal{E}_L = -L\frac{dI}{dt}$ ... (12.28)

where L is the inductance and $\frac{dI}{dt}$ is the time rate of change of current. Therefore to maintain the current from C to D, there must be an impressed voltage V, which is given by

$$V = V_C - V_D = -\mathcal{E}_L = -\left(-\frac{LdI}{dt}\right)$$

$$= L\frac{dI}{dt} \quad \text{... (12.29)}$$

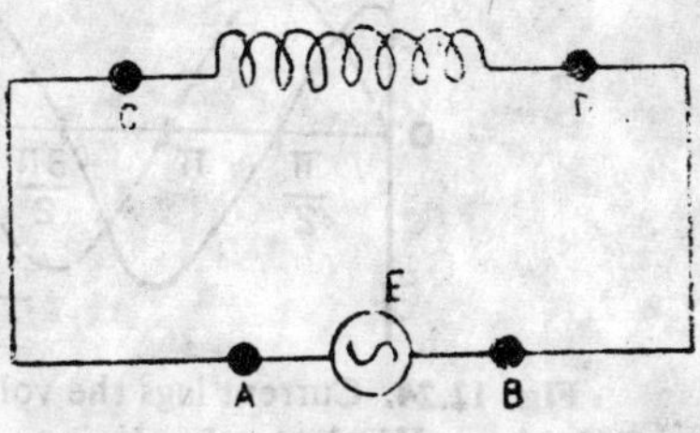

Fig. 12.23. A.C. source with an inductance.

where $(V_C - V_D)$ is the potential difference between the points C and D.

Suppose the current at any instant of time is given by $I = I_0 \sin \omega t$. The applied voltage V at any instant of time that causes such a current is given by

$$V = L\frac{dI}{dt} = L\,\frac{dI_0 \sin \omega t}{dt} = LI_0\omega \cos \omega t \quad \text{... (12.30)}$$

Now the potential difference V between the points C and D must be at every instant of time be equal to the potential difference E across the source terminal A and B. Hence instantaneous value of the applied e.mf. *i.e.* e.m.f. of the a.c. source is given by

$$E = L\omega I_0 \cos \omega t = E_0 \cos \omega t$$
$$= E_0 \sin(\omega t + \pi/2) \quad \text{... (12.31)}$$

where $E_0 = L\omega I_0$ is the amplitude or peak value of the applied e.m.f. comparing with resistive case *i.e.* $E_0 = RI_0$ one can find that ωL plays the same role here as the resistance R in the resistive circuit. This means inductance resists the flow of a.c.

The quantity $\omega L = X_L = 2\pi fL$ is called the inductive reactance of the circuit and has the unit of ohm.

From the expression of the current $I = I_0 \sin \omega t$ and voltage $E = E_0 \sin(\omega t + \pi/2)$ it is evident that current and voltage are 90° or $\pi/2$ out of phase. Fig. 12.24 shows the way in which the instantaneous values I and E vary. I reaches maximum value a quarter cycle later than E. This means, a pure inductance causes the current to lag behind the e.m.f. in phase $\pi/2$ (radian).

Example 12.8. The current through an inductor of inductance 2.0 henry varies simusoidally with an amplitude of 1.0 ampere and a frequency of 50 cycles per second. Calculate the potential difference across the terminals of the inductor.

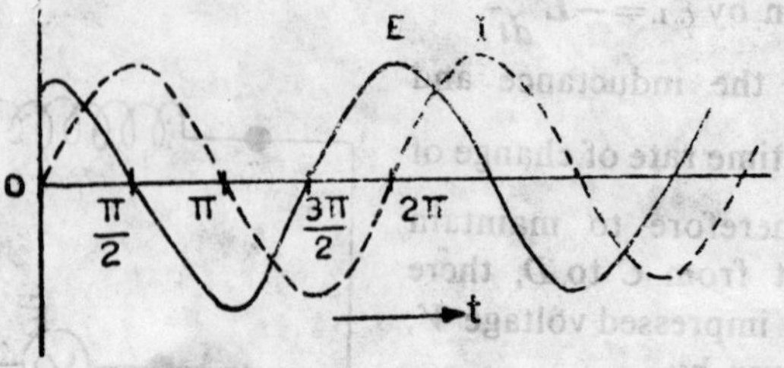

Fig. 12.24. Current lags the voltage E by $\pi/2$ in inductive circuit.

Solution. We have $I = I_0 \sin \omega t = I_0 \sin 2\pi f t$

Potential difference across the inductor is given by

$$E_L = L\frac{dI}{dt} = L\frac{d}{dt}(I_0 \sin 2\pi f t)$$

$$L = 2\pi f L I_0 \cos 2\pi f t$$

Amplitude of the voltage $E_0 = 2\pi f L I_0$

$$= 2 \times 3.14 \times 50 \times 2.0 \times 1$$

$$= 628 \text{ volts.}$$

Therefore the voltmeter will read the effective value

$$E_{\text{eff}} = \frac{E_0}{\sqrt{2}} = \frac{628}{\sqrt{2}} = 448 \text{ volts.}$$

Example 12.9. What is the inductive reactance of a coil if the current through it is 40 mA and voltage across it is 40 v?

Solution. The values of the current and voltage given are the effective values. Hence

$$X_L = \frac{V_{\text{eff}}}{I_{\text{eff}}} = \frac{40 \text{volt}}{40 \times 10^{-3} \text{A}} = 1000 \text{ volt/A}$$

$$= 1000 \text{ ohms}$$

Example 12.10. At what frequency will 0.5 henry inductor have a reactance 1000 ohms?

Solution. $X_L = \omega L = 2\pi f L$

$\therefore$ $1000 = 2\pi f \times 0.5$

$$f = \frac{1000}{\pi} = 318.5 \text{ cycle/sec}$$

$$f = 318.5 \text{ Hz}$$

12.20. A.C. Circuit containing Inductance and Resistance only

An inductance is always associated with resistance. Therefore it is more practical to consider a.c. circuit, which has a series combination of inductance (L) and resistance (R) as shown in Fig. 12.25. In such a circuit, the current is the same in each part of the circuit as the inductance and resistance are in series. Suppose the e.m.f. applied to this circuit is given by $E=E_0 \sin \omega t$. The current in the

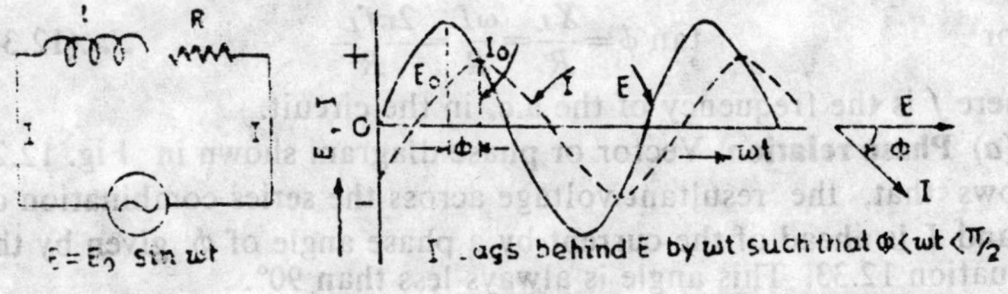

Fig. 12.25. A.C. applied across a coil of resistance R and inductance L.

circuit is a.c. and is therefore opposed by both inductance and resistance. Hence there are voltage drops V_R and V_L across the inductance and the resistance respectively in the circuit. The maximum voltage drop across the resistance is given by $E_{0R}=I_0R$ and the maximum voltage drop across the inductance is $E_{0L}=I_0 \times X_L$ The voltage E_{0R} is in phase with the maximum current I_0 and E_{0L} is ahead of I_0 by a phase angle of 90° or $\pi/2$ radians. Therefore E_{0L} is ahead of E_{0R} by a phase angle of $\pi/2$ radians. Thus there exists a phase difference between the potential drops across the inductance and the resistance. Under this peculiar situation *i.e.* under the existence of phase difference between voltage drops, the voltage drops cannot be added algebraically to give resultant voltage. But mathematical analysis of electrical circuit suggests that to find out the maximum value E_0 of the applied e.m.f. the maximum value E_{0R} and E_{0L} must be treated as vectors and then be vectorially added. For this purpose we draw a vector diagram or phase diagram as shown in Fig. 12.26.

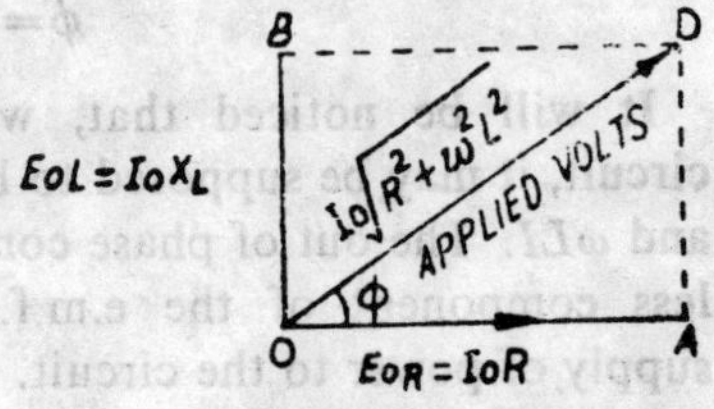

Fig. 12.26. Vector diagram for resistance and Inductor circuit.

Here E_{0R} which is in phase with I_0 is taken along the *x*-axis and is taken equal to *OA* and E_{0L} which is ahead of the current by $\pi/2$ is taken along the *y*-axis and is taken equal to *OB*. Then by the law vectors, *OD* the diagonal of the rectangle *OADB* is equal to the maximum value E_0 of the applied voltage.

$$\therefore \quad E_0=\sqrt{E_{0L}^2+E_{0R}^2}$$

$$=\sqrt{(I_0R)^2+(I_0X_L)^2}$$

$$\text{or} \quad E_0=I_0\sqrt{R^2+X_L^2} \qquad \ldots(12.32)$$

Further, $\tan\phi = \frac{E_{0L}}{E_{0R}} = \frac{I_0 X_L}{I_0 R}$

or $$\tan\phi = \frac{X_L}{R} = \frac{\omega L}{R} = \frac{2\pi f_L}{R} \quad \text{...(12.33)}$$

where f is the frequency of the a.c. in the circuit.

(*a*) **Phase relation.** Vector or phase diagram shown in Fig. 12.26 shows that, the resultant voltage across the series combination of R and L is ahead of the current by a phase angle of ϕ given by the Equation 12.33. This angle is always less than 90°.

(*b*) **Impedance.** On comparing the Equation 12.32 with Ohm's law (*i.e.* $E=IR$), we find that the effective resistance of the circuit denoted by Z is given by

$$Z = \sqrt{R^2 + \omega^2 L^2}$$
$$= \sqrt{R^2 + (2\pi f L)^2} \quad \text{...(12.34)}$$

The effective resistance Z is known as the impedance of the circuit. If R is measured in ohm, f in cycles per second and L is in henry, then Z is expressed in ohm.

(*c*) **Resultant current.** Since the resultant current lags the applied voltage by a phase angle ϕ, we get

$$I = I_0 \sin(\omega t - \phi)$$

where $$I_0 = \frac{E_0}{\sqrt{R^2 + \omega^2 L^2}}$$

$$\phi = \tan^{-1}\frac{\omega L}{R} \quad \text{...(12.35)}$$

It will be noticed that, when a voltage is applied to inductive circuit, it may be supposed to be resolved into two components IR and ωLI. The out of phase component i.e. ωLI is called the watt-less component of the e.m.f. because it does not represent any supply of power to the circuit.

Example 12.11. An alternating voltage of 100 volts at a frequency of 25 c.p.s. or hertz (Hz) is applied to a circuit containing a resistance of 1.5 ohms and inductance of 0.1 henry in series. Find the current flowing, the angle of lag and potential difference across the resistance and the inductance.

Solution. We have

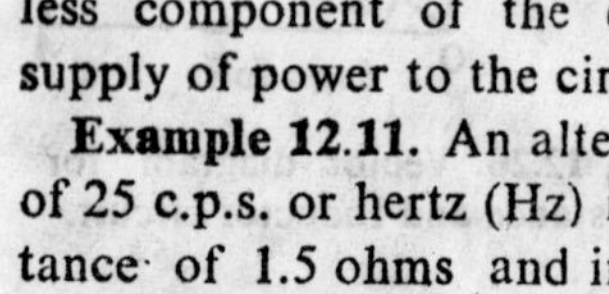

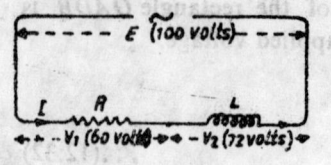

Fig. 12.27

$$\omega = 2\pi f = 2\pi \times 25$$
$$= 157 \text{ radians per second.}$$

Inductive reactance X_L

$$= 2\pi f_L$$
$$= 157 \times 0.01$$
$$= 1.57 \text{ ohms}$$

Resistance $R = 1.5$ ohms

Impedance $= \sqrt{(1.5)^2+(1.57)^2} = \sqrt{4.71} = 2.17$ ohms

Current $I = \frac{100}{2.17} = 46$ amperes.

The angle of lag ϕ is given by

$$\tan\phi = \frac{2\pi fL}{R} = \frac{1.57 \text{ ohms}}{1.5 \text{ ohms}} = 1.047$$

$\therefore$ $\phi = 44° 19'$

Voltage V_R across the resistance

$= RI = 1.5$ ohms $\times 46$ amp $= 69$ volts.

Voltage V_L across the impedance

$= 2\pi fLI = 1.57$ ohms $\times 46$ amp $= 72$ volts.

It seems impossible for the applied 100 volts to supply both the values of voltage of which, the algebraic sum $= 141$ volts, but it must be noted that reactive and resistance voltages have not to get their maximum wants supplied simultaneously. Therefore in this case

$$V_R + V_L \neq E_0 \text{ but } V_R^2 + V_L^2 = E_0^2$$

12.21. A.C. Circuit containing Capacitance only

A capacitor consists of two plates of conducting material separated by an insulator. Its resistance therefore is practically infinite. Hence, we might expect that no current, whether a.c. or d.c. flows in a circuit containing a capacitor. But that is not true. Current flows in a circuit containing capacitor but in a different manner. Let us first consider a capacitor in a d.c. circuit.

Fig. 12.28. Capacitor in d.c. circuit

(*a*) **Capacitor in a D.C. circuit.** Consider a circuit shown in Fig. 12.28. As soon as we press the key electrons begin to flow from the negative terminal of the battery to the plate Q and from P to the positive plate of the battery. The plate P thus begins to acquire positive charge and the plate Q negative charge. This charging of the capacitor continues till the potential difference between the plates becomes equal to that across the battery terminals. Then the flow of charge ceases. This flow of charge is equivalent to a current. This a current does flow in the circuit, during the charging process though not through the capacitor but in the remainder of the circuit. The ammeter would, therefore, show a momentary deflection. The direction of the current is from the plate Q to the plate given P via battery. Its magnitude at any instant is given by the rate of growth of charge on the capacitor.

Hence, $$I=\frac{dQ}{dt} \quad \ldots(12.36)$$

The charging process can be extended in time if a resistance is included in the circuit. It is found that final charge on the capacitor is given by,

$$Q_0=V_0C \quad \ldots(12.37)$$

where V_0 is the battery voltage and C is the capacitance of the capacitor.

(*b*) **Capacitor in an a.c. circuit.** Now consider the capacitor in an a.c. circuit, Fig. 12.29. As the voltage from the source is continually changing, the charge on the capacitor is also continually charging. During a complete cycle, the capacitor is first charged in one direction then discharged and again charged in the reverse direction and discharged. As the charging and discharging of the capacitor is taking place continuously a continuous current exists in the circuit. Let us investigate the nature of this current.

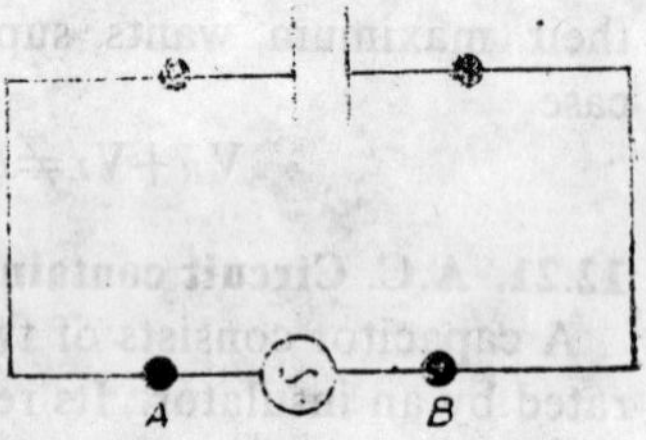

Fig. 12.29 A.C. source with a capacitor.

It is evident that the potential difference across the capacitor between points C and D at every instant has to be exactly the same as that across the source terminals A and B. Therefore the capacitor must charge and discharge in such a manner that the potential difference V across it is sinusoidal and equal to applied e.m.f. at every instant, that is

$$V=E=E_0 \sin \omega t$$

The charge on the capacitor at any instant is given by

$$Q=CV$$

Therefore the current at any instant is given by

$$I=\frac{dQ}{dt}=C\frac{dV}{dt}=C\frac{d}{dt}(E_0 \sin \omega t)$$

$$=E_0C\omega \cos \omega t=I_0 \cos \omega t$$

$$=I_0 \sin(\omega t+\pi/2) \quad \ldots(12.38)$$

where $$I_0=\frac{E_0}{\frac{1}{C\omega}} \quad \ldots(12.38)$$

Thus in this case the current is sinusoidal but 90° ahead of e.m.f. in phase. The Fig. 12.30 shows the way in which the instantaneous

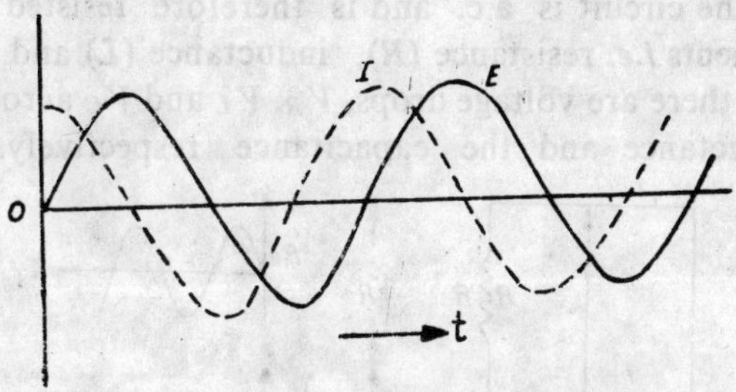

Fig. 12.30. Current I leads the voltage E by $\pi/2$ in a capacitative circuit.

values I and E vary with time.

The quantity

$$X_C=\frac{1}{C\omega}=\frac{1}{2\pi f C} \quad \ldots(12.40)$$

is known as the capacitative reactance of the circuit. It plays the same role as the inductive reactance X_L in the inductive circuit or resistance in resistive circuit. That is it resists the flow of a.c. and its unit is ohm.

Example 12.12. What is the capacitive reactance of 15 f capacitor when it is the part of a circuit, whose frequency is (*i*) 50 c/s. (*ii*) 10 c/s.

Solution. (*i*) Given

$$C=15\ \mu f=15\times10^{-6}\ f$$

frequency $f=50$ c/sec.

We have $$X_C=\frac{1}{2\pi fC}=\frac{1}{2\pi\times15\times10^{-6}\times50}=212.2 \text{ ohms.}$$

(*ii*) Given $C=15\ \mu f=15\times10^{-6}f$

frequency $=10^6$ c/sec

$$X_c=\frac{1}{2\pi fC}=\frac{1}{2\pi\times15\times10^{-6}\times10^6}$$
$$=1.06\times10^{-2} \text{ ohms.}$$

12.22. A.C. through Resistance Inductance and Capacitance in Series or L.C.R Circuit

Consider the circuit in Fig. 12.31 which consists of a resistance (R), inductance (L) and capacitance (C) in series and connected to an a.c. source. In such a circuit, the current is the same in each part of the circuit as it is a series circuit.

Suppose the e.m.f. applied to the circut is $E=E_0 \sin \omega t$. The current in the circuit is a.c. and is therefore resisted by all the circuit elements *i.e.* resistance (R), inductance (L) and capacitance (C). Hence there are voltage drops, V_R, V_L and V_C across the resistance, inductance and the capacitance respectively. But these

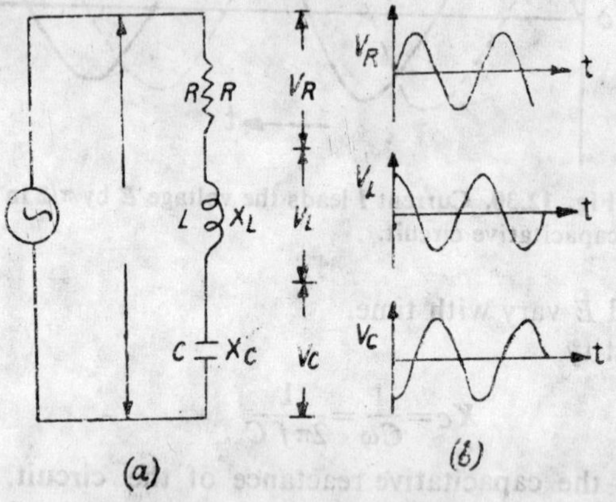

Fig. 12 31. An L.C.R. circuit.

voltages are not in the same phase. Therefore they can not be algebraically added to give the applied voltage. However we can treat them as vectors and add them vectorially.

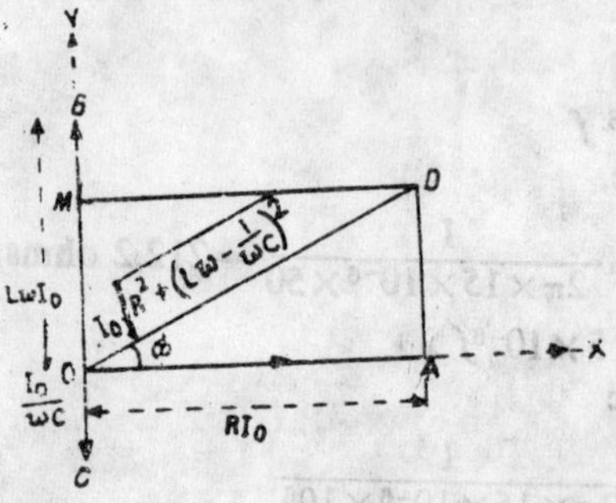

Fig. 12.32 Vector diagram for LCR circuit.

The vector diagram for such a circuit is shown in Fig. 12.32. Here $E_{OR}=I_0R$, in phase with the current is shown along the positive x-axis, $E_{0L}=\omega I_0 L$ which is ahead of the current I_0, and is shown along the positive y-axis, and $E_{0C}=I_0/\omega C$ which lags behind the current by 90°, hence is shown along the negative y-axis.

Let OA, OB and OC in Fig. 12.32 represent, the maximum potential drops across the resistance, inductance and capacitance respectively. If $\omega L > \frac{1}{\omega C}$ then $OB > OC$. Hence $OM = OB - OC$ is the resultant of the components OB and OC. The resistance component OA is E_0R OD, is the resultant e.m.f. E_0 acting across the circuit and hence its magnitude is given by

$$OD^2=(OA)^2+(OB-OC)^2 \qquad \ldots (12.41)$$

$$\therefore \quad E_O=\sqrt{(E_{OR})^2+(E_{OL}-E_{OC})^2}$$

$$=\sqrt{(I_OR)^2+(I_OX_L-I_OX_C)^2}$$

$$=I_0\sqrt{R^2+(X_L-X_C)^2}$$

$$=I_0\sqrt{R^2+\left(\omega L-\frac{1}{\omega C}\right)^2}$$

or
$$I_O=\frac{E_O}{\sqrt{R^2+\left(\omega L-\frac{1}{\omega C}\right)^2}} \quad \ldots (12.42)$$

The quantity $\sqrt{R^2+\left(\omega L-\frac{1}{\omega C}\right)^2}$ is the impedance Z of the circuit.

or
$$Z=\sqrt{R^2+(X_L-X_C)^2} \quad \ldots (12.43)$$

Also the phase angle ϕ is given by $\tan\phi=\frac{DA}{OA}$

$$\frac{X_L-X_C}{R}=\frac{\omega L-\frac{1}{\omega C}}{R} \quad \ldots (12.44)$$

If $\omega L>\frac{1}{\omega C}$, ϕ is positive and the applied e.m.f. leads the current but if $\omega L<\frac{1}{\omega C}$, ϕ is negative. OD lies below OA and e.m.f. lags behind the current.

Resonance. It may be seen if $X_L=X_C$, the impedance

$$Z=\sqrt{R^2+(X_L-X_C)^2}$$

is at its minimum value and equal to R. The current, therefore, in the circuit under this condition is maximum. The circuit is purely resistive. The applied voltage and the current are in phase. This is known as the condition of resonance and frequency f_r at which this occurs is known as resonant frequency.

Here $\quad X_L=X_C$

or
$$2\pi f_rL=\frac{1}{2\pi f_rC}$$

$$\therefore \quad f_r=\frac{1}{2\pi\sqrt{LC}} \quad \ldots (12.45)$$

Example 12.13. A resistor of 100 ohms, an inductance of 0.5 henry and a capacitor of 15 microfarads are connected in series. A 200 volt 50 cycle alternating potential is connected across the group, find (*a*) the impedance of the circuit, (*b*) the current, (*c*) potential difference across each of the three elements and (*d*) the phase angle between the current and the applied voltage.

Solution. (*a*) Impedance of the circuit,

$$Z=\sqrt{R^2+(X_L-X_C)^2}$$

$$X_L=2\pi fL=2\pi\times50\times0.5=157.1 \text{ ohms}$$

$$X_C=\frac{1}{2\pi fc}=\frac{1}{2\pi\times50\times15\times10^{-6}}=212.2$$

$$Z=\sqrt{(100)^2+(212.2-157.1)^2}$$

$$=114.2 \text{ ohms}$$

(*b*) $$I=\frac{E}{Z}=\frac{200}{114.2}=1.75 \text{ amp}$$

(*c*) (*i*) Potential difference across the resistance

$$V_R=IR=1.75 \text{ amp}\times100 \text{ ohm}$$

$$=175 \text{ volts (in phase with current)}$$

(*ii*) Potential difference across the Inductance

$$IX_L=1.75\times157.1$$

$$=274.9 \text{ volts (leading the current by } 90^\circ)$$

Potential difference across the capacitance

$$IX_C=1.75\times212.2$$

$$=371.3 \text{ volts (legging behind the current by } 90^\circ)$$

Since $X_C>X_L$, the current leads the applied voltage by angle ϕ, which is given by

$$\tan\phi=\frac{X_C-X_L}{R}=\tan^{-1}\frac{55.1}{100}$$

$$\therefore \quad \phi=28^\circ51'$$

12.23. Transformer

A transformer is a device for converting large alternating current at low voltage into small current at high voltage and vice-versa. The transformers which convert low voltage into higher ones are called the step up transformers, while those which convert high voltage into lower ones are called step-down transformers.

Principle. The working of a transformer is based on the principle of electro magnetic induction *i.e.* when a magnetic flux linking with a coil, changes an induced e.m.f. and hence induced current set up in it.

Construction. A simple transformer consists of two coils of insulated copper wires, which are wound separately on a continuous soft iron laminated core, Fig. 12.33. One of the coils is called the

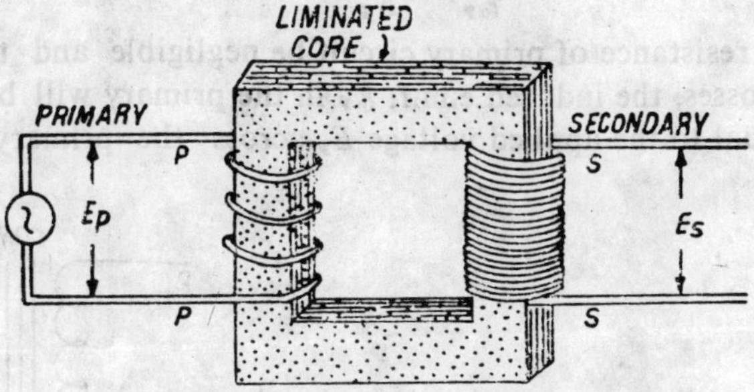

Fig. 12.33. A transformer.

primary and the other one is called secondary.

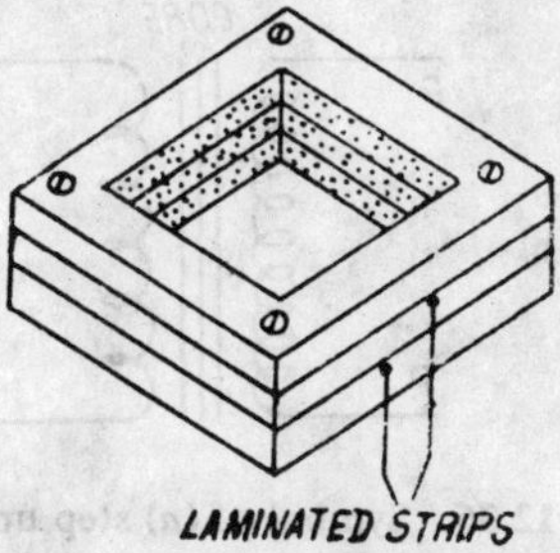

Fig. 12.34. Laminated core of transformer.

The primary is connected to the alternating voltage to be transformed while the secondary is connected to a load, which may be a resistance or any other electrical device to which electric power is to be supplied.

Theory. When an alternating voltage is applied to the primary, the alternating current flows in the coil. This sets up alternating magnetic flux in the core. This changing magnetic flux is linked up with both the primary and the secondary. Hence an induced e.m.f. is caused in the secondary and self induced back e.m.f. in the primary.

Let N_p and N_s be the number of turns in the primary and secondary coils respectively. Let us assume that there is no leakage of magnetic flux so that the same flux passes through each turn of primary and the secondary. Let ϕ be the flux linked with each turn of either coil at any instant of time. Then by Faradays law of electromagnetic induction the self induced back e.m.f. produced in the primary is given by

$$\mathcal{E}_p = -\frac{d(N_p \phi)}{dt} = -N_p \frac{d\phi}{dt} \qquad \ldots (12.46)$$

and the e.m.f. induced in the secondary,

$$\mathcal{E}_s = \frac{d(N_s \phi)}{dt} = -N_s \frac{d\phi}{dt} \qquad \ldots (12.47)$$

$$\therefore \quad \frac{\mathcal{E}_s}{\mathcal{E}_p} = \frac{N_s}{N_p} \quad \ldots (12.48)$$

If the resistance of primary circuit be negligible and there be no energy losses, the induced e.m.f. $\mathcal{E}_p$ in the primary will be numerically equal to the applied voltage E_p across the primary. Further

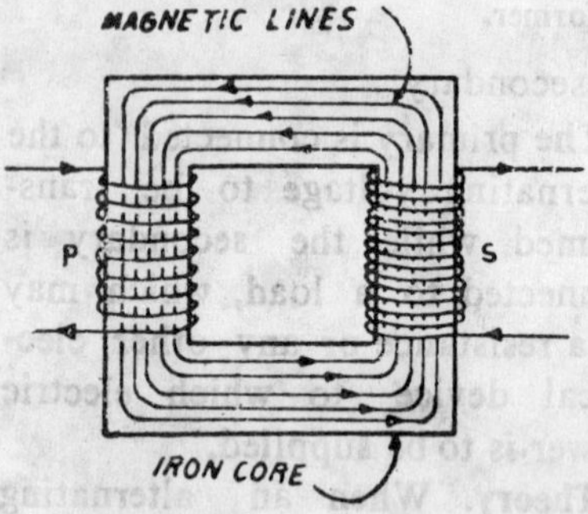

Fig. 12.35. Iron core.

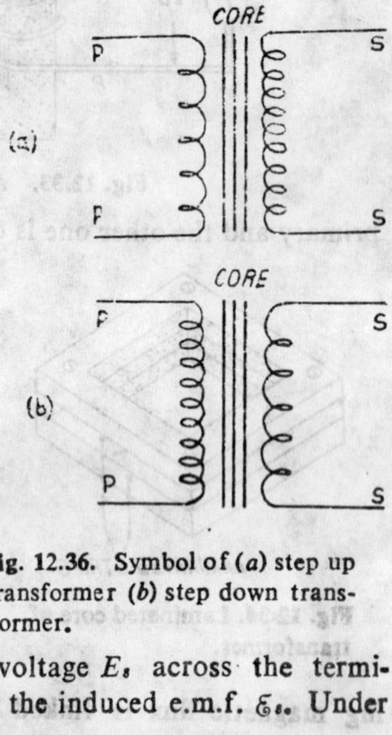

Fig. 12.36. Symbol of *(a)* step up transformer *(b)* step down transformer.

if the secondary circuit be open, the voltage E_s across the terminals of the secondary will be equal to the induced e.m.f. $\mathcal{E}_s$. Under these ideal conditions

$$\frac{E_s}{E_p} = \frac{\mathcal{E}_s}{\mathcal{E}_p} = \frac{N_s}{N_p} = K \quad \ldots (12.49)$$

where K is called the transformation ratio. In Equation 12.48, E_p may be described as the input e.m.f. given to the primary and E_s as the output e.m.f. from the secondary. We then have,

$$\frac{E_s}{E_p} = \frac{\text{output e.m.f.}}{\text{input e.m.f.}} = \frac{N_s}{N_p} \quad \ldots (12.50)$$

When $N_s > N_p$, then $E_s > E_p$, the transformer is known as stepup transformer, if $N_s < N_p$, $E_s < E_p$ then it is known as step down transformer.

If the transformer is ideal transformer, then there is no energy loss due to any cause and the input power must be equal to the output power.

We then have,

$$E_pI_p = E_sI_p \qquad \ldots (12.51)$$

where voltages and currents are effective values. From Equation 12.51.

$$\frac{I_s}{I_p} = \frac{E_p}{E_s} \qquad \ldots (12.52)$$

Thus we see, when voltage is stepped up, the current is correspondingly reduced in the some *ratio* and vice versa.

Energy loss in a transformer. The power cutput of a transformer is always less than the power input because of unavoidable energy losses. These losses are:

(*i*) **Copper loss.** As the alternating current flows through the primary and the secondary heat is developed inside the copper wires wound. The loss of this energy is called copper loss.

(*ii*) **Core loss or iron loss.** During each cycle or a.c., the core is taken through a complete cycle of magnetisation. The energy expanded in this process is finally converted into heat and is therefore wasted. This loss can be minimised by the choice of iron with special magnetic property.

(*iii*) **Loss due to eddy current.** Eddy currents are set up in the iron core of the transformer and these generate heat, with consequent loss of energy. To minimise the losses, the iron core is laminated by making it of number of thin sheets of iron insulated from each other instead of being made of one solid piece of iron.

(*iv*) **Loss due to leakage of flux.** In an actual transformer all the flux linked with the primary does not pass through the secondary but some of them return through the air. So there is an energy loss due to this leakage. This is generally small. It can be minimised by suitably designing the transformer.

Efficiency of a transformer. The efficiency of a transformer η is defined as

$$\eta = \frac{\text{Power output}}{\text{Power input}}.$$

If all the losses mentioned above are minimised then, the efficiency of a transformer is fairly high (90–99%) though not 100 %.

Example 12.14. A step down transformer at the end of a transmission line reduces the voltage from 4800 volts 245 volts. The power output is 9.0 KW and the over all efficiency of the transformer is 92%. The primary winding has 4,000 turns. How many turns has the secondary coil ? What is the power input? What is the current in each of the two coils ?

Solution. We have,

$$\frac{E_P}{E_S}=\frac{N_P}{N_S}$$

If the power efficiency is high, the error introduced by using the terminal voltage is not excessive. Then

$$\frac{4800 \text{ volts}}{240 \text{ volts}}=\frac{4000 \text{ turns}}{N_s}$$

Hence $N_s=200$ turns.

$$\text{Efficiency} = \frac{\text{Power output}}{\text{Power input}} = \frac{P_S}{P_P}$$

$$0.92=\frac{9000 \text{ watts}}{P_P}$$

Therefore $P_P = 9800$ watts

$$I_P = \frac{9800 \text{ watts}}{4800 \text{ volts}} = 2.04 \text{ amp.}$$

$$I_S = \frac{9000 \text{ watts}}{240 \text{ volts}} = 3.7 \text{ amp.}$$

12.24. Long Distance Transmission of Electric Power

Electric energy is produced by the generator at the power station and is then transmitted to different places, which may be situated at a considerable distance from it. The transmission is done by two parallel wires called transmission lines for carrying current from and to the power station. During the transmission the wires get heated and as a result there is a loss of power. This loss has to be minimised so as to obtain greater amount of power at the receiving station. In order to achieve this end, an a.c. system is used in the transmission process. If a d.c. system were used, the losses in transmission would be considerably great.

The following discussions will indicate why an a.c. system is preferred to a d.c. system in long distance transmission.

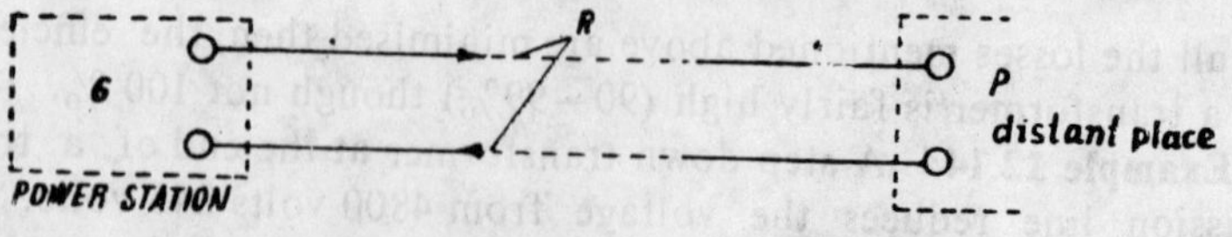

Fig. 12.37. Long distance transmission.

Suppose the power station is situated at G (Fig. 12.37) from where the energy is to be carried to the place P, situated at a distance.

Let R be the resistance of the line wire from G to P and back. Suppose the generator voltage is V and I is the current to be delivered i.e. the generated energy $VI(=W_1)$ k watt. The loss in the transmission line due to production of heat is I^2R k watts. Hence the power available at P is

$$VI-I^2R=I(V-IR)=W \text{ k watt say.}$$

From this it is quite clear that W will be practically equal to VI if I^2R loss is small. This can by achieved in one way by making the transmission lines wires of large diameter. By this, the resistance R of the wires is reduced and energy loss due to heating is minimised. But in this case the cost of outlay is so large that it is almost prohitive. Another way of making I^2R loss small is to make I small. If this method is adopted the voltage must be enormously raised in order that the power supplied may be still equal to W. To make V large means that the insulations in the generator have to be almost perfect, which is not possible. Moreover as the potential drop in the line is small, the voltage at the receiving station, which is equal to $(V-IR)$ is still large. This volage is very unsuitable for use in domestic and industrial purpose in view of the dangers from the shock etc. But on the other hand, if by any means, the low voltage current produced by a generator can be converted into high voltage current and this then transmitted to a great distance where it is again converted into low voltage current, then the whole objective, can be achieved. In d.c. system such changes in voltage cannot readily be made, but can easily be done by means of transformers if an a.c. system is used.

In an a.c. system, the voltage may be increased or decreased by means of a transformer. Hence to avoid heating losses in the line wire, the output voltage of the generator is fist transformed to a much higher value by a step up transformer. It converts the electric power at low voltage and high current to the same power at

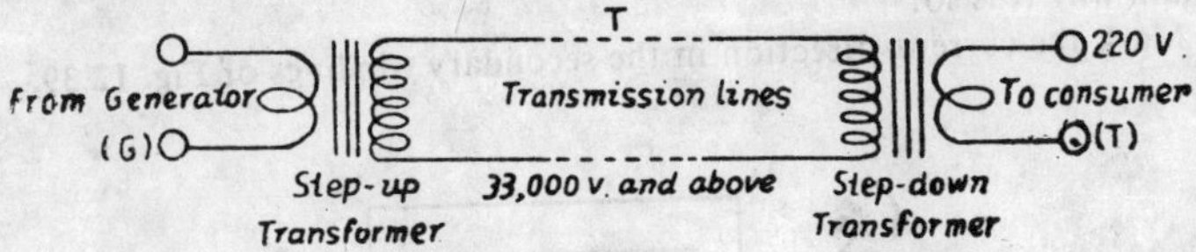

Fig. 12.38. Simple transmission system.

higher voltage and lower current. After the transmission a second transformer is used at the receiving end to step down the voltage to a value which is safe for use (Fig. 12.38). In India for ordinary consumers, the voltage at receiving station is reduced to 220 volts. But by using suitable transformers the voltage may be reduced to any value, which may be needed for industrial and other purposes.

To appreciate the economy in transmitting the power of high voltage and low current, the following example may be illustrative.

Suppose a power generator produces 10 KW power at 25 amp. and 400 volts. It is necessary to deliver this power to a consumer 5 km away on a transmission line whose resistance is 10 ohms. The line loss is given by

$$I^2R = (25)^2 \times 10 \text{ ohm} = 6250 \text{ watts}$$
$$= 6.25 \text{ k watt.}$$

Thus, the line loss constitutes 62.5% of the original power and is wasted. On the other hand if a transformer is used to step up the voltage to 4000 volts, the current will be then only 2.5 amp. Therefore the line loss I^2R is equal to $(2.5)^2 \times 10$ ohms $= 62.5$ watts $= 0.0625$ KW, which is negligibly small.

Thus we see by use of a.c. system and transformers electric energy can be transmitted to very long distance with minimum energy loss.

QUESTIONS

1. What is meant by magnetic flux? State its units in SI system. Is it a scalar or vector quantity?

2. A closed coil of copper wire is moved parallel to itself ia a uniform magnetic field, why is not current induced in it?

3. An ebonite ring and a copper ring of same size both have the same rate of change of flux. How do the induced e.m.f. and induced current in each ring compare?

4. Why do you feel an opposing force, when a metal sheet it pulled outside or pushed inside a magnetic field?

5. A closed copper wire is pulled out with certain velocity from a magnetic field will it be easier to pull it if the ohmic resistance of the coil is increased? Explain why it is so?

6. Mark the current direction in the secondary windings of Fig. 12.39.

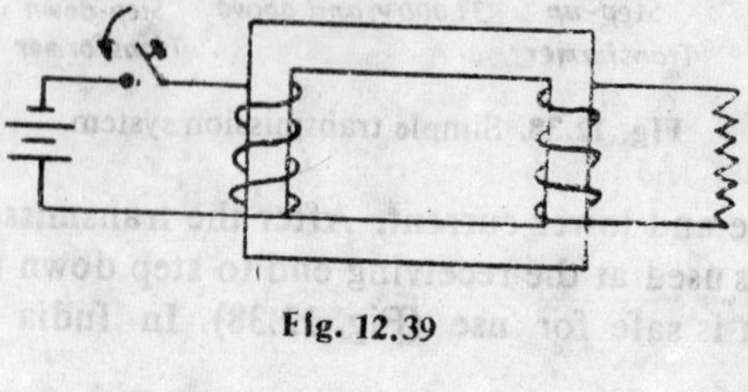

Fig. 12.39

7. State Faraday laws of induction and explain them. on the basis of magnetic flux?

8. Two coils of wire, one connected to a battery and the other connected to a tourch light bulb are placed close to each other. If the current in the battery circuit is rapidly altered, the bulb is found to glow, although no battery is there in that circuit. Why does it happen? Where does the energy come?

9. Does Lenz's law follow from conservation of energy? Explain.

10. Sometimes, when a current carrying circuit is opened a spark is seen at the place, where the circuit is as opened. Why does it happen ?

11. Why are to and fro oscillations are completely absent in a better designed galvanometer?

12. Why is the current in a motor at start is large? What precaution is taken to protect it from damage due to such large current?

13. What do you mean by an alternating current? Can all varying current be called a.c.?

14. Does a transformer convert a.c. to d.c.? Then what does it do? Why are the transformers used in long distance transmission of electric energy?

15. Why is the core of a transformer heated, when alternating current is passed in the coil?

16. If the speed of the generator is low would it affect (*i*) the maximum e.m.f. produced? (*ii*) the frequency of the e.m.f.?

17. What arrangement, will you make to make an electric bulbs glow without the use of a battery?

18. What is the significance of negative value of instantaneous current?

19. What happens to the resistance, inductive reactance, and the capacitative reactance when the frequency applied to a series circuit is doubled?

20. Does the current in a.c. circuit lag, lead or remain in phase with voltage of frequency if appiled to the circuit when,

(*i*) $f=fr$
(*ii*) $f>fr$
(*iii*) $f<fr$

where fr is the resonant frequency

PROBLEMS

1. A 60 cycle a.c. circuit has a voltage of 120 volts and a current of 6.00 amp (effective values).

(*a*) What are the maximum values of these quantities.

(*b*) What is the instantaneous values of the voltage 1/720 sec after the voltage has zero value? *[Ans. 170 volts, 8.50 amp, 85.0 volts]*

2. A coil of 100 turns is perpendicular to the magnetic field, so that the flux passing through the coil is 200×10^{-6} weber. The coil is withdrawn rapidly so that the flux linked drops to zero in 0.1 second. What is the average induced e.m.f. *[Ans. 0.20 V]*

3. A coil of 100 turns and area of 0.002 m^2 is placed at right angles to a magnetic field of flux density 8×10^{-3} wb/m^2. The field is reduced to 10% of its original value in 0.6 seconds. Calculate the average induced e.m.f.

[Ans. $2.8\times10^{-2}V$]

4. A coil is of 100 turns and area 300 cm^2 rotates about an axis prependicular to a magnetic field of 2.0×10^{-2} wb/m^2. If the field of rotation is 50 rev/s, what is average induced e.m.f. in the coil. [*Ans. 12 V*]

5. A copper wire of 30 cm long is perpendicular to a magnetic field of flux density 1.6 wb/m^2 and moves at right angles to the field with speed of 100 cm/s calculate e.m.f. induced in the wire? [*Ans. 0.48 V*]

6. A flate coil of area 10 cm^2 and with 200 turns rotates about an axis in the plane of the coil which is at right angles to a uniform field of flux density of 0.05 wb/m^2. If the speed of rotation of the coil is 25 rad/sec., what are the maximum e.m.f. induced in the coil and the instantaneous value of the e.m.f., when the coil is 45° to the field. [*Ans. 0.25 V, 0.177V*]

7. The electric mains in the house is marked 220 v. 50 c/s. Write the equation for instantaneous voltage. [*Ans. 310 sin 100πt*]

8. An aeroplane with a wing span of 30 meters flies at a horizontal speed of 100 m/s in a region where the vertical component of the magnetic field due to earth is 5.0×10 wb/m^2. What is the potential difference between the tips of its wings? [*Ans. 0.15 V*]

9. A vertical metal disc, radius 8×10^{-2} m is rotating about its centre at 50 revolutions per second with its plane perpendicular to horizontal magnetic field of 0.1 tesla. Find the e.m.f. between the centre and the rim. What is the e.m.f. between the two points on the rim at opposite ends of a diameter. [*Ans. 0.1 V.0*]

10. What e.m.f. will be induced in a 10 H inductor in which the current changes from 10 A to 7 A in 9×10^{-2} sec? [*Ans. 3.33V*]

11. At what frequency will a capacitor of capacitance 5.0 μf have a reactance of 1000 ohms? [*Ans. 32 Hz*]

12. At frequency will a 1 H inductor have a reactance of 1000 ohms? [*Ans. 159.2 Hz*]

13. An inductor is in series with a 100 V, 50 c/s a.c. generator. The current is 10 A and is found to be lagging behind the voltage by 60°. What are the resistance and inductance of the inductor. [*Ans. 5 Ω, 28 mH*]

14. A series circuit consists of a capacitor, a 20 mH inductor of negligible resistance, 125 Ω resistor and an a.c. generator operating at 300 rad/sec. The current is found to lead the voltage by 52°. What is the capacitance of the capacitor. [*Ans. 20.1 μf*]

15. A coil of wire has resistance of 30 ohms and an inductance of 0.10 henry. (*a*) What is its inductive reactance X_L in a 60 cycle circuit? (*b*) Its impedance Z. (*c*) What current there be if the coil is connected to a d.c. source of 120 volts? (*d*) To a 60 cycles a.c. source of 120 volts? [*Ans. 38 ohms, 48 ohms, 4.0 amp, 2.5 amp.*]

16. A resistor of 4.00 ohms, an inductive coil of negligible resistance and inductance of 2.39mH and a good quality 30.0 μf capacitor are connected in series to a source of 500 cycle 110 volt alternating e.m.f. Calculate the reactance of each part of the circuit and the current in the line. [*Ans. 7.50 ohms, 10.6 ohms, 21.8 amp.*]

17. An inductance of resistance 5 and inductance 10 mH is in series with a capacitor and a 10 V 100 c/s a.c. source. The capacitor is adjusted to give resonance in the circuit. Calculate the capacitance of the capacitor and the voltage across the capacitor and the coil? [*Ans. 2.53 μf, 125.7 V, 126.0 V*]

18. A 35.6 ohm rheostat, an inductor of 25.4 ohms resistance and 146 mH inductance and a capacitor of 19.7 μf capacitance are connected in series across a 225 volt, 60 cycle a.c. circuit. Calculate (*a*) impedance of the circuit, (*b*) the current in the circuit, (*c*) what could be done to produce resonance. [*Ans. 100 ohms, 2.25 amp, increase to 358 mH*]

19. How much current is drawn by the primary of a transformer which steps down from 220 volts to 22 volts to operate a device with an impedence of 220 ohm. [*Ans. 0.01 amp.*]

20. An a.c. voltage of 110 V is applied to the primary of a transformer whose efficiency is 99%, the currents in the primary and secondary circuit being 1.60 A and 132 mA respectively. What is the voltage on the secondary and the turn ratio of the transformer? [*Ans. 1320 V, 1 : 12*]

13

OPTICS

From the earliest days man has been familiar with light. It was thought that vision resulted from something sent out from our eye. Later it was believed that light consisted of a stream of tiny elastic particles called corpuscles. This was developed to a theory by Sir Issac Newton towards the middle of the seventeenth century and it is known as Newton's Corpuscular Theory. According to this theory; a luminous body (a burning electric lamp, candle flame or Sun) continuously emits corpuscules in all directions. These corpuscules could (*i*) travel in a straight line without being affected by the earth's gravitation, (*ii*) penetrate the transparent matters, (*iii*) produce the sensation of vision while striking on the retina of the eye and (*iv*) reflect back from a polished surface. Thus rectilinear propagation, reflection, refraction could be accounted satisfactorily but other phenomena like interference, diffraction could not be explained according to this theory. At about same time Christian Huygens proposed wave theory of light in 1678. According to this theory light moves from one point to other just like the motion of waves on the surface of water. This theory could explain the phenomena of interference and, diffraction as well as the reflection and refraction. Later it was found that light is a type of electromagnetic wave which is transverse in nature.

However the modern discoveries like black body radiation photoelectric effect, Compton effect etc, have again established, the particle nature of light. So the dual nature of light is accepted.

13.1. Optics

Light travels in straight line in a homogeneous medium. This rectilinear path of light is represented by a straight line with an arrow head along the direction of propagation of light. But when light reaches a boundary of the medium through which it is travelling, three things happen: Some light is turned (reflected) back in to the first medium and a part is passed (transmitted) through the second and rest is absorbed by the boundary medium. If the surface is polished and glazed like a mirror most of the light is reflected back from the surface. This phenomena is known as reflection and the laws of reflection are stated as: (1) The angle of incidence is equal to the angle of reflection. (2) The incident ray, the reflected ray and the normal to the reflecting surface at the point of incidence all lie in same plane. In case of reflection by a plane mirror the image is always virtual and is formed on the back side of the mirror, the object distance is equal to image distance and size of the image is equal to the size of the object.

13.2. Spherical Mirrors

If the reflecting surface or mirror is curved rather than plane, the same laws of reflection holds but the size and position of the image formed are quite different from those of the image formed by a plane mirror. Curved mirrors may be a part of a sphere, an ellipsoid or a paraboloid and they are named accordingly. Thus there are spherical, ellipsoidal and paraboloid mirrors. Our discussions are limited to spherical mirrors only.

Spherical mirrors are classified as concave when light is reflected from the concave surface. Similarly the spherical mirror is convex when the light is reflected from the convex surface.

Terms associated with the spherical mirror (Fig. 13.1).

1. *Centre of curvature.* The centre of curvature C of a spherical mirror is the centre of the sphere of which the mirror is a part.

2. *Radius of curvature.* Radius of curvature R of a spherical mirror is the radius of the sphere of which the mirror is a part.

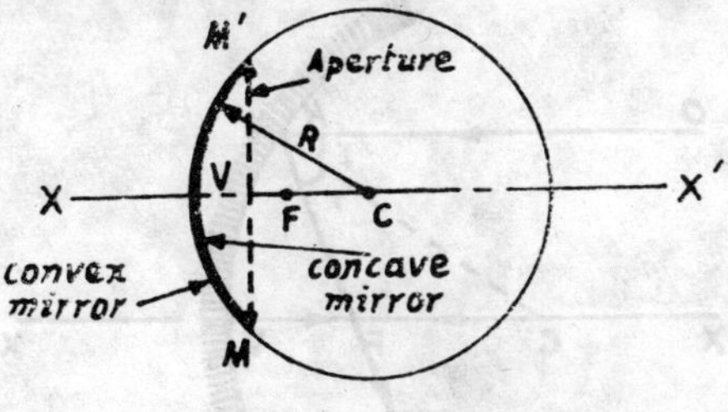

Fig. 13.1. Spherical mirrors.

3. *Vertex or pole of the mirror.* The middle point of the reflecting surface of a mirror is called the pole V of the mirror.

4. *Principal axis.* Principal axis XX' of a spherical mirror is the straight line which passes through the centre of curvature and pole of the mirror.

5. *Principal section.* Principal section $M'VM$ of a mirror is its section by a plane passing through the principal axis of the mirror.

6. *Aperature of a mirror.* The circular periphery of the mirror towards the incident light is called the aperature.

The linear distance *MM'* of the circular outline or periphery of the mirror is called the linear aperature of the mirror and it should be small compared to the radius of curvature of the same mirror.

The angle subtended by this circular periphery of the mirror at its centre of curvature *C* is also known as angular aperature of the mirror. This angle should be less than 4° for all optical purposes.

7. *Principal focus.* If a beam of light parallel to principal axis is incident on a spherical surface, the rays after reflection either converges to or appears to diverge from a common point. This common point *F* is called the principal focus of the mirror.

8. *Focal length f.* The distance *f* of the principal focus from the vertex or pole of the mirror is called the focal length.

9. *Focal plane.* The plane passing through the focus and perpendicular to the principal axis is called the focal plane.

Similarly the distance of the object or the image from the vertex *V* are known as object distance *p* or image distance *q* respectively.

13.3. Relation between Focal Length and Radius of Curvature

For convenience, consider two incident rays *OK* (parallel to the principal axis) and *CV* (coincident with the principal axis) coming

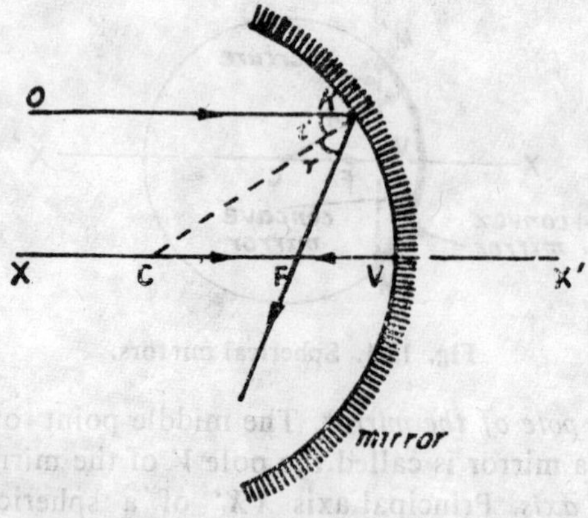

Fig. 13.2. Reflection of a parallel beam of light by a concave mirror.

from a far distant source and falling on a concave mirror. The ray *OK* reflected along *KF* passes through the focus *F* and the other is reflected along *VC*. So an image of the source is formed at the point of intersection of these two reflected rays i.e. at the focus F in front of the mirror.

According to the laws of reflection

$$\angle OKC = \angle CKF$$

But $$\angle KCF = \angle OKC$$

So $$\angle KCF = \angle CKF$$

So $\triangle CFK$ is an isosceles triangle.

Therefore $$CF = KF$$

For small aperature the point K is very close to the point V.

Therefore $$KF \simeq FV$$

Hence $$CF = FV = f.$$

$$f = \frac{CF + FV}{2} = \frac{CV}{2}$$

i.e. focal length = 1/2 radius of curvature.

or $$f = R/2 \qquad \ldots (13.1)$$

The same relation can also be established for a convex mirror.

13.4. Types of Images

Usually two types of images are formed by spherical mirrors, real and virtual.

Real images. Real images are formed by actual intersection of the rays after reflection or refraction. The real images are inverted and can be projected on a screen.

Virtual images. Virtual images are formed by the intersection of imaginary rays (*i.e.* by producing the reflected or refracted rays in the backward direction.) These images are erect and cannot be projected on a screen.

13.5. Location of Image Formed by Spherical Mirrors

The position of an image formed by a spherical mirror can be located in three ways.

1. **By physically observing the image position.** When virtual images are formed by spherical mirror, their position can be located easily by looking through the mirror just like plane mirror whereas real images of an object formed by concave mirror can be projected on a screen.

2. **By drawing the ray diagram.** For this purpose two rays (coming from the object) are chosen —A ray parallel to the principal axis which after reflection passes through the focus or appears to diverge from the focus and another ray through the radius of curvature which is reflected back in the same direction.

The point of intersection of reflected rays is the position of the image of the concerned object point. In the Fig. 13.3 O_1 is the object point, O_1K is the ray parallel to the principal axis, O_1C is the ray passing through the centre of curvature and I_1 is the image point.

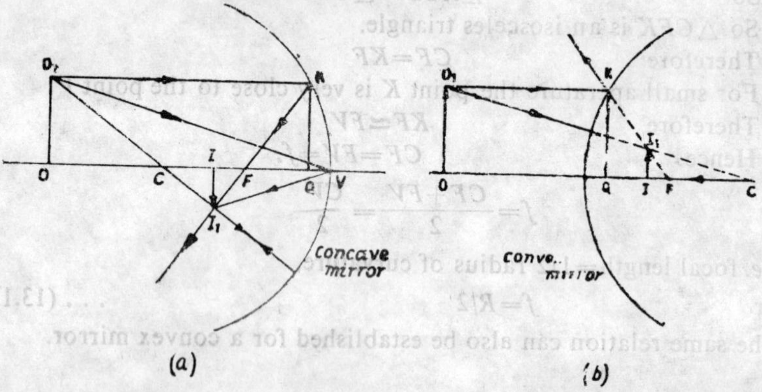

Fig. 13.3. (*a*) Image formation in a concave mirror.
(*b*) Image formation in a convex mirror.

3. **By using the mirror equation.** It is discussed in sections 13.9 and 13.10.

13.6. Mirror Equation

It is a simple relation between the object distance p, image distance q and the focal length f of a mirror. To derive the relation let us consider the Fig. 13.3 (*a*) where OO_1 and II_1 are the length of the object and image respectively. The triangles O_1OC and I_1IC are similar.

Therefore,

$$\frac{OO_1}{II_1}=\frac{OC}{IC}=\frac{OV-CV}{CV-IV} \quad \ldots (13.2)$$

Again the triangles KQF and II_1F are also similar. Therefore,

$$\frac{KQ}{I_1I}=\frac{FQ}{FI} \quad \ldots (13.3)$$

or $$\frac{OO_1}{II_1}=\frac{FQ}{IV-FV} \quad \text{since } KQ=O_1O$$

Now QF is slightly shorter than VF. But if we are dealing with mirrors of small aperature, we can take them to be equal.

So $$QF=VF=f,\ IQ=q,\ OQ=p.$$

Equating the right hand side of the two Equations 13.2 and 13.3 we get,

$$\frac{OV-CV}{CV-IV}=\frac{FQ}{IV-FV}$$

or
$$\frac{p-2f}{2f-q}=\frac{f}{q-f}$$

On cross multiplication we get,

$$pq-2qf-pf+2f^2=2f^2-qf$$

or
$$pq=pf+qf,$$

Dividing through out by pqf we get, $\frac{1}{p}+\frac{1}{q}=\frac{1}{f}$ . . . (13.4)

This is the mirror equation.

Same relation can also be deduced using the Fig. 13.3(*b*) for a convex mirror using the sign convention described in the next section. It is left as an exercise for the students.

Special case. If the surface is plane, $r=\infty$. Here the mirror equation reduces to

$$\frac{1}{p}+\frac{1}{q}=\frac{1}{\infty}=0$$

or
$$p+q=0$$

or
$$p=-q$$

Thus in the case of plane mirror, the image distance and the object distance are equal and the image is behind the mirror, and image distance is negative. Hence we may say that the plane mirror is a special case of spherical mirror.

13.7. Sign Convention

1. All distances are measured from the vertex V or pole of the mirror as the origin.

2. The distances measured in the same direction as that of incident light are taken as negative.

3. The distances measured against the direction of the incident light are taken as positive.

In this book, we shall always draw the incident ray of light as going from left to right as shown in ray diagram. In such case, the distance measured from the pole to the left is positive and to the right negative. This is just reverse to our familiar sign convention used in Cartesion (rectangular) coordinate system.

On the basis of this convention, the focal length (and the radius of curvature) of a convex mirror is negative and that of concave mirror is positive.

From mirror equation it is evident that if two quantities are known the third can be calculated. Equation 13.4 is valid for both concave and convex mirror when proper sign conventions are followed. The conventions are (1) f and R are taken as positive for concave mirrors and negative for convex mirrors. (2) The object distance p and image distance q are taken to be positive for real objects and images, negative for virtual objects and images. While working out problems the symbols in mirror equation are always written with positive sign. Negative signs are introduced only when numerical values are substituted for symbols.

13.8. Magnification in Spherical Mirrors

The image of an object formed by a spherical mirror, in general, is of a size different from that of the object. In some cases, it is smaller than the object and in others it is larger. The ratio of the size of image to the size of the object is called magnification and is usually represented by a symbol M.

$$M=\frac{\text{Image size}}{\text{Object size}}$$

In the Figure 13.3 triangles OO_1V and II_1V are similar.

Therefore $$M=\frac{II_1}{OO_1}=\frac{q}{p} \qquad \ldots(13.5)$$

or $$M=\frac{\text{Image distance}}{\text{Object distance}}$$

This is true for any spherical mirror and also for both virtual and real images. From Equation 13.4 we can easily get,

$$M=\frac{f}{p-f}=\frac{q-f}{f} \qquad \ldots(13.6)$$

13.9. Position and Nature of Images Formed by a Concave Mirror

We will use the mirror equations for this purpose.

1. **Object at infinity**. When the object is placed at infinity we have,

$$\frac{1}{\infty}+\frac{1}{q}=\frac{1}{f}$$

$$\boxed{q=f}$$

So the image is formed at its focus or focal plane.

$$M=\frac{q-f}{f}=\frac{f-f}{f}=0.$$

As q is positive the image is real and inverted and highly diminished.

2. Object at C. When object is placed at C, then

$$P=R=2f$$

$$\frac{1}{2f}+\frac{1}{q}=\frac{1}{f}$$

or $$\frac{1}{q}=\frac{1}{2f} \qquad \text{or} \qquad \boxed{q=2f}$$

Thus image is formed at C and the image distance is positive

The magnification:

$$M=\frac{q}{f}-1$$

$$=\frac{2f}{f}-1$$

$$M=1$$

Hence the size of the image is equal to the size of the object and the image is real and inverted.

3. Object at F. When the object is situated at F, then $P=f$

$$\frac{1}{f}+\frac{1}{q}=\frac{1}{f}$$

$$\frac{1}{q}=0 \qquad \text{or} \qquad \boxed{q=\infty}$$

Thus image is formed at ∞. And the magnification,

$$M=\frac{q}{f}-1$$

$$=\frac{\infty}{f}-1=\infty$$

Thus image is real, inverted and highly magnified.

4. Object is between C and ∞. For convenience let us take the object is at $5f$. Then,

$$\frac{1}{5f}+\frac{1}{q}=\frac{1}{f}$$

or $$\frac{1}{q}=\frac{4}{5f}$$

or $$q=\frac{5}{4}f$$

The image is formed between C and f and the image distance is positive.

$$M=\frac{\frac{5}{4}f-f}{f}=\frac{f}{4f}=.25$$

Therefore the image is real inverted and diminished.

5 **Object between F and C.** When the object is placed between F and C, say at $3/2\,f$,

Then, $2/3f+1/q=1/f$.

or
$$\frac{1}{q}=\frac{1}{f}-\frac{2}{3f}=\frac{1}{3f}$$

so $\boxed{q=3f}$ i.e., q is greater than $2f$ and positive.

The image is formed beyond C.

Thus magnification:

$$M=\frac{q}{f}-1$$
$$=\frac{3f}{f}-1$$
$$=2$$

The image is real, inverted and magnified.

6. Object between pole and focus. When the object is placed between f and the pole of the mirror say at $f/2$.

We have
$$\frac{1}{f/2}+\frac{1}{q}=\frac{1}{f}$$

or
$$\frac{2}{f}+\frac{1}{q}=\frac{1}{f}$$

or
$$\frac{1}{q}=-\frac{2}{f}+\frac{1}{f}=-\frac{1}{f}$$

or
$$\boxed{q=-f}$$

Thus image is formed behind the mirror and q is negative.

Magnification:

$$M=q/f-1 \qquad \text{or } M=-f/f-1=-2$$

Thus the image is erect, virtual and magnified.

Example 13.1. An object 5 cm high is placed 30 cm from a convex mirror whose focal length is 20 cm. Find the position, size and nature of the image?

Solution. Size of the object $=OO'=5\text{cm}$

Object distance $=p=30\text{cm}$

Focal length $=f=-20\text{cm}$

Image distance $=q=?$

Size of the image $=II=?$

Now
$$\frac{1}{f}=\frac{1}{p}+\frac{1}{q}$$

$$\therefore \quad \frac{1}{-20}=\frac{1}{q}+\frac{1}{30} \quad \text{or} \quad q=-12 \text{ cms}$$

Image is 12 cm, from the mirror and is virtual. (Since sign is negative) We know, magnification,

$$=\frac{\text{Size of the image}}{\text{Size of the object}}=\frac{II'}{OO'}=\frac{q}{p}$$

or $$\frac{II}{5}=\frac{12}{30}$$

or $$II'=2 \text{ cms}$$

Image is 2 cm high.

Example 13.2. A man has a concave shaving mirror whose focal length is 20 cm. How far should the mirror be held from his face in order to give an image of two fold magnification?

Solution. An erect virtual magnified image is desired.

Magnification $=M=q/p=2$

Since the image is virtual, p and q have opposite signs or

$$q=-2p.$$

$$\frac{1}{f}=\frac{1}{p}+\frac{1}{q}$$

or $$\frac{1}{p}+\frac{1}{-2p}=\frac{1}{20}$$

or $$p=10 \text{ cm}$$

13.10. Position and Nature of Image Formed by a Convex Mirror

1. **Object at infinity.** f is negative for a convex mirror. We have $p=\infty$ in this case.

so $$\frac{1}{\infty}+\frac{1}{q}=-\frac{1}{f}$$

or $$\frac{1}{q}=-\frac{1}{f} \quad \text{or} \quad \boxed{q=-f}$$

Image is at the focus, and image distance is negative.

Magnification: $$M=q/f-1$$
$$=f/f-1$$
$$=0$$

Thus image is virtual, and deminished.

2. **Object is placed at a distance f.** When an object is placed in intimate contact with the mirror.

Here $$p=f$$

Then we have:

$$\frac{1}{f}+\frac{1}{q}=-\frac{1}{f}$$

or $$\frac{1}{q}=-\frac{2}{f}$$

or

$$q = -f/2.$$

Thus image is formed at $f/2$ and image distance is negative.

$$\text{Magnification}=M=-\frac{q}{f}-1=\frac{f/2}{f}-1=\frac{1}{2}-1=-\frac{1}{2}$$

The image is virtual, erect and diminished and formed behind the mirror. Other cases are not being discussed.

It can be easily shown that when an object is placed anywhere infront of a convex mirror, the image is formed behind the mirror in between focus and the mirror. The image is always virtual, erect and diminished.

13.11. Refraction at Plane Surfaces

The apparent bending of a stick partly immersed in water and the apparent loss of depth of a coin placed inside a breaker containing liquid are due to the phenomenon of refraction. Similar to that of the reflection, there are two laws.

First law. The incident ray, the refracted ray and the normal to the refracting surface at the point of incidence all lie in the same plane.

Second law. For the particular colour of light, the ratio of the sine of the angle of incidence to the sine of the angle of refraction is constant for a given pair of media.

In mathematical form $\frac{\sin i}{\sin r}=\eta$ (constant)

This second law is known as Snell's law in honour of Willebred Snell Van Royen.

13.12. Refraction through Prism

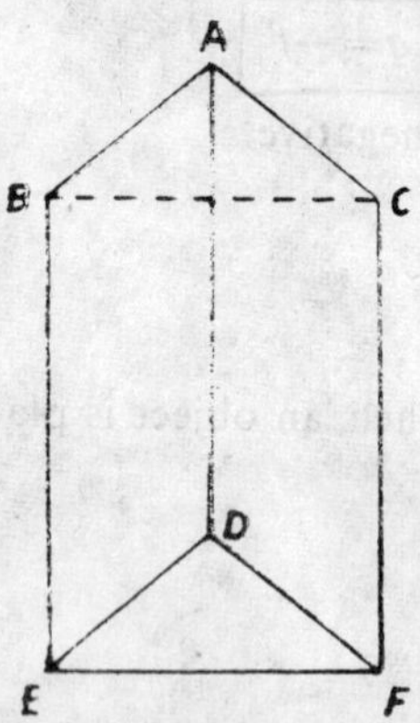

Fig. 13.4. Diagram of a prism.

A block of glass having three vertical faces and two horizontal faces is called Prism and is shown in Figure 13.4.

In the figure *ADEB*, *ADFC* and *BCFE* are three vertical faces and *ABC* and *EDE* are the two horizontal faces.

Refracting edge (AD). The line in which the two refraction faces meet is called the refracting edge of the prism.

Refracting angle or simply angle of the prism. The angle between the two refracting faces is the angle of the prism.

Principal section (ABC). A section of the prism by a plane perpendicular to the refracting edge is called a principal section.

13.13. Deviation due to Prism

A ray of light PQ is incident on the face AB. It gets refracted along QR inside the prism and at R it emerges along RS. In Fig. 13.5 PQ and RS are the initial and final directions of the ray. Angle between them is called the angle of deviation δ. From the Fig. 13.5 we can see that

$$\delta = \delta_1 + \delta_2$$

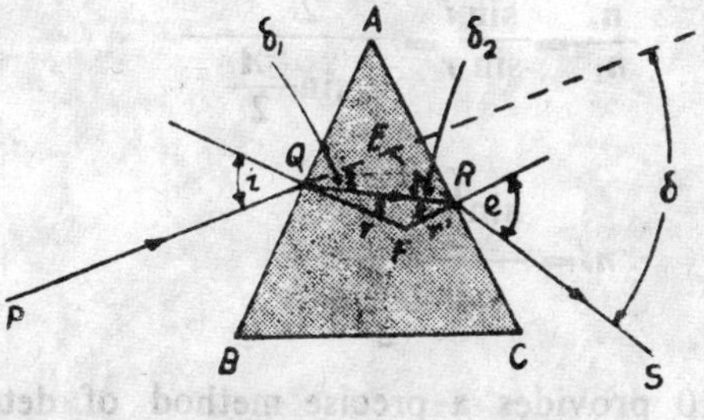

Fig. 13.5. Deviation of a ray by a prism.

Where δ_1 = Deviation due to refraction at the face AB,

δ_2 = Deviation due to refraction at the face AC.

If i, r are the angle of incidence and refraction at the face AB and r' and e are the angle of incidence and angle of emergence at the face AC, we have

$$\delta = (i - r) + (e - r') \qquad \ldots (13.7)$$

In the quadrilateral $AQFR$,

$$\angle AQF = \angle ARF = 90°$$

Hence, $A + \angle QFR = 180°$

$$A = 180° - \angle QFR,$$

Again from traiangle QFR

$$\angle QFR + \angle r + \angle r' = 180°$$

Therefore $\quad r + r' = A \qquad \ldots (13.8)$

Thus $\quad \delta = (i + e) - (r + r')$

$$= i + e - A \qquad \ldots (13.9)$$

or $\quad \delta + A = i + e$

It is found from experiments that the angle of deviation cannot be smaller than a particular value. That angle of deviation is called the angle of mininum deviation D. Minimum deviation occurs only when the ray of light through the prism is parallel to the base . The ray passes symmetrically through the prism under this condition *i.e.* $i = e$ and $r = r'$

Substituting these values in Equations 13.8 and 13.9 we get respectively,

$$r=A/2,$$

and

$$i=(A+D)/2.$$

Relation between refractive index and angle of minimum deviation. We know $n_1 \sin i = n_2 \sin r$,

where, n_1=Refractive index of the surrounding medium

n_2=Refractive index of the material of the prism

$$\frac{n_r}{n_1}=\frac{\sin i}{\sin r}=\frac{\sin\frac{(A+D)}{2}}{\sin\frac{A}{2}}$$

or

$$n_r=\frac{\sin\frac{(A+D)}{2}}{\sin\frac{A}{2}} \quad \ldots (13.10)$$

Equation 13.10 provides a precise method of determining the index of refraction of a transparent material in the form of a prism.

13.14. Thin Lenses

A lens is a transparent refracting medium bounded by two surfaces of regular geometrical form such as spherical or plane or cylindrical surfaces. Lenses whose surfaces are portions of spheres are called spherical lenses and those with cylindrical surfaces are called cylindrical lenses.

Lenses are divided into two types.

1. Convex and 2. Concave. They are shown in the Fig. 13.6.

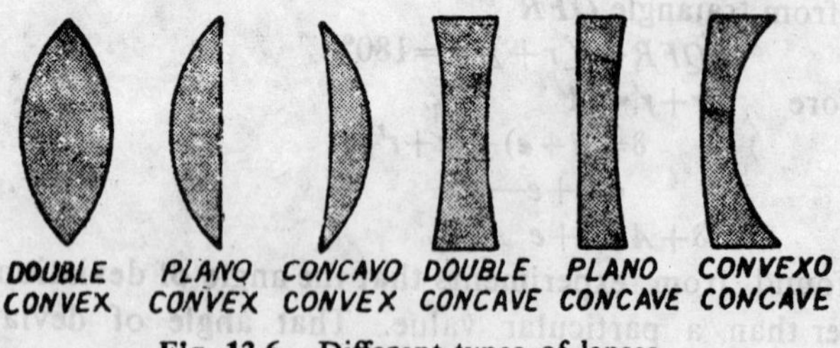

Fig. 13.6. Different types of lenses.

Centre of curvature (C_1 and C_2). The centre of the spheres of which a surface of a lens is a part is called the centre of curvature of lens.

In case if one of the surface is flat, its centre of curvature is considered to be at ∞. Thus each lens has two centres of curvature and hence two radii of curvature as shown in Fig. 13.7(a). It should however, be noted that both the radii of curvature in a lens need not necessarily be always equal.

Principal axis (XX′). The straight line passing through both the centres of curvatuie (C_1 and C_2) is called the principal axis of the lens.

Here, we will consider the cases of spherical lenses only.

Aperature of the lens. When the boundary of the lens is circular, the diameter of the boundary is called aperature of the lens.

Principal section. A lateral section of the lens passing through the principal axis is called the principal section.

Optical centre (O). Inside a lens there is a point so situated on the principal axis that the rays passing through that point are not deviated. This point is known as the optical centre of the lens.

Pole (A_1 and A_2). The points of intersection of the principal axis with the bounding surfaces of a lens are called the poles of those surfaces and the distance between the two poles of the surfaces is the thickness of the lens.

Convex lens. This lens is thicker at the middle than the edges. It is analogous to two prisms placed with their bases in contact as shown in Fig. 13.7.

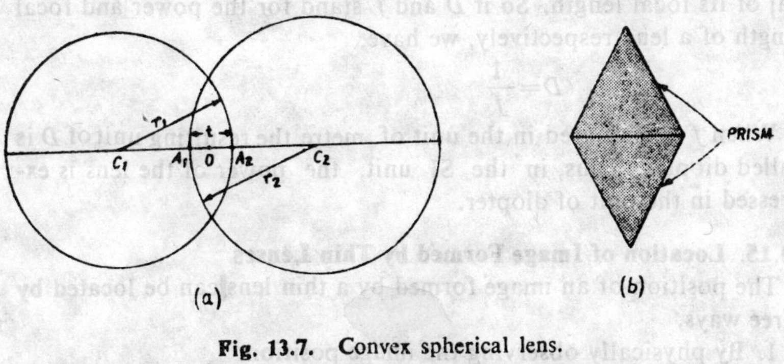

Fig. 13.7. Convex spherical lens.

A prism bends the light rays towards base. Hence, this combination will bend the rays towards the principal axis of the lens. Hence a beam of light passing through a convex lens will converge. Therefore a convex lens is also known as a converging lens.

Concave lens. A concave lens is thicker at the edge than at the middle. So it is analogous to two prisms placed with their vertices on the two opposite face of a rectangular glass slab as shown in Fig. 13.8.

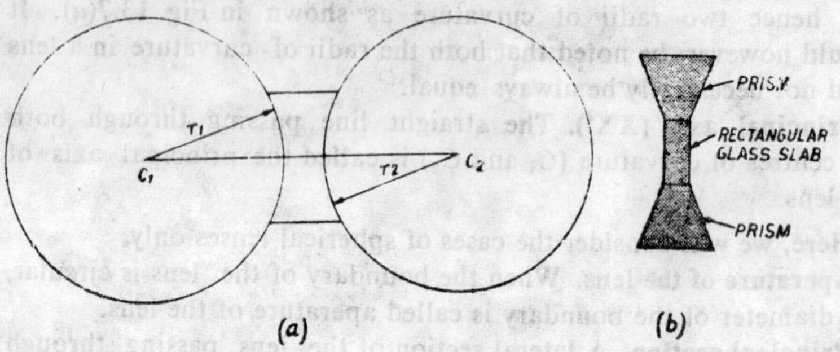

Fig. 13.8. Concave spherical lens.

This arrangement is capable of diverging a light beam. Lens having diverging action is called a diverging lens.

Focus. A point at which rays incident parallel to the principal axis of a lens converge or appear to diverge after refraction through the lens is known as its focus.

Focal length (f). The distance of the focus from the optical centre of the lens is called the focal length of the lens. Each lens has two focii, one on each side. A lens having more converging or diverging power has shorter focal length.

Power of a lens. The power of a lens is expressed as the reciprocal of its focal length. So if D and f stand for the power and focal length of a lens respectively, we have,

$$D=\frac{1}{f}$$

When f is measured in the unit of metre the resulting unit of D is called diopter. Thus, in the SI unit, the power of the lens is expressed in the unit of diopter.

13.15. Location of Image Formed by Thin Lenses

The position of an image formed by a thin lens can be located by three ways.

1. By physically observing the image position.

2. By drawing the ray diagram. For this purpose a ray coming from the object parallel to the principal axis is chosen. This ray after refraction passes through the focus or appears to diverge from the focus. Another ray passing through the optical centre of the lens is also chosen. This ray after refraction go undeviated. The

point of intersection of the two rays is the position of the image of the point object. In the Fig. 13.9(*a*) O_1 is the position of the object point and O_1K is the ray parallel to the principal axis. This ray pass through F_2 after refraction. The ray O_1O passing through the

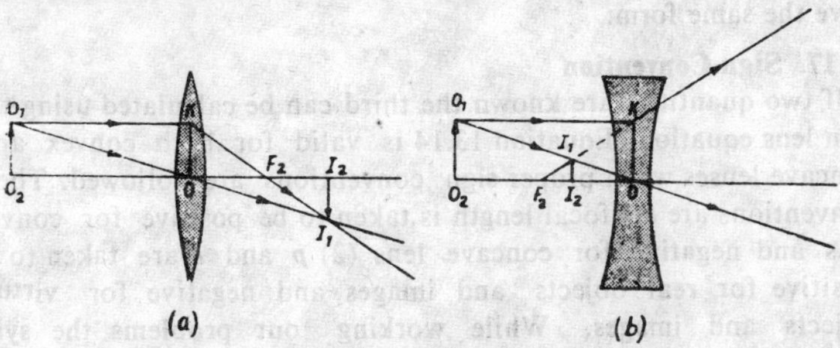

Fig. 13.9. Image formation by convex and concave lenses.

optical centre pass undeviated. I_1 is the position of the image point.

3 By using the thin lens equation is to be described in Sections 13.18 and 13.19.

13.16. Thin Lens Equation

In the Fig. 13.9(*a*) object is represented by O_1O_2. The image formed by the lens is I_1I_2. Let the object distance, the image distance and the focal length of the lens be denoted by *p*, *q* and *f*, respectively. The triangles O_1O_2O and I_1I_2O are similar, we have,

$$\frac{I_2 I_1}{O_2O_1} = \frac{OI_2}{OO_2} \qquad \text{... (13.11)}$$

Triangles KOF_2 and $I_1I_2F_2$ are also similar therefore,

$$\frac{I_2I_1}{OK} = \frac{I_2F_2}{OF_2} \qquad \text{... (13.12)}$$

Since $O_1O_2 = OK$. Using Equations 13.11 and 13.12 we get,

$$\frac{OI_2}{OO_2} = \frac{I_2F_2}{OF_2} = \frac{OI_2 - OF_2}{OF_2} \qquad \text{... (13.13)}$$

or

$$\frac{q}{p} = \frac{q-f}{f}$$

or

$$qf = pq - pf$$

Dividing by pqf we get,

$$\boxed{\frac{1}{p} + \frac{1}{q} = \frac{1}{f}} \qquad \text{... (13.14)}$$

Above equation is the thin lens equation. Same relation can also be deduced using the Fig. 13.9 (*b*) for a concave lens using the proper sign convention discussed in the next section. It is given as an exercise to the students. Students should note that Equation 13.4 for spherical mirrors and Equation 13.14 for spherical lenses have the same form.

13.17. Sign Convention

If two quantities are known the third can be calculated using the thin lens equation. Equation 13.14 is valid for both convex and concave lenses when proper sign conventions are followed. These conventions are (1) focal length is taken to be positive for convex lens and negative for concave lens (2) p and q are taken to be positive for real objects and images and negative for virtual objects and images. While working out problems the symbols in the equation are always written with positive sign. Negative signs are introduced only when numerical values of the symbols are used.

13.18. Magnification

Sizes of image of an object formed by a lens are different from that of the object. The ratio of the size of the image to the size of the object is called magnification. If M is the magnification, from Equation 13.11 we can write,

$$M=\frac{\text{Image size}}{\text{Object size}}=\frac{q}{p} \qquad \ldots(13.15)$$

This is valid for both thin convex and concave lenses.

13.19. Position and Nature of Image Formed by a Convex Lens

1. Object at infinity. When the object is placed at ∞, from Equation 13.14 we get,

$$\frac{1}{\infty}+\frac{1}{q}=\frac{1}{f}$$

Thus the image is real and is formed at the focus on the other-side of the lens.

Hence magnification,

$$M=\frac{f}{\infty}=0$$

The image is highly diminished.

2. **Object at 2 F**. When the object is placed at 2 F, we have,

$$\frac{1}{2f}+\frac{1}{q}=\frac{1}{f}$$

or $$\frac{1}{q}=\frac{1}{2f}, \quad \boxed{q=2f}$$

The image is real and formed at 2 *F* on the otherside of the lens. Hence magnification

$$M=\frac{q}{p}=\frac{2f}{2f}=1$$

Thus image is inverted and of the same size as the object.

3. **Object beyond 2F.** When the object is placed beyond 2 F, for convenience let us take that, the object distance is 5 *f*, we have

$$\frac{1}{5f}+\frac{1}{q}=\frac{1}{f}$$

or $$\frac{1}{q}=\frac{4}{5f}$$

or $$q=\frac{5f}{4}=1.25\,f$$

The image is real inverted and formed between F and 2 F on the otherside of the lens.

Magnification,

$$M=\frac{q}{p}=\frac{1.25f}{5f}=0.25$$

Thus image is diminished.

4. **Object at F.** When the object is placed at F, we have,

$$\frac{1}{f}+\frac{1}{q}=\frac{1}{f}$$

or $$\frac{1}{q}=0$$

or $$\boxed{q=\infty}$$

The image is real and inverted and formed at ∞, on the otherside of the lens.

The magnification

$$M=\frac{q}{p}=\frac{\infty}{p}=\infty$$

Thus image is highly magnified.

5. **Object between F and 2F.** Let the object be placed between *F* and 2*F* say at 3/2*f*. Then we have

$$\frac{1}{\frac{3}{2}f}+\frac{1}{q}=\frac{1}{f}$$

or
$$\frac{1}{q}=\frac{1}{f}-\frac{2}{3f}=\frac{1}{3f}$$

$\therefore$
$$\boxed{q=3f}$$

Thus image is real and formed beyond $2F$ on the other side of lens.

The magnification,

$$M=\frac{3f}{\frac{3}{2}f}=2$$

Thus image is magnified.

6. **Object between F and the lens.** Let the object be placed between F and the lens say at $f/2$, we have,

$$\frac{2}{f}+\frac{1}{q}=\frac{1}{f}$$

or
$$\frac{1}{q}=-\frac{1}{f} \quad \text{or} \quad \boxed{q=-f}$$

Thus image is erect virtual and is formed on the same side of the object.

Here the magnification

$$M=q/p=-2$$

Thus image is erect and magnified.

13.20. Position and Nature of Image Formed by a Concave Lens

1. **Object at ∞.** When an object is placed at ∞ then we have

$$\frac{1}{\infty}+\frac{1}{q}=-\frac{1}{f}$$
$$q=-f$$

Thus image is virtual erect and formed at the focal plane on the same side of the object.

Hence magnification, $M==\frac{q}{f}-1=+\frac{f}{f}-1=0$

Thus image is highly diminished.

2. **Object at 2F.** When the object is placed at $2F$, we have

$$\frac{1}{2f}+\frac{1}{q}=-\frac{1}{f}$$

or
$$\frac{1}{q}=-\frac{1}{f}-\frac{1}{2f}=-\frac{3}{2}f$$

or
$$\boxed{q=-\tfrac{2}{3}f}$$

The image is virtual erect and formed at a distance of $2/3\,f$ on the same side of the object.

Hence magnification,

$$M=\frac{q}{p}=-\frac{\frac{2}{3}f}{2f}=-\frac{1}{3}$$

Thus image is diminished.

3. **Object at F.** When the object is placed at a distance of f we have

$$\frac{1}{f}+\frac{1}{q}=-\frac{1}{f} \quad \text{or} \quad \frac{1}{q}=-\frac{2}{f} \quad \text{or} \quad \boxed{q=-\frac{f}{2}}$$

The image is virtual and formed at a distance $f/2$ on the same side of the object.

$$M=\frac{f/2}{f}=\frac{1}{2}$$

The image is diminished to half of its size.

13.21. Wave Theory of Light (Huygen's Principle)

This theory was suggested by Chistian Huygen. According to this theory a luminous body is a source of disturbance. This disturbance travels through space in the form of waves in a hypothetical medium called ether.

Huygen's principle. The waves will move in all possible directions and it will reach simultaneously at all part of a spherical surface. All the waves at the surface are in the same state of vibration (*i.e.* same phase) and hence that surface represents the wave front at any time. To find the wave front at any subsequent interval, Huygen suggested a method, known as Huygen's principle. Each point on the wave front acts as a source of new disturbance. The new disturbances are called secondary wavelets and spread out with velocity of light. In Fig. 13.10 *AB* is the wave front at certain time, and *a, b, c, d* . . . are the origin of secondary wavelets. To find out the position of the wave front after t seconds describe a series of circles with radii vt taking *a, b, c, d* . . . etc. as centre. These are the origin of the secondary wavelets. The envelope A' B' of these secondary wavelets is the new-wave-front. These secondary wave front spread in the forward direction only.

At a great distance from the source, the radii of curvature of the wave front will be large and it can be regarded as plane. As the rays of light are perpendicular to the wave front, they will be taken parallel.

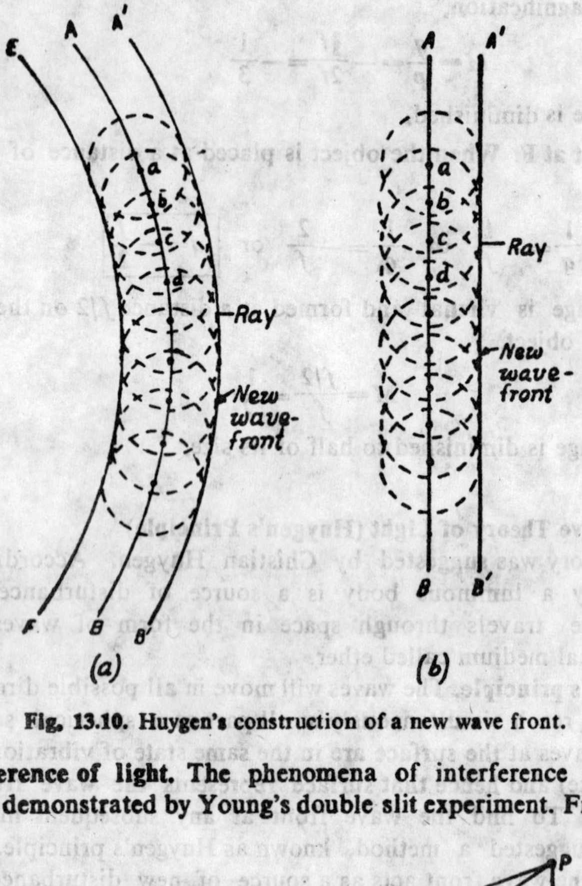

Fig. 13.10. Huygen's construction of a new wave front.

Interference of light. The phenomena of interference of light was first demonstrated by Young's double slit experiment. Fig. 13.11

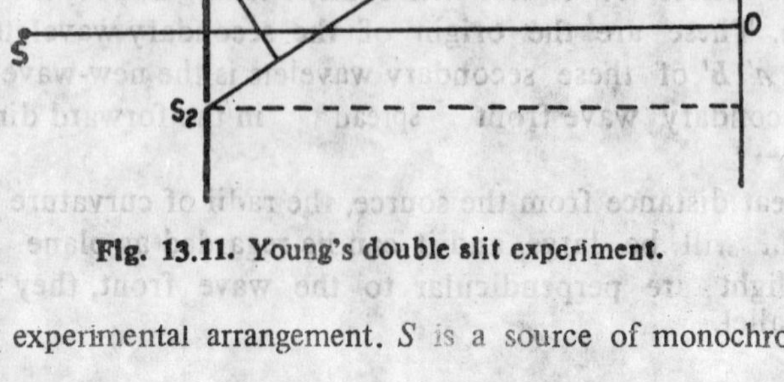

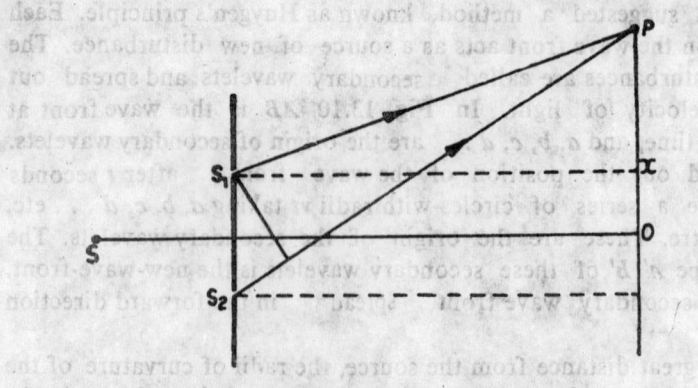

Fig. 13.11. Young's double slit experiment.

shows the experimental arrangement. S is a source of monochromatic light. S_1 and S_2 are two parallel rectangular slits. The separation between the two slits is about 2 mm, the width of each slit is about 0.3 mm. OP is the plane of observation where light from S_1 and S_2 reach. The distance S_1S_2 is 10 cm and S_1O is 2 metre. It is so arranged that the line of sight pass through middle of the two slits. If one observes through an eye piece placed on OP, he observes the following.

1. If either S_1 or S_2 alone is open, the intensity of light is uniform on OP.

2. If both S_1 and S_2 are open, the intensity alternately increases and decreases as one moves along OP *i.e.* bright and dark fringes are seen.

3. The fringe width w is directly proportional to the distance D between slits S_1 S_2 and OP and inversely proportional to the separation between S_1 and S_2.

To explain the observations let us assume that light moves in the form of waves having wave length λ. In that case, secondary wave lets from S_1 and S_2 will start and move towards OP. The secondary waves from S_1 and S_2 would interfere each other as discussed earlier in dealing with superposition of waves. If we consider a point P in the observation plane, the nature of interference between the two waves reaching there depends upon the path difference p. If $OP=x$, $S_1S_2=d$ then,

$$p=S_2P-S_1P$$

$$=\left[D^2+\left(x+\frac{d}{2}\right)^2\right]^{1/2}-\left[D^2+\left(x-\frac{d}{2}\right)^2\right]^{1/2}$$

As D is much greater than x and d

We can write,

$$p=D+\frac{\left(x+\frac{d}{2}\right)^2}{2D}-D-\frac{\left(x-\frac{d}{2}\right)^2}{2D}$$

$$\boxed{p=\frac{xd}{D}} \quad \ldots (13.16)$$

If a_1, a_2 are the amplitudes of the disturbances at p due to S_1 and S_2 respectively then the resultant amplitude of the two super imposed waves differing in phase is given by,

$$A^2=a_1^2+a_2^2+2a_1a_2 \cos \phi \quad \ldots (13.17)$$

The intensity will be maximum when A^2 is maximum *i.e.* $\cos \phi=1$, which means that the phase difference is zero or even multiplies of π. The above condition is satisfied if the path difference is zero or integral multiplies of λ. Therefore condition for maximum intensity at p can mathematically be expressed as

$$\frac{x_n d}{D} = n\lambda \quad \ldots (13.18)$$

Where x_n=is the distance of the *n*th fringe from *O*. Under this condition $A = a_1 + a$.

On the other hand if the path differs at *p* is an odd multiplies of $\frac{\lambda}{2}$, then

$$\frac{x_n d}{D} = \left(n + \frac{1}{2} \right) \lambda \quad \ldots (13.19)$$

Here both the disturbances at *p* are in opposite phase.

Hence,

Phase difference $= (2n+1)\pi$

$$A^2 = a_1^2 + a_2^2 - 2a_1a_2$$

or $$A = (a_1 - a_2)$$

Thus we observe minimum intensity of light at that point. Hence there will be alternate brightness and darkness.

Therefore, the separation between the successive maxima i.e. the fringe width *w*,

$$w = X_{n+1} - X_n = \frac{D}{d}(n+1-n)\lambda$$

or $$w = \frac{D}{d}\lambda \quad \ldots (13.20)$$

If one assumes that light travels in the form of waves the conclusion is that there should be alternately bright and dark fringes. This is exactly what is observed experimentally. So the assumption is taken to be correct. It is also further observed that there is quantitative agreement between the experimental and theoretical result as given in Equation 13.20 *i.e.* $w \propto D$ and $w \propto 1/d$.

It is to be noted that the interference experiments gives us no information whether the wave are longitudinal or transverse.

Usually the wave length of light is expressed in Angstrom Unit (Å) which is 10^{-10} metre. For example the wave length of different colours of visible lights are given below.

Violet	—	4000 Å
Green	—	5600 Å
Yellow	—	5900 Å
Red	—	7500 Å

But the light having $\lambda < 4000$Å called ultraviolet are not visible to our eye, and also light having $\lambda > 7500$ Å called infra-red are also not visible to our eye.

13.22. Coherent Sources

It is found that in Young's double slit experiment if two slits S_1 and S_2 are illuminated by two separate sources of monochromatic light of same wave length no interference pattern is observed. This is because the phase difference between the two sources are not constant.

Two sources are said to be coherent if they emit light waves of the same frequency and are always in phase with each other. In actual practice, it is not possible to have two independent sources which are coherent. It is found that two sources of light can have constant phase difference if they are the secondary sources of a single source. Such sources are called coherent.

The following methods have been developed to produce two coherent sources.

1. The real source and its virtual image formed by a plane mirror serve as two coherent source. (Lloyd's single mirror.)

2. Two virtual images of the same sources formed by two plane mirrors (bimirror).

3. Two virtual images of the same source formed by a biprism —refraction.

Conditions for interference. The following conditions must be satisfied for getting an interference pattern.

1. Two sources of light must be coherent.

2. The source must emit monochromatic light. If a source of white light is used to produce interference pattern the maximum of one colour may coincide with minimum of another colour. As a result clear dark and bright fringes cannot be seen.

3. The two sources should be narrow otherwise clear dark and bright fringes will not be produced, because maxima produced by one region of the sources may coincide with minima produced by the other region of the sources.

4. Separation between the two sources must be very small, otherwise fringe width will be so small that the fringes can not be distinguished.

13.23. Conditions for Maximum and Minimum Intensity

Conditions for maximum. Maximum intensity is observed at a point where the phase difference between the two waves reaching the point is an even multiple of π. The path difference between the two waves is an integral multiple of the wave length.

Conditions for minimum. For minimum intensity at a point the phase difference between the two waves reaching the point should be an odd multiple of π or

the path difference between the two waves should be an odd multiple of half wavelength.

13.24. Llyod Mirror

This is an arrangement for getting two coherent sources. S is a slit illuminated by a monochromatic source. The light through the slit is allowed to fall on a mirror M. This mirror M is either a flat polished metal or a piece of black glass so that no reflection take place from the back of the mirror. Thus image S' is formed by reflection and it acts as another source. These two sources S and S' are coherent and very close to each other. Interference fringes are produced by the two coherent sources (*see* Fig. 13.12).

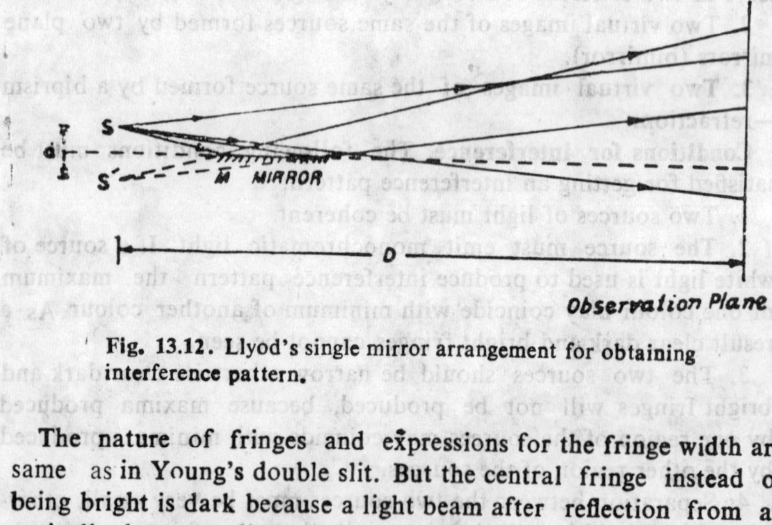

Fig. 13.12. Llyod's single mirror arrangement for obtaining interference pattern.

The nature of fringes and expressions for the fringe width are same as in Young's double slit. But the central fringe instead of being bright is dark because a light beam after reflection from an optically denser medium undergoes a phase change of π. So the position of maximum and minimum are interchanged (*see* Equation 4.53).

13.25. Fresnel Biprism

Fresnel used a biprism to observe interference phenomenon. The biprism ABC consists of two acute angled prisms placed base to base. Actually, it is constructed as a single prism of obtuse angle of about 179°. The acute angle on both sides is about 30′. The prism is placed with its refracting edge parallel to the line source *S* (slit) (*see* Fig. 13.13).

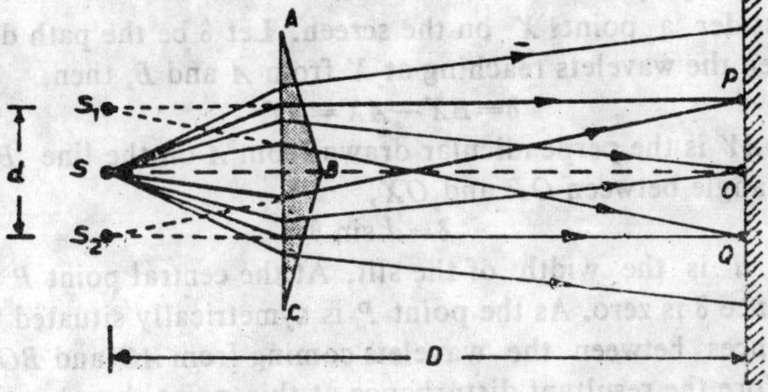

Fig. 13.13. Fresnel's biprism arrangement for obtaining interference.

When light from *S* falls on the lower portion of the prism it is bend upwards and appears to come from the virtual sources S_2. Similarly light falling on the upper portion of the prism is bent downwards and appears to come from the virtual source S_1. Therefore S_1 and S_2 act as two coherent sources and produce interference pattern.

The nature of fringes and the expression for the fringe-width are the same as discussed in Young's double slit experiment.

13.26. Single Slit Diffraction

It is found from experiments that light bends round an obstacle and encroaches into the geometrical shadow to some extent. Due to this bending bright and dark fringes appear at the boundary of the geometrical shadow. This phenomena is called the diffraction of light. The fringes are called the diffraction fringes. The appearance of dark and bright fringes can only be explained by wave theory of light.

Diffraction fringe patterns are different depending upon the nature and size of the source and obstacles. Here we will discuss only about the diffraction pattern produced by a narrow slit.

Let *AB* be a section of a slit perpendicular to the plane of the paper. The slit has been shown magnified many times. Let a plane wave front of light of wave length λ be incident normally on the slit. Each point on the wave front in *AB* can be regarded as a source of secondary wavelets whose radius increases at the speed of light. The resultant amplitude at any point on the screen is obtained by the superposition of the displacedment amplitudes of secondary wavelets reaching there. Let O be the centre of the slit and OP be perpendicular to the slit.

Consider a point X on the screen. Let δ be the path difference between the wavelets reaching at X from A and B, then,

$$\delta = BX - AX = BY$$

where AY is the perpendicular drawn from A on the line BX. If θ is the angle between OP and OX,

$$\delta = d \sin \theta$$

where d is the width of the slit. At the central point P the path difference δ is zero. As the point P is symetrically situated the path differences between the wavelets coming from AO and BO is zero. Therefore the resultant disturbance at this point has large amplitude, hence large intensity.

Let us consider the point X on the screen above P such that the path difference between BX and AX is λ. Since $d \ll OP$ the path difference between OX and AX is $\lambda/2$ to a good approximation. Similarly the path differences between OX and BX is also $\lambda/2$. The whole wave front AB can be considered to be of two halves OA and OB. For every point on the upper half of OA there is a corresponding point on the lower half of OB and the path difference between the secondary waves originating from them is $\lambda/2$. Thus destructive interference will take place at X. So X will be of minimum intensity. There will be a symmetrical point X' below P where the intensity will be minimum.

As one move away from X the above condition will not be satisfied so the point above X will have some intensity. Let us consider a point E such that the path difference between AE and BE is $3\lambda/2$. Now the aperature AB can be divided into three zones. such that the path difference between the wavelets coming from any two corresponding points in the two adjacent zones and reaching at E is $\lambda/2$. So the intensity at E due to the light coming from two adjacent zones is zero. Only light coming from the third zones will contribute to the intensity at E. So the point E will be bright but the brightness will be less than that at P.

Similar argument can be used to explain the appearance of bright and dark diffraction fringes. In general the point X_n is dark if the path difference between AX_n and BX_n is integral multiplies of λ. In other word the nth minimum is obtained in a direction θ_n. When

$$d \sin \theta_n = n\lambda \quad \ldots (13.21)$$

Similarly the point X_n is bright if the path difference between AX_n and BX_n is odd multiplies of $\lambda/2$ or in other words the nth maximum is obtained when,

$$d \sin \theta_n = (2n+1)\,\lambda/2 \qquad \ldots (13.22)$$

Width of the central maximum. The first minimum occurs when,

$$d \sin \theta_1 = \lambda$$

Since θ_1 is very small we can write,

$$\theta_1 = \lambda/d$$

If distance of the first minimum from the slit is y and from the point P is x_1 respectively,

$$\tan \theta_1 = \theta_1 = \frac{x_1}{y}$$

Therefore
$$\frac{\lambda}{d} = \frac{x_1}{y}$$

or
$$x_1 = y\,\frac{\lambda}{d}$$

So width w of the central maximum is given by $w = 2\,\dfrac{\lambda y}{d}$

$$\ldots (13.23)$$

Since the central maximum extends on both sides of the centre from Equation 13.23 it is clear that if d is large the width of the central maximum is large and if d is large the width is small. This result agrees well with the experimental findings. The schematic plot of the intensity distribution on the screen is given in Fig. 13.14(a-b).

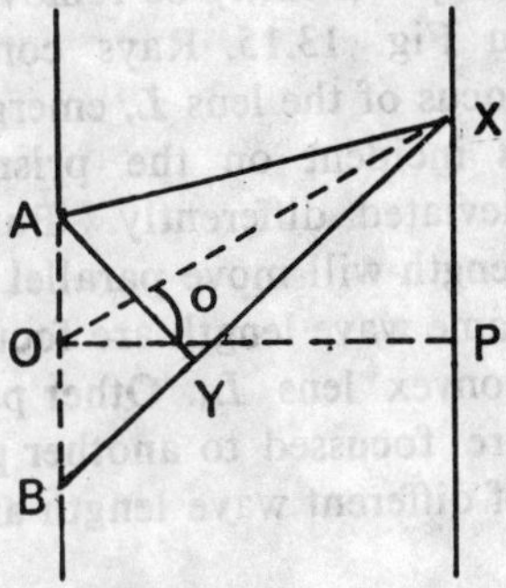

(a)

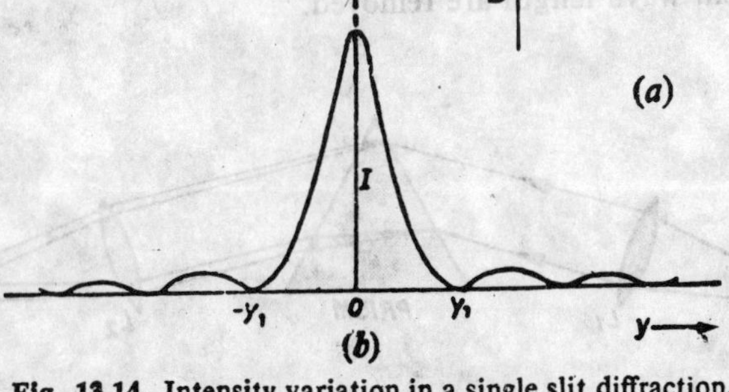

(b)

Fig. 13.14. Intensity variation in a single slit diffraction.

13.27. Difference between Interference and Diffraction

There is no basic difference between interference and diffraction. The change of intensity distribution brought about due to the superpositions of wavelets from a finite but small number of separate sources is called interference. On the other hand, intensity pattern produced owing to the superposition of the wavelets originating from a continuous portion of the wave front of a coherent source is called diffraction. Two non-contiguous parts of the wavefront can be regarded as discrete point sources but the contiguous elements of a wavefront constitute a continuous distribution of coherent sources. Therefore, the intensity pattern due to superposition of wavelets in the former case is usually termed as interference pattern and in the latter case diffraction pattern.

13.28. Production of Pure Spectra

When white light is passes through a prism, it spreads out into different colours. This is called dispersion. We say a spectrum has been formed.Since the colours are known to be related with wavelengths, we shall say that a spectrum is the wave length wise distribution of light from a given source.

The spectrum produced only by a prism is not pure due to the overlaping of lights of different wave length at a point. This difficulty can easily be removed by using two convex lenses as shown in Fig. 13.15. Rays coming from the source S placed at the focus of the lens L, emerge as a parallel beam. The parallel beam is incident on the prism. The light of different wave length are deviated differently. But the emergent rays having same wave length will move parallel to each others. These parallel rays having same wave length are focussed to a point on the screen by the other convex lens L_2. Other parallel rays having different wave lengths are focussed to another point. In this way the overlapping of light of different wave length are removed.

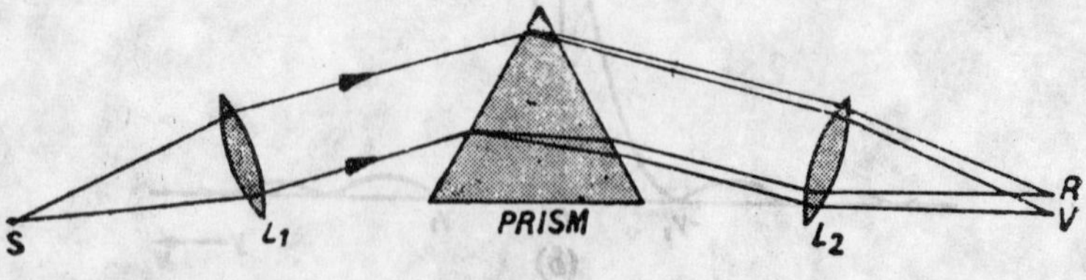

Fig. 13.15. Arrangement for obtaining a pure spectrum.

Spectrometer. The instrument which can produce a pure spectrum and can be used for the study of spectra is called a spectrometer. We shall discuss the main features of a spectrometer.

Construction. A spectrometer has three main parts (*i*) a collimetor, (*ii*) a prism and a graduated table and (*iii*) a telescope.

Collimetor. A collimetor is a tube with a convex lens at one end and a vertical slit at the other end. The slit can be slided into the tube so that it may be adjusted to lie in the focal plane of the lens. Arrangements are provided to make the collimetor axis horizontal.

Prism table. It is a circular horizontal base for the prism, rotating about a vertical axis. The horizontal axis of the collimetor passes through the vertical axis of the prism tube. The prism can be placed over it as desired. The prism table can be made horizontal by three levelling screws.

Telescope. The telescope has its axis horizontal and pass through the axis of the prism table. It is so mounted that it can be rotated about the prism table axis and its angular position can be known from a graduated circle. The axis of the telescope can be made horizontal by adjusting suitable levelling screws.

Fig. 13.16 gives a schematic top view of a spectrometer. *S* is the slit, perpendicular to the plane of the paper. *C* is the collimetor

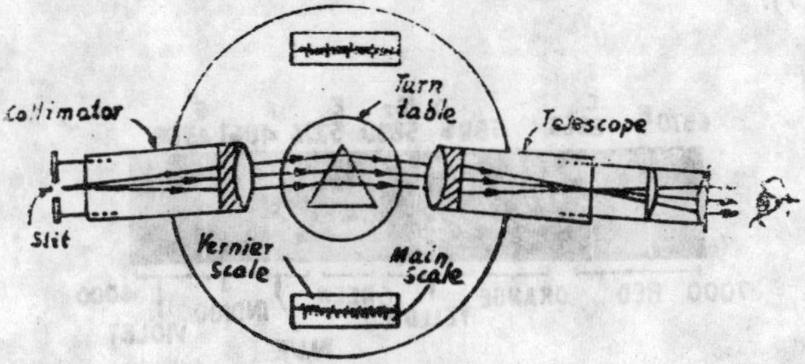

Fig. 13.16. Schematic diagram of a spectrometer.

lens, *P* is the prism. *T* is the telescope objective and *F* is the eye piece. There is a vertical wire *X* in this focal plane of the eye piece and by rotating the telescope one can bring any part of the spectrum on this wire (called a crosswire).

Types of spectra. Spectra in general, can be classified into two major groups, viz. (*i*) emission spectra and (*ii*) absorption spectra.

Emission spectra. The spectrum of a source of light is called emission spectrum. It is the wave length wise distribution of light emitted by that source.

The light emitted from (*i*) incandescent solid or vapour (*ii*) an arc discharge (*iii*) spark discharge (*iv*) a sodium vapour lamp (*v*) a mercury vapour lamp (*vi*) the discharge of electricity through a gas or vapour contained in a discharge tube are familiar examples of emission spectra.

Absorption spectra. When light containing of different wave lengths is passed through a transparent substance some light corresponding to a particular region of the spectrum is absorbed by that material and rest is transmitted. The resulting spectrum of the transmitted light consists of darklines or bands called the absorption lines or bands. Thus the distribution of the wavelength wise absorption of light by a substance is called the ***absorption spectrum*** of the substance. When the light from sun after passing through a narrow slit is dispersed by a prism solar spectrum is produced. It is found that some dark lines are present in the spectrum. These lines are called Fraunhofer lines called *B*, *C*, D_1, D_2, *E* etc. lines, having wave lengths 6870 Å, 6563 Å, 5890 Å 5890 Å, 5270 Å etc. respectively. These lines originate due to the absorption of light of said wave lengths by gases present on the outer layer of the sun. For example the wave length of *C* lines is equal to the wave length of the 1st member of Balmer-Series lines of Hydrogen (*see* Fig. 13.17).

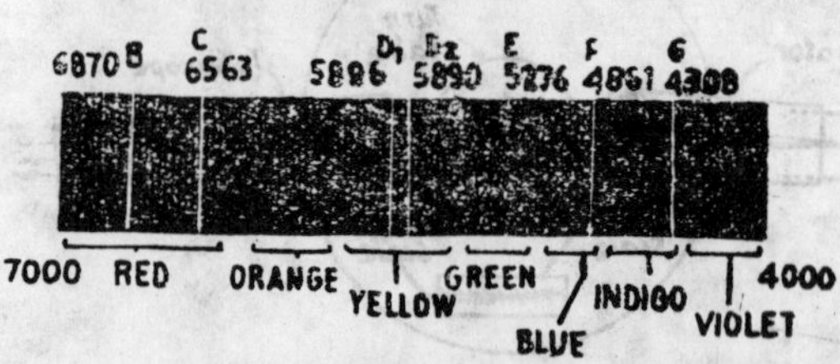

Fig. 13.17. Absorption spectra.

If a beam of white light is passed through cool sodium vapour, light corresponding to the wave lengths of D_1 and D_2 lines are absorbed and rest are transmitted. Thus the spectrum of transmitted light consists of all colours except those corresponding to the absorbed wave lengths. Similarly a piece of glass coloured with copper oxide (tube glass) transmits red light and absorbes all the rest of the wave lengths.

Classification of emission spectra. 1. *Continuous spectra.* It consists of radiations of all wave lengths without break. Incandescent solids, liquids, or gases under high pressure are sources of continuous spectra.

2. *Band spectrum.* In this spectrum we observe some bright bands as in Fig. 13.18. Each band has one sharp head and the other end with decresing intensity. In fact, under a good spectrograph

Fig. 13.18. Band spectra.

each band shows several separate lines which crowd together at the sharp edge but are well spaced at the other end unlike continuous spectrum. The edge may be towards the violet or the red end.

A band spectrum is emitted by a molecular gas e.g. a discharge tube containing CO_2 or NH_3 or candle light. Each molecule has its own characteristic system of bands.

3. *Line spectrum.* In this spectrum, we observe some sharp lines as in Fig. 13.19. These lines may be weak or intense. A line spectrum is emitted by free atoms in the gaseous stage. For example

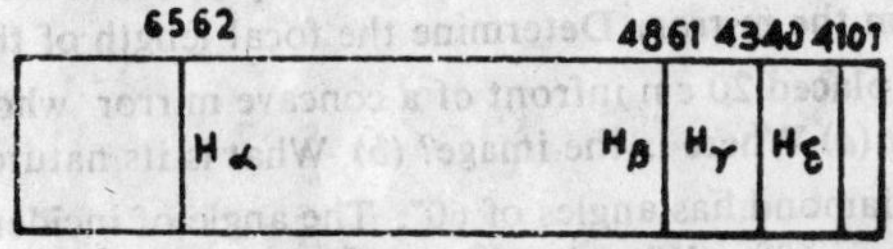

Fig. 13.19. Line spectra.

neon, sodium, mercury etc. in discharge tubes gives line spectra. Each element has its own characteristic spectra. Some salt break up into atoms when put in a bunsen flame and the metal atoms gives a line spectra. Chemists use this phenomenon in the flame test to identify the metal element. From the study of the spectrum of an ore the presence of different elements can be known.

QUESTIONS

1. How can a real image be distinguished from a virtual image?
2. What types of mirrors might possibly be used to make a 'burning glass'?
3. Does a convex mirror ever form an inverted image? Why?
4. Why concave mirror is used as a shaving mirror?

5. Why convex mirror is used as viewfinder in automobiles?

6. What are the two factors that define whether a lens is converging or diverging?

7. Under what conditions does a double convex lens becomes a diverging lens?

8. What is the physical explanations of the formations of a spectrum by a glass prism?

9. What is the evidence to show that all frequencies of light have the same speed in vacuum but different speeds in other medium?

10. When two light waves interfere at some point to produce darkness, what becomes of the energy?

11. If light is bent around obstacles, why can we not see around a house?

12. What sort of spectrum is given by moon light?

13. Account for the colours observed in a cut diamond.

14. Account for the gorgous colours of sunrise and sunset?

PROBLEMS

1. Show that an object at a distance *P* from a concave spherical mirror of radius *R* produces an image at a distance *q* given by:

$$q=\frac{RP}{2P-R}$$

2. A convex mirror forms an image that is one quarter the size of an object placed 132 cm from the mirror. Determine the focal length of the mirror.

3. An object is placed 20 cm infront of a concave mirror whose radius of curvature is 50 cm (*a*) Where is the image? (*b*) What is its nature?

4. A prism of diamond has angles of 60°. The angle of incidence of yellow light on one face is 60°. What is the angle of emergence? (μ of yellow light is 1.73 (30°)

5. When an object is placed 200 mm from a lens, its virtual image is formed 100 mm from the lens. Determine the focal length and character of the lens? (20 cm, concave)

6. What is the power of a diverging lens whose focal length is —30 cm?

7. Prove that minimum distance between the object and real image produced by a convex lens is 4 *f*. *f* is the focal length of the lens.

8. The acute angle of a biprism of refractive index 1.55 is 2°. A slit illuminated by a monochromatic light is placed 10 cm from the biprism. If the distance between the two dark fringes observed at a distance of a metre from the biprism is 0.1 mm find the wave length of light. p (3840 Å)

9. In Young's interference experiment the distance between two slits is .2 mm. The distance between the screen and the source is 80 cm. The 3rd dark band is .5 cm from the central image. Find wave length of the light.

10. A slit is of .3 mm width is placed infront of a convex lens and it is illuminated by a monocromote light of wavelength 5890 Å. In the diffraction pattern formed at the focal plane of the lens, the distance between

the 3rd dark fringe on the left and the 3rd dark fringe on the right is .4 mm. What is the focal length of the lens? (3.4 cm)

11. In a single slit diffraction pattern the distance between screen and the slit is one metre. The wave length of the light used is 5890 Å. The width of the slit is .2 mm. Calculate the separation between the central maximum and the 1st minimum? (.3 cm)

14

PHYSICS OF THE ATOM

It was accepted upto the end of 19th century that the smallest indivisible particle of matter was atom. But this concept was found to be wrong in the light of some experimental results. They are passage of electricity through gases, Spectra of hydrogen atom, emission of X-rays and emission of photoelectrons from metal surfaces etc.

14.1. Passage of Electricity through Gases

At ordinary pressure almost all gases are found to be bad conductors of electricity. Normal air is a perfect insulator. However any gas can conduct electricity when they are exposed to ionising agents such as, ultraviolet light, X-rays, high electric fields and radio active substances etc. But when these agents are removed the gas again becomes insulators.

14.2. Ionisation Theory of Conduction

The induced conductivity of gases has been found to be due to the presence of charged particles called ions. Under normal conditions a gas contains very few ions. So they are bad conductors. But when a gas is exposed to ionising agents mentioned earlier a large number of free ions are produced inside the gas and it becomes a conductor. This process is called the ionisation of a gas. In this process one or more electrons are detached from the neutral atom or *molecule* of the gas. The residual atom or molecule is positively charged and is called a positive ion. The detached electron may move freely or attach itself to a neutral atom or *molecule* and produce a negative ion. The energy required to detach an electron from an atom or molecule is supplied by the ionising agents.

14.3. Passage of Electricity through Gas in the Presence of Strong Electric Field

During cloudy weather lightening are often seen. The lightenings are nothing but passage of electricity through the air. This is possible when very high potential difference develops between the cloud and surface of earth or between two clouds. Such phenomena can be observed in a laboratory when the secondary electrodes of a powerful induction coil are separated from each other through a small distance. This phenomena is called disruptive spark discharge. Such spark discharge requires a potential difference of about 30 KV at atomospheric pressure and room temperature when the gap between the secondary electrodes is about 1 cm.

The situation is different when the pressure of the gas is reduced. At lower pressure less potential difference is required to produce discharge. Electric discharge at low pressure provides vital information about the structure and properties of atoms.

14.4. Passage of Electricity through Gas at Low Pressure

Electric discharge at low pressure can be studied easily using a discharge tube. It consists of a strong glass tube about 50 cm long and 4 cm in diameter closed at both ends. Two disc electrodes of metal, are fixed at both ends as shown in Fig. 14.1.

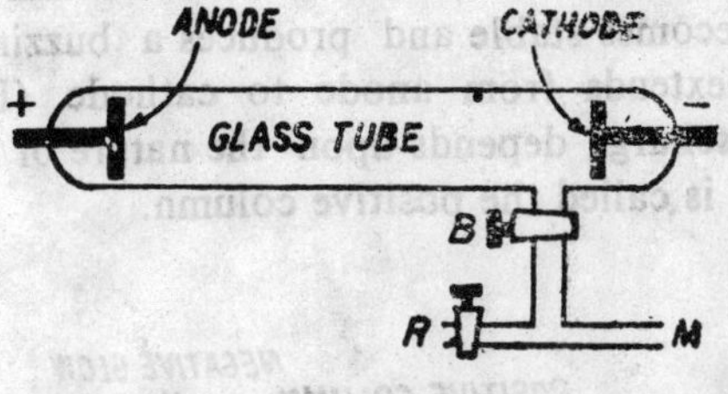

Fig. 14.1. Discharge tube.

One opening B is provided to connect the tube with the pressure gauge M and high vacuum pump R. The two electrodes are con-

Fig. 14.2. Electric discharge when pressure is in between 1 cm and 2 cm of mercury.

nected to the secondary of a powerful induction coil, developing a potential difference of the order of 50,000 volts. When the vacuum pump and the induction coil are allowed to operate the following phenomena are observed.

(1) When the pressure of the gas is about 10 cm of mercury irregular streaks of light appear accompanied by a crackling noise.

(2) When the pressure is between 2 cm and 1 cm of mercury the discharge begins to pass with streamers producing crackling sound. At the beginning the discharge looks like a purplish sinus thread and finally change to a single rope between the cathode and anode. The nature of the discharge is shown in the Fig. 14.2.

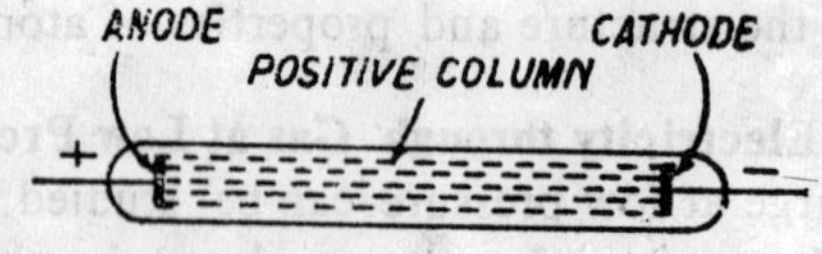

Fig. 14.3. Electric discharge when pressure is in between 1 cm and 5 mm of mercury.

(3) When the pressure is between 1 cm and 5 mm the single rope broadens and becomes stable and produces a buzzing sound. Here the discharge extends from anode to cathode (Fig. 14.3). The colour of the discharge depends upon the nature of the gas. This coloured region is called the positive column.

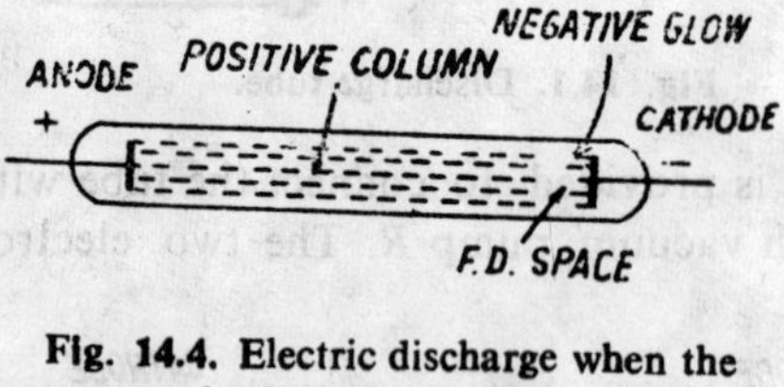

Fig. 14.4. Electric discharge when the pressure is about 1 mm of mercury.

(4) As the pressure reduces or falls to about 1 mm a bluish glow is seen infront of the cathode. This glow is called negative glow. Negative glow is separated from the positive column by a dark space called Farady Dark Space as represented in Fig. 14.4.

(5) As the pressure reduces to about .05 mm the negative glow and the Faraday dark space shifts towards the anode. At this stage a light pinkish glow appears infront of the cathode. This is called

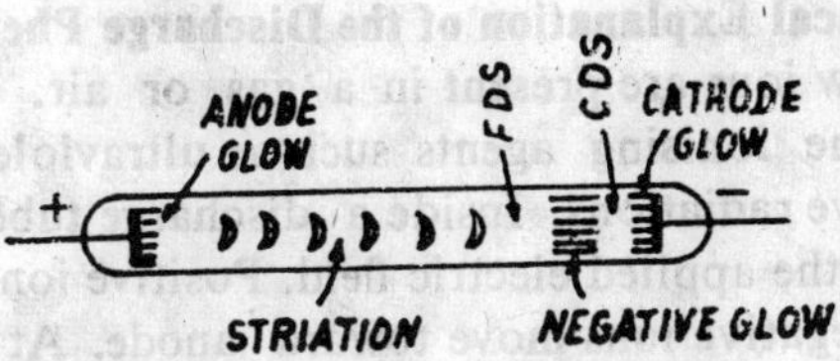

Fig. 14.5. Electric discharge when the pressure is about 0.05 mm of mercury.

the cathode glow, which is separated from the negative glow by a dark space called Crooke's Dark Space. The positive column breaks into alternate bright and dark discs called striations as shown in Fig. 14.5.

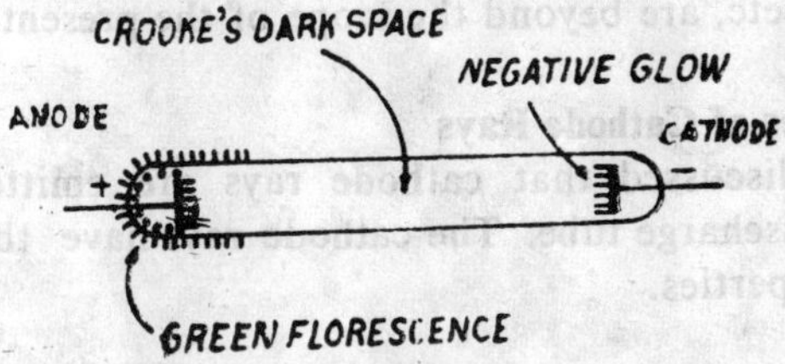

Fig. 14.6. Electric discharge when the pressure is about 0.01 mm of mercury.

(6) As the pressure further falls to .01 mm of mercury the striations recede towards the anode. Faraday dark space and the negative glow disappear. The tube is completely filled with the Crooke's dark space only. At this stage the walls or the tube opposite to the cathode begin to glow with a bluish of greenish light depending upon the glass of the tube. This glow is found to be independent of the gas present inside the tube and also does not depend on the metal electrodes but depends on the type of the glass of the discharge tube. This glow is found to be due to impact of some invisible radiations. or rays emanating from the cathode surface. These rays are called *Cathode rays* (*see* Fig. 14.6).

(7) As pressure falls to about .01 mm of mercury or less the discharge cannot pass through the tube. This is called no discharge stage. At this stage no ions are present inside the tube to maintain the electric current.

14 5. Theoretical Explanation of the Discharge Phenomena

Always a few ions are present in a gas or air. These ions are produced by the ionising agents such as ultraviolet light, cosmic and radio-active radiations. Inside a discharge tube these ions get accelerated by the applied electric field. Positive ions moves towards cathode and negative ions move toward anode. At higher pressure the mean free path of the ions are small. So they cannot acquire sufficient energy to ionise neutral gas molecules by collision. But at lower pressure the mean free path is comparatively large. The ions can acquire sufficient kinetic energy by the same electric field. So these accelerated ions create more ions by collision, as a result of which at lower pressure discharge is possible even at lower potential difference. However at very low pressure when mean free path is of the order of dimension of the discharge tube or more no new ions are created to maintain discharge phenomena. The mechanism of the formation of crooke's dark space, Faraday dark space and Striations etc, are beyond the scope of the present book.

14.6. Properties of Cathode Rays

It has been discussed that cathode rays are emitted from the cathode of a discharge tube. The cathode rays have the following interesting properties.

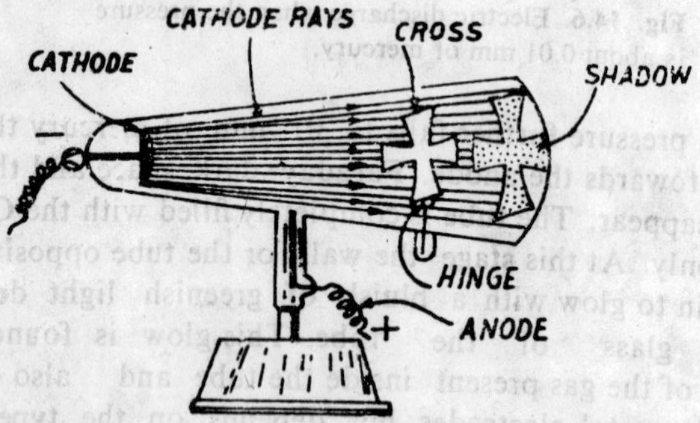

Fig. 14.7. Diagram to demonstrate that cathode rays traval in straight lines.

(*a*) Cathode rays travel in straight lines. This can be experimentally verified. When an obstacle is placed in the path of cathode rays a clear shadow of the obstacle is produced. If the obstacle is removed the shadow vanishes. In Fig. 14.7 formation of a clear shadow of a mica cross is presented.

(*b*) Cathode rays possess considerable kinetic energy. This can be verified by the following experimental arrangement. In Fig. 14.8 cathode rays are focussed on a tungsten plate by a concave cathode inside a discharge tube. Tungsten becomes red hot. As the melting point of tungsten is 3387°C it is evident that cathode rays possess considerable kinetic energy to heat tungsten to such a high temperature.

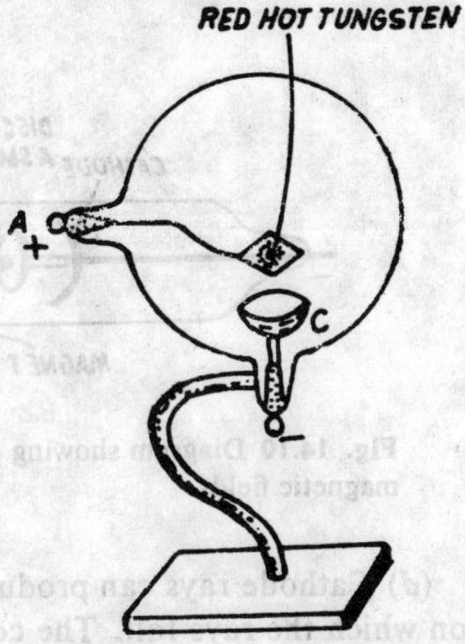

Fig. 14.8. Diagram demonstrating the heating effect of cathods rays.

In Fig. 14.9 along the path of the cathode rays a micavane wheel is placed on two parallel glass rods. When cathode rays are emitted and strike the vanes the wheel rotates and moves away from the cathode. This motion is mainly due to the momentum of the cathode rays. This shows that cathode rays possess considerable amount of kinetic energy. Cathode rays travel with very high speed. Its speed may be up to 1/10th of velocity of the light.

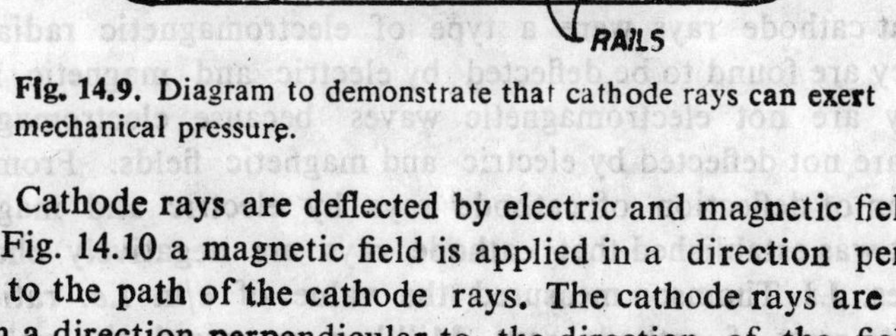

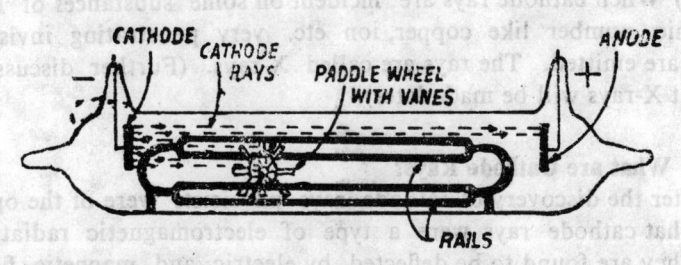

Fig. 14.9. Diagram to demonstrate that cathode rays can exert mechanical pressure.

(*c*) Cathode rays are deflected by electric and magnetic fields.

In Fig. 14.10 a magnetic field is applied in a direction perpendicular to the path of the cathode rays. The cathode rays are deflected in a direction perpendicular to the direction of the field and direction of motion of the cathode rays. If the direction of the field is reversed the direction of deflection will also be reversed. However, when the magnetic field will be applied in the direction of motion of the cathode rays, they will not be affected.

Cathode rays are deflected along the opposite direction of the electric field *i.e.* the beam is deflected towards the positive potential.

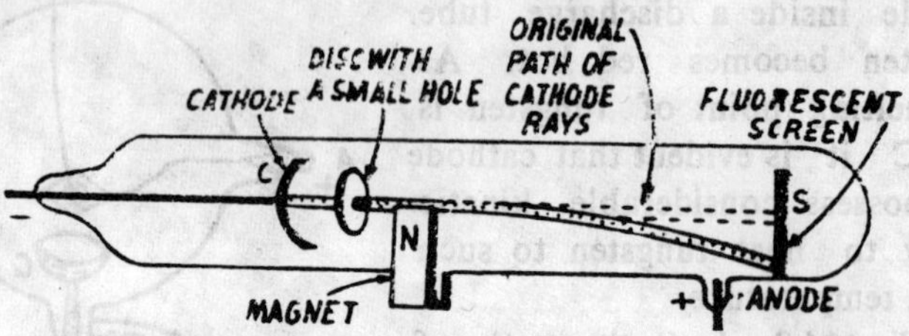

Fig. 14.10 Diagram showing the deflection of cathode rays by magnetic fields.

(*d*) Cathode rays can produce fluorescence in certain substances on which the rays fall. The colour of the fluorescence is different for different substances,for example,colour of the fluorescence produced by soda glass and lead glass are yellowish green and greyish blue respectively.

(*e*) Cathode rays can affect suitable photographic plates.

(*f*) Cathode rays can ionise a gas through which they pass.

(*g*) When cathode rays are incident on some substances of high atomic number like copper, ion etc. very penetrating invisible rays are emitted. The rays are called X-rays. (Further discussion about X-rays will be made later).

14.7. What are Cathode Rays?

After the discovery of cathode rays physicists were of the opin ion that cathode rays were a type of electromagnetic radiation. But they are found to be deflected by electric and magnetic field. So they are not electromagnetic waves because electromagnetic waves are not deflected by electric and magnetic fields. From the direction of deflection of cathode rays by electric and magnetic fields it was established that cathode rays are negatively charged particles. J.J. Thomson measured the value of e/m *i.e.* ratio of charge to mass of cathode rays. Millikan measured the value of negative charge and proved that negative charge is atomic in nature. The two experiments confirmed that the electricity is atomic in nature and cathode rays are nothing but a stream of fast moving electrons.

14.8. Measurement of e/m of Cathode Rays

J J. Thomson measured the e/m of cathode rays. According to his method a narrow pencil of cathode rays is subjected to simultaneous electric and magnetic fields acting at right angles to each other. By measuring the deflections produced by the two separate fields the e/m of cathode rays can be measured.

Apparatus. The apparatus used by Thomson is given in Fig. 14.11. Cathode rays are emitted from the cathode *C*. The rays are reduced to a narrow beam while passing through slits in the anode *A* and metal plug *DE*.The beam is accelerated between cathode and anode due to the electric field and acquires a certain velocity *v* at the anode. The velocity *v* of the electrons remains uniform after it crosses the anode. The beam strikes the fluorescent screen and produces a small luminous spot at *P* on it. By applying a suitable potential difference between the plates *X* and *Y* the luminous spot is displaces through a small distance. Then the magnetic field is applied in a direction perpendicular to the directions of the electric field in such a manner that the deflections due to both the fields are in opposite direction. So by adjusting the magnitude of either the electric or the magnetic field the two deflections can cancel each other so that the spot remains in the original position at *P*.

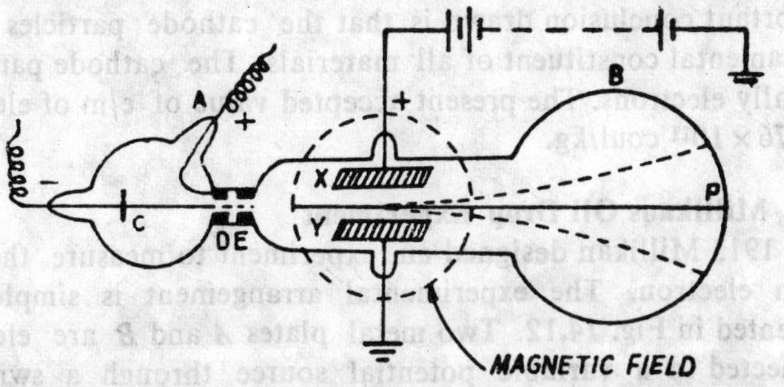

Fig. 14.11. Thomson's apparatus to measure e/m of electrons.

Theory. Let *V* be the potential difference in volts applied between the cathode *C* and the anode *A*. The cathode particles are accelarted by the applied electric field between the cathode and anode. Thus energy gained by the particles is given by

$$eV = \frac{1}{2}mv^2 \qquad \ldots(14.1)$$

where *e* is the charge of the particle in coulumbs, m is mass in kg and the velocity *v* in m/s.

$$\text{Therefore} \quad \frac{e}{m} = \frac{v^2}{2V} \quad \ldots (14.2)$$

If E is the electric field applied between the plates X and Y the cathode particles are deflected towards the positive direction of the Y axis, if plate X is at a higher potential than the plate Y. Let the spot be shifted to another point P_1.

If by application of suitable magnetic field B the spot is brought back to its original position, we have,

$$eE = B\,ev \quad \ldots (14.3)$$

where $e\,E$ is the force due to electric field acting on the cathode particles in the vertically upward direction and Bev is the force due to magnetic field acting in the vertically down ward direction.

$$\text{Therefore } v = \frac{E}{B} \quad \ldots (14.4)$$

Using 14.2 in 14.4 we obtain,

$$\frac{e}{m} = \frac{E^2}{2VB^2} \quad \ldots (14.5)$$

Thus knowing the values of E, V and B for no deflection condition of the beam e/m of cathode rays can be determined. The value of e/m are found to be independent of the cathode material, coating on the cathode and the enclosed gas of the tube. From this, the important conclusion drawn is that the cathode particles are one fundamental constituent of all materials. The cathode particles are actually electrons. The present accepted value of e/m of electrons is 1.76×10^{11} coul/kg.

14.9. Millikans Oil Drop Experiment

In 1913 Millikan designed an experiment to measure the charge of an electron. The experimental arrangement is simple and is presented in Fig. 14.12. Two metal plates A and B are electrically connected to a variable potential source through a switch (not shown in the figure). In the off position of the switch the plates A and B are at same potential. A and B are separated from each other by insulating support S. There are two opening in the insulating support. One is used to illuminate the space between the two plates by a source of light O and a lens L. In the other opening a microscope is used to observe the space between the plates. There is a small hole t at the centre of the plate A. With the help of a sprayer fine shower of oil drops are sprayed over the small hole t. The drops become charged due to friction. When the drops are viewed by the microscope, the small drops are seen to fall towards the plate *B* by the action of gravity. Due to air friction the small drops acquires a constant terminal velocity V_0. The velocity V_0 can be measured easily.

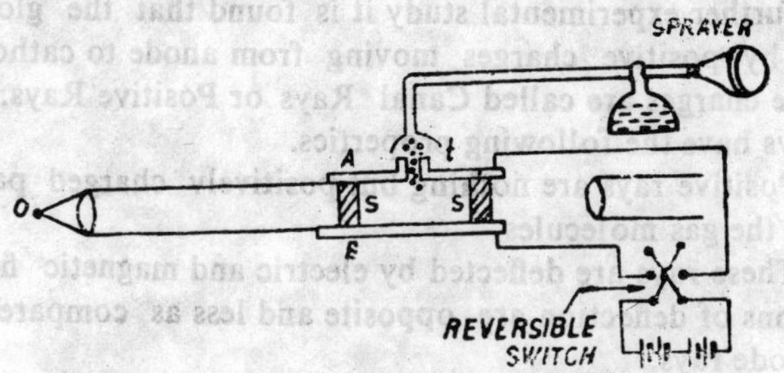

Fig. 14.12. Apparatus for Millikan's oil drop experiment.

Before the drop reaches the bottom plate *B* the switch is made on. The drop instead of falling moves upward under the action of electric field (if negatively charged) and acquires a constant terminal velocity V_1 which again is measured.

If the potential difference is so adjusted that the upward force due to the electric field *E i.e. neE* is balanced by the downward force due to gravity *mg*, under equilibrium condition.

$$mg = neE.$$

Where *n* is the number of electrons attached to the drop of mass ***m***. The mass of the oil drop can be calculated using the formula.

$$\frac{V_0}{V_1} = \frac{mg}{E—mg} \quad \ldots (14.6)$$

The experiment was repeated several times. The charge on the drop was always found to be an integral multiple of a smallest amount of charge 1.6×10^{-19} coulombs. This amount of charge is taken to be the charge of an electron. This experiment established the quantum nature of charge.

14.10. Mass of an Electron

The mass of an electron can be found by using the e/m and *e* value of electrons.

$$\frac{e}{m} = 1.76 \times 10^{11} \text{ coul/kg}$$

and $e = 1.6 \times 10^{-19}$ coul

$$\text{So } m = \frac{1.6 \times 10^{-19} \text{ coul}}{1.76 \times 10^{-11} \text{ coul/kg}} = 9.11 \times 10^{-31} \text{kg}$$

14.11. Positive Rays or Canal Rays

In 1886 Gold Stein found that a portion of the glass tube behind the cathode glows if there are holes in the cathode. The colour of the glow is different from the glow produced by the cathode rays. From further experimental study it is found that the glow is produced by positive charges moving from anode to cathode. These positive charges are called Canal Rays or Positive Rays. The positive rays have the following properties.

(1) Positive rays are nothing but positivelv charged particles *i.e.* ions of the gas molecules.

(2) These rays are deflected by electric and magnetic fields. The directions of deflection are opposite and less as compared to that of cathode rays.

(3) They travel in straight lines.

(4) e/m of canal rays are different for different gases and much less than that of cathode rays.

(5) Their velocities are small compared to that of cathode rays.

(6) They can affect photographic plates.

(7) They can cause fluorescence and phosphorescence.

(8) Their penetrating power is less than that of cathode rays.

14.12. Model of an Atom

An atom as a whole is neutral. But it contains positive and negative charges. The total positive charges are equal to the total negative charges in the atom. But the exact structure of the atom was not known upto 1911.

14.13. Rutherford's Model of an Atom

Rutherford from the experimental results of the scattering of α-particles by tin foils suggested for the first time a model for the structure of an atom. (The scattering experiment was actually performed by Geiger and Morsden).

When two electrons of a helium atom are removed the helium atom is doubly ionised. Its charge is $2e$ (positive) and mass is four times the mass of a proton. Such an ion is called an α—particle.

A schematic diagram of the experimental arrangement is given in Fig. 14.13. A narrow beam of α-particles is allowed to be incident on a screen S at S_0 in an avacuated chamber. F is a tin foil. When the foil is not present the particles of the beam have same energy and strike the screen at S_0. When the tin foil is inserted on the path of the α-particles they strike the screen at diffe-

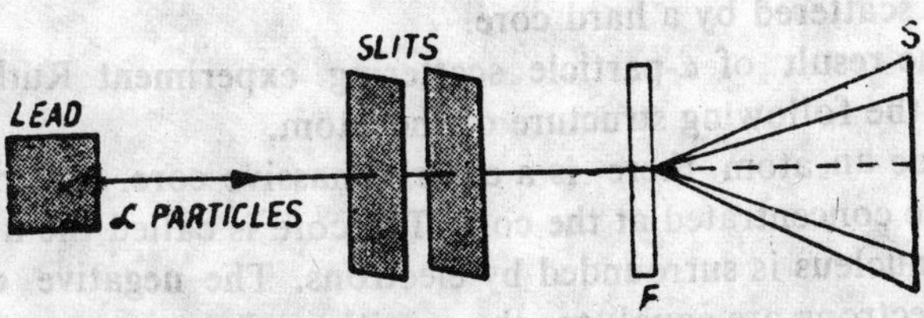

Fig. 14.13. Scattering of α-particles by tin foils.

rent points. This phenomena is called scattering of α-particles. The angles made by the paths of the α-particles with that of the original path is called the scattering angle. Geiger and Marsden measured scattering angles of each particle. The variation of the number $N(\theta)$ of the scattered α-particles with the angle θ is shown in Fig. 14.14. From the graph it is clear that most of the particles are scattered by small angles. Only one particle out of 9000 particles is scattered by an angle more than 90°. In a very few cases the α-particles are reflected back. The types of scattering observed are presented in Fig. 14.15.

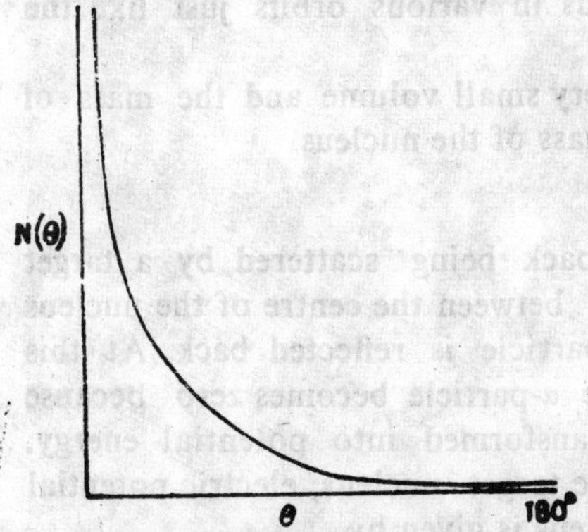

Fig. 14.14. Graph showing the variation of $N(\theta)$ with θ.

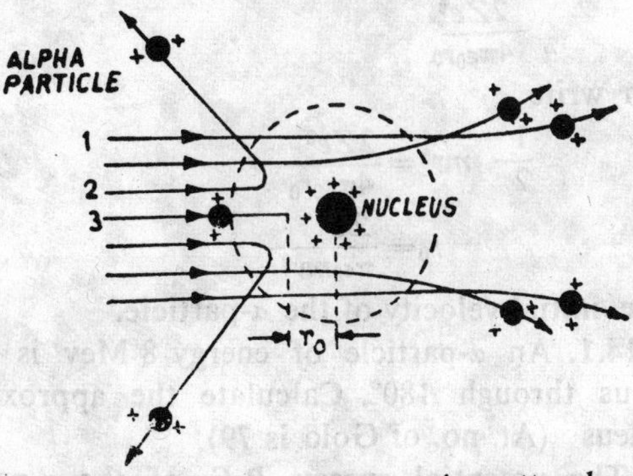

Fig. 14.15. Diagram showing the path of the scattered α-particles.

Since most of the α-particles are scattered by small angles it shows that the foil has empty spaces The little deviation of the particles are due to the columb scattering of charge particles. The large angle scattering of α-particles is only possible when the particle is scattered by a hard core.

From the result of α-particle scattering experiment Rutherford suggested the following structure of the atom.

1. In side an atom there is a central massive core. All positive charges are concentrated at the core. The core is called the nucleus.

2. The nucleus is surrounded by electrons. The negative charges of the electrons are equal to the positive charges present at the nucleus, so that the atom as a whole is neutral Such a system is not stable because the electrons will move towards the nucleus due to coulumb attraction. So later N. Bohr suggested that the electrons are revolving round the necleus in various orbits just like the planets revolve round the sun.

3. The nucleous occupies a very small volume and the mass of the atom is nearly equal to the mass of the nucleus.

14.14. Radius of the Nucleus

Let an α-particle be reflected back being scattered by a target nucleus. Let r_0 be the distance between the centre of the nucleus and the point from which the α-particle is reflected back. At this point the kinetic energy of the α-particle becomes zero because all its kinetic energy has been transformed into potential energy. If Ze is the nuclear charge of the target nucleus, electric potential P at a distance r_0 due to the nucleus is given by

$$P=\frac{Ze}{4\pi\epsilon_0 r_0}$$

So potential energy of the α-particle at that point is

$$\frac{2Ze^2}{4\pi\epsilon_0 r_0}$$

Hence we can write

$$\frac{1}{2}\ mv^2=\frac{2Ze^2}{4\pi\epsilon_0 r_0} \qquad \ldots(14.7)$$

or

$$r_0=\frac{Ze^2}{\pi\epsilon_0 mv^2} \qquad \ldots(14.8)$$

where v is the initial velocity of the α-particle.

Example 14.1. An α-particle of energy 8 Mev is scattered by a Gold nucleus through 180°. Calculate the approximate radius of Gold nucleus (At. no. of Gold is 79).

Solution. The potential energy $P.E.$ of the α-particle at its nearest approach to the nucleus is given by

$$P.E.=\frac{2\times 79\times(1.6)^2\times 10^{-38}\ (\text{coul})^2}{8.83\times 10^{-12}\ r_0}$$

$$=3.65\times 10^{-26}/r_0 \text{ netwon}\times(\text{meter})^2$$

The kinetic energy $K.E.$ of the α-particle is given by

$$K.E. = 8 \times 1.6 \times 10^{-12} \text{ Joules.}$$

So $$r_0 = \frac{3.65 \times 10^{-26}}{8 \times 1.6 \times 10^{-12}} \quad \frac{\text{Joule} \times \text{meter}}{\text{Joule}}$$

$$= 2.85 \times 10^{-15} \text{ meter.}$$

From this calculation it is clear that for α—particles of higher energy r_0 is less. However this calculation gives only the order of nuclear radius. The actual radius of the nucleus is smaller than r_0.

14.15. Bohr's Theory for Hydrogen Atom

Spectra of hydrogen contains different series of spectral lines. A series of spectral lines was first photographed by T. Lyman in the ultraviolet region. According to his name this series was called Lyman series. Another series in the visible region was photographed by Balmer. Similarly according to his name the series was called Balmer series. Subsequently other series in the infrared region were discovered they are called Paschen and Bracket series.

Bohr in 1913 was able to give the theoretical explanation of above observed spectra of hydrogen. To explain this he modified the Rutherford model of atom and accepting the Plank's quantum theory made the following postulates.

14.16. Bohr's Postulates

(1) Modifying the Rutherford's nuclear model of the atom he suggested that the electrons are revolving around the nucleus in different orbits. The central force required for the rotation is provided by the coulomb attraction between the nucleus and electrons.

(2) According to classical picture electron can revolve in any orbit. But Bohr postulated that electron can revolve only in permitted orbits. These orbits are called stationary orbits. When an electron revolves in an orbit it cannot radiate energy. The radius of the orbits are such that the orbital angular momentum of the electron is equal to an integral multiple of $h/2\pi$, h being the Plank's constant. So if r_n is the radius of the nth orbit,

$$mvr_n = \frac{nh}{2\pi} \qquad \ldots(14.9)$$

where m and v are the mass and speed of the electron. This n is a positive integer and is called the principal quantum number. Equation 14.9 is also called Bohr's quantum condition.

(3) An atom radiates energy when the electron jumps from a

higher orbit to a lower orbit and similarly absorbs energy when it goes from a lower orbit to a higher orbit.

If E_1 and E_2 are the energies of two permitted orbits the frequency of radiation emitted or absorbed is given by

$$E_2 \sim E_1 = h\nu$$

The relation is also called Bohr frequency condition, where ν is the frequency of the radiation emitted or absorbed. Bohr using his postulates successfully explained the spectra of hydrogen atom in the following manner.

14.17. Bohr's Theory

In normal state the single electron of the hydrogen atom revolves in the nearest orbit around the nucleus. The principal quantum number of this orbit is 1. When more energy is given to the atom the electron revolves in orbit having larger radius. The principal quantum numbers of these orbit are 2, 3,.........etc. When the electron jumps from an outer orbits to an inner orbit, it radiates energy in the form of electromagnetic radiation. The energy of the radiation depends upon the two concerned orbits.

The electron in the atom revolves continuously around the nucleus in a circular orbit. The centripetal force required for rotation in circular path is supplied by the coulomb attraction between the nuclear charge and the electron. So from 1st postulate we can write

$$\frac{mv^2}{r_n} = \frac{Ze^2}{4\pi\varepsilon_0 r_n^2} \quad \ldots(14.10)$$

From 2nd postulate we can write:

$$I\omega = \frac{nh}{2\pi} \quad \ldots(14.11)$$

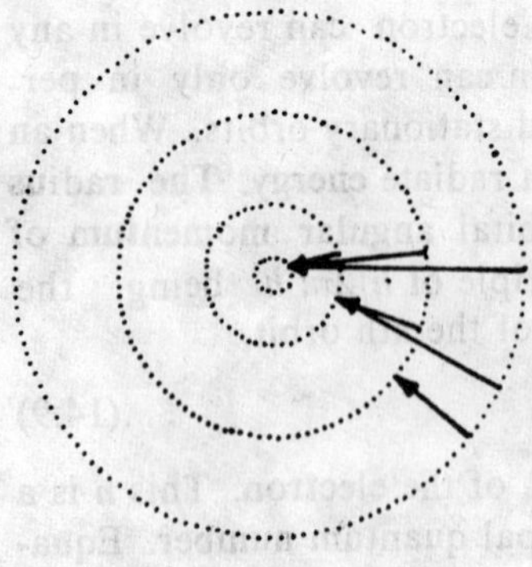

Fig. 14.16. Some orbits of hydrogen.

where I is the moment of inertial and ω is the angular velocity. Equation 14.11 can easily be expressed as:

$$mvr_n = nh/2\pi$$

or

$$v = \frac{nh}{2\pi m r_n} \quad \ldots(14.12)$$

where Z=Atomic number of the nucleus for hydrogen $Z=1$. ϵ_0=Dielectric constant of free space.

If T is the kinetic energy of the electron, we get from Equation 14.10

$$T = \frac{Ze^2}{(4\pi\epsilon_0)2r_n} \quad \ldots(14.13)$$

From Equations 14.12 and 14.13 we can write

$$r_n = \frac{n^2h^2(4\pi\epsilon_0)}{4\pi^2mZe^2} \quad \ldots(14.14)$$

The potential energy of the electron is given by

$$P.E. = \frac{Ze^2}{(4\pi\epsilon_0)r_n} \quad \ldots(14.15)$$

Total energy W of the atom becomes:

$$W = \frac{Ze^2}{(4\pi\epsilon_0)2r_n} - \frac{Ze^2}{(4\pi\epsilon_0)r_n} = -\frac{Ze^2}{(4\pi\epsilon_0)\,2r_n} = T \quad \ldots(14.16)$$

Using 14.14 on 14.16 we get :

$$W = -\frac{2\pi^2mZ^2e^4}{(4\pi\varepsilon_0)^2\,h^2n^2} \quad \ldots(14.17)$$

When electron jumps from the initial to final orbit we can write from 3rd postulate

$$h\nu = (W_i - W_f),$$

$$\nu = \frac{2\pi^2mZ^2e^4}{(4\pi\varepsilon_0)^2\,h^3}\left(\frac{1}{n_f^2} - \frac{1}{n_i^2}\right)$$

where W_i and W_f are the energy of the initial and final orbits,

or
$$\nu = CR\left(\frac{1}{n_f^2} - \frac{1}{n_i^2}\right) \quad \ldots(14.19)$$

where n_i and n_f are the quantum number of the initial and final orbits, c is the velocity of light and ν is the frequency of radiation emitted and R is a constant, called Rydberg constant.

14.18. Explanation of the Experimental Result

In the Equation 14.19 if we take $n_f=1$ and $n_i=2$, we get the value of the frequency of the 1st member of Lyman series. Similarly taking $n_i=3, 4, 5$...we can get frequencies of other members of the series. The calculated values of frequencies agree well with the measured values of the Lyman series.

If we take $n_f=2$ and $n_i=3, 4, 5$...we get the values of the frequencies of the Balmer series. The calculated values and experimental values are same. Similarly if we take $n_f=3$ and $n_i=4, 5, 6$... frequencies of the Paschen series and for $n_f=4$ and n_i—5, 6, 7... frequencies of the Bracket series are obtained. For all the cases the calculated values of frequencies agree well with corresponding experimental values. The formation of different series are shown in Fig. 14 17.

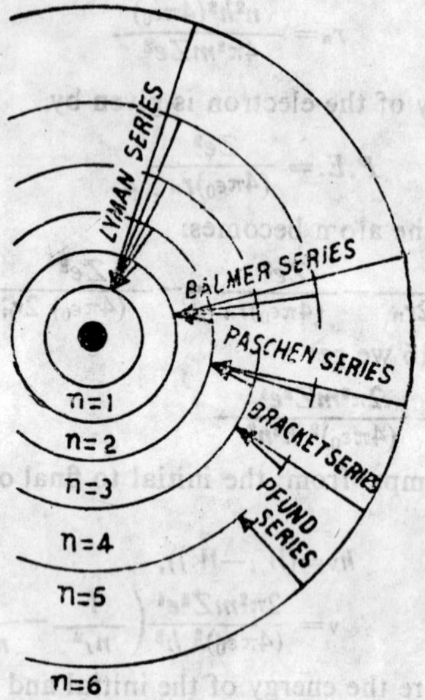

Fig. 14.17. Emission of different series of hydrogen.

14.19. Energy Level Diagram

The diagram given in Fig. 14.17 is not suitable to represent the electron transitions. Because for higher *n* values the radius of the orbits are large. So they cannot be drawn to the same scale. To overcome these difficulties energy level of different orbits are represented by horizontal parallel lines with lowest energy level at the bottom. The advantage of this is that energy levels of higher orbits are closer. So energy levels, can be plotted in a proper scale. The energy level diagram of hydrogen atom is presented in Fig. 14.18.

14.20. Ionisation Potential of an Atom

Ionisation potential of an atom is the amount of energy in ev required to remove an outer most electron completely from the atom. For hydrogen it is the amount of energy required to remove the electron from the 1st orbit. If E_i is the ionisation potential of hydrogen

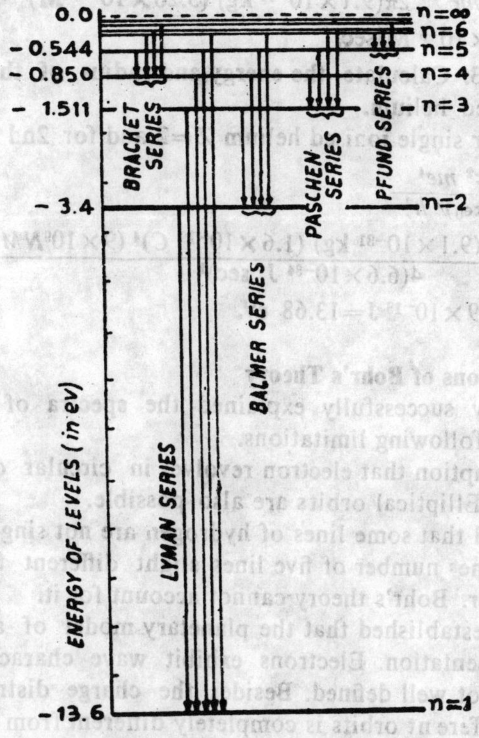

Fig. 14.18. Energy level diagram of hyrogen.

$$E_i = \frac{2\pi^2 me^4}{(4\pi\varepsilon_0)\ h^2}\left(\frac{1}{n^2} - \frac{1}{\infty}\right)$$

$$= \frac{2\pi^2(9.1\times10^{-31}\ \text{kg})(1.6\times10^{-19}\ C)^4\ (9\times10^9\ N-M^2/C^2)^2}{(6.6\times10^{-34}\ \text{J. sec})^2}$$

$$= 2.19\times10^{-18}\ \text{J} = 13.68\ \text{ev}$$

Here the magnitude of the energy of the 1st orbit is equal to the ionisation potential of hydrogen.

Example 14.2. Calculate the radius of the 1st orbit of Hydrogen and speed of the electron moving in that orbit.

Solution. For hydrogen $Z=1$ and for 1st orbit $n=1$. From Equation 14.14 we can write,

$$r_1 = \frac{h^2\ (4\pi\epsilon_0)}{4\pi^2\ me^2} = \frac{(6.6\times10^{-34}\ \text{J sec})^2}{4\pi^2(9.1\times10^{-31}\ \text{kg})(1.6\times10^{-19}C)^2\ (9\times10^9 NM^2/C^2)^2}$$

$$r_1 = 5.26\times10^{-11}\ M = 0.53\ \text{Å}$$

So $v=\frac{h}{2\pi m r}=\frac{(6.6\times10^{-34}\text{ J. sec})}{2\pi(9.1\times10^{-31}\text{ kg})(5.26\times10^{-11}\text{ M})}$

$=2.2\times10^{-6}$ M/sec.

Example 14.3. Calculate the energy and radius of the 2nd orbit of single ionised helium.

Solution. For single ionised helium $Z=2$ and for 2nd orbit $n=2$

$$W_2=\frac{8\pi^2 me^4}{4(4\pi\epsilon_0)^2 h^2}$$

$$=\frac{8\pi^2(9.1\times10^{-31}\text{ kg})(1.6\times10^{-19}\ C)^4(9\times10^9 NM^2/C^2)^2}{4(6.6\times10^{-34}\text{ J. sec})^2}$$

$=2.189\times10^{-18}$ J$=13.68$ ev.

14.21. Limitations of Bohr's Theory

Bohr's theory successfully explained the spectra of hydrogen. But it has the following limitations.

1. The assumption that electron revolves in circular orbit is not fully justified. Elliptical orbits are also possible.

2. It is found that some lines of hydrogen are not single lines but actually contained number of five lines slight different frequencies from each other. Bohr's theory cannot account for it.

3. It is now established that the planetary model of atom is not a correct representation. Electrons exhibit wave character. So the exact orbit is not well defined. Besides, the charge distribution of electrons in different orbits is completely different from the picture given by Bohr's theory.

4. According to Bohr's theory the atomic radius and ionisation potential should vary with atomic number and square of the atomic number respectively. But the above predictions were not experimentally observed.

14.22. Electron Configuration in Atoms and Pauli Exclusion Principle

From further experimental studies it was found that the spectral series of hydrogen have some similarities with those of alkali atoms. Further the spectra of a particular group of elements were also found to be similar with each other. Taking the above facts into account Bohr and Stoner concluded that all the electrons in an atom do not rotate in the same orbit, and maximum number of electrons that can be accommodated in an orbit is given by $2n^2$, where n is the principal quantum number. The orbits with $n=1, 2, 3$ ... etc. are called K, L, M, N etc. shells respectively. The electrons in the outer most incomplete shells or orbits are called the Valence electrons.

Sommerfeld extended the Bohr's model in which circular as well as elliptical orbits are permissible for electrons in an atom. The expression for energy of the atom thus calculated remains same for same principal quantum number. He introduced a new quantum number called the azimuthal quantum number K. The ratio n/K gives the ratio of semi-major and semi-minor axis of the corresponding elliptical orbits. In his theory K could only have values 1, 2,..... n. For $K=0$ semi-minor axis is zero *i.e.* The path of the electron is a line passing through a nucleus, such that an orbit is not permissible. But experimental observation shows that K can have n possible values and can also be zero. This difficulty was removed by introducing another quantum number l called the orbital quantum number. l can have all integral values from o to $n-1$. So $l=K-1$. Orbits specified by different l values are called subshells.

Subsequently it was discovered that Balmer lines are composed of six separate distinct lines. But the frequencies of these lines differ slightly from each other. These lines are called the fine structure. To account for this it was assumed that electron spins about its own axis and a spin quantum number S was assigned for it. The value of S is all ways taken to be $1/2$.

Zeeman discovered that some single spectral lines split up into several lines when the source is placed between pole pieces of an electromagnet. To account for this the revolving electron around the nucleus, is considered as a current carrying coil. So it behaves as a little magnet. The field of the magnet is perpendicular to the plane of the coil. In a strong magnetic field the little magnet try to precess around the field like the precessional motion of a top. The little magnet cannot orient making all possible angles with the applied field, it has only discrete values. The permitted values are described by a magnetic quantum number m_l. can have all possible integral values from $-l$ to $+l$ i.e. $(2l+1)$ m_l values for a particular value of l. For example, for $l=2$, $m=-2, -1, 0, +1, +2$.

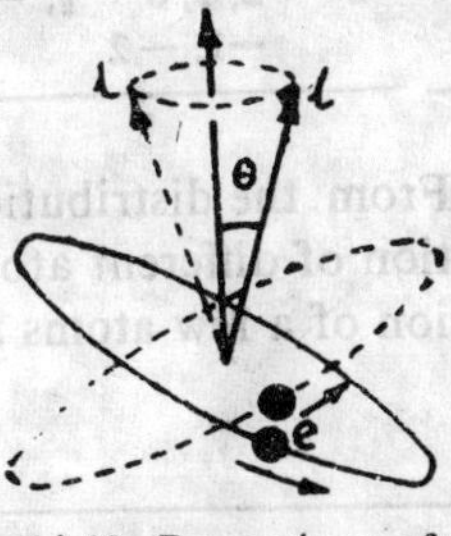

Fig. 14.19. Precession of the little magnet around the field.

Like an orbiting electron a spinning electron is also associated with a magnetic field. In the presence of a magnetic field the field due to spinning electron could take up only two orientations *i.e.* parallel or antiparallel to the field. This gives rise to a spin magnetic quantum number m_s. m_s can have values +1/2 or −1/2. With the help of these four quantum number *n, l*, m_l and m_s the state of an electron in an atom is completely specified.

From systematic study of spectra of elements Pauli stated a principle known as *Pauli exclusion principle*. The principle states that no two electrons will have all the quantum numbers same. Applying this principle maximum number of electrons that can remain in a shell can be known.

Maximum number of electrons that can remain in a shell are given in the table.

TABLE 14.1. Maximum Number of Electrons Permitted in a Shell.

n	*l*	m_l	m_s	*No. of electrons*	*Total No. of electrons*	*Spectroscopic Representation*
1	0	0	$\frac{1}{2}, -\frac{1}{2}$	2	2	$1s^2$
2	0	0	$\frac{1}{2}, -\frac{1}{2}$	2		
	1	1, 0, −1	$\frac{1}{2}, -\frac{1}{2}$		8	$2s^2 - {}^62p^6$
				6		
3	0	0	$\frac{1}{2}, -\frac{1}{2}$	2		
	1	1, 0, −1	$\frac{1}{2}, -\frac{1}{2}$	6	18	$3s^2\ 3p^6\ 3d^{10}$
	2	2, 1, 0 −1, −2	$\frac{1}{2}, -\frac{1}{2}$	10		

From the distribution of electrons in a shell electronic configuration of different atoms can be written easily. Electronic configuration of a few atoms are given below.

TABLE 14.2

Atom	*No. of electrons*	*Spectroscopic rotation*
H	1	$1\ s$
He	2	$1\ s^2$
Li	3	$1\ s^2\ 2s$
Be	4	$1\ s^2\ 2s^2$
Bo	5	$1\ s^2\ 2s^2\ 2p$
Ne	10	$1\ s^2\ 2s^2\ 2p^6$
No	11	$1\ s^2\ 2s^2\ 2p^6\ 3\ s$

14.23. X-ray

In 1895 Rontgen while studying the eletrical discharge phenomena in gas had kept some fluorescent materials near the discharge tube. He observed that the fluorescent material began to glow. The glow was observed even when the tube was covered with black paper in a dark room. The fluorescent material did not glow only when the discharge tube was put out of operation. He concluded, therefore, that some unknown radiations coming out of the tube cause fluorescence. These radiations he named X-rays. Rongten further established that X-rays are produced due to the impact of electrons on walls of the discharge tube.

Now a days X-rays have application in various fields. Different types of X-ray tubes have been developed to produce X-rays for different needs.

14.24. Production of X-rays

A schematic diagram of a Rontgen X-ray tube is presented in the Fig. 14.20. In the diagram *C* represents a concave cathode made of aluminium. *A* represents the anode, *T* is called the anticathode or the target. *T* and *A* are electrically connected outside the tube. The target is made of some substance having high atomic number, high melting point and high electrical conductivity. Some

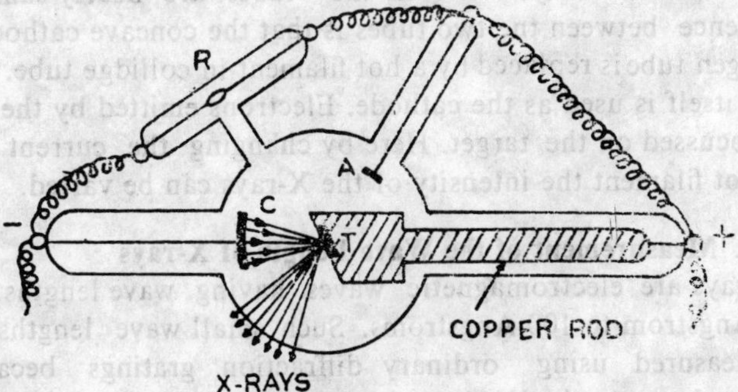

Fig. 14.20. X-ray tube.

such substances are platinum, tungsten, tantalum etc, usually tungsten is used. The target is fixed to the face of a copper rod such that the target makes an angle 45° with the axis of the tube. The target is placed at the centre of curvature of the concave cathode. The copper rod is cooled by water cooling arrangement (not shown in the figure). The cathode, the target and the anode are all kept inside a glass vessel maintained at a pressure of 10^{-2} mm of mercury.

When high potential difference of the order of 100 kv is applied between the cathode and the anode cathode rays are produced and strike the target. X-rays are emitted in the direction shown in the diagram. Only 0.1 to 0.2 % of the kinetic energy of the cathode rays are converted into X-rays. The rest amount of energy is converted to heat.

The intensity of X-rays will be more if more number of electrons strike the target. The number of electrons that strike the target depends upon the pressure of the gas. The energy or quality *i.e.* penetrating power of X-rays depends upon the energy of the electrons. But the energy of electrons depends upon the potential difference between the cathode and anticathode. Therefore X-rays produced by low potential difference have low penetrating power and are called soft X-rays. But X-rays produced by high potential differences have got greater penetrating power and hence are called hard X-rays. Hence in a Rontgen tube the quality of X-rays can be controlled to some extent only by changing the potential differences between *C* and *T*. But there is no control over the intensity of X-rays. These are the two main defects of the tube. To over come these difficulties coollidge tube was invented. The principle of production of X-rays in both the tubes are nearly same. Only difference between the two tubes is that the concave cathode of the Rontgen tube is replaced by a hot filament in collidge tube. The filament itself is used as the cathode. Electrons emitted by the filament are focussed on the target. Here by changing the current through the hot filament the intensity of the X-rays can be varied.

14.25. Measurement of the Wave Length of X-rays

X-rays are electromagnetic waves having wave lengths between one Angstrom to 100 Angstroms. Such small wave lengths cannot be measured using ordinary diffraction gratings because the spacing between the lines are larger than wave length of X-rays. But the spacing of atoms and molecules in a crystal are of the order of a few Angstroms. That is why crystals are used as gratings to measure the wave length of X-rays.

A schematic diagram of X-ray spectrometer is given in Figure 14.21. A narrow beam of X-rays is formed by passing through slits *B* and *C*. The narrow beam in incident on a crystal surface at an angle θ. An X-ray detector is fixed as shown in the diagram. The X-ray tube and the detector are fixed. But the crystal is mounted on a turn table which can be rotated about a vertical axis passing through its centre. If d be the distance between the two neighbouring molecules in the crystal, λ be the wave length of X-rays and θ be the angle of differaction then according to Bragg's rule,

$$2d \sin \theta = n\lambda \qquad \ldots (14.20)$$

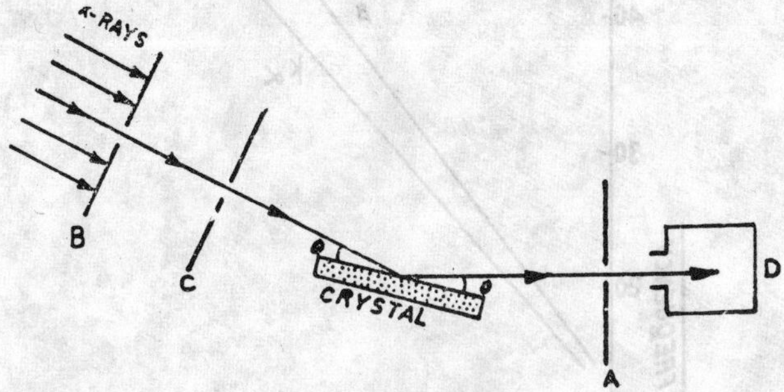

Fig. 14.21. X-ray spectromater

where n is the order of spectra. For a given order, X-rays having different wave lengths are diffracted through different angles θ. For crystals values of d is known. Hence by measuring the values of θ, λ can be known with the help of Equation 14.20.

From measurement of wave length of X-rays it is found that two types of X-rays are emitted from an X-ray tube. They are characteristic X-rays and continuous X-ray. Characteristics X-rays are composed of specific, sharp wave lengths similar to spectral lines of hydrogen. Continuous X-rays contain all possible wave lengths like solar spectrum. Over the continuous X-rays the characteristic X-rays are superimposed.

14.26. Origin of Characteristic X-rays

Characteristic X-rays were first discovered by Barkla. These are called K, L radiations. Moseley from systematic study of K, radiations concluded that they are composed of two radiations called K_α and K_β radiations. He further found that square root of the frequency of the K radiation is proportional to the atomic number of the elements. He plotted square root of the frequency of K radiations against the atomic number and obtained two straight lines (see Fig. 14.22). One for K_α radiation and other for $K\beta$ radiation. According to his name this variation is called Moseley law. From the straight line graph we can write

$$\nu = K\,(Z-a)^2 \qquad \ldots (14.21)$$

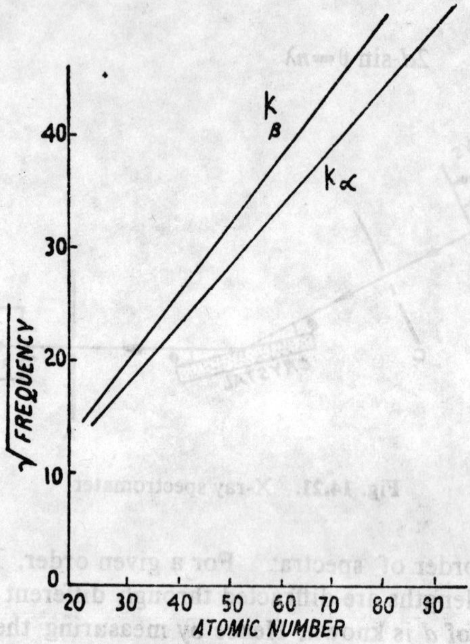

Fig. 14.22. Moseley diagram.

where ν=frequency of the X-rays, a and K are two constants. For K radiation Equation 14.21 can be expressed in the form :

$$\nu = CR\,(Z-1)\left(\frac{1}{1^2}-\frac{1}{2^2}\right) \qquad \ldots (14.22)$$

where $CR\left(\frac{1}{1^2}=-\frac{1}{2^2}\right)=\frac{3}{4}CR=K$ and $a=1$

Comparing (14.22) with (14.19) one can predict the $K\alpha$ radiations are emitted when an electron jumps from orbit $n=2$ to 1st. orbit $n=1$. $(Z-1)$ is written in place of Z in Equation 14.22 because the electron in the orbit $n=1$ effectively reduces the nuclear charge by one. Similarly for $K\beta$ radiation Equation 14.21 can be expressed as

$$\nu = CR\,(Z-9)^2\left(\frac{1}{1^2}-\frac{1}{3^2}\right)$$

where $CR\left(1-\frac{1}{9}\right)=\frac{8CR}{9}=K$ and $a=9 \qquad \ldots (14.23)$

Comparing Equation 14.23 with 14.19 we find that K_β radiations are emitted when electron jumps from orbit $n=3$ to orbit $n=1$. Due to the presence of 1 electron in the orbit $n=1$, 8 electrons in the orbit $n=2$ the effective nuclear charge is reduced by 9. Subsequently different characteristic radiations were discovered. They are called *L, M, N,* radiations. They all obey Moseley law.

From the above discussions the mechanism of characteristic X-radiation can be understood easily. When an energetic electron collides with an electron in the *K* shell of an atom and removes it a vacancy is created in the *K* shell. By this process the atom goes to higher energy state. When another electron from either *L, M* or *N* shells jumps to *K* shell K_α, K_β or K_γ radiations are emitted.

The vacancy created in *L, M* or *N* shells are filled by electron transition from higher orbits and as such different characteristic radiations are emitted. For example transitions from $n=3$, 4, 5 etc. to $n=2$ gives L_α, L , L_γ X-rays respectively. These transitions are similar to the transitions for getting Lyman, Balmer series of hydrogen. The value of *R* is same in both the cases.

14.27. Origin of Continuous X-rays

When an electron moves through an atom it is attracted by the nucleus and repelled by other electrons. The force of attraction due to the nucleus is greater; so the electron is accelerated towards the nucleus. But according to electromagnetic theory an accelerated electron loses energy by radiation. This process is called Bremsstrahlung loss. Due to this process we get continuous X-rays. In this Bremsstrahlung process when the total energy of the electron is transformed to the energy of a X-ray photon we get X-rays having maximum frequency or minimum wave length. This is called the wave length limit. At this limit the energy of the electron is equal to the energy of the X-ray photon. So we can write

$$eV = h\nu_{max}$$

$$\text{or} \quad \lambda_{mn} = hc/eV \quad [\because \lambda = c/\nu]$$

where eV is the energy of the electron in electron volt and *C* is the velocity of light and *h* is the Planks constant. So the cut off wave length is inversely proportional to the applied potential difference between cathode and anode.

14.28. Properties of X-ray

1. X-rays are electromagnetic radiations having wave length between 1Å to 100Å. So they possess all the properties of visible

light. They travel in straight lines with the speed of light. They also exhibit the properties of reflection, refraction, interference and diffraction.

2. They are not affected by electric and magnetic fields.

3. X-rays excite fluorescence in certain substances and affect photographic plates.

4. They emit photoelectrons from certain surfaces on which they fall.

5. They ionise the gas through which they pass.

6. Living tissues are destroyed by X-rays. Hence long exposure to X-rays is harmful for living bodies.

7. X-rays have got high penetrating power. They can penetrate through solid substances. Absorption of X-rays by a substance is governed by the equation

$$I = I_0 e^{-\mu x} \qquad \ldots (14.24)$$

when I_0 is the intensity of incident X-ray and I is the intensity of X-ray after it passes through a matter of thickness x. Intensity of X-rays is defined as the amount of energy flowing per sec through unit area perpendicular to the direction of flow. μ is called the absorption coefficient. μ is found to depend upon the wave length of X-rays and also on the absorbing material. The variation of μ with λ shows sharp discontinuties. But between two consecutive discontinuities μ varies smoothly with λ according to the relation given by

$$\mu = CZ^4\lambda^3\rho \qquad \ldots (14.25)$$

where C is a constant, ρ and Z are the density and atomic number of the absorber respectively. Thus the elements of higher atomic number absorb X-rays more.

14.29. Uses of X-rays

X-rays have many important and useful applications. Some of them are given below:

1. **Radiography.** X-rays can pass through soft substances but are absorbed by hard substances. When a beam of X-rays, after passing through some portions of the human body are incident on a fluorescent screen shadow of bones are clearly visible on the screen. Using this method fractures of bones, presence of bullets, pins etc, are easily detected. They are also used to diagnose diseased lungs and kidneys and tumour in the human body. This method of diagnosis is called Radiography.

2. **Radio-therapy.** X-rays can destroy easily diseased tissues such as in tumours, and cancerous growths. So by using proper X-rays the

cancerous tissues are destroyed. This method of treatment is called radio-therapy.

3. **Engineering and industry**. X-rays are used to detect defects in valves, tennis balls, metal joints and cracks in metals and finished goods.

4. **Crime detection.** X-rays are used to detect the presence of explossives, smuggled goods kept inside leather and wooden boxes at airport and seaports. They are also used to detect the presence of gold hidden inside the body of smugglers or parcels. They can also be used to test the genuiness of documents, pearls, diamonds etc.

5. **Scientific researches.** X-rays are used to study the crystal structure and structure of other substances like glass, fibers etc.

14.30. Photoelectric Effect

It was found from experiment that when light is incident on some substances, electrons are emitted. It was found that ultraviolet light and X-rays can also emit electrons. These electrons are called photoelectrons, and the phenomenon is called photoelectric effect. From detailed experimental study following conclusions were drawn about the photoelectrons and photo electric effect.

1. Number of photo electrons emitted depends only upon the intensity of incident radiation and is directly proportional to the intensity of radiation.

2. The energy of photoelectrons depends only on the frequency of the incident radiation. The kinetic energy of the emitted photoelectrons vary linearly with the frequency of the incident radiation.

3. For a given surface there exists a limiting frequency of radiation below which photoelectrons are not emitted. This limiting minimum frequency is called the threshold frequency.

4. There is practically no time lag between the incidence of radiation and emission of photoelectrons. The time lag is found to be of the order of 10^{-9} sec.

Some of the above conclusions can be experimentally verified in the laboratory.

14.31. Experimental Study of Photoelectric Effect

Fig. 14.23 represents a simple experimental arrangement to study the photoelectric effect. In the diagram *C* represents a photosensitive surface, *A* is the anode placed in front of *C*. Both of them are placed inside a glass bulb maintained at very low pressure *G* is a microammeter and *B* represents available potential source.

When suitable potential difference is applied between *C* and *A* and electromagnetic radiations of suitable frequency is incident on

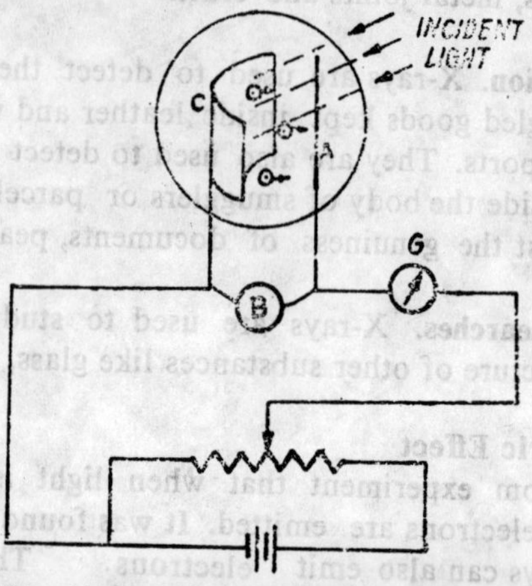

Fig. 14.23. Experimental arrangement for study of photo electric effect.

C, electric current will flow in the circuit. If the incident radiation is obstructed by a screen the current reduces to zero. Thus the incident radiation is causing the emission of photoelectrons. This current is called the photo electric current. This simple experiment establishes the existence of photoelectric effect.

To study the effect of intensity of incident radiation let us use a monochromatic source of ultraviolet radiation whose intensity can be varied. The radiation is incident on *C*. A suitable positive potential is applied to *A* with respect to *C*. Strength of photoelectric current is then measured. Keeping the potential constant intensity of the incident radiation is varied and corresponding photo current is measured. By plotting a graph between intensity of ultraviolet light and photo current a straight line will be obtained. Thus the photocurrent is proportional to intensity of radiation. Since photo current is proportional to the number of photoelectrons emitted, it is evident that number of photoelectrons emitted is proportional to intensity of radiation.

The effect of applied potential on photoelectrons can be studied by measuring photo electric current *I* on applying different potential to the anode. Similar observation are taken for light having different intensity. The variation of photo current as a function of applied voltage are shown in Figure 14.24. The photo current do not increase above a certain value of applied pctential. This limiting current is called the saturation photo current. From the experiment it is found that the intensity of light is proportional to the saturation photo current. Increasing the negative value of the potential photo current decreases. At a certain negative potential photo current reduces to zero. This potential is called the retarding or cut-off potential V_0. When the applied negative potential is more than V_0 no photo current will be emitted even though the intensity of light may be quite strong. The retarding potential is different for different materials and different frequencies of light. If Vo is the retarding potential for a given frequency V_0e measures the maximum kinetic energy of emitted electrons. So we can write,

$$eV_0 = \frac{1}{2}mv^2_m \qquad \ldots (14.26)$$

where v_m is the maximum velocity of electrons.

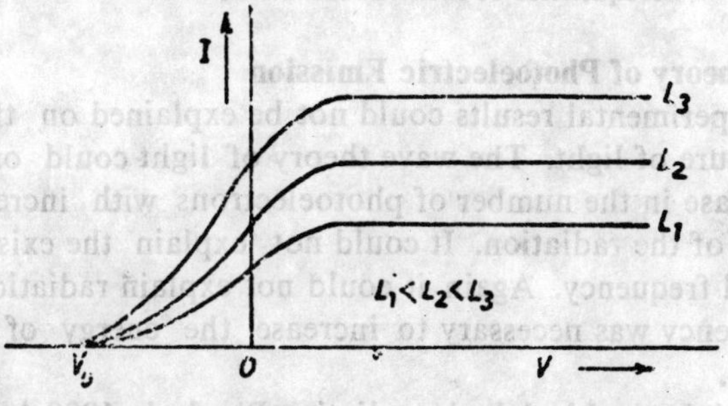

Fig. 14.24. Variation of photo current with applied potential.

Effect of frequency of light on photo emission can easily be studied. Light sources having different frequencies are taken. Radiations of a particular frequency is allowed to be incident on *C*. Cut off potential for the radiations is determined. The same experiment is repeated with radiations of different frequencies and cut off potentials in each case are determined. A graph is plotted between cut off potentials and frequencies of light. A straight line graph (*see* Fig. 14.25) is obtained. This shows that the stoping potential increases with increase of frequency. The graph also indicates that there is a frequency for which stoping potential is zero. This again indicates that no photoelectrons will be emitted if radiation of frequency is less than the threshhold frequency. From this it is clear that when light having greater frequency is used to emit the photoelectrons the energy of electrons is more so larger negative potential is required to stop the emission of photo-electrons.

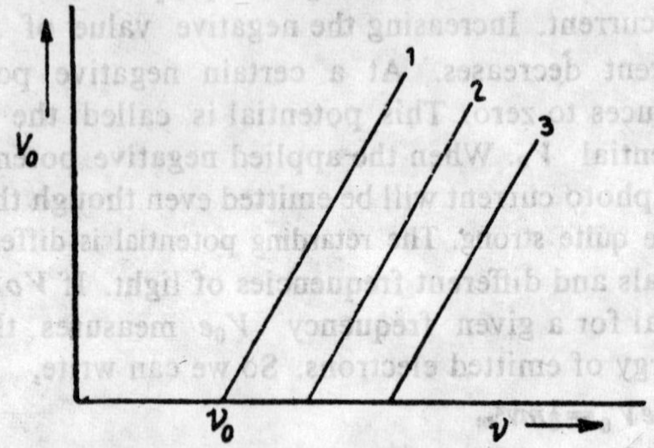

Fig. 14.25. Variation of stopping potential with frequencies of incident radiation.

14.32. Theory of Photoelectric Emission

The experimental results could not be explained on the basis of wave nature of light. The wave theory of light could only explain the increase in the number of photoelectrons with increase in the intensity of the radiation. It could not explain the existence of a threshold frequency. Again it could not explain radiations of greater frequency was necessary to increase the energy of the photo electrons.

To explain the black body radiation Plank, in 1900 A.D postulated that a radiation of frequency ν is associated with a quanta of energy $h\nu$, where h is Planks constant. This quanta of energy is called a photon. In 1905 Einstein using Plank's quantum theory explained the phenomenon of photoelectric effect. According to him, each photon of light absorbed by metallic surface emits photoelectron. A part of the energy is spent to bring the electron from inside to outside. The rest of energy of the photon is given to the electron as kinetic energy. Thus the kinetic energy of the electron is given by,

$$\frac{1}{2}mv^2 = h\nu - w \qquad \ldots (14.27)$$

Where v is the maximum velocity of the emitted electron and w is the amount of energy required to remove an electron from metallic surface. This energy is different for different surfaces and is called the work function of the metallic surface. The incident radiation usually penetrates a distance of about 100 Å into the surface. So photoelectrons are emitted from different depths. As a result, photoelectrons have different energies. This accounts for the variation of photo currents with varying potential. When the frequency of the radiation decreases the energy of photon decreases. This explains the decrease of photocurrent with frequency. When the energy of the incident photon is equal to the work function of the metal no photo electrons will be emitted even though the intensity of light is high. This clearly explains the existence of threshold frequency. Thus all experimental results are explained by Einstein's Theory. If ν_0 is the threshold frequency.

$$\tfrac{1}{2}mv^2 = h(\nu - \nu_0) \quad \ldots (14.28).$$

Where $w = h\nu_0$

This equation is known as Einstein photoelectric equation.

14.33. Application of Photo Electric Effect

Using the principle of photo electric effect photo electric cells are prepared. Photocells directly convert light energy into electrical energy. These photo tubes are widely used for the following purposes.

1. Photo tubes are used in reproduction of sound in cinematography and also in television.
2. They are used to determine opacity of solids and liquids.
3. The temperature of furnaces and of chemical reactions are controlled by photo tubes.
4. They are used in automatic control devices of street lighting systems for traffic signals and for speed of automobiles.
5. Photo tubes are used in photographic camera for controlling exposure time.
6. Automatic warning devices such as Burglars alarm are prepared using photo tubes.
7 They are used to locate the minor flaws or holes in metallic sheets.
8. Photo electric cells are used in solar battery which finds an important use in space vehicles and artificial satellites.

14.34. Dual Nature of Matter

The phenomena of interference, diffraction and polarisation of radiation established the wave nature of radiation. But the phenomena of black body radiation, compton effect and photoelectric effect establish the quantum nature of radiations. In general some times the radiations exhibit wave character and sometimes exhibit particle character. In 1925 Louil de Broglie thought about the wave nature of matter. He was of opinion that matter should exhibit dual character like radiation.

14.35. Wave Length of Matter Waves

According to de Broglie if a mass m is assigned to a radiation then

$$E = mc^2 = h\nu = \frac{hc}{\lambda}$$

$$\text{or } \lambda = \frac{h}{mc} \quad \ldots (14.29)$$

where mc is the momentum of the radiation. If a wave length λ is assigned to a particle then we can similarly write

$$\lambda = \frac{h}{mv} = \frac{h}{p} \quad \ldots (14.30)$$

where p is the momentum of the particle.

So a moving electron can be considered as a wave packet. Let an electron acquires a velocity v when it is acclerated by a potential different of V volts.

Then:

$$1/2mv^2 = V \text{ (e.v)}$$

$$\text{or } \quad m^2v^2 = 2mV \text{ (e.v.)}$$

$$\lambda^2 = \frac{h^2}{2mV(\text{ev})} = \frac{[6.6 \times 10^{-34}\text{J.sec}]^2}{(2 \times 9.1 \times 10^{-31}\text{kg})(V \times 1.6 \times {}^{-19}\text{J})}$$

$$= 1.496 \quad 10^{-18} \frac{\text{Jsec}^2}{V(\text{kg})}$$

$$\therefore \lambda \quad = \frac{1.23 \times 10^{-9}}{V^{\frac{1}{2}}} \text{ meter} = \frac{12.3\text{Å}}{V^{\frac{1}{2}}}$$

From above it is clear that the order of the wave lenght of electrons are equal to that of X-rays.

de Broglie wave length of electron was first measured by Dovisson and Germer.They directed a beam of electrons having energy 54 eV on a crystal of nickel. They measured the intensity of the diffracted beam by an electron detector. The intensity shows a peak value at 50°. The wave length was calculated using Brag's equation

$$n\lambda = 2d \sin \theta$$

For nickel $2d = 2.15$ Å and for $n = 1$

$$\lambda = 1.65\text{A}$$

The theoretical value is given by

$$= \lambda \frac{12.3\text{Å}}{(54)^{\frac{1}{2}}} = 1.66\text{A}$$

From this it was established that at times electrons behave as waves. Similarly the protons and neutrons show also wave character.

QUESTIONS

1. Why gases are bad conductors at ordinary pressure and good conductors at low pressure?
2. How it was established that cathode rays are nothing but electrons?
3. Discuss some important properties of cathode rays.
4. What important conclusion was drawn from the Millikan's oil drop experiment?
5. What are Canal rays?
6. Distinguish between Canal rays and Cathode rays?
7. Discuss about the experimental results obtained from scattering of α– particles by tin foil.
8. Discuss about the postulates of Bohr.
9. Give a clear idea about quantum numbers.
10. In what respect the spectra of hydrogen differs from the spectra of single ionised helium?
11. What is ionisation potential?
12. Discuss about the limitations of Bohr's theory.
13. State Pauli exclusion principle? How, by applying this principle, maximum number of electrons that can remain in a shell be known?
14. What is Zeeman effect? Discuss about the origin of such effect?
15. What are X-rays?
16. Discuss the origin of emission of characteristic X-rays.
17. Discuss the origin of the emission of continuous X-rays.
18. Discuss some important uses of X-rays?
19. Why ordinary grating is not suitable for X-ray diffraction?
20. What are photo electrons?
21. Discuss the origin of emission of photo electrons.
22. How the energy of photo electrons emitted from a surface can be increased?
23. How the number of photo electrons emitted from a surface can be increased?
24. What are de Broglie Waves?
25. Discuss under what circumstances matter behaves as waves and waves behave as matter.
26. What do you mean by dual nature of matter and radiation?

PROBLEMS

1. An α-particle is scattered by a Gold nucleus through 180°. If the approximate radius of the Gold nucleus is one fermi what is the initial velocity of α-particles? (1.05×10^8 m/sec)
2. Calculate the radius of the 2nd orbit of doubly ionised Lithium. (.007 Å)
3. Compare the speed of the electron moving in the 1st orbit of hydrogen with that of electron moving in the 1st orbit of single ionised helium. (0.5)

4. If a potential difference of 400 volt is applied between cathode and anode and total energy of the accelerated electron is converted to X-ray photon what is the wave length of X-rays? (31Å)

5. When an electron revolves in the third orbit of hydrogen find the frequency of revolution? (1.97×10^{12}/sec)

6. If the wave length of the first member of the Balmer series of hydrogen is 6563 A. Calculate the wave length of the first member of the Lyman series. (1215 Å)

7. If the threshold frequency of potassium is 3×10^{14}. Calculate its work function. (19.9×10^{-20}J)

8. If the work function of sodium is 2.13 ev, what is the maximum velocity of photo electrons emitted by the light of wave length 4000 A? (5.85×10^5 m/sec)

9. If the velocity of an electron is 0.8 C. What will be its wave length? (·03 Å)

10. If the wave length of an electron is 0.5 Å, what is the energy of the electron? (3.97×10^{-15} J)

15

THEORY OF RELATIVITY

In the 1st part of 20th century Einstein proposed the Theory of Relativity. This theory contradicts to some extent the Newton's simple theory of mechanics. According to this theory velocity of any material particle cannot be more than the velocity of light and there is no difference between mas and energy. The mathematical theory of relativity is complicated. So we will only discuss about the physical idea involved in it. Before going to develop the theory we will first discuss about some concepts.

15.1. Event, Observer and Frame of Reference

An event is a simple or complicated physical phenomenon taking place in space at certain instant of time. For example lightning in the sky, collision between two cars, falling of a fruit from a tree etc. are called events. A person who observes an event and take measurement is called an observer. Any measuring equipment that can take measurements of an event is also called an observer.

To locate a point in space we require a coordinate system. The position of the point can easily be expressed by three coordinates.

Different coordinate systems are used for different situations. Out of thesefor convenience we use cartesian coordinate system. To get precise information about an event an observer requires a co-ordinate system and a time measuring instrument. For example, an observer saw the falling of a fruit. To express the position of the fruit before it falls he requires three coordinates x, y, z. But for convenience the observer can choose the coordinate system in a particular manner. For example if the x-y plane of the coordinate system coincides with the surface of the earth and the observer is at the origin the new coordinates of the position where the fruit hits the ground is x, y, o. To express the time when it falls from the tree and when it hits the ground the observer requires a time measuring instrument. So in general to record the position of occurrence of an event an observer must set up a coordinate system or a frame of reference in which the observer himself is located.

When bodies in a frame of reference follow Newtonian mechanics, the frame of reference is called *inertial frame*. An observer in the inertial frame is referred to as inertial observer. Any other frame in rectilinear and uniform motion relative to the inertial frame is referred as inertial system.

15.2. Principle of Relative Motion

Suppose one train X is standing at a platform and another train Y is moving with an uniform velocity parallel to the train X. There are two observers A and B in the trains X and Y respectively. The observers A while looking at the train Y cannot say with certainty whether he is moving or the train Y is moving. Similarly observer B also cannot be certain as to whether he is moving or the train X is moving. But when they will see some stationary objects in the platform they can definitely say whether they are moving or at rest. On the other hand if both the trains are moving in the same direction with same velocity both the observers will feel that both of them are at rest. But actually they are moving.

From the above example one can conclude that the state of rest or of uniform rectilinear motion are relative terms. The state of rest or of uniform motion of an object can only be defined relative to other objects.

For further discussion let us imagine that the observer A in the train at rest is throwing a ball upwards and catching it on its descent. Similarly the observer B in the moving train is also throwing a ball upwards and catching it on its descent. From experience we can say that both the observes will have no difficulty to perform the experiment. Again suppose both observers are pouring tea into a cup from a kettle then both of them can do it easily. So experiments conducted in a moving train will not give different results enabling us to distinguish the state of rest or of iuniform rectilinear motion.

From the above examples we can conclude that the laws of Physics are the same in a frame at rest or at uniform rectilinear motion. The above statement is called the principle of relative motion or Newton's Relativity Principle.

15.3. Galilean Transformation

Suppose two observers moving relative to each other take observation of an event occurring in space. The results of the observation will be different. A set of equations can be obtained to correlate the results of the observation. These set of equations are called the Galilean transformation.

To make the ideas clear let us consider two observers O and O'

located at the origins of the inertial frames S and S'. The primed system S' moves with a velocity v parallel to the X-axis relative to the unprimed system or the unprimed system moves with a velocity $-v$ parallel to X-axis relative to the prime system. Let us assume that their stop-watches, meter scales are calibrated to record same time and same length and at $t=0$, origins of both the frames coincide. At $t=0$ let both observers record the coordinates of a point P in space and the coordinates are X_1, Y_1, Z_1 at $t=0$. After time t_1 let them again record the coordinates of the point P.

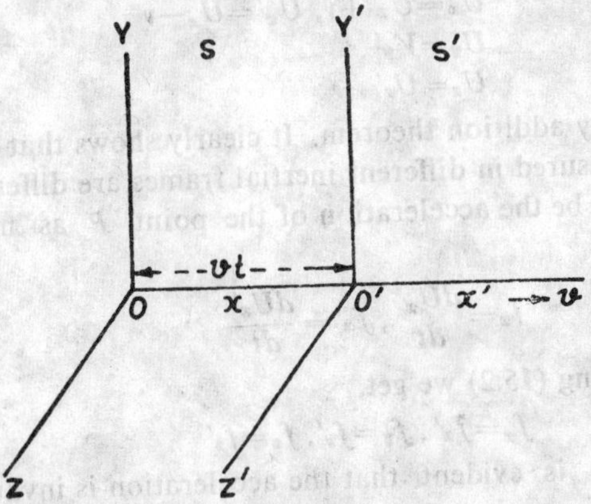

Fig. 15.1. Coordinates of a point in differe̲nt fr..mes

The coordinates recorded by observer O is X_1, Y_1, Z_1 and that by the observer O' is X_2, Y_2, Z_2 and t_2. As there is no relative motion along Y and Z axis, $Y_1=Y_2=Y$ and $Z_1=Z_2=Z$. As the two stop-wathes are calibrated to record same time, $t=t_1=t_2$. Therefore the x-coordinate of the point P as measured by the two observers are related by: $X_1=X_2+vt$ as the primed system is moving with a veloctty v.

Thus writing the above transformation together,

$$\begin{aligned} X_1&=X_2+vt \\ Y_1&=Y_2 \\ Z_1&=Z_2 \\ t_1&=t_2 \end{aligned} \qquad \text{or} \qquad \begin{aligned} X_2&=X_1-vt \\ Y_2&=Y_1 \\ Z_2&=Z_1 \\ t_2&=t. \end{aligned} \qquad \ldots (15.1)$$

These transformation equations are called Galilean transformation. In the above discussion space and time intervals are same in all inertial frames. Thus in Newtonian mechanics both space and time are treated as absolute.

15.4. Newtonian Relativity Principle

Using Galilean transformation the velocity and acceleration of a body in the two frames of reference can easily be related. Let the point P move with time and its velocity components be U_x, U_y, U_z and U_x', U_y', U_z' measured by the observer O and O' respectively. So we can write

$$U_x = \frac{dx_1}{dt},\ U_x' = \frac{dx_2}{dt}$$

Differentiating the Galilean transformation equations we get,

$$U_x = U_x' + v,\ U_x' = U_x - v$$
$$U_y = V_y'$$
$$U_z = U_z' \quad \dots (15.2)$$

This is velocity addition theorem. It clearly shows that the velocity of a body measured in different inertial frames are different.

If f_x and f_x' be the acceleration of the point P as measured in two frames

$$f_x = \frac{dU_x}{dt},\ f_x' = \frac{dU_x'}{dt}$$

Differentiating (15.2) we get,

$$f_x = f_x',\ f_y = f_y',\ f_z = f_z' \quad \dots (15.3)$$

Therefore it is evident that the acceleration is invariant under Galilean transformation. In Newtonian mechanics mass is taken to be invariant under *Galilean Transformation*. Therefore force is also invariant. Similarly the principle of conservation of momentum and energy are the same in different frames. Hence the fundamental laws of mechanics are independent of the velocity of the observer. So in general we can write 'The fundamental laws of mechanics are invariant under Galilean trasformation.' The above statements is called the Newtonian Relativity Principle.

15.5. Application of Newtonian Relativity Principle to Light

According to present concept light waves are electromagnetic waves moving in ether medium. According to wave theory velocity of light C is considered to be the velocity relative to the frame of reference attached to the ether. Suppose C is the speed of light in a frame S in which the ether is at rest. Suppose an observer in a frame S' moving with a velocity parallel to the path of light measures the velocity of light. According to the Galilean velocity transformation the measured speed will be either $c+v$ or $c-v$ depending upon the direction of motion of the observer relative to the direction of propagation of light. In other words the velocity of the moving frame can be estimated by measuring the velocity of light in the moving frame.

Scientists performed number of experiments to verify the above fact. Of all the experiments the most important one is Michelson and Morley experiment. The aim of the experiment was to measure the velocity of earth in ether by measuring the velocity of light in two mutually perpendicular directions *i.e.* parallel and perpendicular to the direction of the velocity of earth. In 1887 Michelson and Morley performed the experiment using a Michelson interferometer. The result of the experiment actually leads to the formulation of the theory of relativity. We will discuss the experiment in brief.

15.6. Michelson and Morley Experiment

The experimental arrangement is presented in Fig. 15.2. A narrow beam of light coming from the source *A* is divided into

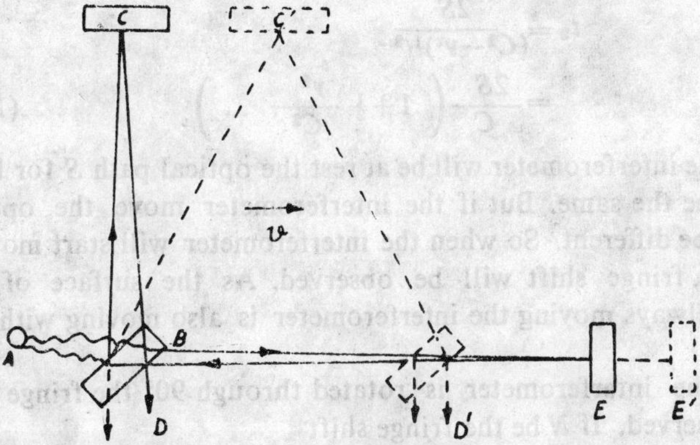

Fig. 15.2. Experimental arrangement of Michelson and Morley experiment.

two parts by the half silvered mirror *B*. The two beams moves at right angles to each other and are reflected by two mirrors *E* and *C*. The two reflected beams meeting again at *B* produce interference pattern in the direction of *D*, depending upon the optical path difference between the two beams.

The *BE* arm of the interferometer is placed parallel to velocity of the earth. Let v, C and S be the velocity of earth, velocity of light and optical path of both *BC* and *BE* respectively. If light takes time t_1' to move from *B* to *E* and t_1'' from *E* to *B*, we can write,

$$Ct_1' = S + vt_1$$

or

$$t_1' = S/(C-v) \qquad \ldots(15.4)$$

as during time interval t_1' the mirror *E* moves through a distance vt_1'.

Similarly,

$$Ct_1'' = S - vt_1'' \qquad \ldots(15.5)$$

or

$$t_1'' = S/(C+v)$$

If t_1 is the time required by light to travel from *B* to *E* and back again from *E* to *B*.

$$t_1 = t_1' + t_1''$$

$$= \frac{S}{C-v} + \frac{S}{C+v} = \frac{2S}{C}\left(1 + \frac{v^2}{C^2} \ldots\right)$$

as

$$v \ll C \qquad \ldots(15.6)$$

If t_2 is the time taken by light to travel from *B* to *C*, and back, we have :

$$\left(\frac{Ct_2}{2}\right)^2 = S^2 + \left(\frac{vt_2}{2}\right)^2 \qquad \ldots(15.7)$$

or

$$t_2 = \frac{2S}{(C^2 - v^2)^{1/2}}$$

$$= \frac{2S}{C}\left(1 + \frac{1}{2}\frac{v^2}{C^2} \ldots\right) \qquad \ldots(15.8)$$

When the interferometer will be at rest the optical path *S* for both rays will be the same. But if the interferometer move the optical path will be different. So when the interferometer will start moving from rest a fringe shift will be observed. As the surface of the earth is always moving the interferometer is also moving with the earth.

When the interferometer is rotated through 90° the fringe shift will be observed. If *N* be the fringe shift

$$N = \frac{\text{Path difference}}{\text{Wave length}}$$

$$= \frac{2Sv^2}{C^2\lambda} \qquad \ldots(15.9)$$

Value of these quantities in Michelsons experiment were

$S = 10$ m

$v = 30$ m/sec. (Revolutional velocity of earth)

$C = 3 \times 10^8$ m/sec.

$\lambda = 5000$ A.

Using the numerical values of S, ν and λ we get,

$$N=0.4 \qquad \ldots(15.10)$$

The fringe shift of 0.4 can easily be measured by the Michelson interferometer. To measure the fringe shift Michelson and Morley kept the interferometer on a stone slab which was floating over mercury. They tried to observe the fringe shift by continuously rotating the interferometer during day and night and at different places on earth. But could not observe any fringe shift. Their final conclusion was that there was no fringe shift at all. This unexpected result contradicts the Newtonian mechanics. From 1887 to 1905 scientists were not able to explain the contradictory result of the experiment. In 1905 Einstein tried to explain the result and developed the Theory of Relativity.

15.7. Postulates of the Special Theory of Relativity

In 1905 Einstein was able to interpret the negative result of Michelson-Morley experiment. According to him relative velocity of a body can only be measured and general laws of Physics will depend upon the velocity of the observer.

For example, length and mass of a body measured by an observer at rest are different from that measured by an observer moving with some velocity. Similarly time interval between two events measured by an observer at rest is different from the time interval measured by a moving observer. To have a clear idea let us imagine that a train is moving with a very high velocity while a parallel beam of light from a torch is incident vertically on a plane mirror fixed at the roof of the train. The beam of light is reflected by the mirror. An observer in the train will see that the light beam is reflected back in the same direction. But an observer in the platform will notice that the light beam will take a longer path as shown in the Fig. 15.3. But from negative result of Michelson and Morley experiment it is established that the velocity of light is invariant. So the observer at rest will see that light takes more time as it has to cover greater distance. On the other hand the observer in the train will see that light takes less time as it covers shorter distance. So from the above discussion it is evident that the time intervals are different in frames moving with different velocities. However, according to classical concept the time intervals are invariant.

From above considerations Einstein made two postulates to develop the theory of relativity. They are:

(1) Fundamental Physical Laws have the same mathematical

form in all intertial frame of reference.

(2) The velocity of light in empty space is constant, and is independent of the motion of the observer and source.

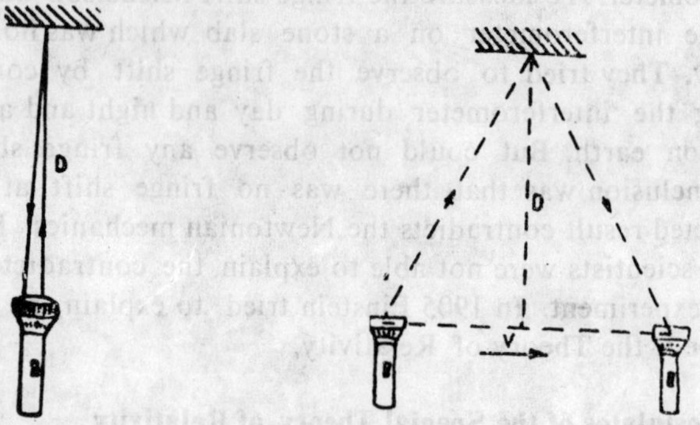

Fig. 15.3. Path of light observed by two observers.

Using theory of relativity Einstein derived relations between the results obtained by two observers in different frames of reference. Here we will give only the final expressions. Let an observer moving with a velocity relative to the earth take some measurements and another observer on earth also takes the same measurements. The equations that correlate the two measurements are given below.

1. **Measurement of length**. Let l_0 and l be the lengths of a rod measured by observer at rest and the observer moving with velocity v with respect to the rod. Then

$$l=l_0\left(1-\frac{v^2}{C^2}\right)^{1/2} \quad \ldots(15.11)$$

There is contraction of length called Lorentz contraction.

2. **Measurement of time**. Let t_0 and t represent the time intervals between two events recorded by stationary observer and moving observers respectively, then

$$t=t_0\Big/\left(1-\frac{v^2}{C^2}\right)^{1/2}, \ t>t_0 \quad \ldots(15.12)$$

This phenomenon is called time dilation.

3. **Velocity of an object**. Let v_0 and v be the velocity of an object measure by the two observers, then.

$$v=\frac{v_0+v}{1+vv_0/C^2} \quad \ldots(15.13)$$

4. **Mass of a body**. If m_0 be the mass of the body at rest and m be the mass of the body when moving with a velocity, then :

$$m=\frac{m_0}{(1-v^2/C^2)^{1/2}},\ m>m_0 \qquad \ldots(15.14)$$

5. **Mass energy relation**. Einstein showed that mass is a form of energy and they are related by the equation given by :

$$E=mC^2 \qquad \ldots(15.15)$$

In Equations 15.11 and 15.15 when $v \ll C$, the expressions reduces to the results obtained by Newtonian mechanics. Thus Newtonian mechanics is a special case of theory of relativity.

Example 15.1. Using Equation 15.13 prove that velocity of light is invariant.

Solution. Let C be the velocity as measured in a frame at rest and C_1 be the velocity of light as measured in frame moving with a velocity v. Using Equation 15 13 we can write :

$$C_1=\frac{C+v}{1+v\ C/C^2}=C$$

Example 15.2. What will be the velocity of the electrons when its mass is two times the rest mass?

Solution. If m_0 is the rest mass and m be the mass,

$$\frac{m}{m_0}=2=\frac{1}{(1-v^2/C^2)^{1/2}}$$

or $$4=\frac{1}{1-v^2/C^2} \quad \text{or} \quad v=\sqrt{\frac{3}{2}}C$$

Example 15.3. A man in a rocket moving with a velocity of 0.5 C relative to earth take rest after every 6 hours of work. What is the time interval according to the time scale of earth?

Solution. From Equation 15.12 we can write :

$$t=6\left(1-\frac{.25\ C^2}{C^2}\right)^{1/2}$$

or $$t^2=\frac{36}{0.75} \qquad \text{or} \qquad t=6.9 \text{ hours.}$$

QUESTIONS

1. A ball is thrown vertically by a person sitting in a train. The ball returns to its starting point. What inferences can be drawn about the velocity of the train?

2. In the above question the ball does not return to its starting point. What inferences can be drawn about the velocity of the train?

3. Identify the following physical quantities which are variant and invariant under Galilean transformation. Space, time, mass, velocity, force, acceleration, momentum and kinetic energy.

4. What is the important conclusion drawn from the negative result of Michelson-Morley ether drift experiment?

5. State and explain postulates of special theory of relativity.

6. What do you know about time dilation and length contraction?

7. What will be the consequences if the velocity of light is 100 km/hour? Discuss

8. Explain, why the effect of special theory of relativity is not observed in daily life.

PROBLEMS

1. The length of a rocket on earth is 200 metres. If the rocket moves with a velocity of 3×10^5 metres/sec what will be its length as observed on earth? (109.9 m)

2. The rest mass of an electron is 9×10^{-28} gm. If its mass is 13×10^{-28} gm when it is moving, what is the velocity of the electron? (2.17 m/sec)

3. If the kinetic energy of an electron is 1 Mev, find its velocity and mass. (2.8×10^8 m/sec, 2.51×10^{-30} kg)

4. Earth receives 2 calories of heat from sun per sec. What is the amount of energy in terms of mass received by the earth during one year? (1.47×10^{-9} kg)

5. A man moving with a velocity 0.8 C relative to earth observes that a train on earth is moving with a velocity of 60 km/hour. What is the velocity of the train as recorded by an observer on earth? p (166.7 km/hour)

6. The half life of a particle is 10^{-7} sec. What will be its half life if it moves with a velocity .098 C? p (1.005×10^{-7} sec)

16

NUCLEAR PHYSICS

Rutherfords scattering experiment had established that at the centre of an atom there is a hardcore called nucleus The radius of the nucleus was estimated to be of the order of 10^{-15} m whereas the radius of an atom is of the order of 10^{-10} m. In Bohr's model of the atom the uucleus was taken to be positively charged. In the smallest nucleus of hydrogen there is only one proton having positive charge. The nuclear charge of helium is two times that of hydrogen. But it was observed that the nuclear mass of helium is four times that of hydrogen or of a proton. Hence the nuclear mass of helium is two times the total mass of the protons in it. In some elements *i.e.* carbon, nitrogen, oxygen, neon etc. nuclear mass is found to be nearly two times the total mass of protons in the nucleus. But in other elements the nuclear mass is found to be more than two times the total mass of the protons. So it was inferred that some other particles, besides protons, are present in the nucleus and due to the presence of these particles protons can remain inside a nucleus. When number of protons in a nucleus increases more number of these particles are required to keep the protons inside the nucleus. In 1932 James Chadwick discovered this new particle which is present in a nucleus. The particle is called neutron. Subsequently other experiments also established that the basic building particles of nucleus are protons and neutrons.

16.1. Structure of Nucleus

Inside a nucleus there are some protons and some neutrons. Both protons and neutrons are called nucleons. Suppose in a nucleus number of protons is Z, number of neutrons is N, and A is the total number of nucleons then :

$$A = Z + N \qquad \ldots (16.1a)$$

Z is the atomic number of the atom and A is called the mass number of the nucleus. An atom having Z protons and A nucleons is represented by a symbol ${}_ZX^A$ where X is the chemical symbol of the element. The symbol ${}_ZX^A$ is called a nuclide. For example a nucleus ${}_6C^{13}$ represents a nucleus containing 6 protons, 7 neutrons or 13 nucleons. Since the chemical symbol specified the atom whose atomic number is known sometimes the pre-subscript is omitted. So the above nuclide at times is presented by the C^{13}.

16.2. Isotopes and Isobars

In general hydrogen nucleus contains only one proton. But at times one or two neutrons are also present along with the proton in the hydrogen nucleus. In such cases the mass number of the atom becomes 2 or 3. When the mass number of hyrogen is 2 it is called deuterium and when the mass number is 3 it is called tritium. Deuterium and tritium are called the *Isotopes* of hydrogen. So hydrogen has three isotopes. They are represented by the symbols $_1H^1$, $_1H^2$, and $_1H^3$. Similarly lithium has two isotopes *i.e.* $_3Li^6$ and $_3Li^7$. In general atoms having same atomic number and differrent mass numbers are called isotopes.

On the other hand, nuclei having same mass number but different atomic number are called *Isobars*. For example $_1H^3$ and $_2He^3$ are isobars. Between two isobars if atomic number of one is equal to the number of neutrons present in the other the two isobars are called mirror nuclides. The isobars, $_1H^3$ and $_2He^3$ are mirror nuclides as number of protons in $_1H^3$ is equal to the number of neutrons in $_2He^3$.

16.3. Nuclear Mass and its Measurement

Different nuclei have different mass. Even mass of two isotopes are also different. Different experimental methods and techniques were developed to measure the mass of ions. The instrument used to measure the mass of ions are called mass spectrographs. Here it is not possible to discuss about the different types of mass spectrographs. However, we will discuss about one of them;the Brain bridge mass spectrograph. The principle of measurement involved in it is simple and it gives accurate results.

Brain-bridge mass spectrograph. The sectional diagram of a Brain-bridge spectrograph is presented in Fig. 16.1. In the diagram the ions from a source (not shown in the figure) are converted to a narrow beam after passing through two slits S_1 and S_2. The beam then passses through a 'velocity selector' which consists of an electric field parallel to the plane of the paper and a magnetic field perpendicular to the plane of the paper. The electric field is produced by applying suitable potential difference between the plates X and Y. The magnetic field is produced by an electromagnet. The direction of the two fields are such that the forces acting on the ion beam oppose each other. So the deviation of the ion beam produced by one field is opposite to that produced by the other field. The two fields are then so adjusted that the ions moving with a certain velocity v are not deviated and pass through the slit S_3. Other ions moving with velocities other than v are deviated

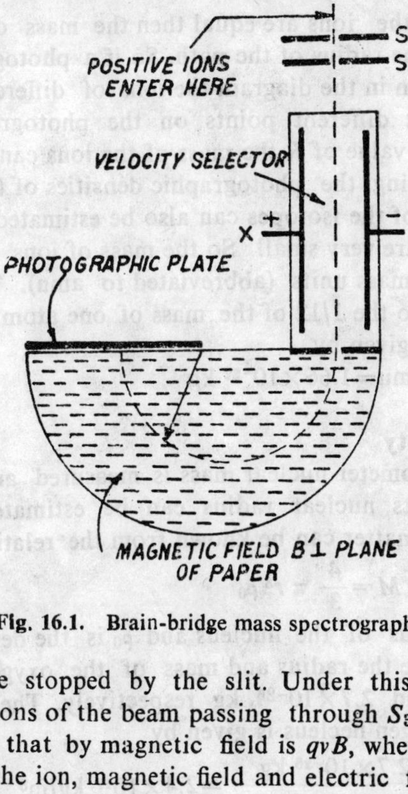

Fig. 16.1. Brain-bridge mass spectrograph.

and hence are stopped by the slit. Under this condition force acting on the ions of the beam passing through S_3 by the electric field is qE and that by magnetic field is qvB, where q, B and E are the charge of the ion, magnetic field and electric field. As the two forces are equal and opposite,

$$qE = qvB$$

or
$$v = \frac{E}{B} \qquad \ldots (16.1b)$$

The beam of ions after passing through the slit S_3 pass through a uniform magnetic field B' perpendicular to the plane of the paper. So the ions move under the influence of a constant force whose direction is always perpendicular to the velocity of the ions. Therefore an ion moves in a circular path with a constant tangential velocity v. If the radius of curvature of the path is R, and mass of ion is M, we can write :

$$qvB' = M\frac{v^2}{R}$$

or
$$R = \frac{Mv}{B'q} \qquad \ldots (16.2)$$

From Equations 16.1 and 16.2 we find:

$$M = \frac{RB'q}{v} = \frac{RB'Bq}{E} \qquad \ldots (16.3)$$

Hence, if charge of the ions are equal then the mass of each ion is proportional to the radius of the path. So if a photograhic plate *P* is placed as shown in the diagram the ions of different isotopes will be incident at different points on the photographic plate. Therefore, from the value of *R* the mass of the ions can be estimated. Besides,comparing the photographic densities of the lines the relative abundance of the isotopes can also be estimated.

Mass of the ions are very small. So the mass of ions are usually expressed in atomic mass units (abbreviated to amu). The amu is defined to be equal to the 1/12 of the mass of one atom of carbon $_6C^{12}$ and its value is given by

$$1 \text{ amu} = 1.66 \times 10^{-27} \text{ kg} \qquad \ldots (16.4)$$

16.4. Nuclear Density

Using mass spectrometer nuclear mass is measured and from the scattering experiments nuclear radius can be estimated. So the density of nuclear matter can be known from the relation,

$$M = \frac{4}{3}\pi r^3 \rho_0$$

where r is the radius of the nucleus and ρ_0 is the density of the nucleus. For example the radius and mass of the oxygen nucleus are 3×10^{-15} m, and 2.7×10^{-26} kg respectively. Therefore the density ρ_0 of the oxygen necleus is given by

$$\rho_0 = \frac{2.7 \times 10^{-26} \text{ kg}}{\frac{4}{3}\pi(3 \times 10^{-15} \text{ m})^3} = 2.4 \times 10^{17} \text{ kg/m}^3 \qquad \ldots (16.5)$$

Nuclei are nearly spherical and their radii are expressed fairly well by the empirical relation given by

$$r = r_0 A^{\frac{1}{3}} \qquad \ldots (16.6)$$

where r_0 is an empirical constant equal to 1.2×10^{-25} m, and A is the mass number of the nucleus. Density of nuclear matter becomes

$$\rho_0 = \frac{m_p A}{\frac{4}{3}\pi(r_0)^3 A} = \text{constant} \qquad \ldots (16.7)$$

where m_p is the mass of the proton.

From this it is evident that the aensity of all nuclei are approximately equal.

16.5. Binding Energy and Mass Defect

In atoms the electrons are bound to the nucleus by long range coulomb force. Similarly two nucleons in a nucleus are bound together by the nuclear forces. The characteristic of nuclear forces are not similar to that of coulomb force. Some aspects of nuclear force are not yet understood completely. However, they have the following characteristics :

1. **Strong attractive force.** Nuclear forces are very strong attractive. They are different from electro static or gravitational forces.

2. **Charge independence.** The nuclear forces do not depend upon the charge of the nucleons. The nuclear force of attraction between two protons, two neutrons or proton and neutron are very nearly same. This is called charge independence of the nuclear forces.

3. **Short range.** The range of nuclear forces are very small. The forces are effective only up to a distance of the order of 10^{-15} m. It falls rapidly with distance.

4. **Saturation.** A nucleon cannot interact simultaneously with all other nucleon but only with those in its immediate neighbourhood. This is in contrest to the nature of coulomb forces. This limitation of nuclear forces is called saturation.

Due to the nuclear force of attraction between nucleons some amount of energy is required to separate the nucleons. This energy is called the binding energy. The binding energy of a nucleus is defined as the energy required to break up the nucleus into its constituents protons and neutrons and to keep them at rest at infinite distances from each other.

The binding energy is always equal to the product of mass defect and square of velocity of light. To have a clear idea about mass defect let us take the example of deuteron nucleus. A deuteron nucleous contains one proton and one neutron.

The rest masss of proton	= 1.007276 amu
The rest mass of neutron	= 1.008665 amu
The total rest mass	= 2.015941 amu
But the rest mass of deuteron	= 2.013554 amu
Difference of two masses	= 0.002387 amu

The mass defect is the difference between the mass of a nucleus and the sum of the mass of its constituent nucleons. From the above calculation the mass defect of deuteron is 0.002387 amu. Using Einstein's mass energy relation the mass defect of deuteron can be expressed in terms of energy units. We have :

$$\text{Mass defect} = .002387 \times 1.66 \times 10^{-27} \text{ (kg)} \times (3 \times 10^{8} \text{ m/sec})^{2}$$
$$= 3.57 \times 10^{-13} \text{ Joules} = 2.23 \text{ Mev.} \quad \ldots (16.8)$$

When a neutron and a proton combine together to form a deuteron nucleus there will be a decrease of mass by .002387 amu. So an energy of 2.23 Mev will be liberated. This mass of 0.002387 amu is converted in to energy. This energy is called binding energy.

In general the binding energy of a nucleus $_ZX^A$ is given by

B.E=(Mass of protons+mass of neutrons−mass of the nucleus)C^2,

where C is the velocity of light. If m_p, m_n and M are the rest mass of proton, neutron and nucleus respectively in amu, then

$$B.E. = [Zm_p + (A-Z)m_n - M] 1.66 \times 10^{-27} \text{ kg} \times (3 \times 10^{8} \text{ m/s})^{2}$$
$$= [Zm_p + (A-Z)m_n - M] 1.494 \times 10^{-10} \text{ Joules.}$$
$$= [Zm_p + (A-Z)m_n - M] 931 \text{ Mev.} \quad \ldots (16.9)$$

If we will take into account the electrons each having rest mass m_e in amu Equation 16.9 becomes.

$$B.E. = [Z(m_p + m_e) + (A-Z)m_n - (M + Zm_e)] \text{ 931 Mev.} \quad \ldots (16.10)$$

When Equation 16.9 is divided by the number of nucleons we get the binding energy per nucleon. This binding energy per necleon is not same for all elecments. For example the binding energies of $_{17}Cl^{35}$ and $_{91}U^{235}$ are 298.1 and 1782.2 Mev. respectively. So the binding energy per nucleon for the two elements become 8.5 and 7.5 Mev. If the binding energy per nucleon of different elements are plotted against mass number a regular curve is obtained as shown in the Fig. 16.2. Following important conclusions can be drawn from the curve.

1. Binding energy per nucleon of all the elements except $_2He^4$, $_6C^{12}$ and $_8O^{16}$ are on the curve. The binding energy per nucleon of $_2He^4$, $_6C^{12}$ and $_8O^{16}$ are comparatively large. This indicates that the nucleons of these elements are strongly bound.

2. The curve shows a maximum at A=56 which corresponds to the mass number of iron. So the iron nucleus $_{26}Fe^{56}$ is strongly bound.

3. Binding energy per nucleon of most of elements is about 8. For lighter and heavier elements it is less than 8.

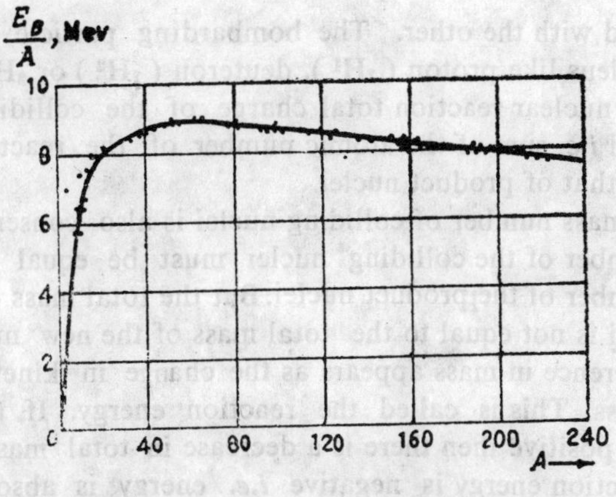

Fig. 16.2. Variation of binding energy per nucleon with mass number.

Example 16.1. Calculate the binding energy per nucleon of $_6C^{12}$.

Solution. The atomic mass of carbon $_6C^{12}$ is 12 amu.

So B.E. $= 931\ (6 \times 1.007276 + (12-6) \times 1.008665 - 12)$ Mev

$= 89.05$ Mev.

Or $\frac{B.E.}{A} = \frac{89.05}{A} = 7.4$ Mev.

16.6. Nuclear Reactions

Nuclear reaction is the process of formation of a new nucleus or a number of nuclei by collision of two other nuclei. For example two deuteron nuclei combine together to form a He nucleus. This process is called a nuclear reaction. The above reaction can be expressed in the form of an equation given by

$$_1H^2 + {_1H^2} = {_2He^4} \qquad \ldots (16.11)$$

Suppose a helium nucleus strikes a nitrogen nucleus to form a flourine nucleus. Subsequently the flourine nucleus being unstable breaks up into a hydrogen and an isotope of oxygen. The above process is also called a nuclear reaction and can be expressed as

$$_2He^4 + {_7N^{14}} \rightarrow [{_9F^{18}}] \rightarrow {_8O^{17}} + {_1H^1} \qquad \ldots (16.12)$$

In general the formation of the unstable compound nucleus is overlooked. So the Equation 16.12 can be written as

$$_2He^4 + {_7N^{14}} \rightarrow {_8O^{17}} + {_1H^1} \qquad \ldots (16.13)$$

In order to effect a nuclear reaction between two nuclei one of them must be accelerated to a high velocity so that it will be able to overcome the electrostatic repulsion between the two nuclei and bombared with the other. The bombarding particle is usually a light nucleus like proton ($_1H^1$), deuteron ($_1H^2$) or $_2He^4$.

In any nuclear reaction total charge of the colliding nuclei is conserved *i.e.* sum of the atomic number of the reacting nuclei is equal to that of product nuclei.

Total mass number of colliding nuclei is also conserved *i.e.* total mass number of the colliding nuclei must be equal to the total mass number of the product nuclei. But the total mass of the colliding nuclei is not equal to the total mass of the new nuclei formed. This difference in mass appears as the change in kinetic energy in the process. This is called the reaction energy. If this reaction energy is positive then there is a decrease in total mass. Similarly if the reaction energy is negative *i.e.* energy is absorbed in the reaction the final mass is greater than the initial mass.

Besides, like other physical processes the total momentum of the nuclei before reaction must be equal to the total momentum of the products after the reaction.

16.7. Discovery of Neutron

In 1930 Bothe and Becker found a very penetrating radiation while bombarding beryllium with α-particles. These radiations were not affected by electric and magnetic fields and having energy of the order of 10 Mev agreed satisfactorily with the mass change in the assumed reaction

$$_4Be^9 + {_2He^4} = {_6C^{13}} + h\nu$$

So they were taken to be γ radiations. In 1932 Curie and Joliot found that when these radiations passed through paraffin or other materials containing hydrogen protons are knocked out of them by the incident radiation. If these protons were knocked out by γ rays the energy of the γ-ray protons should be of the order of 50 Mev. Other experimental results established that these particles are not γ rays.

In 1932 Chadwick showed that these difficulties could be removed if the radiations are assumed to be uncharged particles having mass nearly equal to about that of a proton. This particle was named as ***neutron***. The nuclear reaction of Bothe and Becker would then be

$$_4Be^9 + {_2He^4} = {_6C^{12}} + {_0n^1} \quad \ldots (16.14)$$

where $_0n'$ represents a neutron.

16.8. Properties of Neutron

Neutron is an unchanged particle having mass slightly more than that of protons. Neutrons are present in all nuclei except hydrogen. Inside the nuclei of heavier elements there are more number of neutrons as compared to the number of protons.

Inside a nucleus the neutrons remain in stable condition. But free neutrons are not very stable. Its half life [all radioactive substances disintegrate in time. During the time in which half of a certain amount of the radioactive substance disintegrate, is called the half life of that substance] is about 11 minutes. It is established experimentally that neutrons change into protons in course of time.

As neutrons are uncharged particles they do not experience electrostatic repulsion. So they can penetrate easily into the target nucleus and initiate nuclear reaction. We know, to initiate nuclear reactions by bombarding charged particles, the bombarding particles have to be accelerated to high velocity. But some type of nuclear reactions can also be initiated by comparatively slow moving neutrons.

Example 16.2. Calculate the amount of energy liberated when one gram of helium is formed due to fusion of deuteron neclei.

Solution. When two deuteron nuclei combine together to give a helium nucleus, the mass defect is 0.0241 a mu. The number of helium nuclei present in one gram is given by,

$$N = 1/[4.003 \times 1.66 \times 10^{-24}]$$

Total amount of energy E liberated, becomes

$$E = \frac{(0.0241 \times 1.66 \times 10^{-24} \text{gm})\ (3 \times 10^{8}\ \text{m/s})}{4.003 \times 1.66 \times 10^{-24}}$$

$= 5.418 \times 10^{11}$ Joules $= 3.39 \times 10^{24}$ Mev $= 1.505 \times 10^{5}$ KWH.

Example 16.3. Calculate the amount of energy liberated in the nuclear reaction :

$$_0n^1 + {}_{92}U^{235} \rightarrow {}_{56}Ba^{141} + {}_{36}K^{92} + 3_0n^1$$

Solution. Mass of $_0n^1$ $= 1.0087$ amu

Mass of $_{92}U^{235}$ $= 235.0439$ amu

Total mass $= 236.0526$ amu

Mass of $_{56}Ba^{141}$ $= 140.9139$ amu

Mass of $_{36}Kr^{92}$ $= 91.8973$ amu

Mass of 3 neutrons $= 3.0261$ amu

Total $= 235.8373$ amu

Mass defect $= 236.0526 - 235.8373 = 0.2153$ amu

Energy released $= 931 \times 0.215$ Mev. $= 200.35$ Mev

$= 3.21 \times 10^{-11}$ J $= 8.92 \times 10^{-18}$ KWH

16.9. Radioactivity

In 1896 Becquerel while studying the effect of X-rays on a fluorescent salt of uranium, found that the salt emitted radiation even without being irridiatedby X-rays. The radiation could penetrate dark papers, tinfoil, and other substances. With further experimental study he found that the radiation was characteristic of uranium. It was found that thorium also emit similar radiations. Pierre and Marie Curie also discovered two new elements radium and polonium. These elements also emit radiations like uranium. Such elements are called radioactive elements and the phenomenon is called radioactivity.

It was observed that these radiations coming from radioactive substances affect photographic plates, ionise gases and also produce fluorescence. When these radiations were subjected to a magnetic field it was found that a group of them were deflected slightly in a direction in which a moving positive charge would move. These rays are called α-rays, Another group also was deflected by the magnetic field but deflection is more than that of α-rays and in the opposite direction. These group of rays are called β-rays. Rest of them are not affected by the magnetic field. They are called γ-rays (Fig. 16.3). From subsequent experiment it was established that α-rays are positively charged helium nuclei, β-rays are electrons and γ-rays are electromagnetic radiations. These radiations have different properties given below.

16.10. Properties of α, β and γ Rays

1. α and β-rays are deflected by magnetic fields. The deflections of α-rays are comparatively less and are in a direction opposite to that of β-rays. But γ-rays are not deflected by magnetic and electric field.

2. α, β and γ rays affect photographic plates and fluorescent screen.

3. They can penetrate through different materials. But the penetrating power of α-rays is very small. They are even stopped by a sheet of paper. They can penetrate only a few cm of air. The penetrating power of α-rays emitted by different radioactive elements are different. The penetrating power of β-rays are more than that of α-rays. Their range in air is about several meters. The γ-rays are highly penetrating than β-rays. They are practically not absorbed in air and even can pass through several inches of lead.

4. α and β-rays can ionise the gas through which they move whereas γ-rays practically do not affect the gas. When an α or a β-particle moves through gas it makes number of collisions with the molecules of the gas. Each collision produces a pair of ions. An α or β-particle with more energy can make more number of collisions. The number of ion pairs created by the particle is a measure of the energy or ionising power of the particle. The ionising power of α-particles is much more than that of β-particles and ionising power of β-particles is much greater than γ-rays. The ionisation produced by these rays are in the ratio of 10,000: 100:1

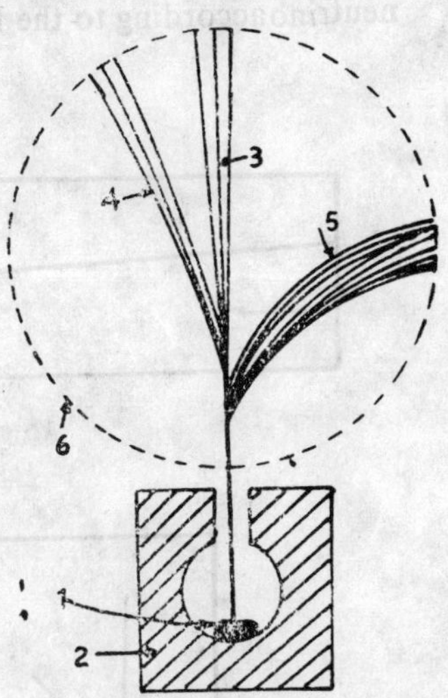

Fig. 16.3. Deflection of α, β. and γ-rays.

The path of the ionising particles can be photographed with the help of a cloud chamber or a bubble chamber. From the photographs it is clear that the path of the α-particles are not changed by the collision because they are comparatively heavy. But at the end of the path the direction may change suddenly due to its reduced energy (Fig. 16.4). As the β-rays are light the direction of their path changes number of time by collision.

16.11. Artificial Radioactivity

In 1934 J.C. Joliot and F. Joliot bombarded aluminium plates by α-particles emanating from polonium. In this experiment they found that some kind of penetrating radiation was coming out of the target aluminium even after removing the radioactive source. Further they found that intensity of this radiations decrease exponentially with time. From further experimental studies they also established that when α-particles bombarded an aluminium target a neutron is emitted according to the reaction.

$$_2He^4 + _{13}Al^{27} \rightarrow _{15}P^{30} + _0n^1 \qquad \ldots (16.15)$$

The isotope of phosphorus $_{15}P^{30}$ is an unstable nucleus. Its half life is 2.55 minutes. It decays into silicon emitting a positron and neutrino according to the following equation.

$$_{15}P^{30} \rightarrow {}_{14}Si^{30} + {}_{1}e^{0} + \gamma \qquad \ldots (16.16)$$

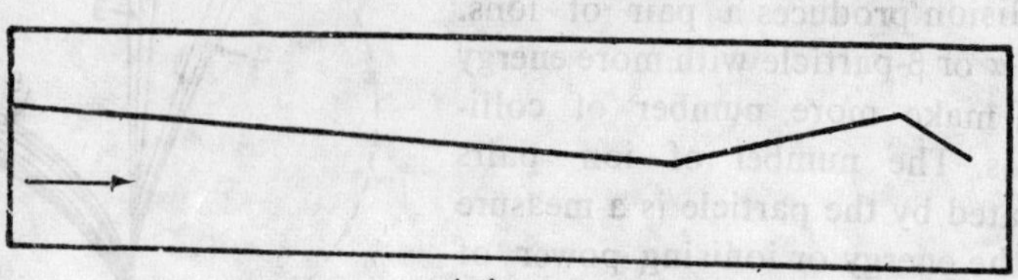

(a)

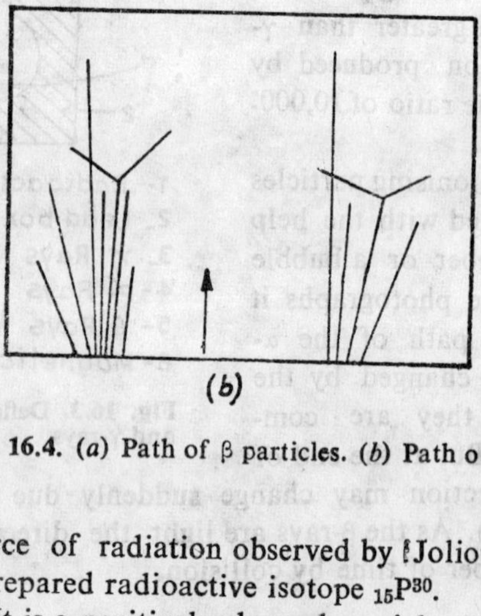

(b)

Fig. 16.4. (*a*) Path of β particles. (*b*) Path of α particles.

So the source of radiation observed by Joliot and Joliot was the artificially prepared radioactive isotope $_{15}P^{30}$.

Positron. It is a positively charged particle. Its mass is equal to that of electron but its charge is equal to that of a proton.

Neutrine. It is an interesting elementary particle with no charge and no mass but having some energy. It always moves with a velocity equal to the velocity of light. The artificially produced radioactive isotope of phosphorus is called a radioisotope. Similar type of nuclear reactions were also observed when magnesium and boron were bombarded by α-particles. The reactions are presented below.

$$_{12}Mg^{24} + {}_{2}He^{4} \rightarrow {}_{14}Si^{27} + {}_{0}n^{1}\ ;\ {}_{14}Si^{27} \rightarrow {}_{13}Al^{27} + {}_{1}e^{0}$$

and

$$_{5}B^{10} + {}_{2}He^{4} \rightarrow {}_{7}N^{12} + 2{}_{0}n^{1}, \qquad \ldots (16.17)$$

$$_{7}N^{12} \rightarrow {}_{6}C^{12} + {}_{1}e^{0} \qquad \ldots (16.18)$$

In the above reactions $_{14}Si^{27}$ and $_{7}N^{13}$ are unstable isotopes of silicon and nitrogen respectively. They disintegrate and become stable isotopes $_{13}Al^{27}$ and $_{6}C^{13}$.

Subsequently Fermi and his co-workers found that radioactive isotopes of elements are easily produced when the elements are bombarded by neutrons because neutrons can easily penetrate into the nucleus and initiate the reaction. Some such reactions are given below.

$$_{13}Al^{27}+{}_{0}n^{1}\rightarrow{}_{11}Na^{24}+{}_{2}He^{4} \quad \ldots (16.19)$$

$$_{11}Na^{23}+{}_{0}n^{1}\rightarrow{}_{11}Na^{24} \quad \ldots (16.20)$$

$$_{26}Fe^{56}+{}_{0}n^{1}\rightarrow{}_{25}Mn^{56}+{}_{1}H^{1} \quad \ldots (16.21)$$

$_{11}Na^{24}$ and $_{25}Mn^{56}$ further disintegrate to emit radiation. In the above reactions $_{11}Na^{24}$ and $_{25}Mn^{56}$ are two radioisotopes. These isotopes, formed artificially show radioactivity as indicated in equations. Radio isotopes have wide applications.

16.12. Uses of Radio Isotopes

Radio isotopes have wide applications in different fields such as in Industry, Agriculture and Medicine etc.

1. **Educational**. Different properties of α, β and γ rays can easily be studied with the help of radio isotopes emitting the particular radiations. For example, the absorption co-efficient of different substances for these different rays can easily be measured. Radioactive indicators have been used on the determination of the solubility of sparingly soluble substances.

2. **Industry**. We know that cracks and defective welding etc. can be detected by X-rays. Similarly using the γ-rays emitted from $_{27}Co^{60}$ can also be put to similar use. Radioactive $_{51}Sb^{24}$ is used to trace the arrival of fresh stock of petroleum transported through pipe lines. In this method a small quantity of $_{51}Sb^{124}$ is introduced at the source when a fresh stock of oil is sent through the pipe line. A detector kept outside the pipe line at the receiving end will indicate the arrival of new stock by detecting radiation emitted by the radio isotope. A mixture of radium and thorium with zinc sulphide is used to coat the pointers and figures of clocks and watches for rendering them visible in darkness.

3. **Medicine**. Ratio isotopes are used in diagnosis of diseases where other conventional methods have failed. For example the position and size of a tumor in the brain can easily be known by the following method. Tumor tissues absorb more radio isotopes than normal tissues. So a small quantity of radio isotope is injected

into the body with organic dyes. After some time radio isotopes will accumulate at the tumor, which can be detected by detectors.

Similarly, radio isotope of iodine are absorbed more by thyroid gland. So by injecting these isotopes the functioning of thyroid gland can be known. Radiation from radio isotopes of $_{27}Co^{60}$ are used for controlling the development and growth of some types of cancer-cells. Leukemia can be treated by the radiation from radio isotopes of phosphorus and iodine respectively.

4. **Agriculture**. Certain types of fertilizers are more suitable for particular types of soil. By using fertilizers mixed with radio phosphorus it is possible to choose the most suitable fertilizer for a particular soil and crop. Irradiating seeds by radiations from radioisotopes in controlled manner it has been possible to develop high yielding and disease resistant varieties of crops.

To protect the crops in insects infected area, a large number of male insects are bred in laboratory and they are sterilised with α-particles. These insects are then released in the infected area. This decreases the population of insects rapidly.

5. **Radio carbon dating.** Radio carbon dating issan useful method for determining the age of geological samples and archealogical remains.

Cosmic rays colliding with atmospheric nitrogen produce radio isotopes of carbon, according to the reaction:

$$_0n^1 + {_7N^{14}} \rightarrow {_6C^{14}} + {_1H^1} \qquad \ldots(16.22)$$

Living organisms assimilate this radioactive carbon along with normal carbon. When the organism dies it cannot further absorb either $_6C^{12}$ or $_6C^{14}$ which was already present in the dead organism decays according to the reaction:

$$_6C^{14} \rightarrow {_7N^{14}} + {-e^{\circ}} + \nu \qquad \ldots(16.23)$$

The half life of $_6C^{14}$ is 5580 years. Knowing original percentage of $_6C^{14}$ the living organism and the present percentage it is possible to determine the age of the archeaological and geological sample. This process of determining the age of the geological samples is called the carbon dating. The carbon dating has been made on the assumption that the production of $_6C^{14}$ has been constant. The age of the geological samples have been fairly accurate for periods extending back to about 40,000 years.

16.13. Nuclear Fission

In 1939 Ottohahn showed that an uranium nucleus disintegrates into two parts by capturing a neutron. It was further found from experiments that a nucleus may disintegrates in different ways resulting in different product nucle. Some of these disintegration processes are given in the following reactions.

$$_0n^1+{}_{92}U^{235}\rightarrow{}_{47}Ag^{113}+{}_{54}Rh^{120}+3_0n1 \quad \text{...(16.24)}$$

$$\rightarrow{}_{54}Xe^{140}+{}_{38}Sr^{94}+2_0n^1 \quad \text{...(16.25)}$$

$$\rightarrow{}_{44}Ru^{115}+{}_{48}Ca^{118}+3_0n^1 \quad \text{...(16.26)}$$

$$\rightarrow{}_{56}Ba^{141}+{}_{36}Kr^{92}+3_0n^1 \quad \text{...(16.27)}$$

In these reactions the product nuclei are of nearly the same size. When a heavy nucleus splits up in to two parts of approximately equal size it is called a nuclear fission. So the reactions given in Equations (16.24-16.27) are called fission reactions.

Besides uranium, thorium, ($Z=90$), protoactintum ($Z=91$) and plutonium ($Z=94$) nucleus can also undergo fission reaction by capturing a neutron.

16.14. Energy Released in Fission Reaction

In a fission reaction it is found that sum of the mass of nucleus and neutron is greater than the sum of the mass of the product nuclei. This mass defect is converted into energy. We have seen in Equation 16.8 that an energy of 200.38 Mev. is released when uranium nucleus disintegrates.According to Equation 16.27 we can have an idea about the amount of energy released in fission reaction if we calculate the amount of energy released when one gram of uranium nucleus disintegrates.

Mass of uranium nuclues $=235.049\times1.66\times10^{-27}$kg.

Therefore number of uranium atom in one gram$=2.56\times10^{21}$.

So total energy released$=2.56\times10^{21}\times3.21\times10^{-11}$J.

$$=2.3\times10^4 \text{ KWH.} \quad \text{...(16.28)}$$

16.15. Chain Reaction

We have seen that an enormous amount of energy of the order 200 mev is liberated during fission of a molecule. Besides two or more neutrons are also ejected in a fission reaction. Let us imagine an ideal case where two neutrons are ejected during the fission of a single uranium nucleus. In a bulk of uranium molecules, suppose one neutron from some source splits an uranium nucleus. As a result two more neutrons are produced. These two neutrons will further split two more nuclei and as such 4 new neutrons will be produced. These four neutrons will initiate fission reaction in four more nuclei and the number of fissionable nuclei will increase in geometric progression. Such a reaction is called a chain reaction. (*see* Fig. 16.5).

It is found that the time required between the emission of a neutron and fission reaction initiated by it is only of the order of 10^{-8} sec. Therefore within a fraction of a secon l many grams of uranium nuclei undergo fission reaction and hence huge amount of energy is released. This principle is utilised in atom bomb. On the

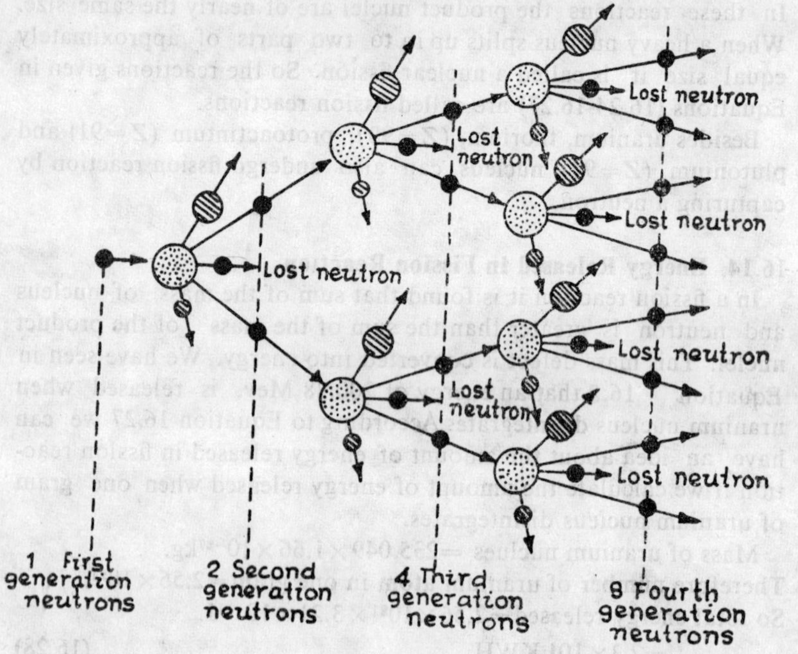

(a)

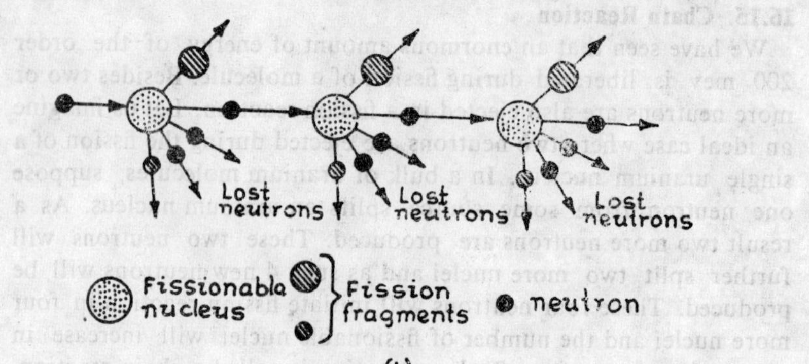

(b)

Fig. 16.5. Chain reaction of $_{92}U^{235}$.

other hand, if rate of fission reaction is decreased by some means and can be controlled then unclear fission energy can be used for peaceful purposes. This type of reaction is called controlled chain reaction and is used in nuclear reactors. Though the principle of chain reaction is simple yet it is difficult to maintain self sustained chain reaction. In actual practice the following facts are to be taken into account.

1. **Absorption by uranium-238.** It is found from experiment that nuclear fission is more effectively initiated by slow or thermal neutrons (energy of the order of 1 ev) as compared to first neutrons (energy of the order of 1 mev). But thermal neutrons can initiate fission reaction only in U^{235} and are absorbed by U^{238}. Uranium consists 3 isotopes *i.e.* U^{238}, U^{235} and U^{234}. In naturally occurring uranium, the proportion of U^{238} is highest. The amount of U^{235} isotope is 140 times less than that of the U^{238} isotope and amount of U^{234} is 100 times less than that of U^{235}. As the percentage of U^{235} is much less than that of U^{238}, there is greater probability of collision of neutrons with U^{238}. By such collision neutrons are lost without producing fission. So it is not possible to sustain chain reaction in natural Uranium.

In actual practice to increase the efficienty of the chain reaction the first neutrons are slowed down by mixing hydrogen–containing substances like water or paraffin with uranium. The hydrogen in the mixture slows down the neutrons. The slow neutrons are absorbed by U^{235} and will initiate fission reaction.

2. **Absorption by impurities.** Some impurities are always present in the fissionable materials. These impurities absorb the neutrons. This loss of neutrons can be reduced by using pure fissionable material.

3. **Leakage from the system.** Some of the secondary neutrons do not take part in fission reaction but escape out of the system. This leakage can be reduced by designing the geometry of the system.

Therefore a self sustaining chain reaction is possible in a sample of pure U^{235} or P_O^{239}, if the loss of neutrons due to leakage is small as compared to the number of neutrons produced during the fission process. From experiment it was found that if the mass of the block of uranium is greater than a minimum size, sustained chain reaction is possible. This mass is called the critical mass.

We introduce a term called reproduction factor K defined as the ratio of the production rate to the leakage rate of neutrons. Since the fission reactions take place throughout the volume of the block the production rate will be proportional tc $4/3\pi r^3$ and the loss of neutrons will be proportional to the surface are *i.e.* $u\pi r^2$ where r is the radius of the spherical block. Therefore we can write,

$$K=\frac{(4/3)\pi r^3}{4\pi r^2}\sim T \qquad \ldots(16.29)$$

If $K=1$, the chain reaction will be sustained. If $K<1$ the reaction will not be sustained and the reaction will stop. If $K>1$ production rate of neutrons will increase and the fission reaction is uncontrolled and may lead to an explosion. In a fission reaction K is always made equal to one. But in an explosive device like atom bomb two or three blocks of uranium each smaller than the critical mass are kept close to each other, but separated from each other. When by a triggering device these blocks are brought together to from a single mass K becomes greater than 1. Fission chain reaction will then proceed rapidly and tremendous quantity of energy will be liberated

16.16. Nuclear Reactor

A device which supply energy from controlled fission reaction is called nuclear reactor. In a nuclear reactor steps are taken to keep the reproduction factor equal to one. The first nuclear reactor or atomic pile was operated successfully in Chicago in 1942. The first nuclear reactor of India was constructed in Trombay in 1956 by the departement of atomic energy. The purpose of a reactor is to generate power for peaceful uses and to prepare radio isotopes for use in various fields.

A schematic diagram of nuclear power plant is presented in the Fig. 16.6. We describe below about ingredients required for a reactor.

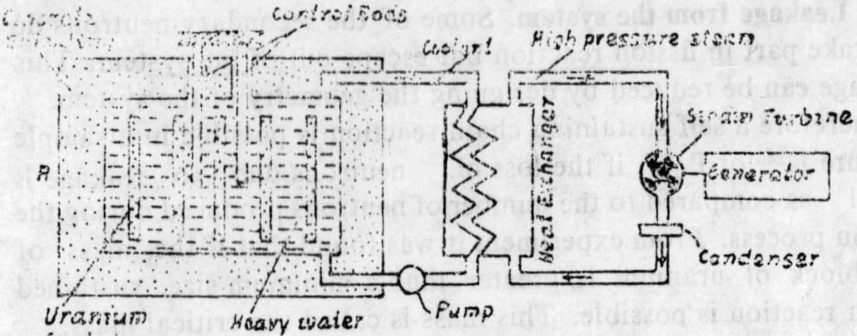

Fig. 16.6. Sectional diagram of a nuclear reactor.

Fuel. In the diagram F represents the fuel which is a fissionable material like U^{235}. The fuel is in the form of rods. After few months the fuel get contaminated due to its own fission products and hence it has to be replaced.

Moderator. The energy of neutrons liberated during fission of a nucleus is of the order of 2 mev. But neutrons having energy of the order 1 ev is required for fission reaction. So the fast neutrons are slowed down by moderator. Usually heavy water, carbon, graphite etc. are used as moderators. In Apsara heavy water is used as moderator.

Control rods. At times ratio of fission reaction increase or decreases rapidly. To keep the reaction at the desired rate the controlled rods are lowered or raised automatically. These rods can absorb neutrons. Usually boron, steel or cadmium are used in control rods.

Coolant. The energy liberated by fission reaction raises the temperature of the reactor. The heat generated is removed by using a suitable coolant such as carbon dioxide. Coolant circulated through the system by pumps gets heated and in its turn converts water to steam which rotates the steam turbine. The turbine is coupled with a generator from which electricity is generated. During this process the steam is condensed to water which is again converted to steam by the coolant to rotate the turbine.

Shielding. During the fission reaction neutrons are emitted. The neutrons are harmful for human body. To protect the persons working near the reactor a thick concreate wall is provided around the reactor. This wall works as a shield to stop the radiations.

Example 16.4. If 200 mev of energy is released per fission of U^{235} (*a*) how many fissions take place in a reactor giving 100 megawatts of energy, (*b*) what is the total mass of U^{235} consumed in one year.

Solution. Energy released per fission = 200 Mev.

$$=200\times10^{6}\times1.6\times10^{-19}\text{J}=3.2\times10^{-11}\text{J}$$

Power generated = 100MW = 100×10^{6}W

(*a*) No. of fission take place per sec $=\dfrac{100\times10^{6}\text{Wsec}}{3.2\times10^{-11}\text{J}}$

$$=31.3\times10^{17}$$

(*b*) Mass of one nucleus U^{235} = 235.0526 amu

$$=235.0526\times1.66\times10^{-27}\text{ kg}$$

Therefore, the total mass U^{235} consumed in 1 year

$$=31.3\times10^{17}\times60\times60\times24\times365\times235.0526\times1.66\times10^{-27}\text{kg}$$

$$=38.5\text{ kg}$$

16.17. Radiation Hazards

In nuclear reactions high energy radiations such as α, β, γ rays and neutrons are emitted. Out of these γ-rays and neutrons are highly penetrating. By these radiations the complex organic molecules break up due to ionisation of its atoms. As a result normal functioning of the biological systems are disturbed. If the radiations are very strong it causes the death of organisms.

To measure the dose of radiation an unit called rontgen (abbreviated to *R*) is introduced. It is defined as the amount of radiation which produces 1.6×10^{12} ion pairs in 1 gm of air. Damage to human organism caused by radiation depends upon dose, the rate of the dose and the part of the body exposed to radiation. Radiations produce two types of harmful effect *i.e.* pathological effects and genetic effects. If the radiation dose exceed 600 *R* the living organisms of the body will die. Even a smaller dose of the order 100 *R* may cause leukemia and cancer. Genetic effect are more dengerous because they are carried over from generation to generation. When these radiations pass through genetic cells they cause mutations of the chromosomes of the cellular nuclei. These mutations are transmitted from generations to generations. So it cannot be said definitely about the safe limit of the radiation to cause genetic damage. One can only say that the genetic damage is less if the dose is small.

16.18. Nuclear Fusion

In fission reaction large nucleus splits into two medium nuclei. But in fusion reaction two small nuclei fuse together to give a larger nuclei. Some of the reactions are given below.

$$_1H^1 + {_1H^1} = {_1H^2} + {_1e^0} \qquad \ldots(16.30)$$

$$_1H^2 + {_1H^2} = {_1H^3} + {_1H^1} \qquad \ldots(16.31)$$

$$_1H^2 + {_1H^2} = {_2He^4} \qquad \ldots(16.32)$$

In these reactions total mass of the product nuclei is less than the total mass of the reacting nuclei. The mass defect is converted into energy. In example of the Section 16.1 it was shown that 1.5×10^5 KHW of energy is liberated when one gram of helium is formed due to the fusion reaction of deuterons. If self sustained fusion reaction can be achieved easily the energy crisis of the world can be solved.

16.19. Self Sustained Fusion Reaction

At ordinary temperature deuteron is in gaseous state and their thermal velocities are comparatively small. In order that two deuteron nuclei may fuse together the velocity of the two nuclei must be sufficiently large to overcome the electrostatic repulsive force and there should be large number of collisions within a short time. The ions can acquire the required velocity when the temperature is of the order of 10^8 K. The number of collision will be adequate when there should be more than 10^{15} ions per c.c. If the conditions are satisfied self sustained fusion reaction is possible. At such high temperature the deuteron atom will be in the ionised state. This state of matter is called the fourth state of matter or the plasma state. But the deuteron plasma cannot be confined by any material wall at this temperature;the wall itself will be vapourised. So scientists are trying to confine the plasma by using electric and magnetic fields. They are trying to confine the deuteron plasma but are not yet completely successful. When they will be successful to confine the deuteron plasma the energy crisis of the world will be solved. In a hydrogen bomb the main source of energy is fusion reaction. The temperature of the ingredients are increased to the required value by fission reaction.

16.20. Energy Generation in Sun and Stars

The temperature of the sun is about 10^7 K. This temperature cannot be achieved on earth easily. In 1939 H. Bethe suggested that due to fusion reaction inside the centre of the sun such high temperature is possible, and these reactions are the immediate source of energy given out by the sun. The expected types of nuclear reactions in the sun are as follows:

$$_1H^1 + {_1H^1} = {_1H^2} + e^0 + \nu \qquad \ldots (16.33)$$

$$_1H^1 + {_1H^2} = {_2He^3} \qquad \ldots (16.34)$$

$$_2He^3 + {_2He^3} = {_2He^4} + {_1H^1} + {_1H^1} \qquad \ldots (16.35)$$

16.21. Cyclotron

We have seen a nuclear reaction can only be initiated when an energetic charged particle collides against a nucleus. Thus the production of high energy particle is essential in nuclear reaction.

There are different devices by which high energetic charged particles can be produced. The cyclotron is one of them. This was invented by E.O. Lawrence in 1930. The cyclotron consists of two semi-circular hollow metal boxes insulated from each other placed in a chamber. The chamber is placed in between the poles of a very strong electromagnet.

An ion introduced at the centre will travel in a semi-circular path in one box. When it goes to the second box it will be accelerated. So it will cover a larger circular path in the second box.

If B is the magnetic field,

q is the charge of the particle,
v is the velocity of the particle.

Then the force exerted on the particle by the magnetic field is :

$$B\,q\,v.$$

Again if m is the mass of the particle.
and r be the radius of the path.

The centripetal force is $\frac{mv^2}{r}$

when the particle moves in a circular path the two forces must be equal. Therefore :

$$B\,q\,v=\frac{mv^2}{r}$$

or
$$r=\frac{mv}{Bq} \quad \ldots (16.36)$$

If time required for one ion to complete a circle is T, then

$$\frac{T}{2}=\frac{\pi r}{v}=\frac{\pi m}{Bq} \quad \ldots (16.37)$$

So the time is independent of the radius of the path and the velocity of the ion.

A high frequency alternating potential can be used to accelerate the ion each time they cross from one D-shaped chamber to the other.

The ions continue to move on a spiral path gaining kinetic energy at the time it moves from one box to another. But lastly the high energy ion came out through a window.

Large cyclotron may produce 10 mev protons, 20 mev deuterons and 40 mev α-particles.

When the velocity of a particle increases, its mass will increase according to the theory of relativity. Therefore the time required to complete a half cycle will also increase. As a result the frequency of the A.C. potential will be unsuitable to accelerate the ion further. To overcome this difficulty synchro cyclotron or frequency modulated cyclotrons are designed. Initially the cyclotrons operates with a frequency $Bq/2\pi m$ The frequency decreases automatically as the time required to complete the half cycle increases due to the increase in mass.

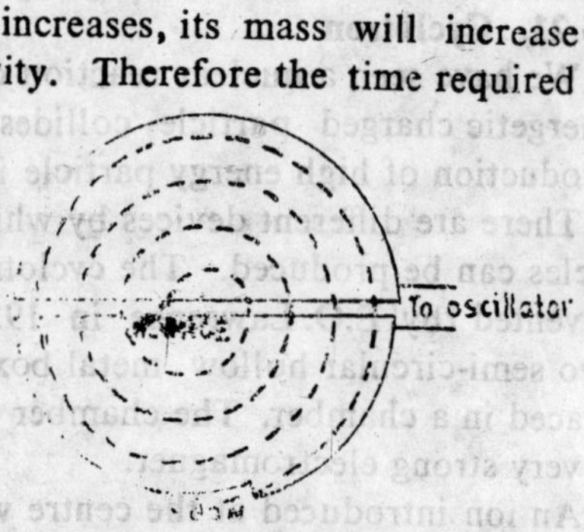

Fig. 16.7. Sectional diagram of a cyclotron.

QUESTIONS

1. Compare the atomic and nuclear radius of hydrogen.
2. Show that the nuclear densities are nearly equal.
3. Distinguish between isotopes and isobars.
4. Discuss the characteristics of nuclear forces.
5. What is mass defect ? How is it related to the binding energy?
6. Distinguish between chemical reaction and nuclear reaction?
7. Discuss the general characteristics of a radioactive atom.
8. How the different rays emitted by radioactive nuclei are identified?
9. Why it is easier to produce more energetic α rays than protons in a cyclotron?
10. Why it is difficult to initiate nuclear reaction by heavier nuclei than lighter nucleil?
11. Which radio active rays can produce more ionisation and why?
12. How neutron was discovered?
13. Compare the properties of α, β and γ rays?
14. What is artificial radioactivity?
15. Discuss the uses of radio isotopes
16. How the age of the geological sample can be estimated?
17. What is fusion reaction?
18. What is controlled fusion reaction? What is its importance?
79. What is fission reaction?
20. What is chain reaction and self sustained chain reaction?
21. Under what conditions self sustained chain reaction is possible?
22. Describe the working principle of a nuclear reactor?
23. Discuss about the hazards of the nuclear reaction?
24. Describe the working principle of a cyclotron?

PROBLEMS

1. If the velocity of an α particle is 3×10^9 cm/sec express its energy in Joules and electron volts? ($\cdot 2.99\times10^{-12}$ J)

2. Find the magnetic field required to rotate an α-particle moving with a velocity of 2×10^9 cm/sec, in a circular path of radius 30 cm.? (1.38)

3. Fill up the blank spaces in the following nuclear reactions?

(*a*) ${}_6C^{12}+{}_2He^4\rightarrow(\qquad)+{}_{+1}e^0$

(*b*) ${}_{17}Cl^{36}+{}_1P^1\rightarrow(\qquad)+{}_2He^4$

(*c*) ${}_{29}Cu^{65}+{}_0n^1\rightarrow(\qquad)+{}_1P^1$

(*d*) ${}_7N^{14}+{}_0n^1\rightarrow(\qquad)+{}_2He^4$

4. The atomic number and mass of radium is 88 and 224 respectively. If it disintegrates by emitting an α-ray, express the reaction in equations?

5. Prove that the velocity of a charged particle moving in a circular path in a magnetic field is given by Bqr / m. Where q=charge of the particle r=radius of the circular path, m=mass of the particle. B=Magnetic field.

6. In the fission of U^{235} 300 mev energy is liberated. What is the mass defect? (5.3×10^{-28} kg)

7. When two deuteron nuclei mix together to give a helium nucleus there is decrease in mass by 0.7%. What is the amount of energy liberated from 1 gm of deuteron? (6.3×10^{14} J)

8. If 3 helium nuclei mixed together to give a carbon nucleus what is the amount of energy liberated in the above reaction? (1.34×10^{-12} J)